ADVANCED PROBABILITY THEORY

PROBABILITY: PURE AND APPLIED

A Series of Textbooks and Reference Books

Editor

MARCEL F. NEUTS

University of Arizona
Tucson, Arizona

1. Introductory Probability Theory, *by Janos Galambos*
2. Point Processes and Their Statistical Inference, *by Alan F. Karr*
3. Advanced Probability Theory, *by Janos Galambos*
4. Statistical Reliability Theory, *by I. B. Gertsbakh*
5. Structured Stochastic Matrices of M/G/1 Type and Their Applications, *by Marcel F. Neuts*
6. Statistical Inference in Stochastic Processes, *edited by N. U. Prabhu and I. V. Basawa*
7. Point Processes and Their Statistical Inference, Second Edition, Revised and Expanded, *by Alan F. Karr*
8. Numerical Solution of Markov Chains, *edited by William J. Stewart*
9. Probability: The Mathematics of Uncertainty, *by Dorian Feldman and Martin Fox*
10. Advanced Probability Theory: Second Edition, Revised and Expanded, *by Janos Galambos*

Other Volumes in Preparation

ADVANCED PROBABILITY THEORY

Second Edition, Revised and Expanded

JANOS GALAMBOS
Temple University
Philadelphia, Pennsylvania

CRC Press
Taylor & Francis Group
Boca Raton London New York

CRC Press is an imprint of the
Taylor & Francis Group, an **informa** business

CRC Press
Taylor & Francis Group
6000 Broken Sound Parkway NW, Suite 300
Boca Raton, FL 33487-2742

First issued in paperback 2019

© 1995 by Taylor & Francis Group, LLC
CRC Press is an imprint of Taylor & Francis Group, an Informa business

No claim to original U.S. Government works

ISBN-13: 978-0-8247-9332-6 (hbk)
ISBN-13: 978-0-367-40163-4 (pbk)

This book contains information obtained from authentic and highly regarded sources. Reasonable efforts have been made to publish reliable data and information, but the author and publisher cannot assume responsibility for the validity of all materials or the consequences of their use. The authors and publishers have attempted to trace the copyright holders of all material reproduced in this publication and apologize to copyright holders if permission to publish in this form has not been obtained. If any copyright material has not been acknowledged please write and let us know so we may rectify in any future reprint.

Except as permitted under U.S. Copyright Law, no part of this book may be reprinted, reproduced, transmitted, or utilized in any form by any electronic, mechanical, or other means, now known or hereafter invented, including photocopying, microfilming, and recording, or in any information storage or retrieval system, without written permission from the publishers.

For permission to photocopy or use material electronically from this work, please access www.copyright.com (http://www.copyright.com/) or contact the Copyright Clearance Center, Inc. (CCC), 222 Rosewood Drive, Danvers, MA 01923, 978-750-8400. CCC is a not-for-profit organization that provides licenses and registration for a variety of users. For organizations that have been granted a photocopy license by the CCC, a separate system of payment has been arranged.

Trademark Notice: Product or corporate names may be trademarks or registered trademarks, and are used only for identification and explanation without intent to infringe.

Library of Congress Cataloging-in-Publication Data

Galambos, Janos
     Advanced probability theory / Janos Galambos. — 2nd ed., rev. and expanded.
          p.   cm. — (Probability, pure and applied ; 10)
          Includes bibliographical references (p. -) and index.
          ISBN 0-8247-9332-3 (hardcover : alk. paper)
          1. Probabilities. I. Title. II. Series.
     QA273.G3126   1995
     519.2—dc20
                                                               95-32193
                                                               CIP

Visit the Taylor & Francis Web site at
http://www.taylorandfrancis.com

and the CRC Press Web site at
http://www.crcpress.com

Preface to the Second Edition

It is with pleasure that I note that many of my colleagues have used the first edition of this book as a text for their courses. Their comments, and the comments of my own students, guided me in preparing the second edition.

The purpose of the book is the same as that of the first edition: to serve as a textbook for a first advanced course in probability theory. I expanded those sections that have generated a large interest in courses offered worldwide, I introduced new topics, and made the book more accessible to those readers with no prior knowledge of probability theory. For this last purpose I added practical examples in Chapter 1 and new sections on distribution theory, elementary at the end of Chapters 2 and 3, and more advanced in Chapters 5 and 6. These sections emphasize model building and are intended to train probabilists who will understand that distributions with their basic characteristics form the core of probability theory. Both exact models and approximate ones are presented, and more than usual emphasis is placed on multivariate distributions. The sections on distributions at the end of Chapters 2 and 3 can be read immediately after the concept of random variables is introduced in Chapter 1, and are useful for further reference.

The sections on martingales, extreme value theory, and Brownian motion have been expanded, and a new section on continuous-time Markov chains has been added. These sections, as any other, are offered not as competitive with specialized books, but rather as encouragement for further study of special topics in probability theory.

In the second edition I have corrected errors, and made several proofs easier to read by adding details which occasionally amount to just a few words while in some places fill several lines. These changes may be just as significant as the new sections in their contributions to making the present volume a better textbook.

The book now covers sufficient material for three semesters. It remains a source for basic concepts of several branches of probability theory, but is not a reference for any special topic. In particular, I did not cover the literature. The selection of references for further reading is, in fact, quite arbitrary. For the material covered, however, the original source is cited.

Many mathematicians from all over the world helped me to make this a better textbook and my sincere thanks go to them. My own students in the past several years gave me very valuable advice through their questions, which led to several improvements in the proofs even if by just a few lines. They should know that I appreciate their attention to detail. For her encouragement to write a second edition of this book, and for her continued interest in its progress, I am indebted to Ms. Maria Allegra of Marcel Dekker, Inc.

Janos Galambos

Preface to the First Edition

This is a textbook for a two-semester introductory course in probability theory at the graduate level. This book is also intended for self-study and as a reference to the basic concepts of and main results in modern probability theory. These latter aims led me to make the book self-contained. Thus, all statements (except those in Section 8.6) are proved in detail, including those theorems that are customarily regarded as results of measure theory. These results are developed and proved as needed for the relevant material in probability theory. I wish to stress that the book is intended to be a reference for the basic results in probability theory only; for specialized topics and for historical background the reader is referred to books dealing with those subjects.

The only prerequisites for the book are a sound knowledge of college calculus and an introductory course in complex analysis. In particular, it is not required that the reader has taken an elementary course in probability theory. However, I assume that the reader will be sufficiently motivated to take an advanced course in probability theory, either via other fields of science such as physics, chemistry, engineering, operations research, and statistics, or by mere interest in the subject as a mathematician who would like to understand a mathematical theory whose results are used so frequently in everyday life. In either case, the student will be aware that the results have direct applications, and thus I did not feel the need to motivate every subject and every group of theorems by listing direct applications. The examples and remarks, however, make numerous references to applications; the book is thus oriented both to applications and to a rigorous development of the mathematical theory. Indeed, I strongly advise reading the discussions (several times formulated as remarks) which appear after the proofs or when theorems are explained before beginning their proofs; these discussions emphasize possible applications, both scientific and mathematical.

The first chapter is a review of the basic concepts, but this is self-contained as well. However, if its pace is too fast, the reader may want to consult my *Introductory Probability Theory* book, which is written for undergraduates (and also published by Marcel Dekker in this series). An occasional reference is made to this book as the "introductory volume," but this is done only to point out that a particular result can also be established by very elementary methods. Such references are kept to a minimum.

To present all the theorems with detailed proofs, more than two semesters would be required. However, I expect that several proofs and, in some cases, even entire topics will be skipped. Also, if all students have taken an elementary course in probability theory, or if students are familiar with Lebesgue integration, then a large chunk of material can be covered rapidly. The rest will still suffice for two semesters. The text has been tried over several years; indeed, the book grew out of my lectures to graduate students at Temple University. I am indebted to my students, whose questions led to several improvements in the presentation.

Each section ends with exercises of various kinds. There is a section at the end of the book that contains detailed outlines of the solutions of the exercises. The exercises are part of the material, and occasional references are made to them. Some exercises are such that they are challenging when stated, but might become very easy after a subsequent section. Therefore, the exercises are to be assigned when the section either is being or has just been covered.

I greatly appreciate the help provided by the staff of Marcel Dekker, Inc. In particular, my thanks are due to Ms. Vickie Kearn, Executive Editor, for her interest and assistance.

Janos Galambos

Contents

ADVANCED PROBABILITY THEORY

1

Review of Basic Concepts

1.1 EVENTS

The three most important concepts in probability theory are *experiment*, *event*, and *probability*. In this section the concepts of experiment and event are introduced and the basic rules of set theory as they relate to these two concepts are summarized.

From the practical point of view, "experiment" has the same meaning as in everyday life: something which is being performed (either in actuality or hypothetically). In the mathematical theory of probability, however, experiment is identified with a set Ω (also called the sample space), whose elements must represent all outcomes (of the experiment). This set is not unique: the experiment we deal with determines the minimum number of elements Ω must have, but that is all we need to know about Ω. Thus, in rolling a die whose faces are distinct, Ω must have at least six elements, but it may have more. This freedom in choosing Ω is convenient both from the mathematical and the practical point of view. The reader will immediately find this obvious by trying to replace $\Omega = [1, 2, 3, \ldots\}$ with any finite set for the experiment of predicting the age of the oldest person in a given society.

When Ω has been chosen, subsets of Ω represent restricted possibilities of the experiment which we call events. Clearly, events are meaningful only if the experiment has been defined. It is meaningless to ask whether a number is even or odd before we have specified where the number came from (e.g., we might have rolled a die with six faces numbered 1 through 6, or picked one number out of the integers 1, 3, and 5; since Ω could be the same set for these two experiments, we clearly need some help in weeding out the superfluous elements—in fact, probability will do that).

It would be convenient if all subsets of Ω could be considered events. But since probability is a number assigned to every event by some rule, contradictions would arise if we insisted that every subset of Ω be an event. For this simple mathematical reason, events are defined as a limited collection $\mathscr{A}$ of subsets of Ω.

Definition 1   For a given experiment Ω, events $\mathscr{A}$ are subsets of Ω, which form a σ-field. That is, the collection $\mathscr{A}$ of events is defined by the following properties. If $A \in \mathscr{A}$, then $A \subset \Omega$. Furthermore, (i) $\Omega \in \mathscr{A}$; (ii) if A_j, $j = 1, 2, \ldots$, all belong to $\mathscr{A}$, then their union is in $\mathscr{A}$; and (iii) if A belongs to $\mathscr{A}$, then so does A^c.

Definition 1 does not explicitly say anything about intersections. However, since the complement of a union is the intersection of the complements of the terms in the union (this is known as de Morgan's law; see below), Definition 1 in fact says that events form an abstract collection which is closed under the operations of union, intersection, and complement for denumerable sequences.

Before proceeding with the consequences of Definition 1, let us list here, as a reminder, the basic rules of the union, intersection, and complement operations.

General Rules:

Commutative law: $A \cup B = B \cup A$ and $A \cap B = B \cap A$.
Associative law: $(A \cup B) \cup C = A \cup (B \cup C)$
              and $(A \cap B) \cap C = A \cap (B \cap C)$.
Distributive law: $A \cap (B \cup C) = (A \cap B) \cup (A \cap C)$.

The special events Ω and the empty set (impossible event) $\varnothing$:

$$A \cup \Omega = \Omega \qquad A \cup \varnothing = A \qquad A \cap \Omega = A \qquad A \cap \varnothing = \varnothing$$

The algebraic definition of the complement:

$$A \cup A^c = \Omega \qquad A \cap A^c = \varnothing$$

De Morgan's law:

$$(\cup A_j)^c = \cap A_j^c \qquad (\cap A_j)^c = \cup A_j^c$$

Operations with a single event:

$$A \cup A = A \qquad A \cap A = A \qquad (A^c)^c = A$$

Implication (one event is the subset of the other):

$$A \subset B \qquad \text{if and only if } A \cap B = A$$

We now return to Definition 1. Note that (i) and (iii) imply that $\varnothing \in \mathscr{A}$. In the formal definition (ii), only denumerably infinite unions are required to remain in $\mathscr{A}$. However, a finite sequence of events can always be extended into an infinite one by including $\varnothing$ infinitely many times. Such an extension does not affect unions. So (ii) is about finite or denumerably infinite sequences of events.

When every condition in Definition 1 holds, except that, in (ii), only finite unions are required to remain in the collection of subsets of a set, the collection is called a *field*.

Clearly, a σ-field is a field. The converse, however, is not true. For example, if $S = (0, 1]$, and if $\mathscr{F}$ is the collection of all finite disjoint unions of right semiclosed intervals $(a, b] \subset S$, then $\mathscr{F}$ is a field. But it is not a σ-field, because

$$\bigcup_{n=2}^{+\infty} \left(0, \frac{1}{2} - \frac{1}{n}\right] = (0, \tfrac{1}{2})$$

is not a member of $\mathscr{F}$.

Fields are convenient intermediate steps in constructing σ-fields. We start with an arbitrary collection $\mathscr{B}$ of subsets of a set S. Let us enlarge $\mathscr{B}$ as follows: (i) we form the class $\mathscr{B}_1$ containing S, $\varnothing$, A, and A^c for all $A \in \mathscr{B}$; (ii) we next form the collection $\mathscr{B}_2$ of finite intersections of elements of $\mathscr{B}_1$; and finally, (iii) we define $\mathscr{B}_3$ as the collection of all finite unions of (pairwise disjoint) members of $\mathscr{B}_2$. Clearly, $\mathscr{B} \subset \mathscr{B}_1 \subset \mathscr{B}_2 \subset \mathscr{B}_3$, and the reader can easily verify that $\mathscr{B}_3$ is a field. It is also the smallest field containing $\mathscr{B}$, because if a single element is dropped from it (repeated elements count only once), it is no longer a field. It is customary to use the notation $\mathscr{F}(\mathscr{B})$ for the smallest field containing $\mathscr{B}$; we have just seen that $\mathscr{F}(\mathscr{B}) = \mathscr{B}_3$.

The further expansion of $\mathscr{F}(\mathscr{B})$ into the smallest σ-field containing $\mathscr{B}$ (and thus $\mathscr{F}(\mathscr{B})$ as well) is not simple; there is, in fact, no general method of constructing it except the following abstract method. Consider all σ-fields containing $\mathscr{B}$; let these be denoted $\sigma_\gamma(\mathscr{B})$, $\gamma \in \Gamma$. The set Γ is not empty because the set of all subsets of S is a σ-field. Then the smallest of these

$$\sigma(\mathscr{B}) = \bigcap_{\gamma \in \Gamma} \sigma_\gamma(\mathscr{B})$$

It is easily verified that the right-hand side is indeed a σ-field, which then is clearly the smallest possible.

In some special cases the following extension of $\mathscr{F}(\mathscr{B})$ into a σ-field is more fruitful than the abstract method described above. Assume that, by some method, we have determined that a collection $\mathscr{B}^*$ of subsets of S satisfies (i) $\mathscr{B}^*$ is a field, (ii) $\mathscr{F}(\mathscr{B}) \subset \mathscr{B}^*$, and (iii) for every denumerably infinite sequence $A_j \in \mathscr{B}^*$ with $A_j \subset A_{j+1}$, $\lim A_j \in \mathscr{B}^*$ (for such special A_j, $\lim A_j$ is both the set of those points which belong to all but a finite number of A_j and the union of A_j). Then $\mathscr{B}^*$ is a σ-field containing $\mathscr{B}$; one has only to observe that if C_j is an arbitrary sequence of sets, then, with

$$A_n = \bigcup_{j=1}^{n} C_j \qquad A_n \subset A_{n+1} \qquad \text{and} \qquad \bigcup_{j=1}^{+\infty} C_j = \bigcup_{n=1}^{+\infty} A_n.$$

Now, if, by excluding a single element from $\mathscr{B}^*$, not all three properties (i)–(iii) above hold, then $\mathscr{B}^* = \sigma(\mathscr{B})$.

We conclude this discussion with an important example. Let S be an open set in the d-dimensional Euclidean space. Let $\mathscr{B}$ be the collection of open subsets of S. The smallest σ-field containing $\mathscr{B}$, $\sigma(\mathscr{B})$, is called the collection of *Borel sets* of S. The set $\mathscr{B}$ is not a field, and the reader is advised to go through the steps in the construction of $\mathscr{F}(\mathscr{B})$. In particular, when S is the whole real line, then starting with all open intervals, or with all closed intervals, leads to the same smallest σ-field containing either of these two groups, namely, to the Borel sets of the real line. The same conclusion is drawn if the initial collection of subsets of the real line is the set of all right semiclosed intervals. Other equivalent definitions of Borel sets are also possible.

Exercises

Let A, B, and A_j belong to the same σ-field of events.

1. Show that
   (i) $A \cup B = A \cup (A^c \cap B)$.
   (ii) $A_1 \cup A_2 \cup \cdots \cup A_n = A_1 \cup \{\bigcup_{j=2}^{n} (A_j \cap A_{j-1}^c \cap \cdots \cap A_1^c)\}$.
   Simplify the right-hand side if $A_1 \subset A_2 \subset \cdots \subset A_n$.
2. Define $A \bigtriangleup B = (A \cap B^c) \cup (B \cap A^c)$. Show that
   (i) $A \bigtriangleup B = (A \cup B) \cap (A^c \cup B^c)$.
   (ii) $B \bigtriangleup (A_1 \cup A_2) = [(B \bigtriangleup A_1) \cap (B \bigtriangleup A_2)] \cup [B^c \cap (A_1 \bigtriangleup A_2)]$.
   (iii) $(A \bigtriangleup B) \cup (A \bigtriangleup B^c) = \Omega$.
3. Show that

$$\bigcup_{n=1}^{+\infty} \bigcap_{j=n}^{+\infty} A_j \subset \bigcap_{n=1}^{+\infty} \bigcup_{j=n}^{+\infty} A_j$$

with equality occurring in each of the following cases: (i) $A_j \subset A_{j+1}$, (ii) $A_{j+1} \subset A_j$, and (iii) $A_j = (1 - 1/j, 2 - 1/j)$.

4. Let $\mathscr{B}$ be a collection of subsets of S, and let A be a fixed subset of S. Set $\mathscr{B} \cap A = \{B \cap A : B \in \mathscr{B}\}$. Show that

$$\sigma(\mathscr{B} \cap A) = \sigma(\mathscr{B}) \cap A$$

where, on the left-hand side, the extension is over subsets of A.

5. Let $A_1, A_2, \ldots, A_n$ be pairwise disjoint subsets of Ω. Set $\mathscr{B} = \{A_1, A_2, \ldots, A_n\}$. Characterize $\mathscr{F}(\mathscr{B})$ and $\sigma(\mathscr{B})$.

6. Let $C_1, C_2, \ldots, C_k$ be subsets of Ω and let the sets

$$C_{i_1} \cap C_{i_2} \cap \cdots \cap C_{i_m} \cap C_{i_{m+1}}^c \cap \cdots \cap C_{i_k}^c \qquad 1 \le i_1 < \cdots < i_m \le k$$

be arranged into a sequence by some rule and denoted by $A_1, A_2, \ldots, A_n$. Show that the smallest field containing all C_j is the same as the smallest field containing all A_r. Hence, if D is a set expressible by a finite number of the union, intersection, and complement operations, then there is a subset T of $\{1, 2, \ldots, n\}$ such that

$$D = \bigcup_{t \in T} A_t$$

Write this representation explicitly if (i) D is the event that exactly s of $C_1, C_2, \ldots, C_k$ occur and (ii) D is the event that at least s of $C_1, C_2, \ldots, C_k$ occur.

1.2 PROBABILITY

Given a sample space Ω and a σ-field $\mathscr{A}$ of subsets of Ω (the events), a probability P is a real-valued function on $\mathscr{A}$ satisfying

(A1) $0 \le P(A) \le 1$.

(A2) $P(\Omega) = 1$.

(A3)   (Additivity) If T is either a finite or a denumerably infinite set of positive integers and if the events A_t, $t \in T$, are mutually exclusive (disjoint), that is, $A_i \cap A_j = \varnothing$ for $i \ne j$, then

$$P(\cup A_t) = \sum P(A_t)$$

The triplet $(\Omega, \mathscr{A}, P)$ is called a probability space.

The properties (A1)–(A3) defining probability are called the *Kolmogorov axioms of probability*, named after the Soviet mathematician A. N. Kolmogorov, who laid down the foundations of modern probability theory in his work of 1933.

Some elementary consequences of the axioms are contained in the following theorems. Notice in the proofs of Theorems 1 through 5 that axiom (A3) is always applied with finite T.

Theorem 1 $P(A^c) = 1 - P(A)$. Consequently, $P(\varnothing) = 0$.

Proof. From $\Omega = A \cup A^c$ and $A \cap A^c = \varnothing$, axioms (A2) and (A3) imply the formula for $P(A^c)$. From this formula, with $A = \Omega$, $P(\varnothing) = 0$ follows. ∎

Theorem 2 $P(A^c \cap B) = P(B) - P(A \cap B)$.

Proof. Since $B = (A \cap B) \cup (A^c \cap B)$, where the terms on the right-hand side are disjoint, the theorem is a direct consequence of axiom (A3). ∎

Theorem 3 (Monotonicity) If $A \subset B$, then $P(A) \leq P(B)$.

Proof. Apply Theorem 2 and axiom (A1). ∎

Theorem 4 $P(A \cup B) = P(A) + P(B) - P(A \cap B)$.

Proof. Apply axiom (A3) and Theorem 2 in Exercise 1.1.1 (i). ∎

Theorem 5 For $n \geq 2$,

$$P\left(\bigcup_{j=1}^{n} A_j\right) \leq \sum_{j=1}^{n} P(A_j)$$

Proof. The inequality is proved by induction over n. For $n = 2$, it follows from the proceding theorem by axiom (A1). Assume now that the inequality holds for n and set A equal to the union of A_j, $1 \leq j \leq n$, and $B = A_{n+1}$. Then apply the inequality with $n = 2$. The assumption of induction thus implies the inequality for $n + 1$, which completes the proof. ∎

Because none of the proofs above uses axiom (A3) for infinite unions, P could have been defined only on a field without change in any of the preceding statements. This is summarized in the corollary below. But first a new concept.

Definition 2 Let $\mathscr{F}$ be a field of subsets of a set Ω. Then a real-valued function μ defined on $\mathscr{F}$ is called an additive set function if axioms (A1)–(A3) apply with μ for P, except that T is assumed to be finite in axiom (A3).

It is not a generally accepted custom to include axiom (A2) in the definition of an additive set function. However, if μ is an additive set function, then so is $c\mu$ for constants $c > 0$, so a finite additive set function can always be transformed into one that satisfies (A2). The only exception is if μ is identically zero, which will be treated separately when it comes up.

The property in axiom (A3) will be called σ-additivity if it applies with infinite T.

Now, as was remarked earlier, we have established the following result.

Corollary Let μ be an additive set function on a field $\mathscr{F}$ of subsets of a set Ω. Then Theorems 1 through 5 apply with μ for P.

With this little preparation we can already solve several problems of practical importance. As an example, we give the following estimate.

Assume an instrument is assembled from 10 components, each of which has a warranty of one year. The manufacturer of the components estimates from past experience that the probability of the return of a component under the warranty equals 0.005. Let us show that, whatever structure the instrument has, the probability p that all components properly function for one year satisfies $p \geq 0.95$.

Indeed, let A_j, $1 \leq j \leq 10$, be the event that component j fails within one year. Then, $P(A_j) = 0.005$. Hence,

$$p = P(A_1^c \cap A_2^c \cap \cdots \cap A_{10}^c) = 1 - P(A_1 \cup A_2 \cup \cdots \cup A_{10})$$
$$\geq 1 - [P(A_1) + P(A_2) + \cdots + P(A_{10})] = 1 - 0.05 = 0.95.$$

Here, we first applied Theorem 1 and then Theorem 5.

Note that Theorems 1–5 entail that, given two events A and B with specific $P(A)$ and $P(B)$, then there are strict limitations in choosing $P(A \cap B)$ and $P(A \cup B)$. For example, if $P(A) = 0.6$ and $P(B) = 0.7$, then, by monotonicity, $P(A \cap B) \leq 0.6$. On the other hand, by Theorem 4 and Axiom 1, $P(A \cap B) = 1.3 - P(A \cup B) \geq 1.3 - 1 = 0.3$. Similarly, $P(A \cup B) \geq P(B) = 0.7$ by monotonicity. In Exercise 3 a general statement is made about the preceding relations.

We now continue the investigation of the basic properties of probabilities.

Theorem 6 (Continuity) Let the events A_1, $A_2, \ldots$ form an increasing $A_j \subset A_{j+1}$ (a decreasing $A_j \supset A_{j+1}$) sequence, and set $A = \cup A_j$ $(A = \cap A_j)$. Then $\lim P(A_n)$ exists and

$$\lim P(A_n) = P(A)$$

Proof. By the monotonicity of probability, $\lim P(A_n)$ exists. To find the limit, we write

$$A = A_1 \cup \left\{ \bigcup_{j \geq 2} (A_j \cap A_{j-1}^c) \right\}$$

and apply axiom (A3). We get

$$P(A) = P(A_1) + \sum_{j \geq 2} P(A_j \cap A_{j-1}^c)$$

$$= P(A_1) + \lim_{n = +\infty} \sum_{j=2}^{n} P(A_j \cap A_{j-1}^c) = \lim_{n = +\infty} P(A_n)$$

where, in the last step, Theorem 2 is applied with A_{j-1} and A_j for A and B, respectively, where $A_{j-1} \subset A_j$. The theorem is thus proved for increasing sequences of events. When the A_j decrease, then the A_j^c increase; therefore, the second part follows from the first part by an appeal to Theorem 1 and de Morgan's law.   ■

The continuity of probability is equivalent to σ-additivity.

Theorem 7 Let μ be an additive set function on the field $\mathscr{F}$ of subsets of a set Ω. Assume that μ is continuous at $\varnothing$; that is, if $A_1, A_2, \ldots$ is a decreasing sequence of $\mathscr{F}$ such that $\cap A_j = \varnothing$, then, as $n \to +\infty$, $\lim \mu(A_n) = 0$. Then μ is σ-additive.

Proof. First note that if μ is continuous at $\varnothing$, then it is continuous at every set of $\mathscr{F}$ in the sense that if A_j^* in a increasing (decreasing) sequence of $\mathscr{F}$ with $A = \cup A_j^*$ $(A = \cap A_j^*)$ then $\lim \mu(A_n^*) = \mu(A)$. Indeed, defining $A_j = (A_j^*)^c \cap A$ $(A_j = A_j^* \cap A^c)$, A_j is decreasing with $\cap A_j = \varnothing$. Thus, by assumption, $\lim \mu(A_n) = 0$, which is equivalent to the previous limit on account of Theorem 2 (see also the Corollary). Now let $B_1, B_2, \ldots$ be pairwise disjoint members of $\mathscr{F}$. Put

$$C_j = \bigcup_{t=1}^{j} B_t \qquad j = 1, 2, \ldots$$

Then the sequence C_j is increasing with $\cup C_j = \cup B_t$. Furthermore, by additivity,

$$\mu(C_j) = \sum_{t=1}^{j} \mu(B_t)$$

which, by continuity, as $j \to +\infty$, converges to the μ-value of the common union of C_j or of B_t on the one hand, and, on the other hand, to the infinite sum of the terms on the right-hand side. Hence, σ-additivity, as well as the theorem, is established. ∎

The continuity of probability holds in the following stronger sense. For a sequence of events $A_1, A_2, \ldots$, define

$$\liminf A_n = \bigcup_{k=1}^{+\infty} \bigcap_{j=k}^{+\infty} A_j \qquad (1.1)$$

and

$$\limsup A_n = \bigcap_{k=1}^{+\infty} \bigcup_{j=k}^{+\infty} A_j \qquad (1.2)$$

If $\liminf A_n = \limsup A_n$, then this common set is called $\lim A_n$. Clearly, $\lim A_n$ exists if the sequence A_j is monotonic. It fact, $\lim A_n = \cup A_j$ if A_j increases and $\lim A_n = \cap A_j$ if A_j decreases, as was seen in Exercise 1.1.3. We now prove the following extension of Theorem 6.

Theorem 8 For every sequence A_j of events on a probability space,

$$P(\liminf A_n) \le \liminf P(A_n) \le \limsup P(A_n) \le P(\limsup A_n)$$

Consequently, if $\lim A_n$ is defined, then $P(\lim A_n) = \lim P(A_n)$.

Proof. Define

$$B_k = \bigcap_{j=k}^{+\infty} A_j$$

Then the sequence B_k is increasing and

$$\lim B_k = \liminf A_n$$

Hence, by Theorem 6,

$$P(\liminf A_n) = \lim P(B_k) \qquad (1.3)$$

On the other hand, by monotonicity, $P(B_k) \le P(A_j)$ for all $j \ge k$, that is,

$$P(B_k) \le \inf\{P(A_j): j \ge k\} \qquad (1.4)$$

By letting $k \to +\infty$, (1.3) and (1.4) imply the first inequality of the theorem. The last inequality is proved in a similar manner, the details of which are omitted. The theorem is established. ∎

Note the following meaning of the events at (1.1) and (1.2). The event at (1.1), that is, $\liminf A_n$, occurs if and only if all but a finite number of the A_j occur (the finite number of exceptions, of course, varies with the elements of Ω; that is, its value is random). On the other hand, lim sup A_n occurs if and only if infinitely many of the A_j occur. One can now see the strong requirement for the existence of lim A_n.

The events lim inf A_n and lim sup A_n frequently come up in the course of proofs of limit theorems. Two of the theorems below provide powerful tools in those proofs. The next theorem is a preparation for these.

Theorem 9   For an arbitrary sequence A_j of events on a probability space,

$$P\left(\bigcup_{j=1}^{+\infty} A_j\right) \leq \sum_{j=1}^{+\infty} P(A_j)$$

Proof.   Set

$$B_k = \bigcup_{j=1}^{k} A_j$$

Then the sequence B_k is increasing, and, by Theorem 5,

$$P(B_k) \leq \sum_{j=1}^{k} P(A_j) \leq \sum_{j=1}^{+\infty} P(A_j)$$

Since the extreme right-hand side does not depend on k, the above inequality continues to hold if $k \to +\infty$ on the left-hand side. By continuity (Theorem 6), the limit of the left-hand side exists and equals $P(\cup B_k) = P(\cup A_j)$. This completes the proof.   ■

Theorem 10 (Borel-Cantelli lemma)   If the events A_j are such that

$$\sum_{n=1}^{+\infty} P(A_n) < +\infty \tag{1.5}$$

then $P(\text{lim sup } A_n) = 0$.

Proof.   From the definition at (1.2), monotonicity and Theorem 9 yield

$$P(\text{lim sup } A_n) \leq P\left(\bigcup_{j=k}^{+\infty} A_j\right) \leq \sum_{j=k}^{+\infty} P(A_j) \qquad \text{for all } k \geq 1$$

By letting $k \to +\infty$ on the extreme right-hand side, the theorem is immediate from the assumption at (1.5).   ■

We finally formulate a complement to Theorem 10, due to Erdös and Rényi (1959).

Theorem 11 Assume that for the events A_j

$$\sum_{n=1}^{+\infty} P(A_n) = +\infty \tag{1.6}$$

Furthermore, upon setting

$$b_{1,n} = \sum_{j=1}^{n} P(A_j) \qquad b_{2,n} = \sum_{1 \le i < j \le n} P(A_i \cap A_j) \tag{1.7}$$

assume that, as $n \to +\infty$,

$$\limsup b_{1,n}^2 / b_{2,n} = 2 \tag{1.8}$$

Then $P(\limsup A_n) = 1$. In particular, condition (1.8) is always satisfied if $P(A_i \cap A_j) \le P(A_i)P(A_j)$ for all $i \ne j$, and thus the conclusion holds under the sole assumption (1.6).

The proof of Theorem 11 will be delayed until Section 1.5. A trivial example, $A_j = A$ for all j, shows that it is possible that $0 < P(\limsup A_n) < 1$.

Exercises

1. Let $\Omega = \{1, 2, \ldots, N\}$ and let $\mathscr{A}$ be the set of all subsets of Ω. Let the numbers $p(k)$, $1 \le k \le N$, satisfy (i) $p(k) \ge 0$ for all k and (ii) $\sum_{k=1}^{N} p(k) = 1$. Define $P(A) = \sum_{a_j \in A} p(a_j)$ for all $A \in \mathscr{A}$. Show that P is a probability on $\mathscr{A}$.
2. Let Ω be the positive real line and let $\mathscr{A}$ be a σ-field of subsets of Ω, containing all intervals. Assume that the set function P is defined on $\mathscr{A}$, satisfying axiom (A3). Furthermore, if $I \in \mathscr{A}$ is an interval of positive length, then $P(I) > 0$ and, for $I_k = (k, k+1)$, $P(I_k) = 1/k(k+1)$, $k \ge 1$. Show that P is not a probability.
3. Let p_1, p_2, and p_{12} be nonnegative real numbers with (i) $p_{12} \le \min(p_1, p_2)$ and (ii) $p_1 + p_2 \le 1 + p_{12}$. Show that there is a probability space $(\Omega, \mathscr{A}, P)$ and two events $A, B \in \mathscr{A}$, such that $p_1 = P(A)$, $p_2 = P(B)$, and $p_{12} = P(A \cap B)$ if and only if (i) and (ii) hold. Generalize the result to the case of three events.
4. For the operation $A \triangle B$, defined in Exercise 1.1.2, show that (i) $d(A, B) = P(A \triangle B)$ satisfies the triangular inequality and (ii) $|P(A) - P(B)| \le d(A, B)$.

5. Let $A = \{a_1, a_2, \ldots\}$ denote a set of distinct positive integers. We say that A has density $d(A)$ if

$$d(A) = \lim_{N = +\infty} \frac{1}{N} \sum_{a_k \leq N} 1$$

exists. Show that the collection $\mathscr{F}$ of those A whose density exists is closed under complementation and that $d(A)$ is an additive set function on $\mathscr{F}$, but $d(A)$ is not σ-additive.

6. Let A_j be the interval $(1/(j + 1), 1/j], j \geq 1$. Find lim inf A_n and lim sup A_n.

7. Prove the following identities
   (i) $(\text{lim inf } A_n)^c = \text{lim sup } A_n^c$.
   (ii) $\text{lim sup}(A_n \cup B_n) = (\text{lim sup } A_n) \cup (\text{lim sup } B_n)$.

8. In the next section we shall prove that there is a probability on the Borel sets of the unit interval $(0, 1)$ such that the probability $P(I) = b - a$ for the interval $I = (a, b)$ (as well as for $I = [a, b]$). We use this probability on the above-mentioned Borel sets. Expand every $0 < x < 1$ into decimal expansion and let $A_n^{(T)}$ be the event that each of the n digits x_j, $T < j \leq T + n$ of x is different from, say, 2. Show that, for every $T \geq 0$, $P(\text{lim sup } A_n^{(T)}) = 0$. Interpret the result.

1.3 A METHOD OF CONSTRUCTING PROBABILITIES*

In Exercise 1 of the preceding section, a general method is given for constructing probabilities when Ω is finite and $\mathscr{A}$ can thus contain all subsets of Ω. However, when the construction of $\mathscr{A}$ is not itself explicit, the construction of P on $\mathscr{A}$ is not simple.

The problem we face is exemplified by the following question. Assume that we want to pick a point at random on the unit interval $(0, 1)$ in such a manner that the probability that the selected point lies in a subinterval (a, b) of $(0, 1)$ is the length $b - a$ of (a, b). Does such a probability exist on the smallest σ-field containing all subintervals of $(0, 1)$ (i.e., on the Borel sets of $(0, 1)$)? If yes, is this probability unique? The answer to both questions is yes, since the choice of the initial values of the probability (in this particular case, of the intervals) is consistent with the axioms of probability.

In general terms, given a set Ω and a collection $\mathscr{B}$ of subsets of Ω, we are looking for a probability on the smallest σ-field $\sigma(\mathscr{B})$ containing $\mathscr{B}$ that has preassigned values on $\mathscr{B}$; $P(B)$ for $B \in \mathscr{B}$. It is assumed that the values $P(B)$,

---

* The details of this section may be skipped at first reading.

$B \in \mathscr{B}$, are consistent with the axioms of probability. The extension of $P(B)$ from $\mathscr{B}$ to $\sigma(\mathscr{B})$ will be done in several steps, each step being of some independent interest.

Step 1: The extension of $P(B)$ from $\mathscr{B}$ to the smallest field $\mathscr{F}(\mathscr{B})$ containing $\mathscr{B}$.

This extension is not necessarily unique (but see the next paragraph); in fact, we have only a guide for it when $\mathscr{B}$ does not have a limited structure. In spite of this limitation, we call it a method of extending $P(B)$ from $\mathscr{B}$ to $\mathscr{F}(\mathscr{B})$, which is as follows. Follow the three simple steps in the construction of $\mathscr{F}(\mathscr{B})$ described in Section 1.1, and the axioms, together with Theorems 1, 2, and 4 (see the Corollary of Section 1.2), as well as the consistency requirement, are to be observed.

Notice that if finite intersections in $\mathscr{F}(\mathscr{B})$ are finite unions of (disjoint) members of $\mathscr{B}$ (this is the case if $\mathscr{B}$ is the collection of all subintervals of an interval), then the procedure above results in a unique extension of $P(B)$ from $\mathscr{B}$ to $\mathscr{F}(\mathscr{B})$.

Step 1 always results in $P(\Omega) = 1$ and $P(\varnothing) = 0$.

Step 2: The extension of $P(B)$ from $\mathscr{F}(\mathscr{B})$ to the collection of $\mathscr{U}(\mathscr{B})$ of all denumerably infinite unions of members of $\mathscr{F}(\mathscr{B})$.

By first forming finite unions, every denumerably infinite union can also be obtained as the union of an increasing sequence; $\mathscr{U}(\mathscr{B})$ can thus be defined as

$$\mathscr{U}(\mathscr{B}) = \{A : A = \cup A_n, A_n \subset A_{n+1}, A_n \in \mathscr{F}(\mathscr{B})\} \tag{1.9}$$

Now, in order to remain consistent with properties of probabilities, in view of the continuity theorem (Theorem 6), our only choice of defining $P(B)$ on $\mathscr{U}(\mathscr{B})$ is the limit relation

$$P(A) = \lim P(A_n) \qquad A \in \mathscr{U}(\mathscr{B}), \qquad (n \to +\infty) \tag{1.10}$$

where A_n is a sequence occurring in the definition of A at (1.9).

Before proceeding we have to show that $P(A)$ is well defined at (1.10). That is, if A_n^* is another increasing sequence in $\mathscr{F}(\mathscr{B})$ such that $A = \cup A_n^*$, then

$$\lim P(A_n) = \lim P(A_n^*) \tag{1.11}$$

To prove (1.11), fix the value of t and form the sequence $C_{t,n} = A_t^* \cap A_n$, $n \geq 1$. Clearly, $C_{t,n}$, $n \geq 1$, is an increasing sequence in $\mathscr{F}(\mathscr{B})$ with

$$C_{t,n} \subset A_n \subset A_{n+1} \subset \cdots \qquad \text{and} \qquad \bigcup_n C_{t,n} = A_t^*$$

Hence,

$$P(C_{t,n}) \leq \lim_k P(A_k) = P(A)$$

thus, by first letting $n \rightarrow +\infty$, and then $t \rightarrow +\infty$,

$$\lim_t P(A_t^*) \leq \lim_k P(A_k) \qquad (1.12)$$

By now reversing the roles of A_n and A_t^*, the reversed inequality of (1.12) is obtained, establishing (1.11).

It is important to record that $\mathscr{F}(\mathscr{B}) \subset \mathscr{U}(\mathscr{B})$ (by taking $A_n = A$ for all n at (1.9)) and the definition of $P(B)$ at (1.10) does not change the value of P for members of $\mathscr{F}(\mathscr{B})$. Step 2 is therefore a real extension of P from $\mathscr{F}(\mathscr{B})$ to the larger set $\mathscr{U}(\mathscr{B})$.

Let us look at the set $\mathscr{U}(\mathscr{B})$ more closely. It contains both Ω and $\varnothing$, and $P(\Omega) = 1$ and $P(\varnothing) = 0$. In addition, $\mathscr{U}(\mathscr{B})$ is closed under finite intersections and finite or denumerably infinite unions. For these operations, the appropriate formulas of Section 1.2 apply to P (such as the ones in the axioms and in Theorems 3, 4, 5, and 6). The proof of this claim is left to the reader as Exercise 1.3.1.

The problem with $\mathscr{U}(\mathscr{B})$ is that it is not closed under complementation. For example, if $\mathscr{B}$ is the collection of all subintervals of $(0, 1)$, then $\mathscr{F}(\mathscr{B})$ contains all closed intervals $[a, a]$ of a single element, and thus $\mathscr{U}(\mathscr{B})$ contains all subsets of $(0, 1)$ with denumerably many elements. In particular, the set of rational points in $(0, 1)$ is in $\mathscr{U}(\mathscr{B})$. But its complement cannot be in $\mathscr{U}(\mathscr{B})$ because the rational points are everywhere dense in $(0, 1)$, and thus the set of irrationals cannot contain as a subset any member of $\mathscr{F}(\mathscr{B})$ other than individual points or their finite unions. But the set of irrationals is not denumerable, so it cannot be represented as a set A at (1.9) with finite sets A_n (i.e., with sets with a finite number of elements).

In spite of this drawback, it is shown in the next step that P on $\mathscr{U}(\mathscr{B})$ essentially determines P on the smallest σ-field $\sigma(\mathscr{B})$ containing $\mathscr{B}$.

Step 3:   The extension of P from $\mathscr{U}(\mathscr{B})$ to $\sigma(\mathscr{B})$.

Let C be an arbitrary subset of Ω. Define

$$P^*(C) = \inf\{P(A): C \subset A, A \in \mathscr{U}(\mathscr{B})\} \qquad (1.13)$$

Because of the properties of P on $\mathscr{U}(\mathscr{B})$ (Exercise 1.3.1),

$$P^*(A) = P(A) \qquad \text{if} \quad A \in \mathscr{U}(\mathscr{B}) \qquad (1.14a)$$

and

$$0 \le P^*(C) \le 1; \qquad P^*(C_1) \le P^*(C_2) \qquad \text{if} \quad C_1 \subset C_2 \tag{1.14b}$$

Next, we prove that, for arbitrary C_1 and C_2,

$$P^*(C_1 \cup C_2) + P^*(C_1 \cap C_2) \le P^*(C_1) + P^*(C_2) \tag{1.15a}$$

and

$$P^*(C_1) + P^*(C_1^c) \ge 1 \tag{1.15b}$$

Indeed, by definition, for every $\varepsilon > 0$, we can find A_1 and A_2 in $\mathcal{U}(\mathcal{B})$ such that $C_1 \subset A_1$, $C_2 \subset A_2$, and $P^*(C_i) \ge P(A_i) - \frac{1}{2}\varepsilon$, $i = 1, 2$. Hence,

$$P^*(C_1) + P^*(C_2) + \varepsilon \ge P(A_1) + P(A_2) = P(A_1 \cup A_2) + P(A_1 \cap A_2)$$
$$\ge P^*(C_1 \cup C_2) + P^*(C_1 \cap C_2)$$

from which (1.15a) follows. In the middle, we again applied a property of P on $\mathcal{U}(\mathcal{B})$ that is proved in Exercise 1.3.1. Now, since Ω and $\varnothing$ are in $\mathcal{U}(\mathcal{B})$, (1.15b) follows from (1.15a) and (1.14a).

Finally, we establish

$$\text{if} \quad C_n \subset C_{n+1}, C = \cup C_n \qquad \text{then} \quad \lim P^*(C_n) = P^*(C) \tag{1.16}$$

Fix $\varepsilon > 0$ and choose $A_n \in \mathcal{U}(\mathcal{B})$ such that $C_n \subset A_n$ and $P^*(C_n) \ge P(A_n) - \varepsilon/2^n$, $n \ge 1$. We first show

$$P^*(C_n) \ge P\left(\bigcup_{k=1}^{n} A_k\right) - \varepsilon(1 - 1/2^n) \qquad n \ge 1 \tag{1.17}$$

The proof of (1.17) is by induction over n. For $n = 1$, (1.17) is valid by the choice of the sets C_1 and A_1. Next, we assume that (1.17) holds for n. Hence,

$$P\left(\bigcup_{k=1}^{n+1} A_k\right) = P\left(\bigcup_{k=1}^{n} A_k\right) + P(A_{n+1}) - P\left(\bigcup_{k=1}^{n} (A_k \cap A_{n+1})\right)$$

$$\le P^*(C_n) + \varepsilon\left(1 - \frac{1}{2^n}\right) + P^*(C_{n+1}) + \frac{\varepsilon}{2^{n+1}} - P^*(C_n)$$

$$= P^*(C_{n+1}) + \varepsilon\left(1 - \frac{1}{2^{n+1}}\right)$$

which establishes (1.17). In estimating the negative term above, we used the fact that, by the choice of the sequences C_k and A_k and since $C_n \subset C_{n+1}$, $C_n \subset A_n \cap A_{n+1}$.

The limit in (1.16) now follows easily from (1.17). On the one hand, since $C_n \subset C$, (1.14b) implies $P^*(C_n) \le P^*(C)$ for every n, and thus $\lim P^*(C_n) \le P^*(C)$ ($\lim P^*(C_n)$ exists by the monotonicity of C_n and by (1.14b)). On the other hand, upon observing that

$$A_n^* = \bigcup_{k=1}^{n} A_k \subset A_{n+1}^* \qquad \cup A_n^* \in \mathscr{U}(\mathscr{B}) \qquad C \subset \cup A_n^*$$

the passages to the limit in (1.17) as $n \to +\infty$, and then $\varepsilon \to 0$, yield $\lim P^*(C_n) \ge P^*(C)$. This completes the proof of (1.16).

We are now in a position to formulate our major extension theorem.

Theorem 12 Set

$$\sigma^*(\mathscr{B}) = \{C : C \subset \Omega, P^*(C) + P^*(C^c) = 1\}$$

The set $\sigma^*(\mathscr{B})$ is a σ-field containing $\mathscr{U}(\mathscr{B})$ (and thus $\mathscr{B}$ and $\mathscr{F}(\mathscr{B})$ as well). The set function P^* is a probability on $\sigma^*(\mathscr{B})$ such that $P^*(B) = P(B)$ for $B \in \mathscr{U}(\mathscr{B})$.

Remark 1 Notice that $\sigma^*(\mathscr{B})$ contains all those subsets C of Ω for which $P^*(C) = 0$. Indeed, since $C^c \subset \Omega$, (1.14a) with $A = \Omega$ and (1.14b) imply $P^*(C) + P^*(C^c) \le 1$, which, in turn, because of (1.15b), implies $C \in \sigma^*(\mathscr{B})$. We refer to such a property by saying that P^* is *complete* on $(\Omega, \sigma^*(\mathscr{B}))$ or that the probability space $(\Omega, \sigma^*(\mathscr{B}), P^*)$ is complete.

More generally, we say that the *probability space* $(\Omega, \mathscr{A}, P)$ is *complete*, or that P is complete on $(\Omega, \mathscr{A})$, if for every $A \in \mathscr{A}$ with $P(A) = 0$ it follows that if $B \subset A$, then $B \in \mathscr{A}$.

Proof of Theorem 12. We first show that $\sigma^*(\mathscr{B})$ is a field and P^* is an additive set function on $\sigma^*(\mathscr{B})$.

By definition, if C is in $\sigma^*(\mathscr{B})$, then so is C^c. Furthermore, in Step 2 we observed that Ω and $\varnothing$ are in $\mathscr{U}(\mathscr{B})$, $P(\Omega) = 1$, and $P(\varnothing) = 0$. Hence, (1.14a) implies that both Ω and $\varnothing$ are in $\sigma^*(\mathscr{B})$. Finally, if $A, B \in \sigma^*(\mathscr{B})$, then (1.15a) and the definition of $\sigma^*(\mathscr{B})$ yield

$$P^*(A \cup B) + P^*(A \cap B) \le P^*(A) + P^*(B)$$
$$P^*[(A \cup B)^c] + P[(A \cap B)^c] \le P^*(A^c) + P^*(B^c)$$
$$= 2 - P^*(A) - P^*(B)$$

The sum of the right-hand sides equals two, while, because of (1.15b), the sum of the left-hand sides is at least two. Consequently, in both cases above, equality must hold, and the columnwise sums on the left-hand side must also equal one. These imply that $A \cup B \in \sigma^*(\mathscr{B})$ and thus $\sigma^*(\mathscr{B})$ is a field and P^* is additive on $\sigma^*(\mathscr{B})$ (apply the first inequality, which turns out to be an equation, with mutually exclusive A and B).

Next, we prove that if $C_n \in \sigma^*(\mathscr{B})$, $C_n \subset C_{n+1}$, then, with $C = \cup C_n$,

$$C \in \sigma^*(\mathscr{B}) \qquad \text{and} \qquad P^*(C) = \lim P^*(C_n) \qquad\qquad (1.18)$$

The limit at (1.18) is just a restatement of (1.16). Now (1.18) and the inequality

$$P^*(C^c) \le P^*(C_n^c) \qquad \text{for every} \quad n \ge 1$$

which is valid for the increasing sequence $C_n \in \sigma^*(\mathscr{B})$ because of (1.14b), yield

$$P^*(C) + P(C^c) \le 1$$

Upon appealing to (1.15b) again, equality must hold above, and thus $C \in \sigma^*(\mathscr{B})$.

We have, in fact, proved that $\sigma^*(\mathscr{B})$ is a σ-field (it is a field and closed with respect to denumerable unions; the monotonicity requirement is not a restriction, since it can always be achieved by first forming finite unions). We also proved that P^* is an additive set function on $\sigma^*(\mathscr{B})$ and that (1.18) holds. Hence, by turning to complements in (1.18), we conclude from Theorem 7 that P^* is a probability on $\sigma^*(\mathscr{B})$.

Finally, (1.14a) implies that $\mathscr{F}(\mathscr{B})$ is a subset of $\sigma^*(\mathscr{B})$ and, thus, by (1.18), so is $\mathscr{U}(\mathscr{B})$. Hence, one more appeal to (1.14a) completes the proof. ∎

Theorem 13 (Main Extension Theorem) Let $\mathscr{F}$ be a field of subsets of a set Ω. Let P be a probability on $\mathscr{F}$. Then there is a unique probability P^* on the smallest σ-field containing $\mathscr{F}$ such that P and P^* coincide on $\mathscr{F}$. (Here, and in what follows, the term "probability" is used whenever the axioms hold even if its domain is not a σ-field).

Proof. Evidently, the smallest σ-field $\sigma(\mathscr{F})$ containing $\mathscr{F}$ is a subset of $\sigma^*(\mathscr{F})$ of Theorem 12. Hence, P^* of Theorem 12 is a probability with the desired property. Consequently, only its uniqueness on $\sigma(\mathscr{F})$ requires proof.

Assume that Q is another probability on $\sigma(\mathscr{F})$ such that $Q(A) = P(A)$ for $A \in \mathscr{F}$. But then $Q(A) = P(A)$ on $\mathscr{U}(\mathscr{F})$, and thus, for every $C \in \sigma(\mathscr{F})$ and $A \in \mathscr{U}(\mathscr{F})$ with $C \subset A$, $Q(C) \le Q(A) = P(A)$. Hence, by (1.13), $Q(C) \le P^*(C)$ for every $C \in \sigma(\mathscr{F})$. Now, if there were $C \in \sigma(\mathscr{F})$ such that $Q(C) < P^*(C)$, we would then have

$$1 = Q(\Omega) = Q(C) + Q(C^c) < P^*(C) + P^*(C^c) = P^*(\Omega) = 1$$

an obvious contradiction. This proves the uniqueness part of the theorem.   ■

Remark 2 If we start with an arbitrary collection $\mathscr{B}$ (rather than a field) of subsets of the set Ω, then Step 1 in our extension method is not necessarily unique (as was already pointed out in Step 1; a concrete example is contained in Exercise 1.3.2). However, when $\mathscr{B}$ is the set of all subintervals of an interval of length one, then even Step 1 provides a unique extension of a "probability" from $\mathscr{B}$ to $\mathscr{F}(\mathscr{B})$. On the other hand, Theorem 13 guarantees the possibility of the unique extension of P from $\mathscr{F}(\mathscr{B})$ to $\sigma(\mathscr{B})$. This settles the question posed at the beginning of this section: there is a unique probability on the Borel sets of the interval $(0, 1)$ such that, for every subinterval (a, b) of $(0, 1)$, $P((a, b)) = b - a$. This particular probability is called the *Lebesgue measure* on $(0, 1)$.

The concept of Lebesgue measure is extended to the whole real line as follows. First, let Ω be a finite interval with endpoints $c < d$. Then, by the main extension theorem, there is a unique probability P on the Borel subsets of Ω such that $P((a, b)) = (b - a)/(d - c)$, where $c < a < b < d$. Therefore, there is a unique σ-additive and nonnegative set function $\lambda(\cdot)$ on the Borel subsets of Ω satisfying $\lambda((a, b)) = b - a$ for all $c < a < b < d$. The function $\lambda(\cdot)$ is called Lebesgue measure on the Borel subsets of the interval Ω. Next, let Ω_j, $-\infty < j < +\infty$, be disjoint finite intervals whose union is the whole real line R. Then $A \subset R$ is Lebesgue measurable if, for each j, $A \cap \Omega_j$ is a Borel set, and then the Lebesgue measure $\lambda(A)$ of A is defined as

$$\lambda(A) = \sum_{j=-\infty}^{+\infty} \lambda_j(A \cap \Omega_j)$$

where $\lambda_j(\cdot)$ is the previously defined Lebesgue measure on the Borel sets of Ω_j. If $A \subset \Omega_j$ is a Borel set then $\lambda(A) = \lambda_j(A)$, and thus the subscript j of $\lambda_j(\cdot)$ in the definition of $\lambda(A)$ above can be dropped.

Note that $\lambda(A) = +\infty$ is possible, but, in the definition, $\lambda(A)$ is always decomposed into a sum of finite values. In general, a nonnegative and σ-additive set function $\mu(\cdot)$ on some subsets of a set Ω is called a measure, and $\mu(\cdot)$ is called σ-finite if Ω can be decomposed into a denumerable collection of disjoint subsets Ω_j such that $\mu(\Omega_j) < +\infty$ for each j. Hence, Lebesgue measure is σ-finite.

Remark 3 If $C \subset \Omega$ which is not in $\mathscr{U}(\mathscr{F})$, it is rather difficult to decide whether or not $C \in \sigma(\mathscr{F})$. However, when it has been decided that $C \in \sigma(\mathscr{F})$, then $P^*(C)$ can arbitrarily closely be approximated by $P(A)$, $A \in \mathscr{F}$. Simply combine (1.10) and (1.13).

Exercises

1. Show that $\mathscr{U}(\mathscr{B})$ and P on $\mathscr{U}(\mathscr{B})$ have the following properties: (i) Ω and $\varnothing$ are in $\mathscr{U}(\mathscr{B})$ with $P(\Omega) = 1$ and $P(\varnothing) = 0$; furthermore, $0 \leq P(A) \leq 1$ on $\mathscr{U}(\mathscr{B})$; (ii) if A and B are in $\mathscr{U}(\mathscr{B})$, then so are $A \cup B$ and $A \cap B$ and $P(A \cup B) = P(A) + P(B) - P(A \cap B)$; (iii) if A and B are in $\mathscr{U}(\mathscr{B})$ and if $A \subset B$, then $P(A) \leq P(B)$; and (iv) if each A_j is in $\mathscr{U}(\mathscr{B})$ and if $A_j \subset A_{j+1}$, $j \geq 1$, then $\cup A_j \in \mathscr{U}(\mathscr{B})$ and $\lim P(A_n) = P(\cup A_j)$.

2. Let Ω and $\mathscr{B}$ be as given below. Let P be a set function defined on $\mathscr{B}$ that is consistent with the axioms of probability. Decide in each case below whether the extension of P from $\mathscr{B}$ to $\mathscr{F}(\mathscr{B})$ is unique:
   (i) $\Omega = (0, 1)$ and $\mathscr{B}$ is the set of all open subintervals (a, b) of Ω with either $b \leq \frac{1}{2}$ or $a \geq \frac{1}{2}$.
   (ii) $\Omega = (0, 1)$ and $\mathscr{B} = \{(0, 1/3), (1/4, 1/2)\}$.
   What is the relation of $\sigma(\mathscr{B})$ to the set of Borel subsets of Ω in part (i)?

3. Let B be a Borel set in the interval $(0, 1)$ and let P be Lebesgue measure (on $(0, 1)$). Show that, for every $\varepsilon > 0$, there exists a finite number of (disjoint) intervals $A_j = (a_j, b_j]$, $1 \leq j \leq n$, such that $B \subset \cup A_j$ and

$$\left| P(B) - \sum_{j=1}^{n} (b_j - a_j) \right| \leq \varepsilon$$

4. Show by the following indirect argument that there are non-Lebesgue measurable subsets of the interval $[0, 1]$. Assume every set $A \subset [0, 1]$ is Lebesgue measurable with measure $0 \leq P(A) \leq 1$. Define $B = \{P(A): P(A) \notin A\}$. Then B is a subset of $[0, 1]$, so $P(B)$ exists. We now face a contradiction when we want to decide whether or not $P(B) \in B$.

(A. R. Freedman (1985))

1.4 CLASSICAL PROBABILITY SPACES

Let the experiment Ω be finite or denumerably infinite. Assume that, for each member ω of Ω, $P(\{\omega\})$ is defined and either $P(\{\omega\}) = 0$ or $P(\{\omega\}) = p > 0$, the same number. By axiom (A2), there are only a finite number N of elements of Ω whose probability is p and $p = 1/N$. We can (at least hypothetically) drop those members of Ω whose probability is zero and work with the probability space $(\Omega_N, \mathscr{A}, P)$ where $\Omega_N = \{\omega_1, \omega_2, \ldots, \omega_N\}$, $\mathscr{A}$ is the set of all subsets of Ω_N, and $P(\{\omega_j\}) = 1/N$ for all j. In this reformulation of the model, it is clear that we speak of a *classical probability space* if the number of possible outcomes is finite and all outcomes are equally likely.

Let now $A = \{\omega_{j_1}, \omega_{j_2}, \ldots, \omega_{j_k}\} \subset \Omega_N$. Then $A \in \mathscr{A}$ and, by axiom (A3),

$$P(A) = k/N \tag{1.19}$$

In other words, the probability of the event A in a classical probability space is the ratio of the outcomes belonging to A to all possible outcomes of the experiment.

As an example we reproduce here the *model for the hypergeometric distribution*. Let a lot contain R items, out of which M are red and the rest are white. Select n items from the lot (either in a single pick or one by one without replacement of the items previously selected) in such a manner that all groups of n items have the same chance of being selected. Let B_s, $0 \leq s \leq \min(n, M)$, be the event that among the n items selected there are s red ones. We then have

$$P(B_s) = \binom{M}{s}\binom{R-M}{n-s} \Big/ \binom{R}{n} \qquad 0 \leq s \leq \min(n, M) \tag{1.20}$$

Indeed, by the definition of the model, there are $N = \binom{R}{n}$ possible outcomes of the experiment and $k = \binom{M}{s}\binom{R-M}{n-s}$ are favorable to B_s. Hence, (1.19) yields (1.20).

The preceding computation actually assumes that the selection of n items is done in a single pick. Hence, the outcomes are sets of n distinct items from the original lot of R items, where the ordering of the positions of the n items is irrelevant. Should we select the items one by one and without replacement than a typical outcome would be an n-vector whose components are distinct elements from the original lot of R, and thus their total number whould be $N = R(R-1)\cdots(R-n+1)$. In this case, the number of favorable outcomes to getting exactly s red items is

$$k = \binom{n}{s}M(M-1)\cdots(M-s+1)(R-M)(R-M-1)\cdots$$
$$\times [R - M - (n - s) + 1],$$

where we first counted the number of ways of getting s positions among the n coordinates for the s red items, and then multiplied this by the number of ways of filling in fixed s positions by red items and the rest by white ones. The ratio k/N once again yields (1.20) for $P(B_s)$.

The probabilities in (1.20) are called the *hypergeometric probabilities* or *distribution* (borrowing a name from another branch of mathematics). One can get a better insight into the model for the hypergeometric distribution by looking at the following sequence of events A_j. Let us choose the items one by one as described earlier. Let A_j be the event that the jth item drawn is red. Then, for an arbitrary choice of the subscripts $1 \leq j_1 < j_2 < \cdots < j_r \leq n$,

$$P(A_{j_1} \cap A_{j_2} \cap \cdots \cap A_{j_r}) = \frac{M(M-1)\cdots(M-r+1)}{R(R-1)\cdots(R-r+1)} \qquad (1.21)$$

The proof is an elementary combinatorial argument, by rule (1.19). Its details are left to the reader.

Formula (1.21) is very significant in the foundations of *exchangeable events*. We say that events are exchangeable if the probability of the intersection of any finite number of them does not depend on the actual subscripts, but only on the number of terms, in the intersection. Formula (1.21) shows that the A_j occurring there are exchangeable. We shall demonstrate later that there is a strong relation between arbitrary exchangeable events and formula (1.21).

Let us now change the selection rule. We shall select n items from a lot of M red and $R - M$ white ones, one by one and each selected item is returned to the lot before the next selection. Then the outcomes can again be represented by n-vectors, but the components are no longer distinct. Hence, the number of all outcomes $N = R^n$. Let B_s, $0 \leq s \leq n$, be as at (1.20). If we proceed in the counting of the favorable cases as in the case of selection of n items one by one but without replacement, we get $k = \binom{n}{s} M^s (R - M)^{n-s}$. Consequently,

$$P(B_s) = \binom{n}{s}\left(\frac{M}{R}\right)^s\left(1 - \frac{M}{R}\right)^{n-s} \qquad 0 \leq s \leq n \qquad (1.20a)$$

The probabilities above represent a binomial distribution (see Section 1.7 for the general case).

Exercises

1. Every time game G is played player A wins or loses \$1. Repeat the game n times, and denote the outcome by an n-vector with components $+1$ or -1, representing A's gain at each repetition. Assume that each possible outcome is equally likely. Let $C_{r,n}$ be the event that A's total profit in n games is r dollars, $-n \leq r \leq n$. Find $P(C_{r,n})$. Show that the maximum over r of $P(C_{r,2n})$ is $P(C_{0,2n})$ and that $P(C_{0,2n}) \geq P(C_{0,2n+2})$, $n \geq 1$.

2. Out of the integers 1 through 100 select 5 numbers (i) with replacement and (ii) without replacement. Let p_2 be the probability that exactly 2 even numbers occur. Calculate p_2 for both selection procedures (i) and (ii).

3. Estimate the difference between the right-hand sides of (1.20) and (1.20a). Deduce from your estimate that the preceding difference tends to zero for fixed s and n, as both M and R tend to infinity but M/R remains unchanged.

4. The integers 1 through 6 are put down in a random order meaning that each permutation occurs with the same probability. Find the probability that (i) number 3 gets into position 3, (ii) at least one number j gets into position j, and (iii) numbers 3 and 5 get into positions 3 and 5, respectively. Answer each question (i), (ii), and (iii) when the integers 1 through n are put down in random order, and find the limit in (ii) as $n \to +\infty$.

1.5 SIMPLE RANDOM VARIABLES

Let X be the number of red items obtained in the hypergeometric model. Then the value of X is determined at the completion of the experiment Ω (of selecting n items according to the model). In other words, X is well defined at every point ω of Ω, that is, X is a function on Ω. Additional properties of X are that the number of its possible values is finite and the sets $\{X = s\}$ are mutually exclusive. These properties exemplify the concept that we now introduce.

Definition 3 On a given probability space, a finite collection $C_1, C_2, \ldots, C_T$ of events is called a *partition* (of Ω) if $C_i \cap C_j = \varnothing$ for $i \neq j$ and $P(\cup C_j) = 1$.

Definition 4 On a given probability space, a function X defined on Ω is called a *simple random variable* if there is a partition $C_1, C_2, \ldots, C_T$ of Ω such that X is (a finite) constant on each C_j.

The above definition of a simple random variable implies that it can take only a finite number of values. Note that a partition C_j, $1 \leq j \leq T$, associated with the random variable X is not unique. As a matter of fact, if the partition C_j, $1 \leq j \leq T$, satisfies the requirement in Definition 4, then so do all partitions obtained from the C_j by splitting any or all C_j into disjoint events.

The following special simple random variable is of particular importance in probability theory. Let $A_1, A_2, \ldots, A_n$ be a (finite) sequence of events. Let f_n be the number of those A_j which occur. Then f_n is a simple random variable, which we call the *frequency* of the A_j, $1 \leq j \leq n$. The ratio f_n/n is called the *relative frequency*. The possible values of f_n are $0, 1, 2, \ldots, n$ and a partition of Ω on whose members f_n is constant is given by

$$C_0 = A_1^c \cap A_2^c \cap \cdots \cap A_n^c \tag{1.22}$$

and, for $1 \leq j \leq n$,

$$C_j = \cup(A_{i_1} \cap A_{i_2} \cap \cdots \cap A_{i_j} \cap A^c_{i_{j+1}} \cap \cdots \cap A^c_{i_n}) \qquad (1.23)$$

where the union is taken over all $1 \leq i_1 < i_2 < \cdots < i_j \leq n$. On C_j, $f_n = j$.

Another frequently used simple random variable is the indicator variable. Let A be an event. Then the indicator variable of A is defined as

$$I(A) = \begin{cases} 1 & \text{if } A \text{ occurs} \\ 0 & \text{if } A \text{ fails to occur} \end{cases}$$

Note that the frequency f_n of $A_1, A_2, \ldots, A_n$ can be written as

$$f_n = I(A_1) + I(A_2) + \cdots + I(A_n) \qquad (1.24)$$

Let us also record here

$$I(A^c) = 1 - I(A) \qquad I(A \cap B) = I(A)I(B) \qquad (1.25)$$

The following characteristics of simple random variables will be the center of our investigation.

Definition 5 Let X be a simple random variable with values $x_1, x_2, \ldots, x_T$ and with partition $C_j = \{X = x_j\}$, $1 \leq j \leq T$. The sequence $p_j = P(C_j) = P(X = x_j)$, $1 \leq j \leq T$, with $p_j > 0$, is called a distribution of X. When the x_j are distinct, the corresponding distribution is called the *distribution* of X. The value

$$E(X) = x_1 p_1 + x_2 p_2 + \cdots + x_T p_T$$

is called the *expected value*, or *expectation*, of X.

Obviously, the expected value is the same for all distributions of X. Note that, for an indicator variable $I(A)$, $E(I(A)) = P(A)$. Furthermore, the definition of $E(X)$ implies that

$$E(X) \geq 0 \qquad \text{whenever} \quad X \geq 0 \qquad (1.26)$$

Theorem 14 Let X and Y be two simple random variables on the same probability space. Then

(i) $E(aX + bY) = aE(X) + bE(Y)$ for constants a and b.
(ii)   if $X \leq Y$, then $E(X) \leq E(Y)$.

Proof. Let X and Y take the values of $x_1, x_2, \ldots, x_T$ and $y_1, y_2, \ldots, y_M$, with the corresponding partitions $C_1, C_2, \ldots, C_T$ and $D_1, D_2, \ldots, D_M$, respectively. Then $B_{ij} = C_i \cap D_j$, $1 \leq i \leq T$, $1 \leq j \leq M$, is a common partition for X and Y and $Z(a, b) = aX + bY$ is a simple random variable with the constant value $ax_i + by_j$ on B_{ij}. Thus

$$E(aX + bY) = \sum_i \sum_j (ax_i + by_j)P(B_{ij})$$

$$= \sum_i \sum_j ax_i P(B_{ij}) + \sum_i \sum_j by_j P(B_{ij})$$

$$= a \sum_i x_i \sum_j P(B_{ij}) + b \sum_j y_j \sum_i P(B_{ij})$$

$$= a \sum_i x_i P(C_i) + b \sum_j y_j P(D_j)$$

$$= aE(X) + bE(Y)$$

which completes the proof of part (i).

For part (ii), choose $a = -1$ and $b = 1$. Then $Z(-1, 1) = Y - X \geq 0$, from which (ii) immediately follows upon an appeal to (1.26) and to part (i) of the theorem. ∎

By induction, part (i) of Theorem 14 extends to a finite number of summands. Thus, from (1.24),

$$E(f_n) = P(A_1) + P(A_2) + \cdots + P(A_n)$$

In particular, if each $P(A_j) = p$, the same value, then $E(f_n/n) = p$.

Definition 6   Let A be an event with $P(A) = p$. When we say that the events $A_1, A_2, \ldots, A_n$ are copies of A, it is then required that $P(A_j) = p$ for each j.

Thus, the expected value of the relative frequency f_n/n in n *copies of A* equals $P(A)$. We intend to show more: under some additional assumption, we want to show that the relative frequency f_n/n in n copies of A converges (in some sense) to $P(A)$, thus establishing the practical meaning of $P(A)$. Instrumental in proving such a limit theorem is the following inequality of Chebyshev.

Theorem 15   Let X be a simple random variable with values $x_1, x_2, \ldots, x_T$ and with distribution $p_1, p_2, \ldots, p_T$. Then, for every $a > 0$,

$$P(|X - E(X)| \geq a) \leq V/a^2$$

where

$$V = V(X) = \sum_{j=1}^{T} (x_j - E(X))^2 p_j$$

Proof. Clearly,

$$V \geq \sum_{\substack{j \\ |x_j - E(X)| \geq a}} (x_j - E(X))^2 p_j \geq a^2 \sum_{\substack{j \\ |x_j - E(X)| \geq a}} p_j$$

Upon observing that the extreme right-hand side equals

$$a^2 P(|X - E(X)| \geq a),$$

the desired inequality follows. ∎

Note that V can also be written as

$$V = V(X) = E\{[X - E(X)]^2\} = E(X^2) - E^2(X)$$

the last equation following from Theorem 14(i). When $X = f_n$, the frequency in n events, easy computations (using (1.24), (1.25), and Theorem 14(i)) yield

$$V(f_n) = \sum_{j=1}^{n} P(A_j) + 2 \sum_{i<j} P(A_i \cap A_j) - \left[\sum_{j=1}^{n} P(A_j) \right]^2 \qquad (1.27)$$

We thus have from Chebyshev's inequality the following result.

Theorem 16 Let $A_1, A_2, \ldots, A_n$ be events such that $P(A_j) = p$ for each j, and

$$\frac{2}{n^2} \sum_{i<j} P(A_i \cap A_j) - p^2 \to 0 \qquad \text{as} \quad n \to +\infty$$

Then, for every $a > 0$, as $n \to +\infty$,

$$\lim P(|f_n/n - p| \geq a) = 0 \qquad (1.28)$$

where f_n is the frequency of the A_j.

Proof. Observe that $V(f_n/n) = V(f_n)/n^2$. Thus (1.27) and Chebyshev's inequality (Theorem 15) imply the claimed limit. ∎

We shall refer to (1.28) by saying that f_n/n converges to p in probability.

Theorem 17 If the events $A_1, A_2, \ldots, A_n$ are such that, for each j, $P(A_j) = p$, and

$$\frac{2}{n^2} \sum_{i<j} P(A_i \cap A_j) - p^2 \leq \frac{c}{n} \qquad (1.29)$$

with some constant $c > 0$, then as $n \to +\infty$,

$$P(\lim f_n/n = p) = 1 \tag{1.30}$$

In particular, (1.29), and thus (1.30) as well, hold if, for all $i < j$, $P(A_i \cap A_j) \le p^2$.

Proof. We first prove (1.30) for the subsequence $n = N^2$, $N \ge 1$ integer. By Chebyshev's inequality, and because of (1.27) and (1.29), for arbitrary $a > 0$,

$$P(|f_n/n - p| \ge a) \le V(f_n/n)/a^2 \le (p + c)/a^2 n$$

Now, since $n = N^2$, the sum of the above probabilities over all N converges. Hence, the Borel–Cantelli lemma (Theorem 10) is applicable, yielding that, with probability one, for sufficiently large N, with $n = N^2$, $|f_n/n - p| < a$, where $a > 0$ is arbitrary. This is exactly the statement at (1.30) with $n = N^2$. We thus have that there is an event D with $P(D) = 1$ such that, on D, $f_n/n \to p$ if $n = N^2$.

Let n now be arbitrary. Let N be a unique integer with $N^2 \le n < (N + 1)^2$. Then, from the decomposition (1.24) of the frequency f_n, we get

$$0 \le f_{(N+1)^2} - f_n = I(A_{n+1}) + \cdots + I(A_{(N+1)^2}) \le (N + 1)^2 - n$$

which, by the choice of N, is further majorized by

$$(N + 1)^2 - N^2 = 2N + 1 \le 2\sqrt{n} + 1$$

Hence, upon dividing by n and observing that $(N + 1)^2/n \to 1$ as $n \to +\infty$, the first part of the proof implies that, on D, $f_n/n \to p$ as $n \to +\infty$. The proof is completed. ∎

From Chebyshev's inequality we can also deduce Theorem 11 of Section 1.2.

Proof of Theorem 11. With the notation of (1.7), $E(f_n) = b_{1,n}$ and formula (1.27) becomes

$$V(f_n) = 2b_{2,n} + b_{1,n} - b_{1,n}^2$$

Hence, by Chebyshev's inequality,

$$P(|f_n - b_{1,n}| \ge (1/2)b_{1,n}) \le 4(2b_{2,n} + b_{1,n} - b_{1,n}^2)/b_{1,n}^2$$

By assumptions (1.6) and (1.8), we thus have

$$\liminf P(f_n < (1/2)b_{1,n}) = 0 \qquad (n \to +\infty)$$

We can therefore choose a subsequence n_k such that

$$\sum_{k=1}^{+\infty} P(f_{n_k} < (1/2)b_{1,n_k}) < +\infty$$

Consequently, the Borel–Cantelli lemma is applicable, which yields that, with probability one, there is a finite value k_0 such that, for $k \geq k_0$, $f_{n_k} \geq (1/2)b_{1,n_k}$. By (1.6), the main part of the theorem is established.

In order to complete the proof, we note that, in the special case $P(A_i \cap A_j) \leq P(A_i)P(A_j)$ for all $i \neq j$, $2b_{2,n} \leq b_{1,n}^2$. On the other hand, from $V(f_n) \geq 0$, we always have $2b_{2,n} + b_{1,n} \geq b_{1,n}^2$. Hence, in the special case above, (1.8) holds whenever (1.6) does. This concludes the proof. ∎

Remark Note that we have actually proved that, with probability one, as $n \to +\infty$, $\lim \sup f_n/b_{1,n} \geq 1/2$. With no change in the proof, $1/2$ can be increased to one.

Exercises

1. For the events $A_1, A_2, \ldots, A_n$, define

$$b_{k,n} = \sum P(A_{i_1} \cap A_{i_2} \cap \cdots \cap A_{i_k}) \qquad k \geq 1$$

where the summation is over $1 \leq i_1 < i_2 < \cdots < i_k \leq n$. Show

$$E\left[\binom{f_n}{k}\right] = b_{k,n}$$

where f_n is the frequency of A_j, $1 \leq j \leq n$. Hence, prove

(i) $b_{k,n} \Big/ \binom{n}{k} \geq b_{k+1,n} \Big/ \binom{n}{k+1}$, $1 \leq k < n$.

(ii) for every integer $0 \leq m \leq (n-1)/2$,

$$\sum_{k=0}^{2m+1} (-1)^k \binom{k+r}{r} b_{k+r,n} \leq P(f_n = r) \leq \sum_{k=0}^{2m} (-1)^k \binom{k+r}{r} b_{k+r,n}$$

2. With the notation of the preceding exercise, show that, for all integers $k \geq 1$ and for every $n \geq 1$,

$$P\left(\bigcup_{j=1}^{n} A_j\right) \geq \frac{2}{k+1} b_{1,n} - \frac{2}{k(k+1)} b_{2,n}$$

Find the optimal value of k.

3. From the inequality of the preceding exercise, deduce

$$P\left(\bigcup_{j=1}^{n} A_j\right) \geq b_{1,n}^2/(2b_{2,n} + b_{1,n})$$

4. (The method of indicators) Let B_j, $1 \leq j \leq M$, be Boolean functions of the events $A_1, A_2, \ldots, A_n$, that is, each B_j can be expressed by a finite number of the operations of union, intersection, and complementation. Let c_j, $1 \leq j \leq M$, be real numbers. Show that, for arbitrary $A_1, A_2, \ldots, A_n$,

$$\sum_{j=1}^{M} c_j P(B_j) \geq 0$$

if and only if

$$\sum_{j=1}^{M} c_j I(B_j) \geq 0$$

Give a new proof for the inequalities of Exercises 1–3.

5. Choose a point x from the interval $(0, 1)$, where the underlying probability space is Lebesgue measure on the Borel sets of $(0, 1)$. Let A_j be the event that the jth decimal digit of x is, say, 3. Show that Theorem 17 is applicable, and thus conclude that, for almost all x, the relative frequency of 3 among the decimal digits of x is $1/10$.

1.6 CONDITIONAL PROBABILITY: THE ELEMENTARY CASE

It was shown in the preceding section that, under certain conditions, the relative frequency f_n/n in a sequence of copies of an event C converges, with probability one, to $P(C)$. Now, if A and B are both events on a given probability space, and if n observations are made, in each of which A and B might have occurred, then the information that B did occur reduces the number n of observations to the number $f_n(B)$, the frequency of B, which is to be used for the "prediction" of the number of occurrences of A. That is, the "new" frequency of A is in fact the frequency $f_n(A \cap B)$ in the old sequence of observations. Thus, the "new" relative frequency is $f_n(A \cap B)/f_n(B)$, which, upon dividing both the numerator and the denominator by n, converges to $P(A \cap B)/P(B)$, with probability one, whenever $P(B) > 0$. Thus, the only reasonable concept of conditional probability is as follows.

Definition 7 Let A and B be two events on the probability space $(\Omega, \mathscr{A}, P)$. Assume $P(B) > 0$. The conditional probability $P(A|B)$ of A, given B, is defined by the formula

$$P(A|B) = P(A \cap B)/P(B)$$

The most significant properties of conditional probability are presented in the following three theorems.

Theorem 18 $P_B(A) = P(A|B)$ is a probability on the smallest σ-field $\mathscr{A}_B$ containing all those subsets C of B which are of the form $C = A \cap B$ with $A \in \mathscr{A}$.

Proof. It follows from Definition 7 and from the monotonicity of probability that $0 \le P(A|B) \le 1$ and that $P(B|B) = 1$. Next, recall Exercise 4 from Section 1.1, which implies that $\mathscr{A}_B$ is simply the collection of $A \cap B$ with $A \in \mathscr{A}$. Thus, it only remains to be proven that if $A_i \in \mathscr{A}$ and $A_i \cap A_j = \emptyset$ if $i \ne j$, then

$$P(\cup A_i|B) = \sum P(A_i|B)$$

This, however, immediately follows from Definition 7 and from axiom (A3) of P upon observing that if the A_i are mutually exclusive then so are the events $A_i \cap B$, $i \ge 1$. The proof is completed. ∎

Theorem 19 (The Total Probability Rule) If the events B_j, $j \ge 1$, are such that $P(B_j) > 0$, $B_i \cap B_j = \emptyset$ if $i \ne j$, and $P(\cup B_j) = 1$, then for an arbitrary event A,

$$P(A) = \sum P(A|B_j)P(B_j)$$

Proof. Write $\Omega = D \cup (\cup B_j)$, where $D \cap B_j = \emptyset$. Then $P(D) = 0$ and therefore $P(A \cap D) = 0$ by monotonicity. We thus have

$$P(A) = P(A \cap \Omega) = P(A \cap D) + \sum P(A \cap B_j) = \sum P(A|B_j)P(B_j)$$

which was to be proved. ∎

A combination of the Total Probability Rule and the definition of conditional probability can be utilized to answer questions of significance to society. Decisions on such questions as whether to introduce compulsory AIDS tests for some groups of professionals, and whether to go through expensive and drastic surgery in order to prevent the spread of some kinds of cancer diagnosed earlier by a seemingly accurate test are usually made on moral and political grounds rather than on scientific ones. The shocking surprise in the following very instructive example is that medical advancement in the tests involved would be of very little help.

Assume that a screening test is 98% effective in detecting disease A when present, and that the test shows A for only 1% of those persons who do not suffer from it. Estimates show that 0.9% of the population has disease A. Let us evaluate the probability of the event B that someone is free of A, given that the test is positive for the patient.

Let $T(A)$ be the event that the test is positive. Then the assumptions above can be translated as $P(T(A)|B) = 0.01$, $P(T(A)|B^c) = 0.98$, $P(B^c) = 0.009$, and thus $P(B) = 0.991$. Consequently, by the Total Probability Rule,

$$P(T(A)) = P(T(A)|B)P(B) + P(T(A)|B^c)P(B^c)$$
$$= 0.01 \times 0.991 + 0.98 \times 0.009 = 0.01873$$

which yields

$$P(B|T(A)) = \frac{P(B \cap T(A))}{P(T(A))} = \frac{P(T(A)|B)P(B)}{P(T(A))} = \frac{0.01 \times 0.991}{0.01873} = 0.529$$

That is, 52.9% of those for whom the test is positive do not have A. What is "wrong" with the test is that the percentage of the population in A is close to the percentage of positive test results among those who are not in A. Note that this same test is very accurate when the result is negative (compute $P(B|T(A)^c)$).

Let us generalize the formula we used in the preceding computation. We say that a sequence $B_1, B_2, \ldots$ of events forms a complete system if $P(B_j) > 0$ for every j and $P(\bigcup_j B_j) = 1$; furthermore, $B_i \cap B_j = 0$ for $i \neq j$. Then, for an arbitrary event C with $P(C) > 0$,

$$P(B_j|C) = \frac{P(C|B_j)P(B_j)}{\sum_k P(C|B_k)P(B_k)}$$

The proof is just a simple combination of the Total Probability Rule and the definition of conditional probability as in the special case of the example above. The formula itself is known as Bayes' Theorem, which is a fundamental formula in Bayesian statistics.

Theorem 20 If the events $A_1, A_2, \ldots, A_n$ are such that

$$P(A_2 \cap A_3 \cap \cdots \cap A_n) > 0,$$

then

$$P(A_1 \cap A_2 \cap \cdots \cap A_n) = P(A_n) \prod_{k=1}^{n-1} P(A_{n-k}|A_{n-k+1} \cap \cdots \cap A_n)$$

Proof. By the successive application of Definition 7 on the righthand side above we see that all terms except the last numerator cancel out, which is indeed the left-hand term. ∎

Exercises

1. Choose a point x from the interval $(0, 1)$ according to the probability space $(\Omega, \mathscr{A}, P)$, where $\Omega = (0, 1)$, $\mathscr{A}$ is the set of Borel subsets of $(0, 1)$, and P is Lebesgue measure. Let X be the first decimal digit of x. Let B be the event that $x < 1/2$. Give the distribution and the expectation of the simple random variable X both on $(\Omega, \mathscr{A}, P)$ and on $(B, \mathscr{A}_B, P_B)$, where $P_B(\cdot) = P(\cdot|B)$.
2. Roll two regular dice and let X and Y be the numbers on the dice. Find $P(X = j|X + Y = k)$, $1 \le j \le \min(6, k)$, $2 \le k \le 12$.
3. A random number N is produced by a machine, where the probability that $N = n$ equals $10^n e^{-10}/n!$. If $N = n$, the "gambler" rolls n regular dice (or one die n times), and wins if exactly 2 dice show the number 6. Find the probability that the gambler wins (if $N = 0$, the gambler loses).
4. For the events A_1, A_2 and A_3, $P(A_3) \doteq 0.5$, $P(A_2|A_3) = 0.6$, $P(A_3|A_2) = 0.8$, and $P(A_1|A_2 \cap A_3) = 0.4$. Find $P(A_1 \cap A_3|A_2)$.

1.7 INDEPENDENCE OF EVENTS

Since conditional probability reflects the influence of one event on another, it is natural to define independence by means of conditional probabilities.

Definition 8 We say that two events A and B are independent if either (i) one of them is of zero probability, or (ii) if $P(B) > 0$, then $P(A|B) = P(A)$. Equivalently, A and B are independent if

$$P(A \cap B) = P(A)P(B) \tag{1.31}$$

The fact that both (i) and (ii) are equivalent to (1.31) is a simple consequence of the monotonicity of probability in case (i) and of Definition 7 in case (ii).

Theorem 21 If A and B are independent, then so are A^c and B.

Proof. By Theorem 2 and by independence

$$P(A^c \cap B) = P(B) - P(A)P(B) = P(B)[1 - P(A)] = P(B)P(A^c)$$

that is, (1.31) also applies to A^c and B. ∎

That some calculation is needed to show independence can be demonstrated by the fact that, when rolling two dice, a six showing on one die is independent of the sum of seven on the two dice, but a six showing on one die is dependent on the sum of, say, eight, on the two dice. Just check it through (1.31).

Definition 9 The events $A_1, A_2, \ldots, A_n$ are called independent if, for every $k \geq 1$ and for all choices of the subscripts $1 \leq i_1 < i_2 < \cdots < i_k \leq n$.

$$P(A_{i_1} \cap A_{i_2} \cap \cdots \cap A_{i_k}) = P(A_{i_1})P(A_{i_2}) \cdots P(A_{i_k}) \qquad (1.32)$$

When (1.32) is satisfied for $k = 2$ only, the A_j are called pairwise independent.

The following simple example shows that not all pairwise independent events are independent. Assume that an urn contains four identical balls, in which, on a slip of paper, the following numbers are written: 1, 2, 3 and 123. A ball is chosen at random and A_j denotes the event that the digit j is on the slip. Then $P(A_j) = 1/2$ for $1 \leq j \leq 3$, and $P(A_i \cap A_j) = 1/4$ for $1 \leq i < j \leq 3$. Hence, the 3 events A_1, A_2, and A_3 are pairwise independent. But they are not independent, because (1.32) fails for $k = 3$: $P(A_1 \cap A_2 \cap A_3) = 1/4$ and not 1/8.

Remark Recall that the major assumptions of Theorems 11 and 17, (1.8) and (1.29), respectively, are satisfied for pairwise independent events.

Definition 10 An infinite sequence of events consists of independent events if every finite segment of the sequence forms a set of independent events.

Definition 11 The collections $\mathscr{A}_j$ of events, $j \geq 1$, are independent if, for every possible n and for every choice of the events $A_j \in \mathscr{A}_j, A_1, A_2, \ldots, A_n$ are independent.

This last definition gives an accurate meaning to the concept of repeating an experiment independently of the former trials. As a matter of fact, let $(\Omega, \mathscr{A}, P)$ be a probability space. For each i, $1 \leq i \leq n$, let $(\Omega_i^*, \mathscr{A}_i^*, P_i)$ be a probability space such that there is an isomorphism between Ω and Ω_i^*, which is also an isomorphism between $\mathscr{A}$ and $\mathscr{A}_i^*$, and if $A \in \mathscr{A}$ and $A_i \in \mathscr{A}_i^*$ are isomorphic pairs, then $P(A) = P_i(A_i)$. We call the experiments Ω_i^* with the underlying probability spaces $(\Omega_i^*, \mathscr{A}_i^*, P_i)$ repetitions, or copies, of $(\Omega, \mathscr{A}, P)$. These n repetitions can be combined into a single experiment (with an associated probability space). In what follows, the construction of such a probability space is given, when, in addition, it is required that the repetitions be independent.

We carry out the construction in somewhat more general terms. Let $(\Omega_i, \mathscr{A}_i, P_i)$ $1 \leq i \leq n$, be probability spaces. Define the product sample space

$$\Omega^{(n)} = \{(\omega_1, \omega_2, \ldots, \omega_n): \omega_i \in \Omega_i, 1 \leq i \leq n\}$$

a cylinder set

$$A^{(n)} = \{(\omega_1, \omega_2, \ldots, \omega_n): \omega_i \in A_i \in \mathscr{A}_i, 1 \leq i \leq n\}$$

and the field $\mathscr{F}^{(n)}$ as the set of all finite unions of (disjoint) cylinders. We call the A_i, $1 \leq i \leq n$, in the definition of $A^{(n)}$ the (basic) edges of $A^{(n)}$. Now, for the cylinder $A^{(n)}$ with basic edges A_i, $1 \leq i \leq n$, put

$$P^{(n)}(A^{(n)}) = P_1(A_1)P_2(A_2)\cdots P_n(A_n) \tag{1.33}$$

Finally, $P^{(n)}$ is extended to $\mathscr{F}^{(n)}$ by additivity.

Theorem 22 Let $\mathscr{A}^{(n)}$ be the smallest σ-field containing $\mathscr{F}^{(n)}$. There is a unique probability on $\mathscr{A}^{(n)}$, which is also denoted by $P^{(n)}$, such that, on cylinders $A^{(n)}$, (1.33) holds.

Definition 12 The probability space $(\Omega^{(n)}, \mathscr{A}^{(n)}, P^{(n)})$ is called the product space of the probability spaces $(\Omega_i, \mathscr{A}_i, P_i)$, $1 \leq i \leq n$. If the latter probability spaces are copies of a single probability space $(\Omega, \mathscr{A}, P)$, then the product space represents the independent repetitions of the experiment whose underlying probability space is $(\Omega, \mathscr{A}, P)$.

Proof of Theorem 22. In view of the main extension theorem (Theorem 13), it suffices to prove that $P^{(n)}$ is a probability on $\mathscr{F}^{(n)}$. Since, by definition, $P^{(n)}$ is an additive set function on $\mathscr{F}^{(n)}$, because of Theorem 7, we have to prove that $P^{(n)}$ on $\mathscr{F}^{(n)}$ is continuous at $\varnothing$. Clearly, it suffices to prove that, for a nonincreasing sequence $A_k^{(n)}$ of cylinders, the validity of $P^{(n)}(A_k^{(n)}) \geq \varepsilon > 0$ for all k implies that $\lim A_k^{(n)}$, as $k \to +\infty$, is not empty.

Let $A_k^{(n)}(\omega_n)$ be the section of $A_k^{(n)}$ at ω_n, that is, a subset of the product of the sample spaces Ω_i, $1 \leq i \leq n - 1$, defined by

$$A_k^{(n)}(\omega_n) = \{(\omega_1, \ldots, \omega_{n-1}): (\omega_1, \ldots, \omega_{n-1}, \omega_n) \in A_k^{(n)}\}$$

Let $C_{k,1}^{(n)}$ be the set of all ω_n such that

$$P^{(n-1)}(A_k^{(n)}(\omega_n)) > \varepsilon/2 \tag{1.34}$$

Then, by assumption and by (1.33),

$$\varepsilon \leq P^{(n)}(A_k^{(n)}) \leq P_n(C_{k,1}^{(n)}) + (\varepsilon/2)(1 - P_n(C_{k,1}^{(n)}))$$

and thus

$$P_n(C_{k,1}^{(n)}) > \varepsilon/2$$

Now, since $A_k^{(n)}$ is nonincreasing in k, so is $C_{k,1}^{(n)}$; consequently, as $k \to +\infty$, $\lim C_{k,1}^{(n)} = C_1$ exists and $P_n(C_1) \geq \varepsilon/2$. Therefore, C_1 is not empty, that is, there is an ω_n^* contained in all $C_{k,1}^{(n)}$ and, for every k, (1.34) applies with ω_n^* in place of ω_n. Upon repeating the foregoing argument with sections $A_k^{(n)}(\omega_n^*, \omega_{n-1})$ of $A_k^{(n)}(\omega_n^*)$ at ω_{n-1}, we obtain an ω_{n-1}^* in Ω_{n-1} such that

$$P^{(n-2)}(A_k^{(n)}(\omega_n^*, \omega_{n-1}^*)) > \varepsilon/4$$

Further repetitions of the preceding argument yield, in n steps, that there exists a point $(\omega_1^*, \omega_2^*, \ldots, \omega_n^*)$ contained in all $A_k^{(n)}$, that is, their limit, as $k \to +\infty$, is not empty. This completes the proof. ∎

The foregoing proof can be repeated almost verbatim to prove the following more general result of Andersen and Jessen (1946). Let T be an arbitrary set (in a finite-dimensional real space) and let T_n be an (arbitrary) subset of n elements of T. Let $(\Omega_t, \mathscr{A}_t, P_t)$, $t \in T$, be a family of probability spaces. Define the product sample space Ω^T as the collection of $\omega_T = (\omega_t, t \in T)$ such that $\omega_t \in \Omega_t$. The collection of those points ω_T for which $\omega_t \in A_t \in \mathscr{A}_t$ if $t \in T_n$ and $\omega_t \in \Omega_t$ if t is in the complement of T_n (i.e., in $T - T_n$) is called a cylinder with base A_t, $t \in T_n$. On a cylinder B_{T_n} with base A_t, $t \in T_n$, define P^T by

$$P^T(B_{T_n}) = \prod_{t \in T_n} P_t(A_t)$$

and extend the definition of P^T to the field $\mathscr{F}$ of all finite (disjoint) unions of cylinders by additivity. Finally, let $\mathscr{A}^T$ be the smallest σ-field containing $\mathscr{F}$. Then the following result holds.

Theorem 23   There is a unique probability on $\mathscr{A}^T$ that coincides with P^T on $\mathscr{F}$.

As was remarked, the proof of Theorem 22 is to be repeated. First note that, when arguing with a nonincreasing sequence of cylinders, only a denumerably infinite sequence of t values is taken into account. Thus, instead of an arbitrary set, one deals with a denumerable set $T^* = \{t_1, t_2, \ldots\}$. Next, instead of arguing successively with sections of cylinders at ω_n down to ω_1, take the first section at ω_{t_1}, then proceed to ω_{t_2}, and so on. The only reason for having chosen the reverse order in the proof of Theorem 22 is the simplicity of notation that way. Otherwise, no change is required.

The following statement, although part of elementary probability theory, is very significant.

Theorem 24 Let A be an event in connection with an experiment with underlying probability space (Ω, A, P). Let $P(A) = p$. Independently repeat the experiment n times and let f_n be the frequency of A in these repetitions. Then

$$P^{(n)}(f_n = r) = \binom{n}{r} p^r (1 - p)^{n-r} \qquad 0 \le r \le n$$

Proof. Let A_j be the event that A has occurred at the jth repetition of the experiment. Then A_j is a cylinder set with $P^{(n)}(A_j) = p$ and the events A_j, $1 \le j \le n$, are independent. By Theorem 21, they remain independent if some of them are replaced by their complements. Now, the event $f_n = r$ is the union of those (disjoint) cylinders whose edges contain exactly r of the A_j, which are isomorphic to A, and the rest are isomorphic to A^c. Thus, the $P^{(n)}$ probability of each of the cylinders in the union equals $p^r (1 - p)^{n-r}$ and, since the number of terms in the union is $\binom{n}{r}$, the statement follows. ∎

The frequency f_n is a simple random variable on the product space describing the n repetitions. Thus, Theorem 24 gives the distribution of f_n, which is called the *binomial distribution* (with parameters n and p). Easy calculations yield that

$$E(f_n) = np \qquad \text{and} \qquad V(f_n) = np(1 - p)$$

Thus, the Chebyshev inequality takes a simple and specific form. It is immediately apparent that f_n/n converges in probability, and with probability one in the infinite product space, to p (see Section 1.5). Much more accurate results can, however, be established. See the Poisson and normal approximations in Chapter 2 of the introductory volume; far-reaching generalizations are presented in the succeeding chapters of the present book.

Exercises

1. Let A be an event on a given probability space. Repeat the experiment independently in connection with A and stop as soon as A has occurred exactly n times. Describe the probability space in which the probability of requiring an even number of repetitions for A to occur n times can be evaluated. Find this probability.

2. Let X be a simple random variable on a given probability space. Repeat the experiment independently in connection with X n times. Let X_j be

the value of X at the jth repetition. Prove that, on the infinite product space $(T = \{1, 2, \ldots\})$, the arithmetical mean

$$\frac{1}{n}(X_1 + X_2 + \cdots + X_n)$$

converges, with probability one, to $E(X)$.

3. Let $X_1, X_2, \ldots, X_n$ be simple random variables, each taking the values $+1$ and -1 with $P(X_j = 1) = 1/2$. Assume that the events $\{X_j = 1\}$, $1 \leq j \leq n$, are independent. Set $S_0 = 0$ and $S_j = X_1 + X_2 + \cdots + X_j$, $1 \leq j \leq n$. Connect the points (j, S_j) in the (x, y)-plane by straight lines, and call such a piecewise linear function $L_n(x)$, $0 \leq x \leq n$, a path from $(0, 0)$ to (n, S_n). By reflecting $L_n(x)$ on the x-axis, show that

$$P(S_j > 0 \text{ for all } 1 \leq j \leq n - 1 \text{ and } S_n = k > 0)$$
$$= P(S_1 = +1, S_n = k) - P(S_1 = -1, S_n = k)$$
$$= (1/2)[P(S_{n-1} = k - 1) - P(S_{n-1} = k + 1)]$$

4. Upon choosing n an odd integer, $k = 1$ and $X_{n+1} = -1$, deduce from the preceding result that

$$P(S_j > 0, 1 \leq j \leq 2m - 1, S_{2m} = 0) = \frac{1}{m}\binom{2m-2}{m-1}\frac{1}{2^{2m}}.$$

1.8 RANDOM VARIABLES, DISTRIBUTION FUNCTIONS

In many instances, it is not the actual outcome of an experiment, but rather some functions of it that are of interest. For example, when choosing a member of a population, an insurance company's interest might be the length of time until an accident by the selected person, a tax agency's interest might be the income of this individual, etc. When we assign values to the outcome of an experiment, we speak of random variables.

Definition 13 Given a probability space $(\Omega, \mathscr{A}, P)$, a real-valued function $X = X(\omega)$, defined on Ω, is a random variable if, for every real number z, the set $[\omega: X(\omega) \leq z\}$, which we abbreviate by $X \leq z$, belongs to $\mathscr{A}$. The function

$$F(z) = F_X(z) = P(X \leq z)$$

defined for all real numbers z, is called the distribution function of X.

Remark Notice that if X is a random variable on a complete probability space $(\Omega, \mathscr{A}, P)$ and if X^* is a real-valued function defined on Ω such that $P(X = X^*) = 1$, then X^* is a random variable also and $F_X(z) = F_{X^*}(z)$. In light of this, we adopt the following convention, whether the underlying probability space is complete or not: random variables need be defined with probability one only. In other words, the values of a random variable can be freely changed on a set of probability zero, and thus it may take the values of plus or minus infinity on such sets. However, since this flexibility does not affect computations in terms of a distribution (function), in proofs involving random variables and their distribution functions, we always argue as if all random variables were defined and finite on the whole Ω. Thus, those arguments are valid with probability one only. This convention, however, considerably reduces the lengths of arguments without loss of generality.

Clearly, the simple random variable discussed in Section 1.5 is a random variable. If X is a simple random variable with the value set $x_1, x_2, \ldots, x_s$ and with distribution $p_1, p_2, \ldots, p_s$, then

$$F_X(z) = \sum_{\substack{i \\ x_i \leq z}} p_i \tag{1.35}$$

where an empty sum is taken as zero. Hence, for a simple random variable X, $F_X(z)$ is a step function. This however, is not a characteristic of simple functions.

Definition 14 If there is a nowhere dense, denumerable set $x_1, x_2, \ldots$ of real numbers such that the distribution function $F_X(z)$ of the random variable X is of the form (1.35), then X, or $F(z)$, is called discrete. The sequence x_1, $x_2, \ldots$ is called the set of values of X and the sequence $p_j = F(x_j) - F(x_j - 0)$, $j \geq 1$, is called the distribution of X. If the set of values of X consists of a single element (x_1), then $F_X(z)$ is called degenerate (at x_1).

The basic properties of a distribution function, given in Theorem 25 below, imply that if the sequence $p_j, j \geq 1$, is the distribution of a discrete random variable X, then $p_j \geq 0$ and $p_1 + p_2 + \cdots = 1$. In particular, if $F(z)$ is degenerate at x_1, then $F(z) = 0$ for $z < x_1$ and $F(z) = 1$ for $z \geq x_1$.

Theorem 25 Every distribution function $F(z)$ satisfies

(i) $F(z)$ is nondecreasing.
(ii) $P(a < X \leq b) = F(b) - F(a)$.
(iii) $F(z)$ is right continuous.
(iv) $\lim F(x) = 0$ as $x \to -\infty$; $\lim F(x) = 1$ as $x \to +\infty$.

Furthermore,

$$F(a) - F(a - 0) = P(X = a) \tag{1.36}$$

Proof. Property (i) follows from the monotonicity of probability (Theorem 3), while property (ii) is an immediate consequence of Theorem 2. Properties (iii) and (iv) and formula (1.36) are easy consequences of the continuity of probability (Theorem 6).   ■

Formula (1.36) thus tells us that, for a discrete random variable X, $p_j = P(X = x_j), j \geq 1$.

Another consequence of Theorem 25 is that, for continuity points a and b of $F(x)$, $P(a < X \leq b) = P(a \leq X \leq b) = F(b) - F(a)$, but at discontinuities the combination of (ii) and (1.36) must be used in computations. For example, if the distribution function of the random variable X

$$F(x) = \begin{cases} 0 & \text{if } x < 0 \\ x^2 & \text{if } 0 \leq x < 1/2 \\ 1 - 1/8x^2 & \text{if } x \geq 1/2 \end{cases}$$

then $P(0.1 < X \leq 2) = P(0.1 < X < 2) = P(0.1 \leq X \leq 2) = 1 - 1/32 - 0.1^2$ but $P(0.5 \leq X \leq 2) = F(2) - F(0.5 - 0) = 1 - 1/32 - 0.5^2$ while $P(0.5 < X \leq 2) = F(2) - F(0.5) = 1 - 1/32 - (1 - 1/8(0.5)^2)$.

On the basis of Theorem 25, we can construct a probability associated with a random variable X on the Borel sets of the real line. Namely, for the field $\mathscr{F}$ of finite unions of right semiclosed intervals $(a, b]$, let $P_X((a, b]) = F(b) - F(a)$, and, for finite unions of disjoint $(a_i, b_i]$

$$P_X(\cup(a_i, b_i]) = \sum P_X((a_i, b_i])$$

Then, on $\mathscr{F}$, $P_X(\cdot)$ is consistent with probability in view of Theorem 25. Thus, the main extension theorem (Theorem 13) yields that there is a unique extension of $P_X(\cdot)$ from $\mathscr{F}$ to the set of Borel sets on the real line. This extension will also be denoted by $P_X(\cdot)$, and, when no confusion is possible, we shall also use the notation $F(z) = F_X(z)$ for $P_X(\cdot)$. We shall call $P_X(\cdot)$ the probability associated with, or induced by, X, or $F(z)$, on the real line.

Notice that if $F(z)$ is a function satisfying properties (i), (iii), and (iv) of Theorem 25, then the above construction of a probability on the Borel sets of the real line can be carried out without any reference to a random variable X. Thus, as a consequence, we have that if $F(z)$ is a function with the above-mentioned properties, then $F(z)$ is in fact the distribution function of a random variable. Just take the real line as Ω, its Borel sets as events and the previously described extension of F as probability. Then, on this space, the distribution function of the random variable $X(\omega) = \omega$ is $F(z)$.

What has been said so far can easily be extended to vectors. Let $(X_1, X_2, \ldots, X_n)$ be a vector whose components are random variables on $(\Omega, \mathscr{A}, P)$. The function

$$F(z_1, z_2, \ldots, z_n) = P(X_1 \leq z_1, X_2 \leq z_2, \ldots, X_n \leq z_n)$$

defined for all real vectors $(z_1, z_2, \ldots, z_n)$, is called the multivariate distribution function of the vector $(X_1, X_2, \ldots, X_n)$. By the continuity theorem, if $z_j \to +\infty$,

$$\lim F(z_1, z_2, \ldots, z_n)$$
$$= P(X_1 \leq z_1, \ldots, X_{j-1} \leq z_{j-1}, X_{j+1} \leq z_{j+1}, \ldots, X_n \leq z_n)$$

which is the multivariate distribution function of the vector

$$(X_1, \ldots, X_{j-1}, X_{j+1}, \ldots, X_n).$$

By repeated application of this method, any number of variables can be eliminated from $F(z_1, z_2, \ldots, z_n)$ and the resulting function is the multivariate distribution function of the remaining components. These distribution functions are called the (lower-dimensional) marginals of $F(z_1, z_2, \ldots, z_n)$. In particular, the one-dimensional, or univariate, marginals of F are the distribution functions of the components.

Further properties of multivariate distribution functions are as follows: (i) $F(z_1, z_2, \ldots, z_n)$ is nondecreasing and continuous from the right in each of its variables, (ii) as one, or several, of its variables converges to negative infinity, $F(z_1, z_2, \ldots, z_n) \to 0$, and (iii) for $a_i < b_i$, $1 \leq i \leq n$.

$$P(a_i < X_i \leq b_i, 1 \leq i \leq n) = \sum (-1)^m F(z_1, z_2, \ldots, z_n) \tag{1.37}$$

where $z_i = g_i a_i + (1 - g_i) b_i$ with $g_i = 1$ or 0, $m = g_1 + g_2 + \cdots + g_n$, and the summation is over all possible assignments of $g_i = 1$ or 0, $1 \leq i \leq n$ (thus, the number of terms is 2^n). Formula (1.37) easily follows by repeated application of Theorem 2 upon observing that the event on the left-hand side is the intersection over $1 \leq i \leq n$ of the events $\{X_i \leq a_i\}^c \cap \{X_i \leq b_i\}$. A particular consequence of (1.37) is that the sum on its right-hand side must be nonnegative.

With no change in the one-dimensional argument, we now get that if $F(z_1, z_2, \ldots, z_n)$ is a function satisfying properties (i) and (ii) of the preceding paragraph and if, furthermore, its univariate marginals are distribution functions and the expression on the right-hand side of (1.37) is nonnegative for all $a_i < b_i$, $1 \leq i \leq n$, then there is a unique probability on the Borel sets of the n-dimensional real space such that the probability of the n-dimensional rectangle whose vertices are $(z_1, z_2, \ldots, z_n)$, $z_i = g_i a_i + (1 - g_i) b_i$, $g_i = 1$ or 0, $a_i < b_i$ fixed, $1 \leq i \leq n$, equals the expression on the right-hand side of (1.37). This probability is called the probability associated with, or induced by, $F(z_1, z_2, \ldots, z_n)$. If no confusion arises, both F and its associated probability will be denoted by F.

There is a large number of multivariate distribution functions with given univariate marginals. One notable exception occurs when the components are independent.

Definition 15   The random variables $X_j, j \geq 1$ are called independent if the events $\{X_j \leq z_j\}, j \geq 1$, are independent for all real numbers z_j.

Thus, for independent components, the multivariate distribution function $F(z_1, z_2, \ldots, z_n)$ of the vector $(X_1, X_2, \ldots, X_n)$ satisfies

$$F(z_1, z_2, \ldots, z_n) = F_1(z_1)F_2(z_2) \cdots F_n(z_n)$$

where $F_j(z_j)$ is the distribution function of X_j. Furthermore, since $\{a_j < X_j \leq b_j\} = \{X_j \leq a_j\}^c \cap \{X_j \leq b_j\}$, for independent random variables, the events $\{a_j < X_j \leq b_j\}$, $j \geq 1$, are independent for all real numbers $a_j < b_j, j \geq 1$ (see Theorem 21). Hence, for independent random variables, the right-hand side of (1.37) becomes the product of $F_j(b_j) - F_j(a_j), 1 \leq j \leq n$. The independence of the events $\{X_j \in A_j\}$, A_j being the finite union of disjoint intervals $(a_i, b_i]$, now easily follows and, thus, the main extension theorem yields that, if the random variables X_j are independent and if A_j is a Borel set on the real line, then the events $\{X_j \in A_j\}, j \geq 1$, are independent.

For some important special distribution functions such as the exponential, normal, gamma, Weibull, extreme value, Poisson, geometric, and others, see the last sections of Chapters 2, 3, 5, and 6. Some of these distribution functions appear among the exercises which follow. Other distribution functions will appear in the present book as the theory unfolds. The important hypergeometric and binomial distributions have been introduced in Sections 1.4 and 1.7 of the present chapter.

Exercises

1. Show that, for a random variable X with the continuous distribution function $F(z)$, $Y = F(X)$ and $Z = -\log Y$ are random variables on the same probability space as X and $F_Y(z) = z$, $0 \leq z \leq 1$ (the uniform distribution on $(0, 1)$), and Z is exponentially distributed, i.e., $F_Z(z) = 1 - e^{-z}, z \geq 0$.
2. Show that if $F(z)$ is a distribution function and if $G(z) = 1 - F(z)$ satisfies the Cauchy equation

$$G(z + x) = G(z)G(x), \quad \text{all} \quad x, z \geq 0$$

then either $F(z)$ is degenerate at zero or $F(z) = 1 - e^{-az}$ with some $a > 0$ ($z \geq 0$). Hence, deduce that if X is a random variable with $P(X \geq 0) = 1$ and $P(X > 0) > 0$ and if, for all $x \geq 0$ and $z \geq 0$,

$$P(X > x + z | X > z) = P(X > x)$$

then the distribution of X is exponential: $F_X(z) = 1 - e^{-az}$, $a > 0$, $z \geq 0$.

3. Let $F(x, y) = 1 - e^{-x} - e^{-y} + G(x, y)$ be a bivariate distribution function, where $G(x, y)$ converges to zero when at least one of x and y tends to infinity. Show that both marginals of F are exponential. Show that F is in fact a distribution function if (i) $G(x, y) = \exp(-x - y + Axy)$, (ii) $G(x, y) = (e^x + e^y - 1)^{-1}$, (iii) $G(x, y) = \exp[-(x^m + y^m)^{1/m}]$, and (iv) $G(x, y) = e^{-x-y}[1 + A(1 - e^{-x})(1 - e^{-y})]$. (In all cases, some limitation on A is necessary.)

4. The univariate marginals of the bivariate function

$$F(x, y) = \begin{cases} 1 - e^{-2x} & \text{if } x > 0, y \geq 1 \\ \frac{1}{2}(1 - e^{-x}) & \text{if } x > 0, 0 \leq y < 1 \\ 0 & \text{otherwise} \end{cases}$$

are proper distribution functions and yet it is not a bivariate distribution function: compute $F(\log 6, 2) - F(\log 6, 1/2) - F(\log 4, 2) + F(\log 4, 1/2)$.

5. Let X and Y be discrete random variables with values $x_j, j \geq 0$, and y_j, $j \geq 0$, and with distributions $p_j, j \geq 0$, and $r_j, j \geq 0$, respectively. Show that X and Y are independent if and only if the events $X = x_j$ and $Y = y_t$ are independent for all j and t. Then evaluate the distribution of $X + Y$ if X and Y are independent with nonnegative integer values and with Poisson distributions $p_j = a^j e^{-a}/j!$, $a > 0$, and $r_j = b^j e^{-b}/j!$, $b > 0$.

6. Let $X_1, X_2, \ldots, X_n$ be independent random variables with the common distribution function $F(z)$. Let

$$W_n = \min(X_1, X_2, \ldots, X_n) \qquad Z_n = \max(X_1, X_2, \ldots, X_n)$$

Show that the distribution functions $L_n(z)$ and $H_n(z)$ of W_n and Z_n, respectively, have the form

$$L_n(z) = 1 - [1 - F(z)]^n \qquad H_n(z) = F^n(z)$$

Then, show that, for exponential $F(z) = 1 - e^{-az}$, $a > 0$, $z \geq 0$, $L_n(z/n) = F(z)$ for every $n \geq 1$, while, for every z

$$\lim H_n((z + \log n)/a) = \exp(-e^{-z}) \qquad (n \to +\infty)$$

REFERENCES

Andersen, E. S. and B. Jessen (1946). On the introduction of measures in infinite product sets, *Danske Vid. Selsk. Mat. Fys. Medd. 25*, no. 4.

Borel, E. (1909). Les probabilités dénombrables et leurs applications arith-méthiques. *Rend. Circ. Mat Palermo 27*, 247–271.

Cantelli, F. P. (1917). Su due applicazioni di un teorema di G. Boole. *Rend. Accad. Naz. Lincei 26*, 39–45.

Chebyshev, P. L. (1944–1951). *Complete Collected Works*, Izd. AN SSSR, Moscow. (In Russian.) (The inequality bearing his name appeared in Liouville's *J. Math. Pures Appl. 12*, (2) (1867), 177–184. See the book by Heyde and Seneta below about Chebyshev's time.)

Erdös, P. and A. Rényi (1959). On Cantor's series with convergent $\sum 1/q_n$. *Ann. Univ. Sci. Budapest, Sectio Math. 2*, 93–109.

Freedman, A. R. (1985). Section of Short Examples, *Canadian Math. Soc. Notes 17*, 7, 30.

Galambos, J. (1984). *Introductory Probability Theory*, Marcel Dekker, New York.

Heyde, C. C. and E. Seneta (1977), *I. J. Bienaymé: Statistical Theory Anticipated*, Springer, New York.

Kolmogorov, A. N. (1933). *Grundbegriffe der Wahrscheinlichkeitsrechnung*, Springer, Berlin. (English translation: *Foundations of the Theory of Probability*, Chelsea, New York, 1956.)

Neveu, J. (1965). *Mathematical Foundations of the Calculus of Probability.* Holden-Day, San Francisco.

2

Expectation and Integral, Weak and Strong Convergence

2.1 DEFINITIONS OF EXPECTATION AND INTEGRAL

Recall from Section 1.5 the concept of expectation for a simple random variable: If X is a simple random variable on a probability space $(\Omega, \mathscr{A}, P)$ with values $x_1, x_2, \ldots, x_T$ and with the associated distribution $p_1, p_2, \ldots, p_T$, then the expected value of X

$$E(X) = p_1 x_1 + p_2 x_2 + \cdots + p_T x_T \tag{2.1}$$

It is established in Exercise 2 of Section 1.7 that if $X_1, X_2, \ldots$ are independent copies of X, then

$$P(\lim(X_1 + X_2 + \cdots + X_n)/n = E(X)) = 1 \qquad (n \to +\infty) \tag{2.2}$$

The converse is also true (see Theorem 1 at the end of the present section): if (2.2) holds, then $E(X)$ is necessarily the value in (2.1). We could, therefore, use either (2.1) or (2.2) as the definition of expectation. This equivalence can be interpreted as saying that (2.1) defines the computational form while (2.2) defines the probabilistic meaning of the expected value of a simple random variable. This is what we shall adopt in extending the concept of the expectation of random variables: we shall give both a probabilistic definition and a computational definition. It will then be shown that the two definitions coincide whenever the expectation is finite.

Before proceeding, let us define some important conventions. First, all random variables in the sequel are defined on the same probability space $(\Omega, \mathscr{A}, P)$. Second, recall the Remark following Definition 13 of Section 1.8, in which it is stated that if the values of a random variable are changed on a set of probability zero, it remains a random variable. Now, in connection with the expectation or integral, we always take a version of random variable such that it is finite on the whole of Ω.

Definition 1   Let $X_1, X_2, \ldots$ be independent copies of the random variable X, i.e., independent random variables, each with the same distribution function as X. Assume there is a finite constant c such that

$$P(\lim(X_1 + X_2 + \cdots + X_n)/n = c) = 1 \tag{2.3}$$

as $n \to +\infty$. We then say that X has finite expectation, or expected value, c. In mathematical notation $E(X) = c$.

The computational definition of $E(X)$ is developed in three steps:

Definition 2   (i) If X is a simple random variable, then its expectation $E(X)$ is defined as in (2.1). (ii) Let the random variable X be nonnegative. Let X_n, $n \geq 1$, be a sequence of simple random variables such that with probability one $0 \leq X_n \leq X_{n+1}$ and $X_n \to X$ as $n \to +\infty$. Then the expectation $E(X)$ is defined as

$$E(X) = \lim E(X_n) \tag{2.4}$$

(iii) Let X be a random variable, and put $X^+ = \max(0, X)$ and $X^- = -\min(0, X)$. Assume that at least one of $E(X^+)$ and $E(X^-)$ does not equal infinity. We then define the expectation of X

$$E(X) = E(X^+) - E(X^-) \tag{2.5}$$

We also call $E(X)$ as defined by steps (i)–(iii) the integral of X and write

$$E(X) = \int_\Omega X \, dP \tag{2.6}$$

If $E(X)$ is finite, then X is said to be integrable.

The integral on the right-hand side of (2.6) is extended to the following class of functions on Ω, which we call *measurable functions* (with respect to $\mathscr{A}$). A function X defined on Ω with probability one whose values are either finite real numbers or plus or minus infinity is called measurable if, for every finite real z, the subset of Ω on which $X \leq z$ belongs to $\mathscr{A}$. Now, if X is measurable and finite with probability one, then X is a random variable, so that Definition 2 applies. On the other hand, if $X \geq 0$ is measurable and

$P(X = +\infty) > 0$, we then define

$$\int_\Omega X \, dP = +\infty$$

Finally, if X is an arbitrary measurable function (but not a random variable) then the integral of X is either plus or minus infinity, depending on whether the integral of X^+ or X^- is infinity. If both are infinity, then, just as for random variables, the integral of X is not defined. The convenience of sometimes dealing with measurable functions rather than with random variables lies in the fact that the family of measurable functions is closed under the usual operations of calculus (the arithmetical operations and the formation of sup, inf, limsup, liminf, and limit). See Exercise 3.

We now return to Definition 2. We must prove that $E(X)$ is well defined in part (ii). First note that there exists a sequence X_n satisfying the requirements in (ii). Indeed, for $X \geq 0$, let $X_n = n$ if $X \geq n$, and $X_n = k/2^n$ if $k/2^n \leq X < (k+1)/2^n$, $0 \leq k \leq n2^n - 1$. Then X_n is simple, $0 \leq X_n \leq X_{n+1}$ and, since X is finite, we eventually get $|X - X_n| < 1/2^n$. Next, observe that the limit in (2.4) always exists because, by Theorem 14(ii) of Section 1.5, $X_n \leq X_{n+1}$ implies $E(X_n) \leq E(X_{n+1})$ for simple random variables. It therefore remains to be shown that if $\{X_n\}$ and $\{Y_n\}$ are two sequences of simple random variables with $0 \leq X_n \leq X_{n+1}$ and $0 \leq Y_n \leq Y_{n+1}$ for all n, and if both converge to $X \geq 0$, then

$$\lim E(X_n) = \lim E(Y_n) \tag{2.7}$$

as $n \to +\infty$. For (2.7) it suffices to prove that, for every fixed s, $\lim X_n \geq Y_s$ implies

$$\lim E(X_n) \geq E(Y_s) \tag{2.7a}$$

Namely, if we let $s \to +\infty$ in (2.7a), $\lim E(X_n) \geq \lim E(Y_s)$ follows, and by virtue of the symmetry in the assumptions about the sequences X_n and Y_n, the reverse inequality is similarly obtained, yielding (2.7).

We thus prove (2.7a). Since Y_s is simple, there is a smallest m and a largest M among its values. First assume that $m > 0$. Let $I_n(\varepsilon)$ be the indicator variable of the event $X_n \geq Y_s - \varepsilon$, where $0 < \varepsilon < m$ is arbitrary. Then, by the properties of expectation for simple random variables (Theorem 14 in Section 1.5),

$$E(X_n) \geq E[X_n I_n(\varepsilon)] \geq E[(Y_s - \varepsilon)I_n(\varepsilon)]$$
$$= E[Y_s - Y_s(1 - I_n(\varepsilon))] - \varepsilon E[I_n(\varepsilon)]$$
$$\geq E(Y_s) - MP(X_n < Y_s - \varepsilon) - \varepsilon$$

which, by first letting $n \to +\infty$ and then $\varepsilon \to 0$, yields (2.7a) under the assumption of $m > 0$. But then, if $m = 0$, the above proof yields, as $n \to +\infty$,

$$\lim E(X_n) \geq \lim E(X_n I(Y_s > 0)) \geq E(Y_s I(Y_s > 0))$$
$$= E(Y_s) - E(Y_s I(Y_s = 0)) = E(Y_s)$$

which completes the proof of (2.7a), and thus of (2.7) as well.

We can now claim that the special sequence X_n constructed in the course of the preceding arguments can also be used in (2.4). We thus have that, for $X \geq 0$,

$$E(X) = \lim_{n = +\infty} \left[\sum_{k=0}^{n2^n - 1} \frac{k}{2^n} P\left(\frac{k}{2^n} \leq X < \frac{k+1}{2^n}\right) + nP(X \geq n) \right]$$

$$= \lim_{n = +\infty} \left[\sum_{k=0}^{n2^n - 1} \frac{k+1}{2^n} P\left(\frac{k}{2^n} < X \leq \frac{k+1}{2^n}\right) + nP(X \geq n) \right] \quad (2.8)$$

where the second limit follows from the first one upon observing that the difference of the two sums in (2.8) does not exceed $1/2^n$.

Finally, the equivalence of Definitions 1 and 2 must be established (for finite $E(X)$). This is contained in the following fundamental theorem.

Theorem 1   (Kolmogorov's Strong Law of Large Numbers)   Let $X_1, X_2, \dots$ be independent copies of X. Then the limit relation $(n \to +\infty)$

$$P(\lim(X_1 + X_2 + \cdots + X_n)/n = c) = 1$$

holds with a finite constant c if, and only if, X is integrable, and then

$$c = \int_\Omega X \, dP$$

The proof of Theorem 1 requires some lengthy preparations; thus, it is postponed to Section 2.5.

Exercises

1. By using formula (2.8), show that
    (i) $E(X) = a$ if X is Poisson distributed with parameter a; i.e., X is discrete with $P(X = j) = a^j e^{-a}/j!, j \geq 0$,
    (ii) $E(X) = 1/a$ if X is exponentially distributed with parameter $a > 0$; i.e., $F_X(z) = 1 - e^{-az}, z \geq 0$,
    (iii) $E(X) = 0$ if X is normal with parameters 0 and 1; that is

    $$F_X(z) = \frac{1}{\sqrt{2\pi}} \int_{-\infty}^{z} e^{-t^2/2} \, dt$$

2. By determining $E(X^+)$ and $E(X^-)$, show that $E(X)$ is not defined if $P(X = 2^n) = P(X = -2^n) = 1/2^{n+1}$, $n \geq 1$.
3. Show that (i) if X and Y are random variables (measurable functions) then so are $X + Y$, XY, $\max(X, Y)$, and $\min(X, Y)$; (ii) if X_n is measurable for each n then, as $n \to +\infty$,

$$\limsup X_n \quad \text{and} \quad \liminf X_n$$

as well as

$$\sup_{n \geq k} X_n \quad \text{and} \quad \inf_{n \geq k} X_n$$

are measurable functions. Show, by an example, that (ii) fails if "measurable" is replaced by "random variable."

2.2 BASIC PROPERTIES OF EXPECTATION

Since Theorem 1 has not yet been proved, we must rely on Definition 2 of the expectation. When we interpret a constant c as a random variable we view it as simple, and thus $E(c) = c$.

Theorem 2 (i) For every random variable X and constant c, $E(cX) = cE(X)$ if $E(X)$ is meaningful; (ii) if the expectations $E(X)$ and $E(Y)$ of the random variables X and Y are defined, and if $E(X) + E(Y)$ is not of the form $\infty - \infty$, then $E(X + Y) = E(X) + E(Y)$; (iii) if $X \leq Y$ and if $E(Y)$ is meaningful, then $E(X) \leq E(Y)$.

Proof. The theorem has been proved for simple random variables in Section 1.5 (Theorem 14). Thus, (i) is immediate from Definition 2. Part (ii) of the theorem also follows if $X \geq 0$ and $Y \geq 0$. Therefore, in the sums

$$X^+ + Y^+ + (X + Y)^- = X^- + Y^- + (X + Y)^+$$

of nonnegative terms we can take expectation term by term. Thus, after rearrangement of the terms on the two sides, which is justified under the assumptions in (ii), we get (ii) for the general case. Part (iii) now is a consequence of part (ii) by $Y = X + Z$, where $Z \geq 0$, and thus, by definition, $E(Z) \geq 0$. This completes the proof. ∎

Corollary X is integrable if, and only if, $|X|$ is. Furthermore, $|E(X)| \leq E(|X|)$.

Proof. Because $X = X^+ - X^-$ and $|X| = X^+ + X^-$, the corollary is immediate from the definition of the integrability and from part (ii) of Theorem 2. ∎

Definition 2 yields that if X is discrete with values $\{x_j\}$ and with distribution $\{p_j\}$ then

$$E(X) = \sum x_j p_j \tag{2.9}$$

whenever one of the sums $\sum x_j^+ p_j$ and $\sum x_j^- p_j$ is finite. Indeed, if each $x_j \geq 0$, we can approximate $E(X)$ by $E(X_n)$, where X_n is simple, defined as $X = X_n$ if $X = x_j$ with $j \leq n$ and $X_n = 0$ otherwise. This yields (2.9) for $X \geq 0$. For arbitrary X, apply the method just described both to X^+ and to X^-, and (2.9) obtains.

Theorem 3 The random variable X is integrable if, and only if,

$$D(X) = \sum_{j=0}^{+\infty} jP(j < |X| \leq j+1) = \sum_{j=1}^{+\infty} P(|X| > j) < +\infty \tag{2.10}$$

In particular, if X is integrable then

$$\lim nP(|X| \geq n) = 0 \qquad (n \to +\infty) \tag{2.11}$$

Proof. Let $X_1 = j$ and $X_2 = j+1$ if $j < |X| \leq j+1$. Then $X_1 < |X| \leq X_2$, $E(X_1) = D(X)$, and $E(X_2) = D(X) + 1 - P(X = 0)$. Part (iii) of Theorem 2 and Corollary thus imply that (2.10) is both necessary and sufficient for the integrability of X. Hence, by Cauchy's principle, as $n \to +\infty$,

$$\lim \sum_{j=n-1}^{2n-1} P(|X| > j) = 0$$

which implies (2.11) upon observing that, by monotonicity in j, the sum above is larger than $nP(|X| \geq 2n)$.

Theorem 4 If X and Y are independent random variables with finite expectations $E(X)$ and $E(Y)$, respectively, then $E(XY) = E(X)E(Y)$.

Proof. We first prove the theorem for simple random variables. Let X and Y be simple with values $x_1, x_2, \ldots, x_T$ and $y_1, y_2, \ldots, y_s$, and with associated distributions $p_1, p_2, \ldots, p_T$ and $r_1, r_2, \ldots, r_s$, respectively. Then, by assumption, $P(X = x_i, Y = y_j) = p_i r_j$. Hence,

$$E(XY) = \sum_{i=1}^{T} \sum_{j=1}^{s} x_i y_j p_i r_j = \sum_{i=1}^{T} x_i p_i \sum_{j=1}^{s} y_j r_j = E(X)E(Y)$$

Next, let X and Y be nonnegative. Then X_n for X and Y_n for Y, satisfying the requirements in part (ii) of Definition 2, can be constructed in terms of

X and Y, respectively (e.g., repeat the construction that led to (2.8)). Consequently, X_n and Y_n can be assumed to be independent. Hence, by (2.4),

$$E(XY) = \lim E(X_n Y_n) = \lim E(X_n)E(Y_n) = E(X)E(Y)$$

Finally, if we write

$$XY = (X^+ - X^-)(Y^+ - Y^-) = X^+ Y^+ - X^- Y^+ - X^+ Y^- + X^- Y^-$$

upon taking expectations, and appealing to Theorem 2 and to the first part of the present proof, the theorem follows. ∎

As a measurement of deviation of a random variable from its own expectation, we introduce the concept of *variance*. If X is a random variable with finite expectation $E = E(X)$, then the expression

$$V(X) = E[(X - E)^2] \tag{2.12}$$

is called the variance of X. Upon expanding $(X - E)^2$, Theorem 2 yields

$$V(X) = E(X^2) - E^2(X) \tag{2.13}$$

Because the variance is a particular expectation, its properties follow immediately from those of the expectation. Some important properties are summarized in the next theorem, the proof of which is left to the reader.

Theorem 5 If X is a random variable with finite variance and if c is a constant, then $V(cX) = c^2 V(X)$ and $V(X + c) = V(X)$. Furthermore, if X and Y are independent,

$$V(X + Y) = V(X) + V(Y)$$

From Theorems 2 and 5, we get, by induction,

Theorem 6 If $X_1, X_2, \ldots, X_n$ are independent random variables with finite expectations E_j and variances V_j, $1 \le j \le n$, then, with $Y_n = n^{-1}(X_1 + X_2 + \cdots + X_n)$,

$$E(Y_n) = (E_1 + E_2 + \cdots + E_n)/n \qquad V(Y_n) = (V_1 + V_2 + \cdots + V_n)/n^2$$

In particular, if $E_j = E$ and $V_j = V$ for all j, then

$$E(Y_n) = E \qquad V(Y_n) = V/n$$

Without calling it variance, we have used the preceding properties in Section 1.5 for relative frequencies. Those results will be extended to the arithmetical mean of general random variables in subsequent sections.

Exercises

1. Let X be a Poisson variable with parameter $a > 0$ and let Y be an exponential variable with distribution function $F(z) = 1 - e^{-bz}$, $b > 0$, $z \geq 0$. Find $E(3X - 2Y)$.

2. Give an explicit formula for the variance of a discrete random variable. Compute the variance of X, if the distribution of X is (i) binomial, (ii) Poisson, and (iii) hypergeometric.

3. Let the distribution function of the random variable X be $F(x) = 1/2 + (1/\pi)\arctan x$. Show that $E(X)$ is not defined by determining the limit of $nP(X \geq n)$ as $n \to +\infty$.

4. The distribution function of the random variable X

$$F(z) = 1 - e/\{z(\log z)^a\}, \qquad z \geq e$$

Give the range of a for which $E(X)$ is finite. Interpret the result in the light of (2.11).

5. Show that if $X \geq 0$ and X^r is integrable for some $r > 0$ then $E(X^u)$ is finite for $0 \leq u \leq r$.

6. Upon utilizing the concavity of $\log z$ in the form

$$\log uv = \frac{1}{r}\log u^r + \frac{1}{s}\log v^s \leq \log\left(\frac{1}{r}u^r + \frac{1}{s}v^s\right)$$

where $u > 0$, $v > 0$, $r > 1$ and $(1/r) + (1/s) = 1$, prove Hölder's inequality

$$E(|XY|) \leq E^{1/r}(|X|^r)E^{1/s}(|Y|^s)$$

From this inequality deduce that (i) $\log E(|X|^a)$ is a convex function of a (over the set $\{a: E(|X|^a) < +\infty\}$) and (ii) $E^{1/a}(|X|^a)$ is nondecreasing in $a > 0$. (Remark: The special case $r = s = 2$ is known as the Cauchy–Schwarz inequality.)

2.3 THE INEQUALITIES OF MARKOV, CHEBYSHEV, AND KOLMOGOROV

The three inequalities of the title are fundamental in the study of the expectation.

Theorem 7 (**Markov**) Let $X \geq 0$ be a random variable with positive finite expectation E. Then, for $a > 0$,

$$P(X \geq aE) \leq 1/a$$

Proof. Let I_a be the indicator variable of the event $\{X \geq aE\}$. Then, by Theorem 2,

$$E(X) \geq E(XI_a) \geq E(aEI_a) = aEE(I_a) = aEP(X \geq aE)$$

Upon division by $aE = aE(X) > 0$, the claimed inequality follows. ∎

Theorem 8 (**Chebyshev**) Let Y be a random variable with finite expectation E and variance $V > 0$. Then, for $b > 0$,

$$P(|Y - E| \geq bV^{1/2}) \leq 1/b^2$$

or, equivalently, for $a > 0$,

$$P(|Y - E| \geq a) \leq V/a^2$$

Proof. Define $X = (Y - E)^2$. By assumption, the conditions of Theorem 7 are satisfied with $E(X) = V$. Hence, the first variant of the Chebyshev inequality follows immediately from that of Markov. On the other hand, the substitution of $a = bV^{1/2}$ into the first variant yields the second one. The theorem is established. ∎

By an appeal to Theorem 6, the Chebyshev inequality implies the following weak form of the limit relation stated in Theorem 1.

Corollary (**Weak Law of Large Numbers**) Let X be a random variable with finite expectation E and finite positive variance V. Let $X_1, X_2, \ldots$ be independent random variables, each distributed as X. Then, for every $a > 0$, as $n \to +\infty$,

$$\lim P(|n^{-1}(X_1 + X_2 + \cdots + X_n) - E| \geq a) = 0$$

A more powerful tool than Chebyshev's inequality is the following inequality of Kolmogorov.

Theorem 9 (**Kolmogorov**) Let $X_j, j \geq 1$, be independent random variables with finite expectation E_j and variance V_j. We put $Y_j = X_j - E_j$, so $E(Y_j) = 0$, and $S_k = Y_1 + Y_2 + \cdots + Y_k$. Then, for arbitrary $a > 0$,

$$P\left(\max_{1 \leq k \leq n} |S_k| \geq a\right) \leq (V_1 + V_2 + \cdots + V_n)/a^2$$

Proof. Let $A_1 = \{|S_1| \geq a\}$, and, for $2 \leq k \leq n$, set

$$A_k = \{|S_1| < a, |S_2| < a, \ldots, |S_{k-1}| < a, |S_k| \geq a\}$$

Notice that

$$B = \bigcup_{k=1}^{n} A_k = \left\{ \max_{1 \le k \le n} |S_k| \ge a \right\}$$

Let I_k be the indicator variable of the event A_k, $1 \le k \le n$, and let $I(B)$ be the indicator of B. Since the events A_k are mutually exclusive, $I(B) = I_1 + I_2 + \cdots + I_n$. Now, by Theorems 5 and 2,

$$\sum_{j=1}^{n} V_j = E(S_n^2) \ge E(S_n^2 I(B)) = \sum_{k=1}^{n} E(S_n^2 I_k) \tag{2.14}$$

Next, observe that, since $S_k I_k$ and $S_n - S_k$ are independent and $E(S_n - S_k) = 0$ (the sum of $E(Y_j) = 0$ for $k < j < n$), Theorem 4 implies

$$E[S_k I_k (S_n - S_k)] = 0 \tag{2.15}$$

Therefore, upon writing

$$S_n^2 = [S_k + (S_n - S_k)]^2 = S_k^2 + (S_n - S_k)^2 + 2S_k(S_n - S_k)$$

we get, from (2.15),

$$E(S_n^2 I_k) = E(S_k^2 I_k) + E[(S_n - S_k)^2 I_k] \ge E(S_k^2 I_k) \ge a^2 E(I_k)$$

This, combined with (2.14) (recall that $E(I_k) = P(A_k)$ and that the events A_k are exclusive), yields

$$\sum_{j=1}^{n} V_j \ge a^2 \sum_{k=1}^{n} P(A_k) = a^2 P(B)$$

which was to be proved.   ∎

While Chebyshev's inequality is associated with the kind of convergence occurring in the weak law of large numbers, it will be demonstrated that the Kolmogorov inequality entails convergence with probability one. For easier reference to these two types of convergence, we introduce the following concepts.

We say that the sequence X_n, $n \ge 1$, of random variables *converges in probability* to X, if for every $a > 0$, as $n \to +\infty$,

$$\lim P(|X_n - X| \ge a) = 0$$

If with probability one, as $n \to +\infty$, $\lim X_n$ exists and is finite, X, say, then X_n is said to *converge* to X *almost surely*, to be abbreviated a.s.

The following three theorems establish relations between these two types of convergence.

Theorem 10 The sequence X_n of random variables converges a.s. to a random variable X if, and only if, $\sup_{k \leq n} |X_n - X| = U_k$ converges to zero in probability.

Proof. By definition, $X_n \to X$ a.s. if, and only if, for every $a > 0$,

$$P\left(\bigcap_{n=1}^{+\infty} \bigcap_{j=n}^{+\infty} \{|X_j - X| \geq a\} \right) = 0 \tag{2.16}$$

Now, the events

$$\bigcup_{j=n}^{+\infty} \{|X_j - X| \geq a\} = \left\{ \sup_{j \geq n} |X_j - X| \geq a \right\}$$

are decreasing, so the continuity of probability (Theorem 1.6) yields that the limit of the probability of the above events is the left-hand side of (2.16). The theorem is established. ∎

Theorem 11 The sequence X_n of random variables converges in probability to a random variable X if, and only if, for every $a > 0$, as $n \to +\infty$,

$$p_n(a) = \sup_{n < m} P(|X_n - X_m| \geq a) \to 0 \tag{2.17}$$

Proof. First assume $X_n \to X$ in probability. Then to every $a > 0$ and $\delta > 0$ there is an integer n_0 such that, for $n_0 \leq n < m$,

$$P(|X_n - X| \geq a/2) < \delta \qquad P(|X_m - X| \geq a/2) < \delta$$

Hence, the elementary inequality

$$P(|X_n - X_m| \geq a) \leq P(|X_n - X| \geq a/2) + P(|X_m - X| \geq a/2)$$

yields $P(|X_n - X_m| \geq a) < 2\delta$, implying $p_n(a) \leq 2\delta$. The limit of (2.17) follows.

Conversely, if (2.17) holds, then, to any integer $k \geq 1$ there is an integer n_k such that $P(|X_n - X_m| \geq a_k) < \delta_k$ for all $n_k \leq n < m$, where $a_k > 0$ and $\delta_k > 0$ are arbitrary. Let us choose the numbers a_k and δ_k such that both infinite series $\sum a_k$ and $\sum \delta_k$ converge. Take a subsequence $s_1 < s_2 < \cdots$ of the consecutive integers which increases at least as fast as the sequence n_k, and put $X_k^* = X_{s_k}$. Then, by the preceding choices,

$$\sum_{k=1}^{+\infty} P(|X_{k+1}^* - X_k^*| \geq a_k) < +\infty$$

which yields, via the Borel–Cantelli lemma (Theorem 1.10), that, except for a finite number of the subscript k, $|X^*_{k+1} - X^*_k| < a_k$ (a.s.). Hence, with probability one,

$$\sup_{n<m} |X^*_n - X^*_m| \leq \sum_{k=n}^{+\infty} |X^*_{k+1} - X^*_k| < \sum_{k=n}^{+\infty} a_k$$

which goes to zero as $n \to +\infty$. We thus established that the subsequence X^*_k converges a.s. to a finite function, i.e., to a random variable X. Therefore, by Theorem 10, X^*_k also converges in probability to X. The previously applied elementary inequality

$$P(|X_n - X| \geq a) \leq P(|X_n - X^*_k| \geq a/2) + P(|X^*_k - X| \geq a/2)$$

and (2.17) (since $X^*_k = X_m$ for some m) now complete the proof. ∎

Remark Notice that in the course of the preceding proof it is established that if $X_n \to X$ in probability then every subsequence of X_n has a further subsequence X'_k such that $X'_k \to X$ a.s. The converse is also true; see Exercise 4.

Theorem 12 The sequence X_n of random variables converges a.s. to a random variable X if, and only if, $\sup_{n<m} |X_n - X_m|$ converges to zero in probability as $n \to +\infty$.

Proof. Theorem 10 entails that if $X_n \to X$ a.s., then both $X_n - X$ and $\sup_{k\leq n} |X_n - X|$ converge to zero in probability. Consequently, we get from the inequality

$$P\left(\sup_{n<m} |X_n - X_m| \geq a\right) \leq P\left(\sup_{n<m} |X_m - X| \geq a/2\right) + P(|X_n - X| \geq a/2)$$

that $X_n \to X$ a.s. implies $\sup_{n<m} |X_n - X_m| \to 0$ in probability. Conversely, the latter implies that, as $n \to +\infty$,

$$\sup_{n<m} P(|X_n - X_m| \geq a) \to 0$$

But then, by Theorem 11, there is a random variable X such that $X_n \to X$ in probability. Hence, on account of the inequality

$$P\left(\sup_{n<m} |X_m - X| \geq a\right) \leq P\left(\sup_{n<m} |X_n - X_m| \geq a/2\right) + P(|X_n - X| \geq a/2)$$

one more appeal to Theorem 10 completes the proof. ∎

Kolmogorov's inequality now enables us to establish the following result which will be instrumental in proving Theorem 1.

Theorem 13 Let $X_1, X_2, \ldots$ be independent random variables with finite variances $V_j, j \geq 1$. Assume

$$\sum_{j=1}^{+\infty} V_j < +\infty \tag{2.18}$$

Then the infinite series $\sum (X_j - E(X_j))$ converges a.s.

Proof. Fix the integer n and apply Kolmogorov's inequality to the sequence $X_{n+1}, X_{n+2}, \ldots$. We have that, if $Y_j = X_j - E(X_j)$ and $S_{k,n} = Y_{n+1} + Y_{n+2} + \cdots + Y_{n+k}$,

$$P\left(\max_{1 \leq k \leq m} |S_{k,n}| \geq a \right) \leq \frac{1}{a^2} \sum_{j=1}^{m} V_{n+j} \leq \frac{1}{a^2} \sum_{j=n+1}^{+\infty} V_j$$

for arbitrary $m \geq 1$ and $a > 0$. Notice that the extreme right-hand side does not depend on m and that its convergence to zero, as $n \to +\infty$, is entailed by (2.18). Therefore, by letting $m \to +\infty$ on the left-hand side, the theorem follows from the continuity of probability (Theorem 1.6) and Theorem 12.

Exercises

1. Let X and Y be random variables such that, for some $p > 0$, $|X - Y|^p$ is integrable. Show that, for arbitrary $a > 0$,

   $$P(|X - Y| \geq a) \leq E(|X - Y|^p)/a^p$$

2. Let X be a Poisson variable with parameter 2. Evaluate $P(|X - E(X)| \geq a)$ for $a = 2$ and 5 and compare the result with the estimate obtained from Chebyshev's inequality.
3. Prove (i) if $X_n \to X$ in probability and $X_n \to X^*$ in probability, then $P(X = X^*) = 1$; (ii) if $X_n \to X$ in probability and $g(x)$ is a continuous function on the real line, then $g(X_n) \to g(X)$ in probability; and (iii) if $X_n \to X$ and $Y_n \to Y$ in probability, and if $g(x, y)$ is a continuous function in the plane, then $g(X_n, Y_n) \to g(X, Y)$ in probability.
4. Show that if every subsequence of X_n, $n \geq 1$, has a further subsequence X'_k such that X'_k converges a.s. to the same random variable X, then $X_n \to X$ in probability.

2.4 SEQUENCES OF INTEGRALS

In the current section, we deal with measurable functions rather than with random variables on a given probability space $(\Omega, \mathscr{A}, P)$. Recall that if X is measurable but not a random variable then its integral, if defined, is infinite. Thus, most statements will be trivial if X is not a random variable. However, when dealing with measurable functions, we have the advantage that inf, sup, liminf, and limsup (and thus limit as well if it exists) are all measurable.

Theorem 14 (**Monotone Convergence Theorem**) If $0 \leq X_j \leq X_{j+1}$ with $\lim X_j = X$, then

$$0 \leq \int_\Omega X_j \, dP \leq \int_\Omega X_{j+1} \, dP \to \int_\Omega X \, dP \tag{2.19}$$

Proof. If, for any n, X_n is not a random variable, then, by monotonicity, neither is any X_j with $j \geq n$. Therefore, the integral of $X_j, j \geq n$, equals $+\infty$, as does that of X. Hence, the theorem holds in the trivial case. Let all X_j now be random variables. Then the inequalities of (2.19) have already been established in part (iii) of Theorem 2. Thus, only the limit of (2.19) needs to be proved. First, let X itself be a random variable. We go back to (2.8). By the construction leading to (2.8), it follows that the expression behind the first limit is increasing in n, implying that, for all fixed n,

$$E(X_j) \geq \sum_{k=0}^{n2^n - 1} \frac{k}{2^n} P\left(\frac{k}{2^n} \leq X_j < \frac{k+1}{2^n}\right) + nP(X_j \geq n) \tag{2.20}$$

while the limit in (2.8) implies that, for sufficiently large n,

$$\sum_{k=0}^{n2^n - 1} \frac{k}{2^n} P\left(\frac{k}{2^n} \leq X < \frac{k+1}{2^n}\right) + nP(X \geq n) > K \tag{2.21}$$

whenever $E(X) > K$. Now, by the continuity of probability (Theorem 1.6), as $j \to +\infty$,

$$\lim P(X_j \geq z) = P(X \geq z) \tag{2.22}$$

for any fixed z. Thus, if we write $P(a \leq X < b) = P(X \geq a) - P(X \geq b)$, we get from (2.20), (2.21), and (2.22) that $E(X_j) > K$ for sufficiently large j, establishing the limit of (2.19). It remains to prove that if all X_j are random variables but X is not, and thus its integral is $+\infty$, then the integral of X_j converges to $+\infty$. However, this is immediately clear from (2.22) and the observation that $E(X_j) \geq E(X_j(z)) = zP(X_j \geq z)$, where $X_j(z)$ is a simple random variable that takes the value z if $X_j \geq z$ and $X_j(z) = 0$ otherwise. The theorem is established. ∎

Theorem 15 (**Fatou's Lemma**) Let Y be an integrable random variable. Assume the measurable functions $X_n \geq Y$ for all n. Then

$$\int_\Omega \liminf X_n \, dP \leq \liminf \int_\Omega X_n \, dP$$

as $n \to +\infty$. On the other hand, if $X_n \le Z$, where Z is integrable, then

$$\limsup \int_\Omega X_n \, dP \le \int_\Omega \limsup X_n \, dP$$

as $n \to +\infty$.

Proof. We proceed with the first part of the theorem. The second part follows immediately from the first one by replacing each X_n by $(-X_n)$ and Y by $-Z$. Now, since $X_n \ge Y$ and Y is integrable, $P(X_n = -\infty) = 0$. Consequently, all the integrals which follow are meaningful.

Set $Y_n = \inf_{k \ge n}(X_k - Y)$. Then $X_n - Y \ge Y_n, 0 \le Y_n \le Y_{n+1}$, and $\lim Y_n = \liminf X_n - Y$. The monotone convergence theorem (Theorem 14) and Theorem 2 (which trivially extends to integrals of measurable functions) thus yield

$$\liminf \int_\Omega (X_n - Y) \, dP \ge \liminf \int_\Omega Y_n \, dP = \int_\Omega \lim Y_n \, dP$$

$$= \int_\Omega (\liminf X_n - Y) \, dP$$

Because Y is integrable, the desired inequality follows by the additivity of integrals and the proof is completed. ∎

Theorem 16 **(Dominated Convergence Theorem)** If X_n, $n \ge 1$, and Y are random variables such that Y is integrable and $|X_n| \le Y$ a.s. for all n, then $X_n \to X$ in probability implies

$$\int_\Omega X_n \, dP \to \int_\Omega X \, dP \qquad (n \to +\infty)$$

Proof. By the remark following the proof of Theorem 11, there is a subsequence X'_k of the sequence X_n such that $X'_k \to X$ a.s. Thus, $|X| \le Y$ a.s., entailing that X is integrable. Furthermore, on account of Theorem 2 and the corollary to it,

$$\left| \int_\Omega X_n \, dP - \int_\Omega X \, dP \right| = \left| \int_\Omega (X_n - X) \, dP \right| \le \int_\Omega |X_n - X| \, dP$$

Thus, the theorem is established if we show that, for $Y_n = |X_n - X| \to 0$ in probability, $|Y_n| \le 2Y$, Y integrable,

$$\int_\Omega Y_n \, dP \to 0 \qquad\qquad (2.23)$$

Consider $\lim \sup \int_\Omega Y_n \, dP$. Let $\{n_k\}$ be a subsequence of integers such that

$$\lim_{k=+\infty} \int_\Omega Y_{n_k} \, dP = \limsup_{n=+\infty} \int_\Omega Y_n \, dP \qquad (2.24)$$

By the remark used earlier, $\{n_k\}$ has a further subsequence $\{r_m\}$ such that $Y_{r_m} \to 0$ a.s. But then, by Fatou's lemma (Theorem 15),

$$0 = \int_\Omega \liminf Y_{r_m} \, dP \leq \liminf \int_\Omega Y_{r_m} \, dP \leq \limsup \int_\Omega Y_{r_m} \, dP$$

$$\leq \int_\Omega \limsup Y_{r_m} \, dP = 0$$

i.e., the limit on the left-hand side of (2.24) is zero. Hence, the right-hand side is zero as well, which is equivalent to (2.23) because $Y_n \geq 0$. ∎

Remark 1 Let $X_n \geq 0$, $n \geq 1$. Then

$$\int_\Omega \sum_{n=1}^{+\infty} X_n \, dP = \sum_{n=1}^{+\infty} \int_\Omega X_n \, dP \qquad (2.25)$$

Simply apply Theorem 2 and the monotone convergence theorem to the sequence $S_k = X_1 + X_2 + \cdots + X_k$, $k \geq 1$. An important particular case is when $X_n = X I_n$, where X is a nonnegative measurable function, I_n is the indicator variable of an event A_n, and the events A_n are mutually exclusive. Then the infinite sum $I_1 + I_2 + \cdots$ is the indicator $I(\cup A_n)$ of the union of A_n, $n \geq 1$, and thus, if X is integrable, we have

$$E(X I(\cup A_n)) = \sum_{n=1}^{+\infty} E(X I_n) \qquad (2.25a)$$

Remark 2 The role of Y in the dominated convergence theorem is important. To see this, let $\Omega = (0, 1)$ and let $\mathscr{A}$ be the collection of Borel sets with Lebesgue measure as probability. Let a_n, $n \geq 1$, be a sequence of numbers with $0 < a_n < 1$ and $a_n \to 0$ as $n \to +\infty$. Let X_n be the simple random variable taking the value n on $(a_n, a_n + 1/n)$ and let $X_n = 0$ otherwise. Then $X_n \to X = 0$ for all $0 < x < 1$, but $E(X_n) = 1$; thus, it fails to converge to $E(X) = 0$.

Exercises

1. Prove that if X is integrable, and if A_n is a sequence of events such that $P(A_n) \to 0$ as $n \to +\infty$, then $E(X I_n) \to 0$, where I_n is the indicator of A_n.

2. Let $X \geq 0$ and integrable. By the dominated convergence theorem (or otherwise) prove that, if $t \geq 0$ and real,

$$\frac{d}{dt} \int_{\Omega} e^{-tX} \, dP = \int_{\Omega} \frac{d}{dt} e^{-tX} \, dP = - \int_{\Omega} X e^{-tX} \, dP$$

3. For the inverse tail function $x = U(y)$, where $y = 1 - F(x)$, find

$$\lim_{t \to 0} \frac{U(t) - U(tu)}{U(t) - U(tv)}, \qquad 0 < u, v < 1$$

and set a lower bound on

$$\lim_{t \to 0} \inf \frac{(1/2)U(t) - \int_{1/2}^{1} U(ts) \, ds}{U(t) - U(tv)}, \qquad 0 < v < 1,$$

if (i) $F(x) = 1 - \exp(-x^2)$, $x > 0$, (ii) $F(x) = 1 - e^{-x}$, $x > 0$, and (iii) $F(x) = 1/(1 + e^{-x})$, x real.

4. Prove Fubini's theorem: let $(\Omega^{(2)}, \mathscr{A}^{(2)}, P^{(2)})$ be the product space of the probability spaces $(\Omega_i, \mathscr{A}_i, P_i)$, $i = 1, 2$ (recall Theorem 22 and Definition 12 of Section 1.7). Assume that the random variable $X = X(\omega_1, \omega_2)$ on the product space is integrable. Show that the integrals below are all well defined and finite and that the stated equations hold:

$$\int_{\Omega^{(2)}} X \, dP^{(2)} = \int_{\Omega_1} \left[\int_{\Omega_2} X \, dP_2 \right] dP_1$$

$$= \int_{\Omega_2} \left[\int_{\Omega_1} X \, dP_1 \right] dP_2$$

2.5 THE STRONG LAW OF LARGE NUMBERS: THE PROOF OF THEOREM 1

We are now in a position to prove Theorem 1. We do this in several steps, each containing a result of independent interest.

Step 1: If the random variables X_j, $j \geq 1$, are independent with finite expectation E_j and variance V_j, then

$$\sum_{j=1}^{+\infty} \frac{V_j}{j^2} < +\infty \tag{2.26}$$

implies that $\sum (X_j - E_j)/j$ converges a.s. In particular, if X_j, $j \geq 1$, are i.i.d. with finite expectation E and variance V, then $\sum (X_j - E)/j$ converges a.s.

The first part follows immediately from Theorem 13 upon observing that $E(X_j/j) = E_j/j$ and $V(X_j/j) = V_j/j^2$. The second part, on the other hand, follows from the first part since, for $V_j = V$ for all j, (2.26) automatically holds. ■

The following result of elementary mathematics helps us to conclude the convergence of arithmetical means from the convergence of an infinite series.

Step 2: (Kronecker's Lemma)   If $\sum a_n$ converges to a finite number A and $b_n \le b_{n+1} \to +\infty$ with n, then, as $n \to +\infty$,

$$\frac{1}{b_n} \sum_{k=1}^{n} a_k b_k \to 0$$

Proof. Set $b_0 = A_0 = 0$, and, for $k \ge 1$,

$$c_k = b_k - b_{k-1}, \qquad A_k = \sum_{j=1}^{k} a_j$$

Then

$$\frac{1}{b_n} \sum_{k=1}^{n} a_k b_k = \frac{1}{b_n} \sum_{k=1}^{n} (A_k - A_{k-1}) b_k = A_n - \frac{1}{b_n} \sum_{k=1}^{n} A_k c_k \qquad (2.27)$$

Since $b_n = \sum_{k=1}^{n} c_k$, we know from elementary calculus that, in view of the assumptions on b_n, $A_n \to A$ implies

$$\frac{1}{b_n} \sum_{k=1}^{n} A_k c_k \to A$$

The lemma thus follows from (2.27). ■

We want to get rid of the assumption of finite variance. This is achieved by the next step of the proof, which is called truncation method.

Step 3:   Given the sequence X_j of random variables, define

$$Y_j = \begin{cases} X_j & \text{if } |X_j| \le j \\ j & \text{if } |X_j| > j \end{cases}, \qquad j \ge 1$$

If $X_j, j \ge 1$, are integrable and identically distributed, then

$$\frac{1}{n} \sum_{j=1}^{n} X_j - \frac{1}{n} \sum_{j=1}^{n} Y_j \to 0 \text{ a.s.} \qquad (n \to +\infty) \qquad (2.28)$$

We appeal to Theorem 3. Since X_1 and X_j are identically distributed, the integrability of X_1 is equivalent to

$$\sum_{j=1}^{+\infty} P(|X_1| > j) = \sum_{j=1}^{+\infty} P(|X_j| > j) < +\infty \qquad (2.29)$$

Now, by definition, $P(X_j \neq Y_j) = P(|X_j| > j)$ and thus (2.29) and the Borel–Cantelli lemma (Theorem 1.10) yield that $X_j = Y_j$ a.s. for all but a finite number of j, from which (2.28) is evident.   ■

Step 4:   Under the assumptions of Step 3, $E(Y_j)$ and $V(Y_j)$ are finite, so
(i) $E(Y_j) \to E(X_1)$ as $j \to +\infty$ and (ii) (2.26) holds with $V_j = V(Y_j)$.

By definition, $|Y_j| \leq j$, and thus $E(Y_j)$ and $V(Y_j)$ are finite. To prove (i), consider the random variables $X_j^* = X_1 I(|X_1| \leq j)$, where, as usual, $I(\cdot)$ is the indicator variable of the event in the parentheses. Clearly, $|X_j^*| \leq |X_1|$ and $X_j^* \to X_1$ as $j \to +\infty$. Furthermore, because of identical distributions $E(X_j^*) + jP(|X_1| > j) = E(Y_j)$. Since X_1 is integrable (2.11) and the dominated convergence theorem (Theorem 16) now give (i). For (ii) first observe

$$\sum_{j=1}^{+\infty} \frac{V_j}{j^2} \leq \sum_{j=1}^{+\infty} \frac{E(Y_j^2)}{j^2} = E\left\{ \sum_{j=1}^{+\infty} \frac{(X_j^*)^2}{j^2} \right\} + \sum_{j=1}^{+\infty} P(|X_j| > j) \qquad (2.30)$$

where the last sum is finite because of the integrability of X_1 and (2.29) and interchanging the summation and taking expectation in the penultimate term on the extreme right-hand side is justified by (2.25). Therefore, (2.26) holds if the penultimate term of (2.30) is finite. Now,

$$\sum_{j=1}^{+\infty} \frac{(X_j^*)^2}{j^2} = \sum_{j=1}^{+\infty} \frac{X_1^2 I(|X_1| \leq j)}{j^2} = \sum_{j=1}^{+\infty} \frac{X_1^2}{j^2} \sum_{k=1}^{j} I(k - 1 < |X_1| \leq k)$$

$$\leq \sum_{j=1}^{+\infty} \frac{1}{j^2} \sum_{k=1}^{j} k^2 I(k - 1 < |X_1| \leq k)$$

$$= \sum_{k=1}^{+\infty} k^2 I(k - 1 < |X_1| \leq k) \sum_{j=k}^{+\infty} \frac{1}{j^2}$$

$$\leq 2 \sum_{k=1}^{+\infty} k I(k - 1 < |X_1| \leq k)$$

where the last inequality is obtained by using $1/j^2 < 1/j(j - 1) = 1/(j - 1) - 1/j, j \geq 2$, so its sum over $j \geq k$ is smaller than $1/(k - 1) \leq 2/k$ for $k \geq 2$. The case $k = 1$ then follows from the estimate for $k = 2$. Hence, by taking expectations, (2.10) entails the finiteness of (2.30).   ■

We now have one part of Theorem 1, which is formulated as

Step 5:   If the random variables $X_j, j \geq 1$, are i.i.d. with finite expectation E
then, as $n \to +\infty$,

$$\frac{1}{n} \sum_{k=1}^{n} X_k \to E \text{ a.s.}$$

Indeed, by Steps 1, 2, and 4,

$$\frac{1}{n} \sum_{k=1}^{n} (Y_k - E(Y_k)) \to 0 \text{ a.s.}$$

and

$$\frac{1}{n} \sum_{k=1}^{n} E(Y_k) \to E$$

Hence, Step 3 implies the desired limit.   ∎

The converse, constituting the second part of Theorem 1, is much simpler.

Step 6:   If the random variables $X_1, X_2, \ldots$ are i.i.d. and if there exists a
(finite) constant c such that

$$\frac{1}{n} \sum_{k=1}^{n} X_k \to c \text{ a.s.} \qquad (n \to +\infty) \tag{2.31}$$

then $E = E(X_j)$ is finite and $c = E$.

Proof.   Put $S_n = \sum_{k=1}^{n} X_k$. Then $n^{-1}S_n \to c$ implies

$$\frac{1}{n} X_n = \frac{1}{n} (S_n - S_{n-1}) = \frac{1}{n} S_n - \frac{n-1}{n} \frac{1}{n-1} S_{n-1} \to 0$$

Hence, (2.31) implies that $|X_n| \geq n$ can occur only for a finite number of n
a.s., which, by the assumed independence and with an appeal to Theorem
1.11 (in particular, see its concluding special case), yields

$$\sum_{n=1}^{+\infty} P(|X_n| \geq n) < +\infty$$

In view of (2.29), this amounts to the $E(X_j)$'s being finite. But when
$E = E(X_j)$ is finite, Step 5 is applicable, and thus $c = E$. The proof is
complete.   ∎

The combination of Steps 5 and 6 is Theorem 1. Notice that, in Step 6,
the conclusion would not change if we assumed only that (2.31) holds with

positive probability. The preceding argument shows that Theorem 1.11 implies that it then must hold a.s. One more remark is in order. The independence of the sequence X_j is not fully employed in the current section. Indeed, it is used only to the extent that we appealed to Kolmogorov's inequality, the additivity of the variance (Theorem 5) and the converse (Theorem 1.11) of the Borel–Cantelli lemma. Consequently, the strong law of large numbers is valid in a more general setting than independence. This will be demonstrated for particular (dependent) models in subsequent sections.

Exercises

1. Let $X_j \geq 0$ be i.i.d. random variables with infinite expectation. Prove that the arithmetical mean of X_j, $1 \leq j \leq n$, goes to infinity a.s. ($n \to +\infty$).
2. Let $X_j \geq 0$ be i.i.d. random variables with infinite expectation. Set $S_n = X_1 + X_2 + \cdots + X_n$. Show that

   (i) $P(\lim \sup(X_n/S_{n-1}) = +\infty) = 1$

   by first showing that, for arbitrary $\varepsilon > 0$, the events $A_n = \{S_{n-1} \leq \varepsilon X_n\}$ occur infinitely often a.s. (establish for these A_n a "converse Borel–Cantelli lemma"). Deduce from (i) that, for any sequence $a_n \to +\infty$ with n and such that

   (ii) $P(\lim \sup(S_n/a_n) < +\infty) = 1$

   we have

   $$P(\lim \inf(S_n/a_n) = 0) = 1$$

   (Note that we shall prove in the next section that the probability in (ii) is either 1 or 0.)

                              Y. S. Chow and H. Robbins (1961)

3. Let $X_j \geq 2$ be integer-valued i.i.d. random variables with $P(X_j = s) = 1/s(s-1)$, $s \geq 2$. Then $E(X_j) = +\infty$. Show that $(n \log n)^{-1} S_n \to 1$ in probability as $n \to +\infty$, where $S_n = X_1 + X_2 + \cdots + X_n$. Compare the result with the preceding exercise.
4. Let X_j be i.i.d. random variables with $E(X_j) = 0$. Assume that $1 < d < 2$ is such that $E(|X_1|^d) < +\infty$. Show that the infinite series $\sum n^{-1/d} X_n$ converges a.s.

2.6   ZERO-ONE LAWS

It was remarked at the end of the previous section that if (2.31) is valid with positive probability then it is valid a.s. In other words, for independent random variables, (2.31) is valid either with probability zero or with probability one. The present section is devoted to laws of this nature. One general theorem follows immediately from Theorems 1.10 and 1.11, the Borel–Cantelli lemmas. We formulate the statement for a special case, covering independence.

Theorem 17   Let the events $A_1, A_2, \ldots$ be such that $P(A_i \cap A_j) \leq P(A_i)P(A_j)$ for all $i \neq j$. Then $P(\limsup A_n) = 0$ or 1 according as $\sum P(A_j)$ converges or diverges.

Some applications of Theorem 17 are listed as examples.

Example 1   If the random variables X_n are independent and $b_n X_n \to 0$ with positive probability, where b_n, $n \geq 1$, are some positive numbers, then $b_n X_n \to 0$ a.s. and, for every $c > 0$, $\sum P(|X_n| \geq c/b_n) < +\infty$.

Indeed, if $A_n = \{b_n|X_n| \geq c\}$, then the events A_n are independent and, thus, Theorem 17 is applicable. We now need only to observe that if $b_n X_n \to 0$ then, for every $c > 0$, A_n can occur only a finite number of times, which has probability either zero or one. The assumption is that it is not zero, so it is one, whose complement, $\limsup A_n$, thus has probability zero. Theorem 17 now yields the conclusion of the example.   ∎

Example 2   Let $X_1, X_2, \ldots$ be independent random variables and set $Z_n = \max(X_1, X_2, \ldots, X_n)$. Let $B_n > 0$ be a nondecreasing sequence of real numbers that tends to infinity with n. Then, for any $k > 0$, $P(Z_n \geq kB_n \text{ i.o.}) = 1$ or 0, depending on whether $\sum (1 - F_j(kB_j))$ diverges or converges. Here i.o. stands for "infinitely often" and $F_j(x)$ is the distribution function of X_j such that $F_j(x) < 1$ for all j and all x.

First note that, because $B_n \to +\infty$ with n and all X_j being unbounded, $P(Z_n \geq kB_n \text{ i.o.}) = P(X_n \geq kB_n \text{ i.o.})$. Indeed, since $Z_n \geq X_n$, then, if $X_n \geq kB_n$ i.o., Z_n, too, exceeds kB_n for these same infinitely many values of n. Conversely, if $Z_n \geq kB_n$, then, for some $j \leq n$, $X_j \geq kB_n \geq kB_j$. Let $j(n)$ denote the largest j with the preceding property. Since B_n is unbounded, an infinite sequence of distinct values of $j(n)$ corresponds to an infinite sequence of n above. We can now apply Theorem 17 to the sequence $A_n = \{X_n \geq kB_n\}$, and the conclusion of the lemma follows.   ∎

Example 3   Let $X_1, X_2, \ldots$ be i.i.d. random variables with a common exponential distribution $F(x) = 1 - e^{-x}$, $x \geq 0$. Then $P(\limsup Z_n/\log n = 1) = 1$, where Z_n is as in Example 2.

We have to show that, for every $a > 0$, $Z_n \geq (1 - a) \log n$ i.o. and

$Z_n \geq (1 + a) \log n$ only for a finite number of n, both statements valid a.s. Since $(B_n = \log n)$

$$\sum_{j=1}^{+\infty} (1 - F(kB_j)) = \sum_{j=1}^{+\infty} j^{-k}$$

which diverges for $k \leq 1$ and converges for $k > 1$, the claimed limiting property follows from Example 2.   ∎

For the next statement we introduce the concept of the "tail" of a sequence X_j, $j \geq 1$, of random variables. Using the notation $\sigma(X_k, X_{k+1}, \ldots, X_{k+m})$ for the smallest σ-field containing all members of $\mathcal{A}$ which are of the form $\{\omega: X_j \leq z_j, k \leq j \leq k + m\}$, z_j arbitrary real (i.e., $\sigma(X_k, X_{k+1}, \ldots, X_{k+m})$ is the smallest sub-σ-field of $\mathcal{A}$ with respect to which the random variables X_j, $k \leq j \leq k + m$, are all measurable), we define $\sigma(X_k, X_{k+1}, \ldots)$ as the smallest σ-field over the field of the union of the sequence $\sigma(X_k, X_{k+1}, \ldots, X_{k+m})$, $m \geq 0$, of σ-fields. Then the tail σ-field of the random variables X_j, $j \geq 1$, is the intersection of the σ-fields $\sigma(X_k, X_{k+1}, \ldots)$, $k \geq 1$. Events of the tail σ-field will be referred to as *events in the tail* of the sequence X_j, $j \geq 1$. Random variables that are measurable with respect to the tail σ-field of X_j, $j \geq 1$, are called *tail functions* of X_j, $j \geq 1$. Typical tail events of the sequence X_j, $j \geq 1$, are the events on which X_j itself, or $\sum X_j$ or $(X_1 + X_2 + \cdots + X_n)/n$, converges. Typical tail functions are $\liminf X_j$, $\limsup X_j$, and the $\liminf$ or $\limsup$ of the arithmetical mean of the first n terms (or of any n terms). Hence, Theorem 17 is, in fact, a statement on a particular event in the tail of the indicators of the events A_n, $n \geq 1$ (whether the sum of these indicators is finite or infinite). The following statement is about the tail of independent random variables.

Theorem 18   **(Kolmogorov's Zero-One Law)**   Let $X_1, X_2, \ldots$ be independent random variables. If the event A is in the tail of X_j, $j \geq 1$, then $P(A) = 0$ or 1. The tail functions of X_j are degenerate.

Proof.   Arguing as at the end of Section 1.8, one gets that if X_j, $j \geq 1$, are independent and if J_1 and J_2 are disjoint subsets of the positive integers, then $\sigma(X_j, j \in J_1)$ and $\sigma(X_j, j \in J_2)$ are independent. Since the tail σ-field is the intersection of $\sigma(X_j, j \geq k)$, $k \geq 1$, $A \in \sigma(X_j, j \geq k)$ for all $k \geq 1$. In particular, $A \in \sigma(X_j, j \geq 1)$. Furthermore, by independence, $\sigma(X_1, X_2, \ldots, X_k)$ and $\sigma(X_j, j > k)$ are independent for all k, and since the latter contains A, A is independent of $\sigma(X_j, 1 \leq j \leq k)$ for every k. But then A is independent of $\sigma(X_j, 1 \leq j)$, which has already been established to contain A. That is, A is independent of itself, implying $P(A \cap A) = P(A)P(A)$, which is possible only if $P(A)$ is either 0 or 1. Hence, the smallest σ-field containing all tail events differs from $\{\emptyset, \Omega\}$ only by events of zero probability. Since random variables are equivalent if they differ only on such sets, tail functions are necessarily degenerate. The proof is complete. ∎

In plain language, A is a tail event of $X_j, j \geq 1$, if it is determined by X_j, $j \geq 1$, but a finite number of this sequence has no influence on A. Thus, as mentioned earlier, the sets on which X_j or $\sum X_j$ converge are tail events. Similarly, $\limsup X_j$, when finite a.s., is a tail function, but $\sup_j X_j$ is not. Some important cases are collected in the next statement.

Corollary If the random variables X_j are independent, then (i) X_j either converges a.s. or diverges a.s.; (ii) $\sum X_j$ either converges a.s. or diverges a.s.; (iii) $\limsup X_j$ is either a finite constant a.s., $+\infty$ a.s., or $-\infty$ a.s.; (iv) if $b_n > 0$, $b_n \to +\infty$ with n, and $(X_1 + X_2 + \cdots + X_n)/b_n$ converges (to a finite value (s)) with positive probability, then it converges a.s., and the limit is constant a.s.

Example 4 Let $X_1, X_2, \ldots$ be i.i.d. exponential variables with the distribution function $F(x) = 1 - e^{-x}$, $x \geq 0$. Then $\limsup X_n/\log n = 1$ a.s.

Because for every sequence $c_n > 0$, the event $\{X_n \geq c_n \text{ i.o.}\}$ is a tail event, its probability is either zero or one. By choosing $c_n = (1 - a) \log n$, $0 < a < 1/2$, and then $c_n = (1 + a) \log n$, $a > 0$, Theorem 17 leads to the desired limiting property. ∎

This argument appears to be similar to the one employed in Example 2 for Z_n. In fact, the normalization is the same for X_n and Z_n, giving the false impression that somehow Z_n and X_n are always "comparable" in limit for independent variables. That $\limsup Z_n/c_n$ is not a tail event is demonstrated in Exercise 2. Yet similarities do exist between $\limsup X_n/c_n$ and $\limsup Z_n/c_n$, as was seen in Example 2. This can be viewed as an extension of tail properties to nontail events. A general result of this kind now follows.

Theorem 19 **(Zero-One Law of Hewitt and Savage)** Let $X_1, X_2, \ldots$ be independent and identically distributed random variables. Let A be an event that belongs to $\sigma(X_j, j \geq 1)$ and is invariant under finite permutations of the sequence X_j. Then $P(A)$ is either 0 or 1.

Proof. Since the sequence $\sigma(X_j, 1 \leq j \leq k)$, $k \geq 1$, is increasing, their union is a field. Therefore, by the definition of $\sigma(X_j, j \geq 1)$ (see the paragraph preceding Theorem 18), Section 1.3 applies. In particular, (1.10) and (1.13) imply that, for every event A and for arbitrary $\varepsilon > 0$, there is an n and an event $A_n \in \sigma(X_j, 1 \leq j \leq n)$ such that

$$P(A \cap A_n^c) < \varepsilon \qquad \text{and} \qquad P(A_n \cap A^c) < \varepsilon \qquad (2.32)$$

Now, consider A as a set of infinite-dimensional vectors $\{(x_1, x_2, \ldots)\}$, where x_j is a value of X_j. The elements of A_n are evidently such that $x_1, x_2, \ldots, x_n$ are restricted by A_n, but $x_{n+1}, x_{n+2}, \ldots$ are arbitrary values (of $X_j, j > n$). Next, define A_n^* as the set of vectors obtained from members of A_n by the permutation $(x_1, x_2, \ldots, x_{2n}) \to (x_{2n}, x_{2n-1}, \ldots, x_1)$. Since $x_j, j > n$, are arbitrary, A_n^* depends only on $X_{n+1} = x_n, X_{n+2} = x_{n-1}, \ldots, X_{2n} = x_1$, and thus A_n and A_n^* are independent. Furthermore, by the invariance property of A, a member of $A \cap A_n$ also belongs to $A \cap A_n^*$ after the permutation. Therefore, since the X_j are identically distributed, (2.32) yields, as $n \to +\infty$,

$$\lim P(A_n \cap A_n^*) = \lim P(A_n^*) = \lim P(A_n) = P(A)$$

Finally, using the independence of A_n and A_n^* on the extreme left-hand side, we get $P^2(A) = P(A)$, the solution of which is either 0 or 1. ∎

Events invariant under finite permutations of $X_j, j \geq 1$, are called *symmetric over the sequence* $X_j, j \geq 1$. Tail events are clearly symmetric (they simply do not depend on a finite number of terms), but there are others, two of which are presented below.

Example 5 Let $X_1, X_2, \ldots$ be i.i.d. random variables. Set $Z_n = \max(X_1, X_2, \ldots, X_n)$ and $S_n = X_1 + X_2 + \cdots + X_n$. Then the events $\{Z_n \geq B_n \text{ i.o.}\}$ and $\{a \leq S_n < b \text{ i.o.}\}$ occur with probability either 0 or 1. (Here, B_n, a, and $b > a$ are finite real numbers.)

Indeed, both events are symmetric over $X_j, j \geq 1$; thus, Theorem 19 applies.

Exercises

1. With the notation and assumptions of Example 2, show that $P(\limsup Z_n/B_n = t) = 1$, where $t = 0$ if the series $(*) \sum (1 - F_j(kB_j))$ converges for all $k > 0$ and $t = \sup S$ otherwise, where S is the set of $k > 0$ for which $(*)$ diverges.

   R. Mucci (1977)

2. Let $X_1, X_2, \ldots$ be independent random variables, where X_1 is the unit exponential $(F(x) = 1 - e^{-x}, x \geq 0)$, and, for $j \geq 2$, $P(0 < X_j < 1) = 1$. Then $0 < P(Z_n \geq 2 \text{ i.o.}) < 1$, where $Z_n = \max(X_1, X_2, \ldots, X_n)$.

3. Let the simple random variables X_j with $P(X_j = 1) = P(X_j = -1) = 1/2, j \geq 1$, be independent. Prove

   $$P(X_1 + X_2 + \cdots + X_n = 0 \text{ i.o.}) = 1$$

   Show that the conclusion continues to hold if X_j has a symmetric distribution on the integers: $P(X_j = u) = P(X_j = -u) = p(u)$ for all $j \geq 1$ and all $u \geq 1$, with $\sum_{u=1}^{+\infty} p(u) = 1/2$, and also $E(|X_1|) < +\infty$.

2.7 LEBESGUE–STIELTJES INTEGRALS

The results of the current section will facilitate the actual computation of the expectation. It is clear from (2.8) that the distribution function $F(x)$ of a random variable X uniquely determines $E(X)$. It is therefore not surprising that the integral below, which is developed from a distribution function, will provide a new form for the evaluation of the expectation.

In the present section, we shall extend the concept of the distribution function. We say that a function $F(x)$ defined on the whole real line is a distribution function in the extended sense (or simply "distribution function" in the current section) if (i) $0 \le F(x) \le 1$; (ii) it is nondecreasing; and (iii) it is continuous from the right.

We write $F(+\infty)$ for the limit of F as $x \to +\infty$; $F(-\infty)$ has a similar meaning at $-\infty$. Let $v_F = F(+\infty) - F(-\infty)$. Assume $v_F > 0$. Then we define P_F as the unique probability on the Borel sets of the real line such that $v_F P_F((a, b]) = F(b) - F(a)$ for all numbers $a < b$ of continuity points of $F(x)$. That such a probability exists and is unique follows from Section 1.3 (see particularly the concluding remark in its Step 1 and main extension theorem) and from the following simple lemma (which guarantees that all intervals are included in the extension of the collection of the particular intervals above).

Lemma   The set of discontinuity points of a distribution function $F(x)$ is denumerable.

Proof.   In the $y = F(x)$ coordinate system, project the graph of $F(x)$ into the y-axis. Because of the monotonicity assumption, to every point x of discontinuity of $F(x)$ there corresponds an interval (a_x, b_x) on y that is left blank in the projection. For different x, the intervals (a_x, b_x) are disjoint. Since every interval contains a rational number, the points of discontinuity of F can be associated with a subset of the rationals, which is denumerable. ∎

Definition 3   Let $g(x)$ be a Borel-measurable function on the real line. The Lebesgue–Stieltjes integral of g with respect to F,

$$\int_{-\infty}^{+\infty} g(x)\, dF(x) = v_F \int_R g\, dP_F \qquad \text{if } v_F > 0$$

where R signifies the real line. If $v_F = 0$ (i.e., if F is a constant on R), then the integral on the left-hand side is defined to be zero.

Just as the integral on the right-hand side is not always meaningful, neither is the left-hand side: if both $P_F(g(x) = +\infty)$ and $P_F(g(x) = -\infty)$ are positive, the Lebesgue–Stieltjes integral of $g(x)$ is undefined. If the integral in Definition 3 is finite, we say that $g(x)$ is integrable with respect to $F(x)$. For numbers $-\infty \le a < b \le +\infty$, let $I(a, b)$ be the indicator of the interval $[a, b]$ (if a or b is infinite, then open on that side). We then define

$$\int_a^b g(x)\, dF(x) = \int_{-\infty}^{+\infty} g(x) I(a, b)\, dF(x)$$

A final remark on Definition 3   We can replace condition (i) in the definition of a distribution function $F(x)$ by assuming only that both $F(-\infty)$ and $F(+\infty)$ are finite, and observe that, with suitable constants $c > 0$ and d, $cF(x) + d = F^*(x)$ will then satisfy condition (i). Furthermore, $P_{F^*}(A) = cP_F(A)$ for all Borel sets A. Therefore, if we multiply the integral with respect to F^* in Definition 3 by $1/c$, we can speak of integrals with respect to this further extended F.

Because the "new integral" is defined in terms of an "old type," all statements on integrals continue to hold for Lebesgue–Stieltjes integrals. Thus, g is integrable if and only if $|g|$ is, integration and summation can be interchanged (for a finite number of summands, or even for an infinite number of terms if all summands are positive) and the limiting properties of Section 2.4 apply.

Let $a < b$ be finite numbers and assume that $F(x)$ is continuous at both a and b. Let $g(x)$ be a continuous function on $[a, b]$. We call the set $a = x_{0,k} < x_{1,k} < \cdots < x_{n,k} = b$ of points $x_{j,k}$, $0 \le j \le n$, a division D_k of the interval $[a, b]$. To D_k, we assign the Riemann–Stieltjes sums

$$R_F(D_k; g) = \sum_{j=0}^{n-1} g(z_j)[F(x_{j+1,k}) - F(x_{j,k})]$$

where the points $x_{j,k}$ are assumed to be continuity points of $F(x)$, and z_j are arbitrary numbers satisfying $x_{j,k} \le z_j \le x_{j+1,k}$. Finally, we say that, as $k \to +\infty$, the maximum of D_k, abbreviated $\max D_k$, tends to zero if $\max\{x_{j+1,k} - x_{j,k} : 0 \le j \le n - 1\} \to 0$.

Theorem 20   Under the conditions stated in the previous paragraph, if $\max D_k \to 0$ as $k \to +\infty$,

$$\lim R_F(D_k; g) = \int_a^b g(x)\, dF(x)$$

If, in addition, $F(x)$ is differentiable on $[a, b]$, then, with $f(x) = F'(x)$,

$$\int_a^b g(x)\, dF(x) = \int_a^b g(x) f(x)\, dx$$

where the right-hand side is the Riemann integral of elementary calculus.

Proof. Note that the function $g_k(x) = g(z_j)$ for $x_{j,k} \leq x \leq x_{j+1,k}$, $0 \leq j \leq n - 1$, and $g_k(x) = 0$ outside of $[a, b]$, is a simple random variable on the Borel sets of the real line, and it is uniformly bounded in k because of the continuity assumption on $g(x)$. Furthermore, $g_k(x) \to g(x)$ as $k \to +\infty$ and max $D_k \to 0$. Therefore, by the dominated convergence theorem (Theorem 16), we get the limit in the first part of the theorem. For the second part, apply the mean value theorem of calculus, yielding $F(x_{j+1,k}) - F(x_{j,k}) = f(z_j)(x_{j+1,k} - x_{j,k})$, where z_j is some number with $x_{j,k} \leq z_j \leq x_{j+1,k}$. Now, with these z_j for $g(z_j)$, we can argue as in the first part with the simple function $g_k^*(x)$, taking the value $g(z_j) f(z_j)$ on $[x_{j,k}, x_{j+1,k}]$. The theorem is established. ∎

Since the second part of Theorem 20 is applicable with $F(x) = x$ (recall the paragraph following Definition 3 and the fact from Chapter 1 that P_F with $F(x) = x$, $a \leq x \leq b$, a, b finite, is called Lebesgue measure), we proved that, for continuous g, on finite intervals, Riemann and Lebesgue integrals coincide. Next, we extend the concept of Lebesgue integral over the whole real line in two steps. First, let $g(x)$, x real, be nonnegative and assume it is Lebesgue measurable on all finite intervals. Let $a_j < a_{j+1}$ be finite real numbers with $a_j \to -\infty$ as $j \to -\infty$ and $a_j \to +\infty$ as $j \to +\infty$. Then

$$\int_{-\infty}^{+\infty} g(x)\, dx = \sum_{j=-\infty}^{+\infty} \int_{a_j}^{a_{j+1}} g(x)\, dx$$

Second, if $g(x)$ is Lebesgue measurable on all finite intervals, then, as before, we put $g^+(x) = \max(0, g(x))$ and $g^-(x) = -\min(0, g(x))$, and define

$$\int_{-\infty}^{+\infty} g(x)\, dx = \int_{-\infty}^{+\infty} g^+(x)\, dx - \int_{-\infty}^{+\infty} g^-(x)\, dx$$

whenever the right-hand side is meaningful. The reader can easily verify that all rules of integration established so far continue to hold for this extended Lebesgue integral. Furthermore, if $g(x)$ is defined and finite for all x and if its Lebesgue integral is finite, then the Lebesgue integral and the improper Riemann integral of $g(x)$ coincide. However, improper Riemann integrals are not necessarily Lebesgue integrals.

Example 6 The improper Riemann integral

$$\int_0^1 \frac{\sin(1/x)}{x} \, dx = \int_1^{+\infty} \frac{\sin y}{y} \, dy$$

is not a Lebesgue integral.

Indeed, we know from calculus that the above improper integral is finite but the improper integral of the absolute values is infinite. Hence, it is not a Lebesgue integral. ∎

Remark By agreement, a distribution function $F(x)$ must be defined on the whole real line. Therefore, when we say $F(x) = x$ on $[a, b]$, $a < b$ finite, we shall always mean that $F(x)$ is continuously extended to R as one constant for $x < a$ and another constant for $b < x$ (remember, we speak of extended distribution functions; when we deal with distribution functions of random variables, $F(x) = x$ on $(0, 1)$, for example, automatically implies what we have just said by an earlier convention). Now, with this specification, $F(x) = x$ on (a, b), and then extended to R, is not differentiable on R, nor on $[a, b]$; the two endpoints a and b are exceptions. Since Riemann integrals are not affected if the integrand is changed at a finite number of points, and neither are Lebesgue–Stieltjes integrals if $F(x)$ is continuous, we add one more general rule of terminology: when we speak of the differentiability of $F(x)$ (in the classical, or elementary, sense, as opposed to that of the Radon–Nikodym derivative to be introduced later) and we permit a finite number of exceptions, we always assume that $F(x)$ is continuous on the whole real line. In this way, the contributions $g(z_j)[F(x_{j+1,k}) - F(x_{j,k})]$ to Riemann–Stieltjes sums for a finite number of j always disappear at the limit when g is bounded over finite intervals (which is the case for continuous g). So Theorem 20 continues to hold if we permit a finite number of exceptions to differentiability.

Let us now look further at the relation of improper Riemann integrals and Lebesgue–Stieltjes integrals.

Theorem 21 Assume that $g(x)$ is a continuous function on R and that the distribution function $F(x)$ is differentiable on R, except perhaps at a finite number of points. Let $f(x) = F'(x)$. If g is integrable with respect to P_F then

$$\int_{-\infty}^{+\infty} g(x) \, dF(x) = \int_{-\infty}^{+\infty} g(x) f(x) \, dx$$

where the right-hand side is the improper Riemann integral.

Proof. Let $a < b$ be arbitrary finite numbers. Then, with the indicator $I(a, b)$ employed earlier, we set $g_{a,b}(x) = g(x)I(a, b)$. Clearly, $|g_{a,b}(x)| \leq |g(x)|$ and $g_{a,b}(x) \to g(x)$ as $a \to -\infty$ and $b \to +\infty$. Then the integrability assumption on $g(x)$ ensures that the dominated convergence theorem applies, yielding, as $a \to -\infty$ and $b \to +\infty$,

$$\lim \int_a^b g(x)\, dF(x) = \int_{-\infty}^{+\infty} g(x)\, dF(x)$$

Upon applying Theorem 20 to the left-hand side, the theorem follows from the definition of improper Riemann integrals. ∎

Corollary Let X be a random variable on an arbitrary probability space with distribution function $F(x)$. Assume that $F'(x) = f(x)$ exists, except perhaps for a finite number of x. Then, if $E(X)$ is finite,

$$E(X) = \int_{-\infty}^{+\infty} xf(x)\, dx$$

where the right-hand side is the improper Riemann integral from calculus.

Proof. First observe that, if X is integrable, then (2.11) applies. Next, upon appealing to (2.8) for both X^+ and X^-, one gets

$$E(X) = \int_\Omega X\, dP = \int_{-\infty}^{+\infty} x\, dF(x) \tag{2.33}$$

Since $g(x) = x$ is continuous, Theorem 21 now implies the corollary. ∎

The integral labeled in the corollary as an improper integral is not always improper. Frequently, $f(x) = 0$ outside a bounded set.

Example 7 The distribution function of the random variable X

$$F(x) = \begin{cases} x & \text{if } 0 \leq x \leq 1/2 \\ 2(x - 1/2)^2 + 1/2 & \text{if } 1/2 \leq x \leq 1 \end{cases}$$

Then

$$E(X) = \int_0^{1/2} x\, dx + \int_{1/2}^1 4x(x - 1/2)\, dx = \frac{13}{24}$$

Notice that $F'(x) = f(x)$ exists everywhere except at $x = 0$, $1/2$ and 1. Furthermore, $f(x) = 0$ for $x < 0$ and $x > 1$. Hence, by the corollary, ordinary Riemann integrals can be used to find $E(X)$. The integrals are those of $xf(x)$. ∎

When applying the corollary, we first have to determine whether or not $E(X)$ is finite. But note that, when writing (see (2.33))

$$E(X) = \int_a^b x \, dF(x) + \int_{-\infty}^a x \, dF(x) + \int_b^{+\infty} x \, dF(x) \qquad (2.34)$$

where $a < b$ are arbitrary finite numbers, the first term on the right-hand side is always finite, and thus we in fact have to establish that the last two terms converge to zero as $a \to -\infty$ and $b \to +\infty$. This is exactly the criterion for the absolute convergence of the improper integral in the corollary, implying that we can replace the condition "if $E(X)$ finite" by "if the improper integral $\int_{-\infty}^{+\infty} xf(x) \, dx$ be absolutely convergent."

Example 8 Let X be a normal variable with parameters m and σ; i.e., the distribution function of X

$$F(x) = \frac{1}{\sqrt{2\pi}\sigma} \int_{-\infty}^x e^{-(y-m)^2/2\sigma^2} \, dy$$

Then $E(X) = m$.

Indeed, since $F'(x) = (\sqrt{2\pi}\sigma)^{-1} \exp[-(x-m)^2/2\sigma^2]$, $xF'(x) \leq x^{-2}$, say, for all large x, and thus we get from symmetry about m and from elementary calculus that the corollary is applicable. Hence, upon writing $x = (x - m) + m$, we have

$$E(X) = \frac{1}{\sqrt{2\pi}\sigma} \int_{-\infty}^{+\infty} (x - m) e^{-(x-m)^2/2\sigma^2} \, dx$$

$$+ \frac{m}{\sqrt{2\pi}\sigma} \int_{-\infty}^{+\infty} e^{-(x-m)^2/2\sigma^2} \, dx$$

where the first integral is zero by symmetry and the last term equals $m[F(+\infty) - F(-\infty)] = m$ (for more details on the normal distribution, see Section 2.9). ∎

To evaluate variance, we need $E(X^2)$. This, in principle, requires the computation of the distribution function of X^2. In the theorem that follows it is established that such computation can be avoided; i.e., it is not always necessary to refer to (2.33).

We use a set-up a bit more general than the one just indicated. Observing that (2.33) transforms an integral on Ω with respect to P to an integral on R with respect to P_F (due to the fact that a random variable X maps Ω into the real line R and that this mapping induces P_F on the Borel sets of R, utilizing P), we introduce the following concept. Let Ω and Ω^* be two sets and let $\mathcal{A}$ and $\mathcal{A}^*$ be σ-fields of subsets of Ω and Ω^*, respectively. We say that T is a measurable mapping of Ω into Ω^*, if the inverse mapping

$$T^{-1}(A^*) = \{\omega : \omega \in \Omega, T(\omega) \in A^*\}$$

satisfies (i) $T^{-1}(A^*) \in \mathcal{A}$ for all $A^* \in \mathcal{A}^*$ and (ii) T^{-1} is a homomorphism for the operations of taking complements, and denumerable unions and intersections. That is,

$$T^{-1}(\Omega^*) = \Omega \qquad T^{-1}(\varnothing) = \varnothing \qquad T^{-1}(A^{*c}) = (T^{-1}(A^*))^c$$

and

$$T^{-1}(\cup A_j^*) = \cup T^{-1}(A_j^*) \qquad T^{-1}(\cap A_j^*) = \cap T^{-1}(A_j^*)$$

Theorem 22   Given $(\Omega, \mathcal{A}, P)$ and Ω^* with a σ-field $\mathcal{A}^*$ of subsets of Ω^*, let T be a measurable mapping of Ω into Ω^*. Define P^* on $\mathcal{A}^*$ by $P^*(A^*) = P(T^{-1}(A^*))$. If g is real-valued function on Ω^*, finite a.s. with respect to P^* (i.e., P), and such that the projection of every $A^* \in \mathcal{A}^*$ is a Borel set, then

$$\int_\Omega g(T(\omega))\, dP = \int_{\Omega^*} g\, dP^* \tag{2.35}$$

In particular, if $\Omega^* = R$, $\mathcal{A}^*$ is the set of Borel sets, and $g(x)$ is a continuous real-valued function on R, then, for every random variable X on $(\Omega, \mathcal{A}, P)$ with distribution function $F(x)$,

$$\int_\Omega g(X)\, dP = \int_{-\infty}^{+\infty} g(x)\, dF(x) \tag{2.36}$$

The equations above are valid in the sense that if one side of the equations is meaningful then so is the other.

Proof.   Both $g(T(\omega))$ and $g(\omega)$ are, in fact, random variables on $(\Omega, \mathcal{A}, P)$ and $(\Omega^*, \mathcal{A}^*, P^*)$, respectively, and thus Definition 2 of integrals is applicable on both sides of (2.35). First note that if g is the indicator variable of a set A^* then, by definition, (2.35) holds because it then reduces to $P^*(A^*) = P(T^{-1}(A^*))$. Next, let g be a simple random variable with values x_1, $x_2, \ldots, x_m$ and associated distribution $p_1, p_2, \ldots, p_m$. Then $g(T(\omega))$ is simple as well, taking of course the same values x_j, and, because of what we have just observed for indicators, it has the same (associated) distribution as g on $(\Omega^*, \mathcal{A}^*, P^*)$. Consequently, (2.35) holds when g is simple. If $g \geq 0$, let $g_1 \leq g_2 \leq \cdots$ be nonnegative simple random variables converging to g. The validity of (2.35) for simple random variables and the monotone convergence theorem (or simply Definition 2) yield the desired result. Finally, if g is arbitrary, upon writing $g = g^+ - g^-$, Definition 2 and the already established cases lead to (2.35).

The special case (2.36) follows immediately. Indeed, the first part of the theorem and the concepts involved were geared towards covering (2.36). The proof is complete. ∎

In the special case when $\Omega = \Omega^*$, $\mathscr{A} = \mathscr{A}^*$, and $P = P^*$, the measurable mapping T is called *measure preserving*. In such a case, it is of special interest to study the sequence obtained by repeated application of T: $T^n = T(T^{n-1})$. See Exercise 5 for such an example.

Example 9   With the notation of Example 8 (the normal case), $V(X) = \sigma^2$.

Since $V(X) = E(X^2) - E^2(X)$ and $E(X) = m$, with the choice $g(x) = x^2$ in (2.36), an appeal to the corollary and integration by parts (write $x^2 e^{-x^2/2} = x(-e^{-x^2/2})'$) gives the desired value. ∎

We shall frequently face the need to compute $m_k = E(X^k)$ for k other than one or two. The expectation m_k is called the kth *moment* of X.

Example 10   Let X be a standard normal variable, i.e., its distribution function is as in Example 8 with $m = 0$ and $\sigma = 1$. Then $m_k = 0$ for k an odd integer and $m_{2k} = (2k - 1)(2k - 3) \cdots 3 \cdot 1$ (the product of the odd integers up to $2k$, which can also be written as $(2k)!/k!2^k$).

We again appeal to (2.36), now with $g(x) = x^k$, and to the corollary. By symmetry about zero, $m_k = 0$ if k is odd. On the other hand, repeated integrations by parts (quite similarly as in Example 9) yield $m_k = (k - 1)m_{k-2}$, $k > 2$, from which the desired result follows for k even. ∎

We conclude the present section with the following remarks. In later sections, integration with respect to decreasing functions will come up. If $G(x)$ is nonincreasing, continuous from the left, and both $G(+\infty)$ and $G(-\infty)$ are finite, then $F_G(x) = -G(x)$ is a distribution function (in the extended sense). Lebesgue–Stieltjes integrals with respect to G are therefore defined as the negative of the corresponding Lebesgue–Stieltjes integral with respect to $F_G = -G$. Formally, $dG = -d(-G)$. Furthermore, if $U(x) = F(x) + G(x)$, where F and $-G$ are distribution functions, $dU = dF + dG$, meaning that an integral with respect to U is defined as the sum of two integrals, one with respect to F, the other with respect to G, whenever such sum is meaningful. The results of the present section are applicable to both of these extensions.

Some of the results of the present section are expressed in terms of the derivative of F. Since the current section is for extended distribution functions, F' has no further significance. However, if $F(x)$ is the distribution function of a random variable and if $f(x) = F'(x)$ exists (for all x), then f and F uniquely determine each other, i.e., $f(x)$ is an important characteristic of X. We call $f(x) = F'(x)$ the *density function* of X or of $F(x)$. As in the theorems where it occurs, we again permit F not to be differentiable at a finite number of points. Thus, if F is continuous and $F'(x) = f(x)$ exists for all but a finite number of x then $f(x)$ is called a *density* of X or of F. Whether f is unique or not, the relation

$$F(x) = \int_{-\infty}^{x} f(t) \, dt$$

is valid. Here, the integral is a (improper) Riemann integral. We know this from calculus.

Exercises

1. Find the expectation and variance of X, the density function $f(x)$ of which is as given below ($f(x) = 0$ outside the specified domain):
    (i) $f(x) = Cx^{b-1}e^{-ax}$, $x > 0$, $a > 0$, $b > 0$, and C is uniquely determined by a and b. (Gamma distribution; special cases: $b = 1$ is the exponential distribution; $b > 0$ integer: Erlang distribution; $2b > 0$ integer, $a = 1/2$: chi-square distribution with $2b$ degrees of freedom);
    (ii) $f(x) = ab^a/x^{a+1}$, $x > b$, $a > 0$, $b > 0$ (Pareto distribution);
    (iii) $f(x) = bx^{b-1} \exp(-x^b)$, $x > 0$, $b > 0$ (Weibull distribution; note the special case $b = 1$).
2. Let Y be a standard normal variable and m and $\sigma > 0$ be constants. Then $X = \exp(m + \sigma Y)$ is said to have a lognormal distribution. Find its density and show that

    $$E(X^r) = \exp(rm + r^2\sigma^2/2), \qquad r \geq 1 \text{ integer}.$$

3. Let the distribution function of X

    $$F(x) = \begin{cases} 0 & \text{if } x < 0 \\ x + 1/4 & \text{if } 0 < x < 1/2 \\ 1 & \text{if } x > 1/2 \end{cases}$$

    Find $E(X)$. Generalize your method of evaluating $E(X)$ to $F(x)$ with an arbitrary number (finite, or denumerable but not dense) of discontinuities, which is differentiable between any two discontinuities.
4. Prove the following method of integration, which is called integration by parts. If $a < b$ are (finite) real numbers and $F(x)$ and $G(x)$ are distribution

functions in the extended sense, then

$$\int_a^b G(x)\,dF(x) = F(b)G(b) - F(a)G(a) - \int_a^b F(x-0)\,dG(x)$$

5. Let the probability space $(\Omega, \mathscr{A}, P)$ be defined as follows: $\Omega = [0, 1)$, $\mathscr{A}$ is its collection of Borel subsets, and, for $A \in \mathscr{A}$,

$$P(A) = \frac{1}{\log 2}\int_0^1 \frac{I(A)}{1+x}\,dx \qquad \text{(Gauss's measure)}$$

where, as usual, $I(A)$ is the indicator of A. Define the mapping T by $Tx = \{1/x\}$ if $0 < x < 1$ and $Tx = 0$ for $x = 0$, where $\{y\}$ signifies the fractional part of $y \geq 1$, i.e., $\{y\} = y - [y]$ with $[y]$ denoting the largest integer not exceeding y (its integer part). Show that T is a P-measure preserving transformation of $[0, 1)$ into itself (observe that Lebesgue measure is not preserved by T). (Repeated application of T leads to the continued fraction expansion of x.)
6. Let the function $f(x)$ be defined on the closed interval $[a, b]$ and let $a = x_0 < x_1 < \cdots < x_n = b$ be a partition of $[a, b]$. Then the supremum $v_f(a, b)$ of the sums

$$\sum_{j=1}^n |f(x_j) - f(x_{j-1})|$$

over all partitions $x_0, x_1, \ldots, x_n$ of $[a, b]$ is called the total variation of $f(x)$. Show that if $f(x)$ is of bounded variation, i.e., if its total variation is bounded, then the functions

$$F(x) = (1/2)v_f(a, x) + (1/2)f(x), \qquad a \leq x \leq b$$

and

$$G(x) = (1/2)v_f(a, x) - (1/2)f(x), \qquad a \leq x \leq b$$

are nondecreasing functions. Hence, by the obvious equation $f(x) = F(x) - G(x)$, interpret the term "integration with respect to a continuous function $f(x)$ of bounded variation."

2.8 WEAK CONVERGENCE

One of the most difficult parts of applied probability is determining the exact distribution function $F(x)$ of the random variable X under consideration. When $F(x)$ cannot be determined, approximations might provide the answer.

Definition 4 We say that a sequence $F_n(x)$ of distribution functions converges weakly to the distribution function $F(x)$ if $F_n(x) \to F(x)$ at each continuity point x of $F(x)$ as $n \to +\infty$. If, in addition, $F_n(-\infty) \to F(-\infty)$ and $F_n(+\infty) \to F(+\infty)$, weak convergence becomes complete convergence. If F_n and F are the distribution functions of the random variables X_n and X, respectively, then X_n is said to converge to X weakly (completely) when F_n converges to F weakly (completely).

The definition above is, again, for distribution functions in the extended sense. Thus, if F_n is a probability distribution function (in the strict sense) then, in approximations, complete convergence is of interest. This new concept of weak convergence is not to be confused with weak laws of large numbers, which involve convergence in probability.

The sequence $F_n(x) = 1 - e^{-x/n}$, $x > 0$, of exponential distributions clearly converges weakly to $F(x) \equiv 0$, but the convergence is not complete.

At first, it appears that weak convergence is not a useful tool. Simply, a sequence of identically distributed random variables X_n always converges weakly, and completely, to some random variable X, but there may be no relation between X_n and X. In spite of this involved triviality, weak convergence became a major tool of probability theory. For example, it will be demonstrated that averages are frequently asymptotically normally distributed regardless of their original distribution. This is the kind of approximation that was hinted at in the first paragraph.

The current section, and the whole of Chapter 3, are mainly preparations for establishing (from Chapter 4 on) general weak convergence results.

Theorem 23 Let D be a dense set on R. If the sequence F_n of distribution functions converges on D to $F(x, D)$ then F_n converges weakly to $F(x) = \lim F(x_d, D)$, $x_d \in D$, $x \leq x_d$, $x_d \to x$.

Proof. The reader can easily verify that $F(x)$ of the theorem is a well-defined distribution function. Now, let $x_{1d} < x < x_{2d}$ be two sequences of D, where x is an arbitrary real number. Since, for every n, $F_n(x_{1d}) \leq F_n(x) \leq F_n(x_{2d})$, it follows, upon letting $n \to +\infty$,

$$F(x_{1d}, D) \leq \lim\inf F_n(x) \leq \lim\sup F_n(x) \leq F(x_{2d}, D)$$

and thus, as $x_{1d} \to x$, and $x_{2d} \to x$, for continuity points x of $F(x)$, the two extreme sides become $F(x)$, yielding the desired limit. ∎

Theorem 24 (**Weak Compactness of Distributions**) Every infinite sequence F_n of distribution functions has a subsequence which converges weakly.

Proof. By Theorem 23, it suffices to demonstrate that F_n has a subsequence which converges at every rational number $r_j, j \geq 1$. Consider the sequence $0 \leq F_n(r_1) \leq 1$ of numbers. By the Bolzano–Weierstrass theorem of calculus, it has a subsequence $F_n^{(1)}(r_1)$ that converges. For this particular subsequence, consider now the numbers $F_n^{(1)}(r_2)$. By the same argument, it has a further subsequence $F_n^{(2)}(r_2)$ that converges. Note that $F_n^{(2)}(x)$ converges at both r_1 and r_2. Choosing then $x = r_3$, a further subsequence that converges can be found, and the procedure can be continued, obtaining a sequence $F_n^{(k)}(x)$ of subsequences of $F_n(x)$ such that the kth subsequence converges at $x = r_j$, $1 \leq j \leq k$. Let us select that subsequence $F_k^{(k)}(x)$ of $F_n(x)$ which is obtained by taking just one term, the kth, from the kth subsequence previously selected. Then, in view of the sequential nature of the selections, this last subsequence, with the exception of a finite number of its terms, is contained in $F_n^{(k)}(x)$ for every $k \geq 1$. Consequently, it converges at every $r_k, k \geq 1$. This completes the proof. ∎

We now turn to the investigation of the relation of weak (complete) convergence of distributions to the convergence of their respective Lebesgue–Stieltjes integrals.

Theorem 25 (Helly–Bray Lemma) If $F_n \to F$ weakly, then, for continuity points $a < b$ of F, as $n \to +\infty$,

$$\lim \int_a^b g(x)\, dF_n(x) = \int_a^b g(x)\, dF(x)$$

where $g(x)$ is an arbitrary continuous function on R.

Proof. We apply Theorem 20; thus, we approximate all the integrals involved by Riemann–Stieltjes sums. We use the same division $D_k, a = x_{0,k} < x_{1,k} < \cdots < x_{m,k} = b$ for all F_n and F, and we assume that all $x_{j,k}$ are continuity points of F_n, for all n, and F (recall the Lemma of Section 2.7), and max $D_x \to 0$ as $k \to +\infty$. Then the Riemann–Stieltjes sums

$$R_{F_n}(D_k; g) = \sum_{j=0}^{m-1} g(x_{j,k})[F_n(x_{j+1,k}) - F_n(x_{j,k})]$$

and

$$R_F(D_k; g) = \sum_{j=0}^{m-1} g(x_{j,k})[F(x_{j+1,k}) - F(x_{j,k})]$$

can be viewed as

$$R_{F_n}(D_k; g) = \int_a^b g_k(x)\, dF_n(x)$$

and similarly with F, where $g_k(x)$ is the simple function taking $g(x_{j,k})$ on $[x_{j,k}, x_{j+1,k})$. Thus, since the continuous function $g(x)$ is uniformly continuous on $[a, b]$, as max $D_k \to 0$,

$$\left| \int_a^b g(x)\, dF_n(x) - R_{F_n}(D_k; g) \right| \leq \sup_{a \leq x \leq b} |g_k(x) - g(x)| \to 0$$

and a similar estimate is valid with F as well. In addition, by weak convergence, as $n \to +\infty$,

$$\lim R_{F_n}(D_k; g) = R_F(D_k; g)$$

Hence, since (we suppress the variable x)

$$\int_a^b g\, dF_n - \int_a^b g\, dF = \left[\int_a^b g\, dF_n - R_{F_n}(D_k; g) \right] + [R_{F_n}(D_k; g) - R_F(D_k; g)]$$

$$+ \left[R_F(D_k; g) - \int_a^b g\, dF \right]$$

upon applying the triangular inequality and the estimate above, by which the dependence of the first difference on n is eliminated, the limits above entail that, if first $n \to +\infty$ and then $k \to +\infty$, the left-hand side goes to zero (argue with lim sup on the left-hand side). The theorem is established. ∎

Theorem 26 (**Helly–Bray Theorem**)   Let $g(x)$ be a continuous function on R. If either $g(x) \to 0$ for $x \to +\infty$ and $x \to -\infty$ and the distribution functions $F_n \to F$ weakly or $g(x)$ be bounded on R and $F_n \to F$ completely, then, as $n \to +\infty$,

$$\lim \int_{-\infty}^{+\infty} g(x)\, dF_n(x) = \int_{-\infty}^{+\infty} g(x)\, dF(x)$$

Proof.   Let $a < b$ be continuity points of F. Decompose every integral on R as in (2.34) (replace x by $g(x)$). Then integrals on $[a, b]$ converge by the Helly–Bray lemma, while the integrals on $(-\infty, a)$ and $(b, +\infty)$ are arbitrarily small if $g(x) \to 0$ at the infinities and $a \to -\infty$ and $b \to +\infty$. On the other hand, if g is bounded, $|g| \leq M$, say, then it is integrable, and complete convergence yields

$$\left| \int_{|x| \geq c} g\, dF_n \right| \leq M(F_n(+\infty) - F_n(c) + F_n(-c) - F_n(-\infty))$$

$$\to M(F(+\infty) - F(c) + F(-c) - F(-\infty)) \to 0 \qquad (2.37)$$

as $c \to +\infty$. The proof is complete. ∎

An essential part of the above proof is the estimate in (2.37): the left-hand side is made small for all large n when c is large. This was achieved by the assumption of complete convergence. However, we can simply make (2.37) the assumption, and then the conclusion of Theorem 26 follows. To formalize the content of (2.37), we say that $g(x)$ is uniformly integrable in the sequence F_n, $n \geq 1$, if to every $\varepsilon > 0$ there is a $c_0(\varepsilon)$ and an n_0 such that, for all $c \geq c_0$ and $n \geq n_0$,

$$\int_{|x| \geq c} |g(x)|\, dF_n(x) < \varepsilon$$

Theorem 27   Let $g(x)$ be a continuous function on R and assume it is uniformly integrable in the sequence F_n of distribution functions. Then $F_n \to F$ weakly implies

$$\lim \int_{-\infty}^{+\infty} g\, dF_n = \int_{-\infty}^{+\infty} g\, dF \qquad (n \to +\infty)$$

Proof. Note that g is integrable with respect to F. Indeed, by continuity, it is integrable on finite intervals (a, b), $a < 0$, $b > 0$, while the integral of g with respect to F can be made arbitrarily small on $(-\infty, a)$ and $(b, +\infty)$ by observing that, with $c = \min(-a, b)$, $c_0 \leq c < c_1$,

$$\varepsilon \geq \int_{c \leq |x| \leq c_1} |g|\, dF_n \to \int_{c \leq |x| \leq c_1} |g|\, dF$$

thus, the same inequality continues to hold if $c_1 \to +\infty$.

For the rest of the proof, the second part of the last proof can be repeated without change but with the added advantage that (2.37) is a part of the assumptions of the theorem.   ■

We apply Theorem 27 with $g(x) = x^k$, $k \geq 0$, integer, i.e., with the moment sequence

$$m_{k,n} = \int_{-\infty}^{+\infty} x^k\, dF_n(x)$$

The moments of F are denoted by m_k. The remarkable result in the following theorem is that, even though its proof relies on Theorem 27, it reverses the assumption and conclusion of that theorem.

Theorem 28   **(Method of Moments)**   Let F_n be a sequence of distribution functions. Assume that, for every $k \geq 0$, $m_{k,n} \to m_k (n \to +\infty)$. Then the sequence $\{m_k\}$ is necessarily the sequence of moments of a distribution function F. If it is known that the moments m_k, $k \geq 0$, uniquely determine F, then $F_n \to F$ completely.

Remark   Theorem 28 is also known as the Fréchet–Shohat theorem. It should be noted that it would have sufficed to assume that $m_{k,n} \to m_k$ for all $k \geq k_0$ with some fixed $k_0 > 0$. It then follows that $m_{k,n} \to m_k$ for $0 \leq k < k_0$ as well (see Exercise 3). However, the present author has not yet seen an application of Theorem 28 in which this generalization was an advantage. Finally, the reader should not be discouraged by the extra assumption of requiring that F is uniquely determined by its moments. Easy rules for this will be established in Chapter 3. In particular, it will be seen that the normal and the Poisson distributions (and many others) are determined by their moments, so Theorem 28 can be used to prove asymptotic normality as well as convergence to the Poisson distribution (see Exercises 4 and 5).

Proof.   Let $2s$ be an arbitrary positive even number. Then, for every $0 \leq k < 2s$ and for $c > 1$,

$$\int_{|x| \geq c} |x^k| \, dF_n(x) \leq c^{-(2s-k)} \int_{|x| \geq c} x^{2s} \, dF_n(x) \leq 2m_{2s} c^{-(2s-k)}$$

for all sufficiently large n. Hence, for every $k \geq 0$ (recall that $2s$ was arbitrary), x^k is uniformly integrable in F_n. Now, by the weak compactness of distributions (Theorem 24), F_n has a subsequence $F_{n_j} \to F$ weakly. Let us apply Theorem 27 to this subsequence. By the uniform integrability of x^k in F_n, and thus in F_{n_j}, the limits m_k are the moments of F. But if the sequence m_k, $k \geq 0$, uniquely determines F, then every subsequence of F_n converges to the same F, yielding that $F_n \to F$ weakly. In order to prove that the convergence is, in fact, complete, first observe that $m_{0,n} \to m_0$ is equivalent to

$$F_n(+\infty) - F_n(-\infty) \to F(+\infty) - F(-\infty) \tag{2.38}$$

Next, since $F_n(-\infty) \leq F_n(x) \leq F_n(+\infty)$, for continuity points x of F, when $F_n \to F$ weakly, $\limsup F_n(-\infty) \leq F(x) \leq \liminf F_n(+\infty)$, and thus

$$\limsup F_n(-\infty) \leq F(-\infty) \leq F(+\infty) \leq \liminf F_n(+\infty) \tag{2.39}$$

It follows from (2.38) and (2.39) that $F_n(+\infty) \to F(+\infty)$ and $F_n(-\infty) \to F(-\infty)$; thus, the desired complete convergence is established. ∎

Exercises

1. Show that the sequence $F_n(x)$ of distribution functions is weakly convergent, but not completely, if (i) $F_n(x) = 1 - n/x$, $x \geq n$, (ii) $F_n(x)$ is discrete with the following discontinuities at the nonnegative integers: $F_n(0) - F_n(0-) = 1/2$ for all $n \geq 1$ and $F_n(k) - F_n(k-) = 1/2n$ for $1 \leq k \leq n$, $n \geq 1$.

2. Let

$$F_n(x) = \begin{cases} 1 & \text{if } x > 1 \\ 1/2 + x^2/4n & \text{if } 0 < x < 1 \\ (1/2)e^{nx} & \text{if } x < 0 \end{cases}$$

Find

$$\lim \int_{-3}^{2} \frac{e^x}{1 + x^2} \, dF_n(x) \qquad (n \to +\infty)$$

3. With the notation of Theorem 28, assume that, for some $k_0 > 0$, $m_{k,n} \to m_k$ for all $k \geq k_0$. Prove that $m_{k,n} \to m_k$ for $0 \leq k < k_0$ as well.

4. Let the random variable f_n be binomial, with parameters n and $0 < p < 1$. Upon borrowing the fact from Chapter 3 that the normal distribution is uniquely determined by its moments, prove by the method of moments that, as $n \to +\infty$ and p is kept fixed, $(f_n - np)/(np(1 - p))^{1/2}$ converges completely to U, the distribution of which is standard normal. (Use Stirling's formula $m! = Km^{m+1/2}c^{-m}e^{c(m)/m}$, where $0 < c(m) < 1$ and K is constant, not depending on m, to approximate the factorials of the binomial distribution.)

5. Again assuming that it has been established that the Poisson distribution is uniquely determined by its moments, prove the Poisson approximation to the binomial by the method of moments: if $n \to +\infty$, $p \to 0$, and $np \to \lambda$, then the distribution of f_n converges to the Poisson distribution (with parameter λ) (the notations are the same as in the preceding Exercise).

2.9 SPECIAL DISTRIBUTION FUNCTIONS: PART I*

We have encountered a few distribution functions in the previous sections, but mainly as examples or exercises for stressing a point or computing a characteristic such as expectation, variance or other moments. In the present section we start a systematic development of models in which the choice of the underlying distribution function is the major point. There can be a variety of arguments for selecting the distribution function of a random variable in a particular applied problem, the most common of which are (i) determining the distribution by mathematical arguments which we call characterization theorems, (ii) approximations via limit theorems, and (iii) selecting one that is supported by data from experiments. The ideal way of selecting the

---

* Part II is Section 3.8.

underlying distribution would be (i), but this is available only for a very few cases. Quite acceptable, and in several cases as good as (i), is method (ii), for which several examples will be developed in subsequent chapters. However, the most frequent way is still method (iii), which can be effective if we become familiar with properties of several specific distributions. We can then recognize the similarities between the properties of a distribution and those encountered in a particular problem. For this reason, some distributions will be included in our list without an applied problem being attached to it. The interesting characteristics of these distributions will justify their detailed studies.

We start with univariate distributions. Unless otherwise stated, the basic random variable is X and its distribution function is $F(x)$ throughout this section. Naturally, $F(x)$ may contain parameters. Two parameters A and $B > 0$ can always be introduced by transforming x to $(x - A)/B$, i.e., by shifting the origin and changing the scale on the x-axis. The new distribution function $F_1(x) = F((x - A)/B)$ is said to be of the same type as $F(x)$, and the parameters A and B are called location and scale parameters, respectively. When X is discrete, its possible values are listed as $x_1, x_2, \ldots$ and its distribution is denoted $p_k = P(X = x_k)$. If x_k is a nonnegative integer we alternatively write k for x_k.

In the list below, the first two distributions have fully been covered in Chapter 1.

Hypergeometric distribution   This distribution is associated with, and fully characterized by, the selection model of Section 1.4. n items are selected at random from a lot of R items. Each item is one of two types. Let us call the two types red and white. If the number of red items in the lot is M, the number X of red items among those selected is called a hypergeometric variable, and its distribution is a hypergeometric distribution. In the selection procedure, one either chooses the n items together, or the items are chosen one by one, always at random and without replacement.

The possible values of X are the integers $\max(0, n + M - R) \le k \le \min(M, n)$ and, as established in Section 1.4,

$$p_k = \binom{M}{k}\binom{R - M}{n - k} \bigg/ \binom{R}{n}$$

The parameters M, n, and R must satisfy $0 < M, n < R$. Note that, due to the binomial coefficients occurring in p_k, $p_k = 0$ if a "wrong value" of k is chosen. Hence, in particular in summation formulas, one can always write $0 \le k \le n$.

Let I_k be the indicator of the event that, when the selection is one by one, the kth selection results in a red item. Then

$$X = I_1 + I_2 + \cdots + I_n \tag{2.40}$$

where, by (1.21), $E(I_j) = M/R$ for all j and $E(I_j I_k) = M(M - 1)/R(R - 1)$ for all $j \neq k$. Hence, by Theorem 2 and formula (2.13), $E(X) = n(M/R)$ and $V(X) = n(M/R)(1 - M/R)(R - n)/(R - 1)$.

By simplifying in the binomial coefficients, one gets

$$p_{k+1} = \frac{(M - k)(n - k)}{(k + 1)(R - M - n + k + 1)} p_k$$

from which it follows that p_k, as a function of k, initially increases, reaches a maximum at the largest integer not exceeding $(n + 1)(M + 1)/(R + 2)$ and then steadily decreases thereafter. This can also be expressed that, in the most likely case, $X/(n + 1)$ is $(M + 1)/(R + 2)$, or, for large M, R, and n, X/n is about M/R. This is a remarkable result, since it says that, in the most likely case, the percentage of the red items among the selected items is the same as the percentage of the red items in the whole lot. What makes the model really applicable is the fact that we can eliminate the restriction "in the most likely case" in the preceding statement (this will be done via approximating the hypergeometric distribution by a binomial distribution— see the next entry—which will in turn be approximated by the normal distribution). This provides the foundation for random inspection for quality control, for conducting opinion polls, for checking the extent of the spread of a disease, and for many others. (See Exercise 4.)

Binomial distribution The binomial distributions are the most significant ones in the foundations of probability theory. As discussed in Sections 1.5 and 1.7, if A is an event in connection with a random experiment such that $P(A) = p$, and if the experiment is repeated n times, independently of each other, then the number X of times when A occurred, called the frequency of A in these repetitions, has the binomial distribution

$$p_k = P(X = k) = \binom{n}{k} p^k (1 - p)^{n-k} \qquad 0 \leq k \leq n$$

A representation as at (2.40) applies except that, this time, the I_j are independent. Hence, $E(X) = np$ and $V(X) = np(1 - p)$. We also know by now (Theorem 17) that, in the infinite product space, the relative frequency X/n converges to p a.s. This provides the justification for interpreting abstract theorems of probability in terms of observations; in other words, this limit makes the mathematical theory of probability a very practical discipline.

The binomial distributions can also serve as approximations to the hypergeometric distributions. Upon writing $p_k(M, n, R)$ for the hypergeometric probabilities with parameters M, n, and R, one immediately gets

$$\binom{n}{k}\left(\frac{M-k}{R}\right)^k\left(\frac{R-M-(n-k)}{R}\right)^{n-k} \leq p_k(M,n,R) \leq \binom{n}{k}p^k(1-p)^{n-k}$$

where $p = M/R$ (after simplifying in the factorials, the lower limit is evident, while, for the upper limit, note that if $0 < a \leq U < V$, then $(U-a)/(V-a) < U/V$). The two bounds are very close to each other if both M and R are large and if R is of a much larger magnitude than n (which assumptions are always true both in quality control and in the case of opinion polls).

The following two approximations to the binomial distributions are very practical. (i) If $n \to +\infty$ and $p \to 0$ in such a way that $np \to \lambda$, a finite positive constant, then, for fixed k, $p_k \to \lambda^k e^{-\lambda}/k!$ (Poisson approximation) (we have given a proof for this approximation in Exercise 5 of the preceding Section 2.8, and several others will come up in subsequent chapters, in particular, see Example 9 in Section 3.2). (ii) (normal approximation) if $n \to +\infty$ and p is fixed, then $\lim P(a < (X - np)/(np(1-p))^{1/2} \leq b) = N(b) - N(a)$, where $N(x)$ is the standard normal distribution function; see below for $N(x)$. For the simplest proof, see Theorem 3.10.

Poisson distribution   Let X be a nonnegative integer valued random variable with distribution

$$p_k = \lambda^k e^{-\lambda}/k! \qquad \lambda > 0$$

Then X is called a Poisson variable, and its distribution a Poisson distribution (with parameter λ). The Poisson distribution appears both as an approximation (such as to the binomial) but more significantly as the exact distribution of the number of random points in a fixed interval if the points arrive successively on the positive real line in such a way that the gaps between two consecutive arrivals are independent and identically distributed exponential variables (the origin counts as a point). We call such randomly distributed points a Poisson process. An alternative way of arriving at a Poisson process is to assume that random points on the positive real line satisfy (i) the distribution of points in an interval depends on the length of the interval but not on its location, (ii) the numbers of points in disjoint intervals are independent random variables, and (iii) in an interval of length Δt, the probability of having exactly one point is asymptotic to $\lambda(\Delta t)$ and of having more than one point is bounded by a constant times $(\Delta t)^2$ as $\Delta t \to 0$.

The proofs are put off until Chapter 8, although the proofs are simple and do not rely on the material covered in Chapters 3–7.

The first description of a Poisson process makes it readily applicable because of the lack of memory property of an exponential distribution (see (2.45)). The second alternative, on the other hand, is in practical terms and the applied scientist must decide whether assumptions (i)–(iii) are applicable to a particular problem. The somewhat mathematical assumption (iii) is interpreted for application as "points do not arrive in groups." In most service related applications such as the arrival of customers at a bank, at a service station or at a department store, and the arrival of telephone calls at a busy switchboard the assumptions of a Poisson process are routinely accepted.

Write

$$u_{(t)} = u(u - 1) \cdots (u - t + 1) \qquad t \geq 1 \text{ integer} \tag{2.41}$$

Since, for $k \geq t$, $k_{(t)} \binom{n}{k} = n_{(t)} \binom{n - t}{k - t}$, one immediately gets $E(X_{(t)}) = \lambda^t$.

Hence, $E(X) = V(X) = \lambda$.

Given that the second characterization of a Poisson process has been proved, by splitting up an interval (a, b) at an inner point c, one gets (i) if X_1 and X_2 are independent Poisson variables, then so is $X_1 + X_2$ and (ii) to every Poisson variable X there exist independent Poisson variables X_1 and X_2 such that $X = X_1 + X_2$ (see also Raikov's theorem in Exercise 2 of Section 4.2). Both (i) and (ii) have easy proofs by characteristic functions (see Chapter 3). Property (ii) can be iterated, yielding that every Poisson variable can be decomposed into an arbitrary number of independent and identically distributed Poisson variables (recall that distribution depends on length only in a Poisson process). Hence, by the classical central limit theorem (quoted accurately below at the normal entry), $(X - \lambda)/\lambda^{1/2}$ is asymptotically normal. This is why tables for Poisson distributions are usually given for $\lambda \leq 25$.

It is very rare that the type of a discrete distribution would have any significance. The Poisson distributions are an exception to this remark, which we shall see in subsequent chapters (for example, see Theorem 4 in Section 5.2).

Geometric distribution This distribution is also known as discrete waiting time distribution because of the following model. Let A be an event in connection with a random experiment and let $0 < p = P(A) < 1$. Repeat the experiment independently of each other and let X be the number of repetitions required for A to occur for the first time. Then X takes the integers with

$$p_k = p(1 - p)^{k-1}, \qquad k \geq 1$$

This is called a geometric distribution, and X is called a geometric variable. We have

$$P(X > t) = p \sum_{t < k} (1 - p)^{k-1} = (1 - p)^t \tag{2.42}$$

and

$$P(X > t + s \,|\, X > t) = P(X > s) \qquad t, s > 0 \text{ integers} \tag{2.43}$$

The property at (2.43) is called the lack of memory on the positive integers. The reader is invited to prove that no other distribution can satisfy (2.43). If the restriction "integers" is changed to "arbitrary real numbers," (2.43) becomes a characteristic property of the exponential distribution among all nonnegative and nondegenerate distributions (recall Exercise 2 in Section 1.8). Another close relation of the geometric distribution to the exponentials is the limit relation

$$\lim P(X \le x/p) = 1 - e^{-x} \qquad p \to 0 \tag{2.44}$$

The proof is simple: on the right-hand side of (2.42) write $\exp(t \log(1 - p))$. By Taylor's expansion, $\log(1 - p) \sim -p$, so, with $t \sim x/p$, (2.44) follows. For further relations between the geometric and exponential distributions see the section *Exponential distribution*.

If X_r denotes the number of repetitions required for A to occur exactly r times in the model at the beginning of this section, then X_r is the sum of r independent copies of X. One can calculate this directly to yield

$$P(X_r = u) = \binom{u - 1}{r - 1} p^r (1 - p)^{u-r} \qquad u \ge r \text{ integer}$$

If we write $u = t + r$, then the right-hand side can also be written as $\binom{-r}{t} p^r (-q)^t$, $q = 1 - p$, $t \ge 0$. Due to this form, the above distribution is called a negative binomial distribution.

We now turn to some special univariate continuous distributions.

Normal distribution   We call X a normal variable with parameters m and σ, denoted $X \in N(m, \sigma)$, if

$$F(x) = (2\pi\sigma^2)^{-1/2} \int_{-\infty}^{x} e^{-(1/2\sigma^2)(y-m)^2} \, dy$$

By the substitution $u = (y - m)/\sigma$ in the integral above, we find that all normal variables are of the same type as $N(0, 1)$. From this it also follows that if $X \in N(m, \sigma)$, then $Y = (X - m)/\sigma \in N(0, 1)$. Members of $N(0, 1)$ are called standard normal. We have just established that for most nonstatistical statements it is not a restriction if we limit discussion to standard normal variables.

In Examples 8 and 9 we computed that if $X \in N(m, \sigma)$ then $E(X) = m$ and $V(X) = \sigma^2$. For the moments $E(X^k)$, $k \geq 1$, for $X \in N(0, 1)$ we refer to Example 10.

Normal distributions come up in applications both as exact distributions and, more frequently, as approximations. First, let us look at the way in which properties of the normal distributions can be utilized in evaluating the accuracy of measurements in laboratories. Assume a quantity m is to be measured. Instead of m, we measure U which fluctuates from experiment to experiment, so we view it as a random variable, and $X = U - m$ as a random error. Since, by the nature of measurement errors, U has the same chance of being below m as over it, $E(X) = 0$, but m is unknown. Now, by analyzing the structure of measurement errors, we shall be able to prove in Theorem 3.14 that X is a normal variable. Since its variance is unknown, a single value of U is insufficient for estimating X. However, if we make a few measurements $U_1, U_2, \ldots, U_r$ (as opposed to a large number of them), which we assume to be independent, the following further properties of the normal distribution will help us to calculate the reliability of the mean

$$M_r = \frac{1}{r}(U_1 + U_2 + \cdots + U_r)$$

as a measurement of m. We use (i) U_j is normal with $E(U_j) = m$, (ii) a sum of independent normal variables is normal (to be proved in Section 3.2), so M_r is normal with $E(M_r) = m$ and $V(M_r) = V(U)/r = V(X)/r$, and (iii) M_r and the sample variance

$$s_r^2 = \frac{1}{r-1} \sum_{j=1}^{r} (U_j - M_r)^2$$

are independent for normal variables (see Exercise 2 of Section 3.5), and the ratio $T_r = (M_r - m)/s_r$ is no longer dependent on the parameter $V(U) = V(X)$. The distribution of T_r is known as a t-distribution with $r - 1$ degrees of freedom, which has extensively been tabulated. Therefore, we proceed as follows: we choose a value p, and from a table of the distribution of T_r we determine $c(p)$ such that

$$P(|M_r - m|/s_r \leq c(p)) \geq p$$

We then have that the true value m is in the interval $(M_r - c(p)s_r, M_r + c(p)s_r)$ with probability at least p. Note that every statement above is exact, that is, not an approximation.

We have faced the normal distribution as an approximation in connection with the binomial and Poisson distributions. Both approximations are special cases of the following theorem that is known as the classical central limit theorem: if $Y_1, Y_2, \ldots, Y_n$ are independent and identically distributed random variables with finite expectation E and variance V, then the asymptotic distribution of $(Y_1 + Y_2 + \cdots + Y_n - nE)/\sqrt{nV}$ is standard normal. This theorem will be proved in Section 3.3, but let us explore its consequences here: (i) the symptotic normality of the properly normalized binomial distribution is just a special case of the above statement in light of the decomposition as at (2.40) but with independent I_j; (ii) we also mentioned a decomposition of a Poisson variable into independent summands which can be utilized for applying the central limit theorem. First, carry out the decomposition for an integer valued parameter λ, and then show that the normalization by $\lambda^{1/2}$ makes little difference whether the parameter is an integer or not; (iii) the most significant consequence of the central limit theorem is in applied statistics: if one is interested in the behavior of the averages then it makes no difference whether we assume a priori that the average was made up by normal variables or by others having finite variance. Ultimately the normal distribution is used for evaluating the mean in either case. This justifies the frequent assumptions in statistics that observations come from normal populations. However, if the interest in the statistical evaluation is different from the mean then the assumption of normality without further investigations is not justified (neither can we use the central limit theorem if the number of observations is small).

Exponential distribution   In several engineering applications and in accident insurance, the family of exponential distributions plays a fundamental role. A nonnegative random variable X is called exponential if its distribution function $F(x) = 1 - e^{-\lambda x}$, $x \geq 0$, and $\lambda > 0$ is a scale parameter. It has two fundamental properties (i) no ageing or lack of memory: X is exponential if, and only if, for all real numbers $x > 0$ and $y > 0$,

$$P(X > x + y | X > x) = P(X > y) \tag{2.45}$$

(ii) constant hazard or failure rate: we define the hazard or failure rate $r(t)$ by

$$r(t) = \lim_{\Delta t \to 0} \frac{1}{\Delta t} P(X \leq t + \Delta t | X > t) = \lim_{\Delta t \to 0} \frac{F(t + \Delta t) - F(t)}{(1 - F(t))\Delta t}$$

which limit always exists when $F'(t)$ does and then

$$r(t) = \frac{F'(t)}{1 - F(t)} \tag{2.46}$$

For exponential $F(x)$, $r(t) = \lambda$, a constant. Note that the right-hand side of (2.46) is the derivative of $-\log(1 - F(t))$, so $r(t)$ uniquely determines $F(t)$. In particular, if $r(t)$ is a constant then $F(t)$ is exponential.

We have established the equivalence between (2.45) and exponentiality in Exercise 2 of Section 1.8. Let X be a random life in an extended sense that allows to include time to the first accident, or to the first breakdown, or actual human life. If X satisfies (2.45) then we have both that X is exponential and that, given that X has lasted x units, its future chance for an extra y units is the same as if the measuring of time would have just started. But the latter description is the actual definition of an accident which makes accident insurance and guarantees under warranty (the breakdown of a new item can be viewed as an accident by the manufacturer) quite a simple problem from the scientific point of view.

The assumption of constant failure rate, which implies exponentiality for the distribution of the underlying quantity, is usually made for a machine that has gone through the run-in period but it is still relatively new. During running-in, the failure rate usually decreases, while for older machines, it increases.

Let us collect a few more important properties of the exponential distributions. First, we have computed in Exercise 1(ii) of Section 2.1 that $E(X) = 1/\lambda$. Furthermore, by integrating by parts, one gets $V(X) = 1/\lambda^2$. For $\lambda = 1$, X is called a unit exponential variable.

An interesting property of the exponentials is that if $X_1, X_2, \ldots, X_n$ are independent copies of X, then

$$n \min(X_1, X_2, \ldots, X_n) \overset{(d)}{=} X \tag{2.47}$$

(the notation above means that they equal in distribution). The proof is simple by noting that $P(\min(X_1, X_2, \ldots, X_n) > x) = P(X_j > x, 1 \leqslant j \leqslant n) = P(X_1 > x)P(X_2 > x)\cdots P(X_n > x) = P^n(X > x) = \exp(-n\lambda x)$. A consequence of (2.47) for accident insurance is that if the expected time to an accident is 80 years, say (so most people do not experience it), then a company writing 30,000 policies on a given day will get a claim the very next day.

Recall (2.43) and (2.44), pointing to the close relation between the geometric and exponential distributions. A more accurate relation is exhibited by the observation that if X is unit exponential, then $Y_p = [X/\{-\log(1 - p)\}] + 1$ is geometric with parameter p. Here, $[u]$ signifies the integer part of u.

Let us close this entry by recalling Exercise 1 of Section 1.8: for an arbitrary random variable Y with continuous distribution function $F(x)$, the random variable $X = -\log F(X)$ is unit exponential.

Weibull distribution   Although there are several models leading to Weibull variables (one is extreme value theory, to be discussed in Section 6.5), we introduce this distribution as it was first introduced to the literature: on a purely experimental basis. It had been recognized by several researchers that, in engineering statistical applications, the normal distribution had been inadequate for analyzing data. W. Weibull (1939, 1951) recommended the following family of distributions: $F(x) = 1 - \exp\{-[(x - A)/B]^\beta\}$, $\beta > 0$, and A, $B > 0$ location and scale parameters; $x \geq A$. All members of this family, which we call Weibull distributions, are of the same type as $F_0(x; \beta) = 1 - \exp(-x^\beta)$, $x \geq 0$. Consequently, we discuss $F_0(x; \beta)$ only. The reason for its being widely applicable is that its density function $f_0(x; \beta) = F_0'(x; \beta) = \beta x^{\beta - 1} \exp(-x^\beta)$ has a wide variety of shapes as β varies (β itself is called shape parameter). In addition, its failure rate $r(t) = \beta t^{\beta - 1}$ is monotonic, a property valid in many engineering applications (see the relevant discussion in the section Exponential distribution).

Note that $F_0(x; 1) = 1 - e^{-x}$ and that, for X with $F_0(x; \beta)$, X^β is unit exponential. What is quite noteworthy is the fact that $F_0(x; 3.6)$ is very close to a normal distribution (just draw the graph of its density with the help of a computer).

Uniform distribution on an interval   We say that X is uniformly distributed on the interval (a, b) if the density function $f(x)$ of X is constant on (a, b), and zero outside of (a, b). Hence, $F(x) = (x - a)/(b - a)$, $a \leq x \leq b$. Since X "does not favor any part of (a, b)," X has the role of selecting points at random from (a, b) (the extension of classical models of selecting from finite sets). Easy computations yield $E(X) = (a + b)/2$ and $V(X) = (b - a)^2/12$. In addition to the mentioned selection problem, the most frequent (mathematical) application of a uniform variable stems from the fact that if Y is a random variable with continuous distribution function $F(x)$ then $X = F(Y)$ is uniform on $(0, 1)$ (Exercise 1(i) of Section 2.1).

Next, we turn to vector variables $(X_1, X_2, \ldots, X_d)$, $d \geq 2$. Here we introduce two families of multivariate distributions, which list will be expanded in subsequent chapters.

Multivariate normal distribution   We write

$$R(x_1, x_2, \ldots, x_d) = \{(u_1, u_2, \ldots, u_d): -\infty < u_j \leq x_j, 1 \leq j \leq d\}.$$

Now, we say that the distribution of the vector $(X_1, X_2, \ldots, X_d)$ is d-dimensional normal if its distribution function

$$F(x_1, \ldots, x_d) = C \int_{R(x_1, \ldots, x_d)} \{\exp[-Q(u_1, \ldots, u_d)]\} \, du_1 \ldots du_d \quad (2.48)$$

where C is a suitable constant and $Q(u_1, \ldots, u_d)$ is a positive definite quadratic form of $(u_1, \ldots, u_d)$. This form assumes that the expectations $E(X_j) = 0$. If we replace each u_j by $u_j - E_j$, then $E(X_j) = E_j$, $1 \le j \le d$. We know from analytic geometry that to a positive definite quadratic form Q there is a positive definite matrix A such that

$$Q(u_1, \ldots, u_d) = (u_1, \ldots, u_d)A(u_1, \ldots, u_d)'$$

where $(\ldots)'$ signifies the transpose of the vector $(\ldots)$. It is immediate that the special matrix $I^* = A$, whose main diagonal entries are positive and all others zero, leads to independent normal components X_j, and, upon referring once more to analytic geometry, we know that all other cases can be transformed to I^* by appropriate linear orthogonal transformations. Upon utilizing such a transformation one gets that the entries of the inverse matrix A^{-1} are $E(X_i X_j)$, $1 \le i, j \le d$. The values $\mathrm{cov}(X_i, X_j) = E(X_i X_j)$, $i \ne j$, are called covariances and, of course, $E(X_j^2) = V(X_j)$ since $E(X_j) = 0$. We refer to a matrix with such entries as a variance-covariance matrix. Direct computation (although not routine) yields that the constant C in the definition equals $(2\pi)^{-d/2}|A|^{1/2}$.

In the bivariate case $(d = 2)$, instead of using the matrix A, due to the properties of A^{-1}, we can write the integrand, which is the bivariate density, in the explicit form

$$f(u_1, u_2) = (2\pi\sqrt{1 - \rho^2})^{-1} \exp[-(u_1^2 - 2\rho u_1 u_2 + u_2^2)/2(1 - \rho^2)]$$

where we normalized so that $E(X_1) = E(X_2) = 0$ and $V(X_1) = V(X_2) = 1$, and where $-1 < \rho = E(X_1 X_2) < +1$.

By letting one, or several, $x_j \to +\infty$ in (2.48), we find that all lower dimensional distributions of $(X_1, X_2, \ldots, X_d)$ are normal. In particular, each X_j is a normal variable. The converse, of course, is not true: one can start with dependent normal variables whose joint distribution is not multivariate normal. The simplest example is if we take $X_1 = X_2 = X \in N(0, 1)$, then $F(x_1, x_2) = P(X \le x_1, X \le x_2) = P(X \le \min(x_1, x_2))$ which does not depend on the positive values of x_2 if $x_1 = 0$. So, it cannot satisfy (2.48).

When we learn conditioning with respect to a random variable in Chapter 7, we shall be able to discuss further interesting properties of multivariate normal vectors.

A bivariate exponential distribution Let Y_1, Y_2 and Y_3 be independent univariate exponential variables with respective parameters λ_j, $1 \le j \le 3$. Set $X_1 = \min(Y_1, Y_3)$ and $X_2 = \min(Y_2, Y_3)$. Then, by arguing as at (2.47), we get that X_1 and X_2 are exponential variables with parameters $\lambda_1 + \lambda_3$

and $\lambda_2 + \lambda_3$, respectively. Furthermore,

$$P(X_1 > x_1, X_2 > x_2) = P(Y_1 > x_1, Y_2 > x_2, Y_3 > \max(x_1, x_2))$$
$$= \exp(-\lambda_1 x_1 - \lambda_2 x_2 - \lambda_3 \max(x_1, x_2))$$

The model can be applied to describing the operation of a system which is fed through two parallel lines, both being controlled by a central switch (its life is Y_3, while the two lines operate with lives Y_1 and Y_2 whenever the switch is on). The model was first introduced by Marshall and Olkin (1967).

We close this Part I on distributions by emphasizing that random quantities can be influenced by several components, each having its own distribution function. In such cases a mixture of distribution

$$F = \sum p_j F_j \qquad p_j > 0 \qquad \sum p_j = 1 \tag{2.49}$$

is the appropriate choice for the underlying distribution. For choosing each F_j, and the weight p_j for it, the individual contributing factors have to be analyzed.

Exercises

1. Show that the parameters n and M of the hypergeometric distribution can be interchanged; i.e.,

$$\binom{M}{k}\binom{R-M}{n-k} \bigg/ \binom{R}{n} = \binom{n}{k}\binom{R-n}{M-k} \bigg/ \binom{R}{M}$$

2. If $p_k(M, n, R)$ is the kth term in the hypergeometric distribution with the given parameters, prove the recursive formula

$$p_k(M+1, n, R) = \frac{(M+1)(R-M-n+k)}{(R-M)(M+1-k)} p_k(M, n, R)$$

State the dual of this recursive formula obtained by the result of Exercise 1.

3. Let X and Y be independent binomial variables with parameters (n_1, p) and (n_2, p), respectively. Use a decomposition as at (2.40) but with independent I_j to conclude that $X + Y$ is a binomial with parameters $(n_1 + n_2, p)$. Deduce from this result that the distribution of X, given $X + Y$, is hypergeometric.

4. From a population of $R = 10^6$ people, $n = 1,600$ are selected at random (without replacement) for interview. If $k = 576$ stated to vote for candidate A in a forthcoming election, analyze the statement "if elections were held today, $k/n = 576/1600 = 36\%$ would vote for A." (Use the binomial and normal approximations to the hypergeometric distribution.)

5. Let X be the number of items in a box. Assume that X is random with

a Poisson distribution. Mark each item, independently of each other and of X, with the same probability p. Show that the number of marked items has a Poisson distribution.

6. Starting from the geometric distribution, prove the stated formula for the negative binomial distribution (use induction over r).

7. Switch to polar coordinates in the double integral

$$\int_{-\infty}^{+\infty} \int_{-\infty}^{+\infty} e^{-(x^2+y^2)/2} \, dx \, dy$$

to conclude that the constants involved in a normal distribution guarantee its being a proper distribution function.

8. For a random variable Y with a continuous distribution function, define the conditional density $f(x|y)$ of a random variable X, given $Y = y$, by the double limit

$$\lim_{\Delta x \to 0} \lim_{\Delta y \to 0} P(x < X \le x + \Delta x | y < Y \le y + \Delta y)/\Delta x$$

assuming that $P(y < Y \le y + \Delta y) > 0$. Calculate $f(x|y)$ for a bivariate normal vector (X, Y).

REFERENCES

Bray, H. E. (1919). Elementary properties of the Stieltjes integral. *Ann. Math. 20*, 177–186.

Chebyshev, P. L. (See the reference in Chapter 1.)

Chow, Y. S. and H. Robbins (1961). On sums of independent random variables with infinite moments and fair games. *Proc. Nat. Acad. Sci. USA 47*, 330–335.

Doeblin, W. (1940). Remarques sur la théorie métrique des fractions continues. *Compositio Math. 7*, 353–371.

Fréchet, M. and J. Shohat (1931). A proof of the generalized second limit theorem in the theory of probability. *Trans. Amer. Math. Soc. 33*, 533–543.

Galambos, J. (1982). The role of functional equations in stochastic model building. *Aequationes Math. 25*, 21–41.

Galambos, J. and S. Kotz (1978). *Characterizations of Probability Distributions. Lecture Notes in Math. 675*, Springer, Heidelberg.

Halmos, P. R. (1974). *Measure Theory*, Springer, Berlin.

Hardy, G. H., Littlewood, J. E., and G. Pólya (1934). *Inequalities*, Cambridge University Press.

Helly, E. (1912). Über lineare Funktionaloperationen. *Sitz. Nat. Kais. Akad. Wiss. 121.*

Hewitt, E. and L. J. Savage (1955). Symmetric measures on Cartesian products. *Trans. Amer. Math. Soc. 80*, 470–501.

Johnson, N. L. and S. Kotz (1970–1972). *Distributions in statistics, Vol. I–IV*, Wiley, New York.

Knopp, K. (1924). *Theorie und Anwendung der unendlichen Reihen*, Springer, Berlin.

Kolmogorov, A. N. (See the reference in Chapter 1.)

Markov, A. A. (1951). *Selected Works.* Izd. AN SSSR, Leningrad. (In Russian.)

Marshall, A. W. and I. Olkin (1967). A generalized bivariate exponential distribution. *J. Appl. Probability 4*, 291–302.

Mucci, R. (1977). *Limit Theorems for Extremes.* Ph.D. Thesis, Temple University.

Weibull, W. (1939). *A statistical theory of the strength of material*, Ingenioer Vetenskaps Akademiens Handligar, Stockholm (No. 151).

Weibull, W. (1951). A statistical distribution function of wide applicability. *J. Applied Mech. 18*, 293–297.

3
Transforms of Distribution

This chapter is devoted to establishing the basic properties of the three major transforms of distribution: characteristic functions, Laplace transforms, and generating functions. Here, distribution means either distribution function, or, for discrete random variables X, the distribution of X. These transforms prove to be efficient and simple tools in proving limit theorems and determining some properties of distribution. While several applications will be included in the present chapter, their thorough utilization begins with Chapter 4 and recurs throughout the book.

3.1 CHARACTERISTIC FUNCTIONS: BASIC PROPERTIES

Before going to the definition of characteristic function, some extensions of the results of the previous chapters to complex-valued functions on Ω are necessary. These will be remarks rather than new developments. The probability space $(\Omega, \mathscr{A}, P)$ is assumed to be the same throughout the whole section.

A function Y on Ω, the values of which are complex numbers $U + iV$, $i^2 = -1$, is said to be a random variable if both U and V are real-valued random variables. U and V are called the components of Y; U is called the real and V the imaginary part of Y. By definition, $E(Y) = E(U) + iE(V)$. With these definitions, all results on integrals, independence, and distributions (Y is viewed as a bivariate vector (U, V) from the point of view of distributions), remain unchanged. This is essentially due to the facts that complex numbers are added component by component and that the distance between two complex numbers is majorized by the sum of the distances of the real and imaginary parts. In particular, the distance between two complex numbers goes to zero if and only if the distances between the real and imaginary parts both tend to zero.

The concept of characteristic function is based on the complex-valued exponential function of purely imaginary numbers. That is, we associate the random variable

$$Y = e^{itX} = \cos(tX) + i\sin(tX) \qquad t \text{ real}$$

with a real-valued random variable X. The second equation above is a result of the elements of complex analysis. It thus follows that $|e^{itX}| \leq 1$ and, upon expanding $\sin u$ and $\cos u$ into Taylor series, the Taylor series and all differentiability properties of $g(u) = e^{itu}$ are seen to be the same as those of e^z, z real. This is all, for the most part, that we need from complex analysis.

Definition 1 Let X be a real-valued random variable with distribution function $F(x)$. The function

$$\varphi(t) = E(e^{itX}) = \int_\Omega e^{itX}\, dP = \int_{-\infty}^{+\infty} e^{itx}\, dF(x)$$

defined for all real numbers t, is called the characteristic function of X or $F(x)$. If $F(x)$ is a distribution function in the extended sense, then $\varphi(t)$ is defined by the last equation alone.

Because

$$\varphi(0) = \int_{-\infty}^{+\infty} dF(x) = F(+\infty) - F(-\infty) \tag{3.1}$$

the value $\varphi(0)$ of a characteristic function always determines whether $\varphi(t)$ is the characteristic function of a random variable ($\varphi(0) = 1$) or a distribution function in the extended sense ($\varphi(0) < 1$). If $F(x) = c$, a finite constant, then $\varphi(t) = 0$ for all t, while $\varphi(0) \neq 0$ otherwise (see (3.1)). These are the only differences when $\varphi(t)$ is the characteristic function of a random variable or when F in the definition of φ is a distribution function in the extended sense. These properties will be used routinely in the sequel.

Theorem 1 Every characteristic function $\varphi(t)$ is well defined for all real t. Furthermore, $|\varphi(t)| \leq 1$ and $\varphi(t)$ is continuous on R.

Proof. By the corollary of Section 2.2, the first two statements are immediate. On the other hand, the continuity of φ on R is an easy consequence of the dominated convergence theorem (Theorem 2.16). ∎

It is, in fact, true that every characteristic function is uniformly continuous on the whole real line. Indeed, for arbitrary real numbers t and h,

$$|\varphi(t + h) - \varphi(t)| \leq \int_{-\infty}^{+\infty} |e^{ihx} - 1| \, dF(x)$$

$$= 2 \int_{-\infty}^{+\infty} |\sin(xh/2)| \, dF(x)$$

which estimate is independent of t. Furthermore, the integral on the extreme right-hand side can be made small for small h upon observing that if we cut the path of integration into three parts

$$\int_{-\infty}^{-A} \cdots + \int_{-A}^{A} \cdots + \int_{A}^{+\infty} \cdots$$

the two outermost terms are small for any h if A is large, while the middle term is small for a fixed A if h is small. The uniform continuity of $\varphi(t)$ is thus established.

We have faced the problem of differentiation under the integral sign in Exercise 2 of Section 2.4. With the same argument used there, i.e., with an appeal to the dominated convergence theorem, we can establish the following very important differentiation rules.

Theorem 2 Let X be a real-valued random variable with distribution function $F(x)$ and characteristic function $\varphi(t)$. If, for some integer $k > 0$, the moment $E(X^k)$ is finite, then $\varphi(t)$ is differentiable k times and

$$\varphi^{(j)}(0) = i^j E(X^j) \qquad 0 < j \leq k \tag{3.2}$$

Conversely, if, for an integer $n \geq 1$, $\varphi^{(2n)}(0)$ exists and is finite, then the moments $E(X^k)$, $0 \leq k \leq 2n$, are all finite.

Proof. We have established in Exercise 4 of Section 2.2 that the finiteness of the kth moment implies finiteness for the jth moment for $0 < j \leq k$. Thus, if we form the differential ratio

$$\frac{\varphi(t + h) - \varphi(t)}{h} = \int_{\Omega} \frac{e^{i(t+h)X} - e^{itX}}{h} \, dP$$

then the application of the mean value theorem of calculus to the integrand yields the estimate

$$h^{-1}|e^{i(t+h)X} - e^{itX}| = |iXe^{i(t+\vartheta h)X}| = |X|$$

where $0 < \vartheta < 1$. Hence, the dominated convergence theorem is applicable, yielding the desired differentiation rule for $j = 1$. We now proceed by induction: the combination of the mean value theorem and the dominated convergence theorem, just as above, leads to (3.2) for all $0 < j \le k$.

For the converse, we evaluate the $2n$th derivative of φ through symmetric differential ratios. That is, we successively form the differences

$$\Delta_{1,h}\varphi(0) = \varphi(h) - \varphi(-h) \qquad \Delta_{k,h}\varphi(0) = \Delta_{1,h}\Delta_{k-1,h}\varphi(0)$$

and evaluate

$$\varphi^{(2n)}(0) = \lim_{h=0} \frac{\Delta_{2n,h}\varphi(0)}{(2h)^{2n}} \tag{3.3}$$

Now, note that $\Delta_{1,h}e^{itX}|_{t=0} = e^{ihX} - e^{-ihX} = 2i \sin hX$; thus, (3.3) takes the form

$$\varphi^{(2n)}(0) = \lim_{h=0} \int_\Omega \left(\frac{2i \sin hX}{2h}\right)^{2n} dP$$

By Fatou's lemma (Theorem 2.15) and by $(\sin hX)/h \to X$ as $h \to 0$, the last limit formula implies $|\varphi^{(2n)}(0)| \ge E(X^{2n})$. Hence, $E(X^{2n})$ is finite and, thus, so are all moments of smaller order. The theorem is proved. ∎

We apply formula (3.2) to the Maclaurin–Taylor expansion of $\varphi(t)$. If X is a random variable with characteristic function $\varphi(t)$ and if $m_n = E(X^n)$ is finite, then $m_k = E(X^k)$, $1 \le k \le n$, are all finite, and, for all real t,

$$\varphi(t) = 1 + \sum_{k=1}^{n-1} m_k \frac{(it)^k}{k!} + t^n \int_0^1 \frac{(1-u)^{n-1}}{(n-1)!} \varphi^{(n)}(tu) \, du$$

$$= 1 + \sum_{k=1}^{n} m_k \frac{(it)^k}{k!} + o(t^n) \qquad t \to 0 \tag{3.4}$$

Furthermore, if moments m_n of all orders are finite, then

$$\varphi(t) = \sum_{k=0}^{+\infty} m_k \frac{(it)^k}{k!} \qquad (m_0 = 1) \tag{3.5}$$

for all t for which the right-hand side converges, which is known from calculus to be equivalent to $|t| < 1/\rho$, where $\rho = \limsup(|m_k|/k!)^{1/k}$, $k \to +\infty$.

In a sequence of examples we now evaluate the characteristic functions of some particular distributions and supplement the results of the current section.

Example 1 **(Standard Normal Distribution)** If X is normal with $E(X) = 0$ and $V(X) = 1$, then $\varphi(t) = \exp(-t^2/2)$.

In Example 10 of Section 2.7, we found that $m_k = 0$ if k is odd and $m_{2k} = (2k)!/k!2^k$. Thus, by (3.5),

$$\varphi(t) = \sum_{k=0}^{+\infty} (-1)^k \frac{(t^2/2)^k}{k!} = e^{-t^2/2}$$

which is convergent for all real t. ∎

Example 2 **(Unit Exponential Distribution)** Let X have the distribution function $F(x) = 1 - e^{-x}$, $x > 0$. Then $\varphi(t) = 1/(1 - it)$.

We use the definition of characteristic function and Theorem 2.21 with $f(x) = F'(x) = e^{-x}$. We have

$$\varphi(t) = \int_0^{+\infty} e^{itx} e^{-x} \, dx = \frac{e^{(it-1)x}}{it - 1} \bigg|_{x=0}^{+\infty} = \frac{1}{1 - it}$$ ∎

Example 3 **(Uniform Distribution on $(0, 1)$)** If X is uniformly distributed on $(0, 1)$, i.e., if its density $f(x) = 1$ on $(0, 1)$ and $f(x) = 0$ otherwise, then $\varphi(t) = (e^{it} - 1)/it$.

By definition

$$\varphi(t) = \int_0^1 e^{itx} \, dx = \frac{e^{it} - 1}{it}$$ ∎

Example 4 **(Binomial Distribution)** If the distribution of X is binomial with parameters n and p, then $\varphi(t) = (1 + p(e^{it} - 1))^n$.

By the formula for the expectation of a discrete random variable and the binomial theorem,

$$\varphi(t) = \sum_{k=0}^{n} e^{itk} \binom{n}{k} p^k (1 - p)^{n-k} = (pe^{it} + 1 - p)^n$$

which is the desired formula. ∎

Example 5 (Poisson Distribution) Let X be a Poisson variable with parameter $a > 0$. Then $\varphi(t) = \exp[a(e^{it} - 1)]$.

Indeed,

$$\varphi(t) = \sum_{k=0}^{+\infty} e^{itk} \frac{a^k e^{-a}}{k!} = e^{-a} \sum_{k=0}^{+\infty} \frac{(ae^{it})^k}{k!} = e^{-a} e^{ae^{it}}$$

where the last equation is justified by Taylor's formula for e^z.   ∎

The computation of the characteristic function is not always as simple as in the preceding examples. Many times, complex integration is required.

Example 6 (Cauchy Distribution) The random variable X is called standard Cauchy if its density function

$$f(x) = \frac{1}{\pi(1 + x^2)} \qquad \text{all } x$$

For such an X, $\varphi(t) = e^{-|t|}$.

Note that $E(X^+) = E(X^-) = +\infty$ and, thus, $E(X)$ is not defined. It is therefore not surprising that $\varphi(t)$ is not differentiable at zero (but see the next example). Now,

$$\varphi(t) = \frac{1}{\pi} \int_{-\infty}^{+\infty} \frac{e^{itx}}{1 + x^2} \, dx$$

which can also be considered a complex integral with respect to z, $z = x + iy$, where $y = 0$. With this in mind, we choose an arbitrary value N and evaluate, for $t > 0$,

$$\int_C \frac{e^{itz}}{1 + z^2} \, dz = \int_{-N}^{+N} \frac{e^{itx}}{1 + x^2} \, dx + \int_{C*} \frac{e^{itz}}{1 + z^2} \, dz \qquad (3.6)$$

where C is the closed curve consisting of the real interval $[-N, +N]$ and the semicircle $C^* = \{z : z = x + iy, |z| = N, y > 0\}$. By showing that the second integral on the right-hand side tends to zero as $N \to +\infty$, we obtain $\varphi(t)$ from the limit of the right-hand side of (3.6).

We first show that the last integral of (3.6) tends to zero as $N \to +\infty$. Because $t > 0$ and, on C^*, $y > 0$ and $|z^2| = N^2$, with $|C^*|$ signifying the length of C^*, for $N > 1$,

$$\left| \int_{C*} \frac{e^{itz}}{1 + z^2} \, dz \right| \leq |C^*| \max_{C*} \left| \frac{e^{itz}}{1 + z^2} \right| \leq \frac{\pi N}{N^2 - 1} \max_{y > 0} e^{-yt}$$

which is further majorized by $\pi/(N-1) \to 0$ as $N \to +\infty$. Next, we evaluate the left-hand side of (3.6) by the method of residues (a method borrowed from complex analysis). Since, inside the closed curve C, the integrand has a single pole at $z = i$ and is regular elsewhere,

$$\int_C \frac{e^{itz}}{1+z^2}\, dz = 2\pi i \operatorname{Res}_C \frac{e^{itz}}{1+z^2} = 2\pi i \left. \frac{e^{itz}(z-i)}{1+z^2}\right|_{z=i} = \pi e^{-t}$$

which implies the desired form of $\varphi(t)$ for $t > 0$. For $t < 0$, a similar computation yields $\varphi(t) = e^t$ if the semicircle C^* is replaced by $C^{**} = \{z: |z| = N, y < 0\}$. (Given this choice, the anticlockwise direction inside the closed curve C assigns the direction $N \to -N$ on the real line and the residue theorem yields $(-\pi e^t)$ for the integral over C.) The repetition of the steps in the computation is unnecessary. ∎

In the second part of Theorem 2, we assumed that an even-order derivative of φ at the origin is finite. The following example, essentially due to Zygmund (1947), shows that "even" here is essential.

Example 7 Let X be a discrete random variable taking the integers $\pm j$, $j \geq 1$, with distribution $P(X = j) = P(X = -j) = r(j)/j^2$, where $r(j)$ is a strictly decreasing function tending to zero as $j \to +\infty$ such that

$$2\sum_{j=0}^{+\infty} \frac{r(j)}{j^2} = 1 \quad \text{and} \quad \sum_{j=1}^{+\infty} \frac{r(j)}{j} = +\infty$$

Then the characteristic function of X

$$\varphi(t) = \sum_{j=1}^{+\infty} \frac{2r(j)\cos tj}{j^2}$$

is differentiable at zero, even though $E(X)$ is undefined.

By the last assumption on $r(j)$, $E(X^+) = E(X^-) = +\infty$, yielding that $E(X)$ is indeed undefined. Now, the fact that $\varphi(t)$ is of the given form follows from the symmetry of the distribution of X about zero: the contribution of $(-j)$ and j can be combined to

$$\frac{(e^{it(-j)} + e^{itj})r(j)}{j^2} = \frac{2r(j)\cos tj}{j^2}$$

Finally, to see that $\varphi'(0) = \lim\{\varphi(t) - 1\}/t$, $t \to 0$, is finite, we use $0 \le 1 - \cos u \le \min(2, u^2)$, u real. We get

$$\left| \frac{1 - \varphi(t)}{t} \right| = \left| \frac{1}{t} \sum_{j=1}^{+\infty} \frac{2r(j)(1 - \cos jt)}{j^2} \right|$$

$$\le |2t| \sum_{j=1}^{[|1/t|]} r(j) + \left| \frac{1}{t} \right| \sum_{j>|1/t|} \frac{4r(j)}{j^2} \to 0 \qquad \text{as} \quad t \to 0$$

that is, $\varphi'(0) = 0$. (Here, $[y]$ denotes the integer part of y.) ∎

In all examples, the characteristic functions are evaluated for the standard form of the respective distributions. The following simple theorem helps out when parameters are present.

Theorem 3 Let the random variable X have the characteristic function $\varphi_X(t)$. Then, for arbitrary real numbers a and b,

$$\varphi_{aX+b}(t) = e^{itb} \varphi_X(at)$$

Proof. By definition,

$$\varphi_{aX+b}(t) = E(e^{it(aX+b)}) = e^{itb} E(e^{itaX}) = e^{itb} \varphi_X(at) \qquad ∎$$

Example 8 (Normal Distribution) Let X be a normal variable with $E(X) = m$ and $V(X) = V$. Then its characteristic function

$$\varphi(t) = \exp(itm - Vt^2/2).$$

Since $X = m + V^{1/2}U$, where U is standard normal, Example 1 and Theorem 3 yield the stated form of $\varphi(t)$. ∎

Exercises

Characteristic function below means that of a random variable.

1. Show that $g(t)$ is not a characteristic function if
   (i) $\lim(g(t) + 1)/t = 0$ as $t \to 0$,
   (ii) $\lim(g(t) - 1)/t^2 = 2$ as $t \to 0$,
   (iii) $g(t) = e^{-t^4}$.
2. Assume that the functions $\varphi_1(t), \varphi_2(t), \ldots, \varphi_k(t)$ are characteristic functions and that the numbers $a_j \ge 0$, $1 \le j \le k$, satisfy $a_1 + a_2 + \cdots + a_k = 1$. Show that $a_1\varphi_1(t) + a_2\varphi_2(t) + \cdots + a_k\varphi_k(t)$ is also a characteristic function. Hence, conclude that if $\varphi(t)$ is a characteristic function, then so are $\varphi(-t)$ and the real part $\text{Re } \varphi(t)$ of $\varphi(t)$.

3. The random variable X is said to have a lattice distribution if it is discrete with possible values of the form $a + jb, j = 0, j = \pm k, k \geq 1$, where a and b are some real numbers. Show that X is a lattice variable if and only if its characteristic function $\varphi(t)$ satisfies $|\varphi(t_0)| = 1$ for some real number $t_0 \neq 0$.

4. Deduce from Exercise 3 that if $\varphi(t)$ is the characteristic function of a nondegenerate distribution then $[\varphi(t)]^a$, $a < 0$, cannot be a characteristic function.

5. Determine the interval of convergence of the Taylor series of the characteristic function for (i) the exponential distribution and (ii) the log-normal distribution. What about the Pareto distribution?

6. Find the expectation and variance of X if its characteristic function $\varphi(t) = (1 + t^2)^{-2}$.

3.2 CHARACTERISTIC FUNCTIONS: THE UNIQUENESS AND CONTINUITY THEOREMS

The two major results of the present section are formulated in the following theorems.

Theorem 4 (Uniqueness) Let $\varphi(t)$ be the characteristic function of $F(x)$. Then, for any continuity points $a < b$ of $F(x)$,

$$F(b) - F(a) = \lim_{U = +\infty} \frac{1}{2\pi} \int_{-U}^{+U} \frac{e^{-ita} - e^{-itb}}{it} \varphi(t) \, dt$$

Consequently, if $F(x)$ is the distribution function of a random variable, then $\varphi(t)$ uniquely determines $F(x)$.

Theorem 5 (Continuity) Let F_n, $n \geq 1$, be a sequence of distribution functions with corresponding characteristic functions φ_n. Assume that $F_n \to F$ completely. Then $\varphi_n(t)$ converges to the characteristic function $\varphi(t)$ of $F(x)$. Conversely, if $\varphi_n(t) \to \varphi(t)$, which is continuous at $t = 0$, then $\varphi(t)$ is the characteristic function of the distribution function $F(x)$ and $F_n \to F$ completely.

The following statement is an important consequence of Theorem 5.

Theorem 5a If the sequence $\varphi_n(t)$ of characteristic functions converges to a characteristic function $\varphi(t)$, then the convergence is uniform on any fixed (finite) interval $-T \leq t \leq T$.

Proof of Theorem 5a. Let F_n and F be the distribution functions corresponding to φ_n and φ, respectively. Then, by Theorem 5, $F_n \to F$ completely. Let $a < b$ be real numbers such that

$$1 - F(b) + F(a) < \varepsilon$$

and

$$1 - F_n(b) + F_n(a) \leq 1 - F(b) + F(a) + \varepsilon < 2\varepsilon$$

the latter being valid for all sufficiently large n. Then, for large n,

$$|\varphi_n(t) - \varphi(t)| \leq \left| \int_a^b e^{itx}\, dF_n(x) - \int_a^b e^{itx}\, dF(x) \right| + 3\varepsilon$$

Next, we partition the interval $[a, b]$ into N equal parts and denote the kth division point $(a = x_0)$ by x_k. Then, upon writing

$$\int_a^b e^{itx}\, dF_n(x) - \int_a^b e^{itx}\, dF(x)$$

$$= \sum_{k=0}^{N-1} \left(\int_{x_k}^{x_{k+1}} e^{itx_k}\, dF_n(x) - \int_{x_k}^{x_{k+1}} e^{itx_k}\, dF(x) \right)$$

$$+ \sum_{k=0}^{N-1} \left[\int_{x_k}^{x_{k+1}} (e^{itx} - e^{itx_k})\, dF_n(x) + \int_{x_k}^{x_{k+1}} (e^{itx_k} - e^{itx})\, dF(x) \right]$$

the triangular inequality and the elementary inequality

$$|e^{itx} - e^{itx_k}| \leq |tx - tx_k| \leq T(x_{k+1} - x_k) = T/N$$

yield

$$|\varphi_n(t) - \varphi(t)| \leq \sum_{k=0}^{N-1} \left| \int_{x_k}^{x_{k+1}} dF_n(x) - \int_{x_k}^{x_{k+1}} dF(x) \right| + \frac{2T}{N} + 3\varepsilon$$

This estimate is uniform in $t \in [-T, T]$; the right-hand side becomes small if we first choose N large and fix it and then let $n \to +\infty$. The proof is completed. ∎

The proofs of Theorems 3 and 5 are postponed to the second part of this section. Although both theorems will be among the most frequently used statements in the remainder of this book, the following results alone would be a sufficiently thorough demonstration of the ways in which they are applied.

Example 9 (Poisson Approximation to the Binomial) If the distribution of X is binomial with parameters n and p, then, as $n \to +\infty$, $p \to 0$, and

$np \to a$, a positive finite number, the distribution of X converges to the Poisson distribution with parameter a.

We have shown in Example 4 that the characteristic function of X, $\varphi_n(t) = (1 + p(e^{it} - 1))^n$. In view of our assumptions, we can write $p = (a + \varepsilon)/n$, where $\varepsilon \to 0$ as $n \to +\infty$. Hence, by the elementary relation $(1 + x(n)/n)^n \to e^x$ if $x(n) \to x$ as $n \to +\infty$,

$$\varphi_n(t) \to e^{a(e^{it} - 1)} \qquad n \to +\infty$$

Now, we know from Example 5 that the limit above is the characteristic function of a Poisson distribution and, thus, by Theorem 4, no other distribution can have this as its characteristic function. Finally, Theorem 5 tells us that the above limit relation implies that the distribution function of X converges completely to the Poisson distribution function with parameter a. But, for discrete distributions, the complete convergence of distribution functions is equivalent to the term by term convergence of the distributions, which was to be proved. ∎

For the next statement, we introduce the concept of *convolutions*. Let F_1 and F_2 be distribution functions. Then the function

$$F(z) = \int_{-\infty}^{+\infty} F_2(z - x)\, dF_1(x) \tag{3.7}$$

is called the convolution of F_1 and F_2. Clearly, $F(z)$ is a distribution function (monotonicity is evident, while right continuity and its values at the infinities can be determined by the dominated convergence theorem). It can also be deduced from the definition that the convolution is symmetric (commutative) in F_1 and F_2. Here, however, we choose to deduce this from the uniqueness theorem via the following result.

Theorem 6 Let F, F_1, and F_2 be distribution functions with corresponding characteristic functions φ, φ_1, and φ_2. Then F is the convolution of F_1 and F_2 if and only if

$$\varphi(t) = \varphi_1(t)\varphi_2(t) \tag{3.8}$$

Proof. We formally write

$$\varphi(t) = \int_{-\infty}^{+\infty} e^{itz}\, dF(z) = \int_{-\infty}^{+\infty} e^{itx}\left[\int_{-\infty}^{+\infty} e^{it(z-x)}\, dF_2(z-x)\right] dF_1(x)$$

$$= \int_{-\infty}^{+\infty} e^{itx}\left[\int_{-\infty}^{+\infty} e^{ity}\, dF_2(y)\right] dF_1(x) = \varphi_2(t)\varphi_1(t) \tag{3.9}$$

and the reader is asked in Exercise 1 to justify these steps by turning to Riemann–Stieltjes sums. We thus have shown that (3.7) implies (3.8). But then the converse follows from the uniqueness theorem, which completes the proof. ∎

Since (3.8) is symmetric in F_1 and F_2, then, when more than two terms are involved, it is also associative, and we have that the convolution defined in (3.7) is a commutative and associative operation.

Some other consequences of Theorem 6 are given below. First note that, from the basic properties of integrals (Theorem 2.4), if X and Y are independent, then

$$\varphi_{X+Y}(t) = E(e^{it(X+Y)}) = E(e^{itX})E(e^{itY}) = \varphi_X(t)\varphi_Y(t)$$

We thus have, from Theorem 6,

Corollary 1 If X and Y are independent random variables with distribution functions F_1 and F_2 and characteristic functions φ_1 and φ_2, respectively, then the distribution function $F(z)$ of $X + Y$ satisfies (3.7) and (3.8) applies.

Corollary 2 If X and Y are independent (i) normal, or (ii) standard Cauchy variables, then $X + Y$ has the same type of distribution as X does. In addition, for independent Poisson variables X and Y, $X + Y$ also is Poisson.

Recall that $F(x)$ and $G(x)$ are of the same type if there are some constants a and $b > 0$ such that $F(x) = G(a + bx)$.

Corollary 2 is an immediate consequence of Corollary 1 and the uniqueness theorem. Indeed, since (3.8) applies, one has only to observe that the product of two normal characteristic functions is normal, and similarly for Poisson and Cauchy characteristic functions.

We can also employ the uniqueness theorem, together with the Taylor expansion (3.5), to supplement the method of moments.

Corollary 3 Assume the distribution function $F(x)$ of X is such that all moments $m_k = E(X^k)$, $k \geq 1$, are finite and, as $k \to +\infty$,

$$\lim \sup(|m_k|/k!)^{1/k} = 0.$$

Then the moments m_k, $k \geq 1$, uniquely determine $F(x)$.

Indeed, by the assumption on the growth of m_k, the Taylor series of $\varphi(t)$ in (3.5) is absolutely convergent for all real t, that is, the sequence m_k uniquely determines $\varphi(t)$ and, thus, by the uniqueness theorem, $F(x)$ as well.

Two special cases of Corollary 3 are given separately below because of their importance.

Corollary 4 If $F(x)$ is either normal or Poisson, then its sequence of moments uniquely determines $F(x)$.

Proof. We appeal to Corollary 3. For the case of normal $F(x)$, recall Example 10 of Section 2.7, in which it is established that $m_{2k+1} = 0$ and $m_{2k}/(2k)! = 1/k!2^k$. One can now apply Stirling's formula (see the introductory volume, pp. 69–70) or the elementary inequality

$$\log(k!) = \sum_{j=1}^{k} \log j \geq \int_{1}^{k} (\log x)\, dx = k \log k - k$$

to conclude

$$\left(\frac{1}{k!}\right)^{1/k} \leq \frac{e}{k} \to 0 \qquad \text{as} \quad k \to +\infty \tag{3.10}$$

Hence, for standard normal $F(x)$, Corollary 3 implies Corollary 4. If $F(x)$ is normal with parameters m and σ, then $F(m + \sigma x)$ is standard normal, so Corollary 4 is established for the normal distributions.

We turn to the case when $F(x)$ is Poisson. By (3.10),

$$m_k = \sum_{j=0}^{+\infty} j^k \frac{a^j e^{-a}}{j!} \leq e^{-a} \sum_{j=1}^{+\infty} j^k \frac{(ea)^j}{j^j}$$

$$\leq e^{-a}(ea)^k \sum_{j=1}^{k} j^{k-j} + e^{-a}(ea)^k \sum_{j=k+1}^{+\infty} \left(\frac{ea}{j}\right)^{j-k} \tag{3.11}$$

For large k, the last sum becomes smaller then $2ea/k$. On the other hand, the penultimate sum requires careful estimates. Consider the continuous function $z^{k-z} = \exp((k - z)\log z)$, $1 \leq z \leq k$. Its maximum is at the solution of the equation $-\log z + (k - z)/z = 0$, i.e., $z \log z + z = k$. For large k, this solution lies between $k/\log k$ and $2k/\log k$, and thus, for large k, the penultimate sum on the right-hand side of (3.11) is smaller than $k(2k/\log k)^k$. Therefore, by (3.10) and (3.11),

$$\lim \sup(m_k/k!)^{1/k} \leq \lim \sup(3e^2 a)k^{1/k}/\log k = 0$$

as $k \to +\infty$, which completes the proof on account of Corollary 3. ∎

Another special case of Corollary 3 is the case of bounded random variables.

Corollary 5   Let X and Y be two random variables that are bounded with probability one. Then all of their moments are finite, and if $E(X^k) = E(Y^k)$ for all $k \geq 1$, then X and Y are identically distributed. In particular, if $f(x)$ and $g(x)$ are two density functions on $(0, 1)$ and if

$$\int_0^1 x^k f(x)\, dx = \int_0^1 x^k g(x)\, dx \qquad \text{for all} \quad k \geq 0 \tag{3.12}$$

then $f(x) = g(x)$ a.s. (with respect to Lebesgue measure).

Since, for bounded X and Y, $|E(X^k)|^{1/k}$ and $|E(Y^k)|^{1/k}$ are also bounded, (3.10) and Corollary 3 clearly imply Corollary 5. It thus follows that the uniform distribution $F(x) = x, 0 \leq x \leq 1$, or the power distributions $F(x) = x^a, 0 \leq x \leq 1, a > 0$, are all determined by their moments. A somewhat less obvious example of the application of Corollary 5 is given below. The result itself is due to Chan (1967).

Example 10   Let X be a random variable with distribution function $F(x)$ and finite expectation $E(X)$. Let $X_1, X_2, \ldots, X_n$ be independent copies of X and set $Z_n = \max(X_1, X_2, \ldots, X_n)$. Then $u_n = E(Z_n)$ is finite for every $n \geq 1$, and the sequence u_n, $n \geq 1$, uniquely determines $F(x)$.

We have observed before that the distribution function

$$H_n(x) = P(Z_n \leq x) = P(\text{all } X_i \leq x) = F^n(x)$$

so

$$u_n = \int_{-\infty}^{+\infty} x\, dF^n(x) = n \int_0^1 F^{-1}(y) y^{n-1}\, dy$$

where $F^{-1}(y) = \inf\{x \colon F(x) > y\}$. The last equation is easily seen to be valid by observing that, for continuity points a, b of $F(x)$, with $a = F^{-1}(c)$ and $b = F^{-1}(d)$,

$$a[F^n(b) - F^n(a)] = F^{-1}(c)[d^n - c^n]$$

while, by the mean value theorem, $d^n - c^n = nc_1^{n-1}(d - c)$, where $c \leq c_1 \leq d$. Hence, upon forming the Riemann–Stieltjes sums to the two sides of the equation, the equality follows. Now, if $F^{-1}(y)$ is integrable (the only trouble spots could be $y = 0$ and $y = 1$; take the simple example $F(x) = 1 - 1/x$, $x \geq 1$, when $F^{-1}(y) = 1/(1 - y), 0 < y < 1$), which is assumed in the example ($n = 1$ gives $E(X) = E(Z_1)$), then so are $F^{-1}(y) y^{n-1}$, $n \geq 1$, i.e., u_n is finite for every $n \geq 1$. Next, if $G(x)$ is another distribution function for which

$$u_n = n \int_0^1 G^{-1}(y) y^{n-1}\, dy, \qquad n \geq 1$$

then (3.12) holds with $f(y) = F^{-1}(y)$ and $g(y) = G^{-1}(y)$. Consequently, $F^{-1}(y) = G^{-1}(y)$ a.s., i.e., $F(x) = G(x)$ on a dense set; thus, $F(x) = G(x)$ (see Theorem 2.23). ∎

With somewhat more effort than simply a direct application of the Taylor expansion (3.5), one can generalize Corollary 3. That is, even though characteristic functions in general are not determined by their values on an interval (see Exercise 8), analytic characteristic functions are uniquely determined by their Taylor series if they are convergent for at least one point different from the origin. This can be seen as follows. If we repeat the argument of the initial lines of the proof of Theorem 2 we get, by induction, that if all moments are finite, then

$$\varphi^{(n)}(t) = i^n \int_{-\infty}^{+\infty} x^n e^{itx}\, dF(x)$$

for all n and all t. Consequently, $|\varphi^{(n)}(t)| \leq E(|X^n|)$, assumed finite. Now, if, as $n \to +\infty$,

$$\limsup(|E(X^n)|/n!)^{1/n} = 1/R < +\infty$$

then

$$\varphi(t^*) = \sum_{n=0}^{+\infty} \frac{\varphi^{(n)}(t)}{n!} (t^* - t)^n$$

is absolutely convergent in the circle $|t^* - t| < R$ (where t and t^* can even be taken to be complex). Next, build overlapping circles with centers on the real line (and with radius R) and let one of the centers be at the origin. Then $t = 0$ for this circle, and $\varphi^{(n)}(0) = i^n E(X^n)$; that is, in this particular circle, $\varphi(t)$ is determined by the moments $E(X^n)$, $n \geq 0$. With the circles overlapping, $\varphi(t)$ is "analytically continued" from one circle to the next. Then, by the result of complex analysis which says that analytic continuation leads to a unique function, we have that $\varphi(t)$ is uniquely determined by the initial circle at the origin, and thus by the set of moments, in a strip of the complex plane that contains the real line. That is, $\varphi(t)$ is uniquely determined for all real t by the moments whenever the moments are such that $R > 0$.

As an example when this extension applies but Corollary 3 does not, take the exponential distribution with unit expectation (Example 2 of Section 3.1). Here, $R = 1$ and we can now conclude that the moments of the exponential distribution uniquely determine the distribution.

From the generality of the corollaries, one might wrongly conclude that all widely applied distributions are uniquely determined by their moments. An important counterexample is the lognormal distribution. Its moments have been computed in Exercise 2 of Section 2.7. Now, it is observed in Heyde (1963) that these same moments are obtained for a random variable whose density function

$$cg(x) = f(x)\{1 + (1/2)\sin[2\pi k\sigma^{-2}(\log x - m)]\}$$

where $f(x)$ is the density function of the lognormal distribution as described in the quoted exercise and k is a positive integer. The constant $c > 0$ is there to make $g(x)$ a density function. The value $1/2$ in $g(x)$ is not significant; any value between 0 and 1 would do.

We now turn to the proofs of Theorems 4 and 5. We shall need the following elementary result.

Lemma The improper Riemann integral

$$\int_{-\infty}^{+\infty} \frac{\sin x}{x}\, dx = \pi$$

Proof. First note that $(\sin x)/x = (\sin(-x))/(-x)$; hence, it will suffice to evaluate our integral on the positive real line. By interchanging the order of integration in

$$\iint_T (\sin x)e^{-xy}\, dx\, dy$$

where $T = \{(x, y): 0 \le x \le n\pi, 0 \le y < +\infty\}$, we get

$$\int_0^{n\pi} \frac{\sin x}{x}\, dx = \int_0^{+\infty} \frac{1 - (-1)^n e^{-n\pi y}}{1 + y^2}\, dy$$

Upon letting $n \to +\infty$, this yields

$$\int_0^{+\infty} \frac{\sin x}{x}\, dx = \int_0^{+\infty} \frac{dy}{1 + y^2} = \frac{\pi}{2}$$

and the proof is completed. ∎

Proof of Theorem 4. Set

$$I_U(a, b) = \frac{1}{2\pi} \int_{-U}^{+U} \frac{e^{-ita} - e^{-itb}}{it}\, \varphi(t)\, dt \tag{3.13}$$

where the integrand is defined by continuity to be $(b - a)\varphi(0)$ at $t = 0$. Since the integrand is bounded, we can interchange the order of integrations if we replace the characteristic function $\varphi(t)$ by its integral form. We get

$$
\begin{aligned}
I_U(a, b) &= \frac{1}{2\pi} \int_{-\infty}^{+\infty} \left[\int_{-U}^{+U} \frac{e^{it(x-a)} - e^{it(x-b)}}{it} \, dt \right] dF(x) \\
&= \frac{1}{2\pi} \int_{-\infty}^{+\infty} \left[\int_{-U}^{+U} \frac{\sin t(x-a) - \sin t(x-b)}{t} \, dt \right] dF(x) \\
&= \frac{1}{\pi} \int_{-\infty}^{+\infty} \left[\int_{U(x-b)}^{U(x-a)} \frac{\sin t}{t} \, dt \right] dF(x)
\end{aligned}
$$

where we first used the fact that $(\cos tu - \cos tv)/t$ is an odd function, and thus that its integral over $(-U, +U)$ is zero, and then, for $(\sin tu)/t$, which is even, the integration was changed to be over $(0, +U)$, over which linear substitutions were carried out. Now, by the lemma, the inner integral in the last form for $I_U(a, b)$ is bounded in U. Hence, by letting $U \to +\infty$, the dominated convergence theorem and the lemma yield that, whenever $a < b$ are continuity points of $F(x)$,

$$
\lim I_U(a, b) = F(b) - F(a) \qquad (U \to +\infty) \tag{3.14}
$$

since the limit of the inner integral is 0 if $x < a$ or if $x > b$ and π for $a < x < b$. The proof is complete. ∎

The formula of Theorem 4, which is restated in (3.13) and (3.14), is called an inversion formula. It should be noted that the limit does not necessarily result in an improper Riemann integral. However, when $\varphi(t)$ is absolutely integrable, then the inversion formula just proved does become an improper integral. Even more is true in this case.

Theorem 7 If the characteristic function $\varphi(t)$ is absolutely integrable on R, then the corresponding distribution function $F(x)$ has a density $F'(x) = f(x)$, which is continuous, and

$$
f(x) = \frac{1}{2\pi} \int_{-\infty}^{+\infty} e^{-itx} \varphi(t) \, dt \tag{3.15}
$$

Proof. In the inversion formula (3.13) and (3.14), choose $a = x - h$ and $b = x + h$. We have

$$
\frac{F(x+h) - F(x-h)}{2h} = \lim_{U = +\infty} \frac{1}{2\pi} \int_{-U}^{+U} \frac{\sin th}{th} e^{-itx} \varphi(t) \, dt
$$

Because the integrand is bounded by $|\varphi(t)|$, the integral of which is assumed to be finite, the dominated convergence theorem entails that the limit on the right-hand side is a Lebesgue integral, and thus an improper Riemann integral as well. Next, as a function of h, the integrand on the right-hand side has a limit as $h \to 0$ and is dominated by the integrable function $|\varphi(t)|$. Consequently, the dominated convergence theorem is once more applicable, we get that the limit of the two sides exists as $h \to 0$, and the desired formula follows.

A final appeal to the dominated convergence theorem yields that $f(x)$ is continuous because the integrand in (3.15) is continuous in x. The theorem is established. ∎

The density of a distribution function might exist even though its characteristic function is not absolutely integrable. Such densities can be recovered from the characteristic functions by summability methods, one of which is given below.

Theorem 8 Let $\varphi(t)$ be the characteristic function of the distribution function $F(x)$, the density of which, $f(x)$, exists. Then, for continuity points x of $f(x)$,

$$f(x) = \lim_{U = +\infty} \int_{-U}^{+U} e^{-itx} \left(1 - \frac{|t|}{U} \right) \varphi(t) \, dt$$

The proof relies on the convolution formula ((3.7) and (3.8) and its extension to densities in (3.18) below), and on the fact that (see Exercise 7)

$$\psi(t) = 1 - |t| \quad |t| \le 1 \quad \text{and} \quad \psi(t) = 0 \text{ elsewhere} \tag{3.16}$$

is the characteristic function of the distribution whose density function

$$g(x) = \frac{1}{\pi} \frac{1 - \cos x}{x^2} \tag{3.17}$$

On the convolution formula we remark that it can easily be verified that if F_1 has a density f_1 in (3.7), then so does F, f, say, and

$$f(z) = \int_{-\infty}^{+\infty} f_1(z - x) \, dF_2(x) = \int_{-\infty}^{+\infty} f_1(z - x) f_2(x) \, dx \tag{3.18}$$

the latter equation being valid if F_2 has a density f_2 as well.

Proof of Theorem 8. Consider the function $\eta_U(t) = \varphi(t)\psi(t/U)$. Since $\psi(t)$ is a characteristic function with corresponding density function $g(x)$ of (3.17), $\psi(t/U)$ is the characteristic function of the density function $Ug(Ux)$. Thus, Theorem 6 ensures that $\eta_U(t)$ is a characteristic function as well, the distribution of which is the convolution of $F(x)$ and $G(Ux)$, where

$G'(x) = g(x)$. Furthermore, since $\psi(t/U)$ is evidently absolutely integrable, so is $\eta_U(t)$, implying that Theorem 7 is applicable to $\eta_U(t)$. That is, the distribution determined by $\eta_U(t)$ has a density $h(x; U)$ and can be evaluated by both formulas corresponding to (3.15) and (3.18). We have

$$h(x; U) = \frac{1}{2\pi} \int_{-\infty}^{+\infty} e^{-itx} \eta_U(t)\, dt$$

$$= \frac{1}{\pi} \int_{-\infty}^{+\infty} f\left(x - \frac{v}{U}\right) \frac{1 - \cos v}{v^2}\, dv \qquad (3.19)$$

We now show that, as $U \to +\infty$, $\lim h(x; U) = f(x)$, resulting in the desired formula (just replace $\eta_U(t)$ by its definition in (3.19)). We argue with the second formula for $h(x; U)$ in (3.19). We write

$$h(x; U) - f(x) = \frac{1}{\pi} \int_{-\infty}^{+\infty} \left[f\left(x - \frac{v}{U}\right) - f(x) \right] \frac{1 - \cos v}{v^2}\, dv$$

If we split the integral into two parts—first integrating for $|v| \le U^a$ and then for $|v| > U^a$—it is easily seen that $0 < a < 1$ can be chosen so that the first integral becomes small because of the difference $|f(x - v/U) - f(x)|$ (recall that x is a continuity point of $f(x)$), while the second integral is made small by the presence of v^2 in the denominator. The details are left to the reader. ∎

A more general summability method can be found in Lukacs [1970, pp. 39–40].

The following statement is not directly related to the uniqueness theorem. However, the calculations involved in its proof are very similar to those used in the proof of Theorem 4.

Theorem 9 Let $\varphi(t)$ be the characteristic function of the distribution function $F(x)$. Then, for every x, as $U \to +\infty$,

$$F(x) - F(x - 0) = \lim \frac{1}{2U} \int_{-U}^{+U} e^{-itx} \varphi(t)\, dt$$

Proof. If we write in the integral form of a characteristic function and interchange integrations, we get

$$\frac{1}{2U} \int_{-U}^{+U} e^{-itx} \varphi(t)\, dt = \frac{1}{2U} \int_{-\infty}^{+\infty} \left\{ \int_{-\infty}^{+U} e^{it(z-x)}\, dt \right\} dF(z)$$

$$= \int_{-\infty}^{+\infty} \frac{\sin U(z-x)}{U(z-x)}\, dF(z)$$

where the integrand is one at $z = x$ and goes to zero as $U \to +\infty$. Hence, the dominated convergence theorem completes the proof. ∎

As a particular case of Theorem 9 we have that $F(x)$ is a continuous distribution function if and only if the limit in Theorem 9 is zero for every x.

We now turn to the proof of Theorem 5.

Proof of Theorem 5. The fact that $F_n \to F$ completely implies $\varphi_n(t) \to \varphi(t)$ is an immediate consequence of the Helly–Bray Theorem (Theorem 2.26). Therefore, only the second part of the theorem needs proof. That is, we assume $\varphi_n(t) \to \varphi(t)$, which is continuous at $t = 0$. First note that $|\varphi_n(t)| \le 1$ and $\varphi_n(t) \to \varphi(t)$ imply

$$\int_0^u \varphi_n(t)\,dt \to \int_0^u \varphi(t)\,dt \qquad u \text{ finite,} \quad n \to +\infty \tag{3.20}$$

Furthermore,

$$\int_0^u \varphi_n(t)\,dt = \int_0^u \left\{ \int_{-\infty}^{+\infty} e^{itx}\,dF_n(x) \right\} dt$$

$$= \int_{-\infty}^{+\infty} \frac{e^{iux} - 1}{ix}\,dF_n(x) \tag{3.21}$$

where the last integrand is defined by continuity at $x = 0$. Now, by the weak compactness of distributions (Theorem 2.24), F_n has a subsequence $F_{n_k}(x) \to G(x)$ weakly. By another appeal to the Helly–Bray theorem, (3.21) thus yields that, as $k \to +\infty$,

$$\int_0^u \varphi_{n_k}(t)\,dt \to \int_{-\infty}^{+\infty} \frac{e^{iux} - 1}{ix}\,dG(x) \tag{3.22}$$

Let $\varphi_G(t)$ be the characteristic function of G. Since G is the weak limit of a subsequence of F_n, it might be a distribution function in the extended sense only. Yet, a formula similar to (3.21) applies to φ_G and G. Hence, such a formula and the combination of (3.20) and (3.22) give

$$\int_0^u \varphi(t)\,dt = \int_{-\infty}^{+\infty} \frac{e^{iux} - 1}{ix}\,dG(x) = \int_0^u \varphi_G(t)\,dt \tag{3.23}$$

Equation (3.23) has two implications. First, both extreme sides are differentiable at $u = 0$ because the integrands are continuous at $t = 0$. Thus, since $\varphi_n(0) = 1$ and $\varphi_n(t) \to \varphi(t)$, we have, by differentiation,

$$1 = \varphi(0) = \varphi_G(0)$$

The extreme sides yield $G(+\infty) - G(-\infty) = 1$ and, since $0 \le G(-\infty) \le G(+\infty) \le 1$, necessarily $G(+\infty) = 1$ and $G(-\infty) = 0$. That is, G is a proper distribution function. Second, the right-hand side of (3.23) is differentiable for all u; its derivative is $\varphi_G(t)$. But then the left-hand side, which does not depend on G, is differentiable also. In other words, $\varphi_G(t)$ is uniquely determined by $\varphi(t)$ and, by the uniqueness theorem, so is G. Consequently, the same limit is obtained for every subsequence of F_n; this limit is a proper distribution function. Therefore, the convergence, as well as the proof, is complete. ∎

Exercises

1. Let $g(x, y)$ be a continuous and bounded function in the plane. Let F_1 and F_2 be two distribution functions. Prove

$$\int_{-\infty}^{+\infty} \left[\int_{-\infty}^{+\infty} g(x, y)\, dF_1(x) \right] dF_2(y)$$

$$= \int_{-\infty}^{+\infty} \left[\int_{-\infty}^{+\infty} g(x, y)\, dF_2(y) \right] dF_1(x)$$

2. Let $X_1, X_2, \ldots, X_n$ be independently and identically distributed exponential variables. Show that $X_1 + X_2 + \cdots + X_n$ is a gamma variable.

3. Let $X_1, X_2, \ldots, X_n$ be independent and identically distributed random variables with finite expectation. Let $X_{1:n} \le X_{2:n} \le \cdots \le X_{n:n}$ be a rearrangement of the X_j ($X_{r:n}$ is called the rth *order statistic* of the *sample* X_j, $1 \le j \le n$). Show that, for every $1 \le r \le n$, $E(X_{r:n})$ is finite, and whatever $1 \le r(n) \le n$, the sequence $E(X_{r(n):n})$, $n \ge 1$, uniquely determines the distribution of X_1.

4. Show that if $\varphi(t)$ is a characteristic function then so is $|\varphi(t)|^2$ and, in general, $|\varphi(t)|^{2n}$, $n \ge 1$ integer. With odd exponents, the statement is not true. Show, in particular, that $\varphi(t) = 1/4 + (3/4)e^{it}$ is a characteristic function, but $|\varphi(t)|$ is not.

5. Using characteristic functions, prove the normal approximation to the binomial: if X is binomial with parameters n and $0 < p < 1$, then $[np(1-p)]^{-1/2}(X - np)$ has an asymptotic normal distribution.

6. Let Y be a Poisson variable with parameter $a > 0$. Show that the distribution of $(Y - a)/a^{1/2}$ converges to the standard normal distribution as $a \to +\infty$.

7. Let $F(x)$ be a distribution function with characteristic function $\varphi(t)$. Show that, for arbitrary real numbers a and $h > 0$,

$$\int_0^h [F(a + x) - F(a - x)]\, dx = \frac{1}{\pi} \int_{-\infty}^{+\infty} \frac{1 - \cos ht}{t^2} e^{-ita} \varphi(t)\, dt$$

With the special choice of $F(x) = 1$ if $x > 0$ and $F(x) = 0$ if $x < 0$ (i.e., degenerate at zero), and thus $\varphi(t) = 1$ for all t, and with $h = 1$, deduce that the characteristic function $\psi(t)$ corresponding to the density in (3.17) is given by (3.16).

8. Let X be a discrete random variable taking the values 0 and $\pm(2k + 1)\pi$, $k \geq 0$, with distribution $P(X = 0) = 1/2$ and $P(X = (2k + 1)\pi) = P(X = -(2k + 1)\pi) = 2/(2k + 1)^2\pi^2$. Show that the characteristic function $\varphi(t)$ of X satisfies $\varphi(t) = \psi(t)$ for $-1 \leq t \leq 1$ and $\varphi(t + 2) = \varphi(t)$, where $\psi(t)$ is given by (3.16). (This is Khintchine's example for characteristic functions coinciding on a full interval.)

9. Let $F(x)$ be a distribution function with discontinuities at the points x_j, $j \in J$, where J is possibly empty and, at most, denumerable (Lemma of Section 2.7). Put $p_j = F(x_j) - F(x_j - 0)$. Show that if $\varphi(t)$ is the characteristic function of $F(x)$, then, as $U \to +\infty$,

$$\lim \frac{1}{2U} \int_{-U}^{+U} |\varphi(t)|^2 \, dt = \sum p_j^2$$

where the summation is over $j \in J$ and the empty sum is zero.

10. Prove that if the sequence $\varphi_n(t)$ of characteristic functions converges uniformly to $\varphi(t)$ for $-t_0 \leq t \leq t_0$, $t_0 > 0$, then $\varphi(t)$ is continuous at $t = 0$.

3.3 CLASSICAL FORMS OF THE CENTRAL LIMIT THEOREM, AND A MODEL FOR MEASUREMENT ERRORS

The asymptotic normality of normalized sums has always been a central problem of probability theory. Three of the now classical results are presented in this section. Each provides sufficient conditions for asymptotic normality. Necessary conditions will be dealt with in subsequent chapters. It will also be shown that the central limit theorem can be used to find the exact, not just the limiting, distribution of errors in laboratory measurements.

Throughout this section, $X_1, X_2, \ldots$ are random variables defined on the same probability space and $S_n = X_1 + X_2 + \cdots + X_n$.

Theorem 10 **(Classical Central Limit Theorem)** Let $X_1, X_2, \ldots$ be independent and identically distributed with $E(X_1) = 0$ and $V(X_1) = \sigma^2 > 0$ finite. Then, $S_n/\sigma n^{1/2}$ is asymptotically normally distributed as $n \to +\infty$.

Proof. Let $\varphi(t)$ and $\varphi_n(t)$ be the characteristic functions of X_1 and $S_n/\sigma n^{1/2}$, respectively. Then, by Corollary 1 of Section 3.2 and Theorem 3,

$$\varphi_n(t) = \varphi^n\!\left(\frac{t}{\sigma\sqrt{n}}\right)$$

Furthermore, (3.4) is applicable with $m_1 = 0$ and $m_2 = \sigma^2$, yielding

$$\varphi_n(t) = \left[1 - \frac{t^2}{2n} + o\!\left(\frac{t^2}{n}\right)\right]^n \qquad n \to +\infty$$

The theorem is thus established by the continuity theorem (Theorem 5) and the elementary limit relation

$$\left(1 + \frac{x(n)}{n}\right)^n \to e^x \qquad \text{if} \quad x(n) \to x \qquad \text{and} \qquad n \to +\infty \tag{3.24}$$

$\blacksquare$

Theorem 11 (Liapunov's Condition) Let $X_1, X_2, \dots$ be independent with $E(X_j) = 0$, $V(X_j) = \sigma_j^2 > 0$, $E(|X_j|^3) = E_{3,j}$ finite. Set $s_n^2 = \sigma_1^2 + \sigma_2^2 + \cdots + \sigma_n^2$. If

$$\frac{1}{s_n^3} \sum_{j=1}^{n} E_{3,j} \to 0 \qquad \text{as} \quad n \to +\infty \tag{3.25}$$

then S_n/s_n is asymptotically normally distributed.

Proof. Let $\varphi_j(t), j \geq 1$, and $\psi_n(t)$ be the characteristic functions of $X_j, j \geq 1$, and S_n/s_n, respectively. Then, as in the previous proof,

$$\psi_n(t) = \varphi_1\!\left(\frac{t}{s_n}\right)\varphi_2\!\left(\frac{t}{s_n}\right)\cdots\varphi_n\!\left(\frac{t}{s_n}\right)$$

We again apply the finite Taylor expansion in (3.4)

$$\varphi_j\!\left(\frac{t}{s_n}\right) = 1 - \sigma_j^2\frac{t^2}{2s_n^2} + \vartheta E_{3,j}\frac{t^3}{6s_n^3} \qquad |\vartheta| \leq 1$$

and observe that $\varphi_j(t/s_n) \to 1$ uniformly for $1 \leq j \leq n$, because (recall Exercise 6 of Section 2.2)

$$\max_{1 \leq j \leq n}\left(\frac{\sigma_j}{s_n}\right)^3 \leq \max_{1 \leq j \leq n}\frac{E_{3,j}}{s_n^3} \leq \frac{1}{s_n^3}\sum_{j=1}^{n} E_{3,j} \to 0$$

Thus, upon using

$$\log \varphi_j(u) = \log[1 - (1 - \varphi_j(u))] = \varphi_j(u) - 1 + o(|\varphi_j(u) - 1|) \tag{3.26}$$

valid for $\varphi_j(u) - 1 \to 0$, we get, with some $|\vartheta^*| \leq 1$,

$$\log \psi_n\left(\frac{t}{s_n}\right) = -\frac{t^2}{2}(1 + o(1)) + \vartheta^*|t^3|\frac{1}{s_n^3}\sum_{j=1}^n E_{3,j} \to -\frac{t^2}{2}$$

as $n \to +\infty$. The continuity theorem of characteristic functions now completes the proof. ∎

Theorem 12 (Lindeberg) Let $X_1, X_2, \ldots$ be independent with $E(X_j) = 0$ and $V(X_j) = \sigma_j^2 > 0$ finite. Set $s_n^2 = \sigma_1^2 + \sigma_2^2 + \cdots + \sigma_n^2$. Assume, for every $\varepsilon > 0$,

$$\frac{1}{s_n^2}\sum_{j=1}^n \int_{|x|\geq \varepsilon s_n} x^2\, dF_j(x) \to 0 \tag{3.27}$$

as $n \to +\infty$. Then,

$$\frac{1}{s_n}\max_{1\leq j\leq n}\sigma_j \to 0 \tag{3.28}$$

as $n \to +\infty$ and S_n/s_n is asymptotically normal.

Proof. If we split the integration in σ_j^2 according to whether $|x| \geq \varepsilon s_n$ or $|x| < \varepsilon s_n$ and if x^2 is replaced by $\varepsilon^2 s_n^2$ in the latter case, we get

$$\sigma_j^2 \leq \int_{|x|\geq \varepsilon s_n} x^2\, dF_j(x) + \varepsilon^2 s_n^2$$

Hence,

$$\max_{1\leq j\leq n}\left(\frac{\sigma_j}{s_n}\right)^2 \leq \max_{1\leq j\leq n}\frac{1}{s_n^2}\int_{|x|\geq \varepsilon s_n} x^2\, dF_j(x) + \varepsilon^2$$

where the first term on the right-hand side can be increased further by the expression in (3.27). Consequently, (3.27) ensures the validity of (3.28).

To prove the asymptotic normality of S_n/s_n, we use a truncation of the variables X_j and apply Liapunov's condition to the truncated variables. It will then be observed that the effect of the truncation is negligible, and thus the desired limit law follows. So, let $X_{j,n} = X_j$ if $|X_j| < \varepsilon_n s_n$ and $X_{j,n} = 0$ otherwise, where $\varepsilon_n \to 0$ as $n \to +\infty$ and is such that

$$\delta_n(\varepsilon_n) \to 0 \qquad \delta_n(\varepsilon_n)/\varepsilon_n \to 0 \qquad \delta_n(\varepsilon_n)/\varepsilon_n^2 \to 0 \tag{3.29}$$

where $\delta_n(\varepsilon)$ signifies the expression in (3.27). Note that only the last relation has to be guaranteed, because the previous two follow from it. They are collected in one place for the sake of easier reference. That such ε_n can be found is ensured by (3.27). Indeed, take an arbitrary sequence $a_1 > a_2 > \cdots$

of positive numbers with $a_n \to 0$ as $n \to +\infty$. Since $\delta_n(\varepsilon) \to 0$ for every fixed $\varepsilon > 0$, there is an integer n_j for every $j \geq 1$ such that $\delta_n(a_j) \leq a_j^3$ for all $n \geq n_j$. Clearly, the integers n_j can be chosen to be distinct; then, with $\varepsilon_n = a_j$ for $n_j < n \leq n_{j+1}$, (3.29) is satisfied. With this argument completed, the variables $X_{j,n}$ are well defined. Since they are bounded, all of their moments are finite. We have, in fact,

$$|E(X_{j,n})| = \left| \int_{|x| < \varepsilon_n s_n} x \, dF_j(x) \right| = \left| \int_{|x| \geq \varepsilon_n s_n} x \, dF_j(x) \right|$$

$$\leq \frac{1}{\varepsilon_n s_n} \int_{|x| \geq \varepsilon_n s_n} x^2 \, dF_j(x)$$

where the first equation follows from the definition of $X_{j,n}$, while the second is implied by $E(X_j) = 0$. Consequently,

$$\frac{1}{s_n} \sum_{j=1}^{n} |E(X_{j,n})| \leq \delta_n(\varepsilon_n)/\varepsilon_n \tag{3.30}$$

Next, we observe that, on the one hand, $E[(X_j - X_{j,n})^2] = V(X_j) - V(X_{j,n}) - E^2(X_{j,n})$, while, on the other hand,

$$E[(X_j - X_{j,n})^2] = \int_{|x| \geq \varepsilon_n s_n} x^2 \, dF_j(x)$$

Hence, by independence and $\sum_{j=1}^{n} b_j^2 \leq (\sum_{j=1}^{n} |b_j|)^2$,

$$0 \leq 1 - \frac{1}{s_n^2} \sum_{j=1}^{n} V(X_{j,n}) \leq \delta_n(\varepsilon_n) + \frac{1}{s_n^2} \left(\sum_{j=1}^{n} |E(X_{j,n})| \right)^2 \tag{3.31}$$

Finally, in view of $|X_{j,n}| < \varepsilon_n s_n$ and $|E(X_{j,n})| < \varepsilon_n s_n$,

$$\sum_{j=1}^{n} E(|X_{j,n} - E(X_{j,n})|^3) \leq 2\varepsilon_n s_n \sum_{j=1}^{n} V(X_{j,n}) \tag{3.32}$$

Now, upon setting $S_n^{(n)} = X_{1,n} + X_{2,n} + \cdots + X_{n,n}$, $E^{(n)} = E(S_n^{(n)})$ and $V_n^2 = V(S_n^{(n)})$, (3.29)–(3.32) imply that Liapunov's condition is satisfied by $S_n^{(n)} - E^{(n)}$; thus, the asymptotic distribution of $(S_n^{(n)} - E^{(n)})/V_n$ is standard normal. A second implication of the estimates in (3.29)–(3.32) is that when writing

$$\frac{S_n^{(n)}}{s_n} = \frac{V_n}{s_n} \frac{S_n^{(n)} - E^{(n)}}{V_n} + \frac{E^{(n)}}{s_n}$$

the normalizing constants V_n/s_n and $E^{(n)}/s_n$ converge to 1 and 0, respectively.

Hence, (see Exercise 1) $S_n^{(n)}/s_n$ is also asymptotically standard normal. Now, to complete the proof, we observe that

$$P\left(\frac{S_n^{(n)}}{s_n} \neq \frac{S_n}{s_n}\right) \leq \sum_{j=1}^{n} P(X_{j,n} \neq X_j) = \sum_{j=1}^{n} \int_{|x| \geq \varepsilon_n s_n} dF_j(x)$$

which is further increased by plugging in the term $x^2/\varepsilon_n^2 s_n^2$, yielding the upper estimate $\delta_n(\varepsilon_n)/\varepsilon_n^2$, which converges to zero because of (3.29). Appealing once more to Exercise 1, the asymptotic normality of S_n/s_n obtains, which was to be proved.   ∎

Remark This proof is very instructive, for it shows that, in spite of the drastically different assumptions in the theorems of Liapunov and Lindeberg, these theorems are essentially equivalent. The proof actually shows that Liapunov's theorem implies Lindeberg's. This is achieved by the very powerful method of truncation. The converse case can easily be seen to hold. As a matter of fact,

$$\frac{1}{s_n^2} \int_{|x| \geq \varepsilon s_n} x^2 \, dF_j(x) \leq \frac{1}{\varepsilon s_n^3} \int_{-\infty}^{+\infty} |x|^3 \, dF_j(x)$$

Thus, summation over j yields that (3.27) holds whenever (3.25) does.

The remainder of this section is devoted to applications. We shall give one application from statistics, one from the field of model building, and one from the very distant field of number theory.

First note that the normal approximation to the binomial is a special case of the classical central limit theorem. Namely, let $X_j = 1$ or 0 with $P(X_j = 1) = 1 - P(X_j = 0) = p$ for all j. Then, if the X_j are independent, $X_1 + X_2 + \cdots + X_n$ is binomial. Hence, since $E(X_j) = p$ and $V(X_j) = p(1 - p)$, we have from Theorem 10 that

$$P(X_1 + X_2 + \cdots + X_n \leq np + x(np(1 - p))^{1/2}) \to N(x) \tag{3.33}$$

as $n \to +\infty$, where $N(x)$ is the standard normal distribution function.

From this special form we deduce a result frequently used in statistics. Let $Y_1, Y_2, \ldots, Y_n$ be independent random variables with a common distribution function $F(x)$. Let $Y_{1:n} \leq Y_{2:n} \leq \cdots \leq Y_{n:n}$ be a rearrangement of the Y_j. As noted earlier, the set Y_j, $1 \leq j \leq n$, is called a sample from $F(x)$ and $Y_{r:n}$ is called the rth order statistic of the sample. If $r = [nq] + 1$, where $[y]$ signifies the integer part of y, $Y_{r:n}$ is called the qth quantile of the sample. We now prove the following result on quantiles.

Theorem 13 With the notation of the preceding paragraph, define Q by $q = F(Q)$ for $0 < q < 1$. Assume that $F(x)$ is such that $F'(x) = f(x)$ exists and is continuous at Q and that $f(Q) > 0$. Then

$$P\left(Y_{[nq]+1:n} \leq Q + \frac{x}{f(Q)} \left(\frac{q(1-q)}{n} \right)^{1/2} \right) \to N(x)$$

as $n \to +\infty$.

Proof. Let us set $r(q) = [nq] + 1$ and

$$z_n(q) = Q + \frac{x}{f(Q)} \left(\frac{q(1-q)}{n} \right)^{1/2}$$

Let $X_j = 1$ if $Y_j > z_n(q)$ and $X_j = 0$ otherwise. Then

$$P(Y_{r(q):n} \leq z_n(q)) = P(X_1 + X_2 + \cdots + X_n \leq n - r(q))$$

and the X_j are independent with $p = P(X_j = 1) = 1 - F(z_n(q))$. Therefore, if we establish that

$$n - r(q) = np + x(np(1-p))^{1/2} + o(n^{1/2}) \tag{3.34}$$

the theorem follows from (3.33) and Exercise 1. Now, by the mean value theorem of calculus and by the continuity of $f(x)$,

$$F(z_n(q)) = F(Q) + f(Q^*) \frac{x}{f(Q)} \left(\frac{q(1-q)}{n} \right)^{1/2}$$

$$= q + (1 + o(1))x \left(\frac{q(1-q)}{n} \right)^{1/2}$$

where Q^* is a value between Q and $z_n(q)$ and, thus, $f(Q) - f(Q^*) = o(1)$ as $n \to +\infty$. A simple substitution of $p = 1 - F(z_n(q))$ into (3.34) yields its validity, which completes the proof. ∎

Note that, since (3.34) permits an error term of $o(n^{1/2})$, we proved more than stated: all order statistics whose order deviates from $r(q)$ by less than an order of $n^{1/2}$ can be normalized by $z_n(q)$ and such normalized order statistics are asymptotically standard normal.

Another consequence of the classical central limit theorem is the justification of the fact that errors in laboratory measurements, which are due to inaccurate reading and the inaccuracy of the instrument, are normally distributed. The fact that the average of errors (given a large number of measurements) is asymptotically normal is just a restatement of Theorem 10. The point we make here is that the exact distribution of errors is normal and, yet, is deduced from a limit theorem. The importance of finding the exact distribution is obvious when experimentation is expensive and only a very few repetitions are possible.

We introduce the model through an example. Assume a chemist measures weight by reading if off on a scale. If the true weight is w but he measures W, the error is $X = W - w$. If the material is split into two parts whose weights are w_1 and w_2, $w_1 + w_2 = w$, but whose measurements are W_1 and W_2, respectively, then the measurement error of the same material is $W_1 - w_1 + W_2 - w_2 = X_1 + X_2$. It is the nature of the quantities involved, rather than mathematical reasons, that allow us to make the following assumptions:

(i) X, X_1, and X_2 are identically distributed.
(ii) X_1 and X_2 are independent.
(iii) $X_1 + X_2$ has the same type of distribution as X.
(iv)   The expectation and the variance V of X are finite, $V > 0$.

We therefore adopt as the model for random errors in measurement random variables that satisfy properties (i)–(iv).

Theorem 14   In the model defined by (i)–(iv), the distribution of X is normal.

Proof. Without loss of generality, we may assume $E(X) = 0$; we denote $V(X) = V > 0$. Recall that (iii) translates as the existence of some a and $b > 0$ such that the distribution of $a + bX$ is the same as that of $X_1 + X_2$. Hence, by (i) and (ii), $a + bE(X) = 0$ and $b^2 V(X) = V(a + bX) = V(X_1 + X_2) = 2V$, so $a = 0$ and $b = \sqrt{2}$. Furthermore, if $\varphi(t)$ denotes the characteristic function of X, X_1, and X_2, our conditions imply

$$\varphi^2(t) = \varphi(\sqrt{2}t)$$

But then $\varphi^4(t) = \varphi^2(\sqrt{2}t) = \varphi(2t)$ and, in general, if $\varphi^N(t) = \varphi(ut)$, then $\varphi^{2N}(t) = \varphi^2(ut) = \varphi(\sqrt{2}ut)$. Hence, by induction, for every $N = 2^k$, $k \geq 1$,

$$\varphi^N(t) = \varphi(2^{k/2}t) = \varphi(\sqrt{N}t) \qquad \text{i.e.,} \quad \varphi(t) = \varphi^N(t/\sqrt{N})$$

On the right-hand side of the last equation we recognize the characteristic function of $(X_1 + X_2 + \cdots + X_N)/\sqrt{N}$, where the X_j are independent and each X_j is distributed as X. Hence, we can apply Theorem 10 to the right-hand side of the subsequence, yielding the limit $\exp(-Vt^2/2)$ as $N \to +\infty$, while the left-hand side is not dependent on N, so the equation continues to hold in the limit as well. That is, $\varphi(t) = \exp(-Vt^2/2)$, which is equivalent to the claimed normality.   ∎

We now look at a remarkable application of Liapunov's theorem in pure mathematics, due to Elliott (1970) and Kátai (1969). The problem is from number theory, but the solution requires no knowledge of that subject beyond the definition of a prime number.

For a positive integer n, define $d(n)$ as the number of divisors d of n, abbreviated to $d|n$, i.e., $d(n)$ is the number of solutions in d of $n = dy$, $d \geq 1$ and $y \geq 1$ integers. For calculational simplicity, we consider square-free numbers n only. That is,

$$n = p_1 p_2 \cdots p_r \qquad p_1 < p_2 < \cdots < p_r \text{ primes} \tag{3.35}$$

We set

$$v_n^2 = \frac{1}{4} \sum_{j=1}^{r} (\log p_j)^2$$

Theorem 15 Assume that $n \to +\infty$ through a sequence of integers for which $(\log p_r)/v_n \to 0$. Define

$$F_n(x) = \frac{1}{d(n)} v_n\{d : d|n, \log d \leq \tfrac{1}{2} \log n + xv_n\}$$

where $v_n\{d : \cdots\}$ signifies the number of integers $d \leq n$ that satisfy the conditions in the dotted space. Then

$$F_n(x) \to N(x)$$

as $n \to +\infty$.

Remark The dependence of v_n on n comes through both r and p_j, $1 \leq j \leq n$. Implicit among the assumptions is that v_n and r tend to infinity with n. It should be noted that, under (3.35), $d(n) = 2^r$ and

$$4v_n^2 \leq (\log p_r) \sum_{j=1}^{r} \log p_j = (\log p_r)(\log n)$$

and so $v_n = o(\log n)$.

Proof of Theorem 15. Define the discrete random variables X_j, $1 \leq j \leq r$, as independent, where X_j takes the values 0 and $\log p_j$, each with probability $1/2$. The underlying probability space is immaterial. Writing k_j for 0 or 1,

$$Y_r = X_1 + X_2 + \cdots + X_r = \sum_{j=1}^{r} k_j \log p_j = \log(p_1^{k_1} p_2^{k_2} \cdots p_r^{k_r})$$

and thus the set of values of the sum Y_r is the set $\{\log d : d|n\}$. Furthermore, $P(Y_r \le z) = (1/d(n))v_n\{d : d|n, \log d \le z\}$, where P is the probability in the (undetermined) probability space for the X_j. Therefore, with $z = (\log n)/2 + xv_n$, $P(Y_r \le z) = F_n(x)$, and the theorem will be proved if we show that $E(Y_r) = (\log n)/2$, $V(Y_r) = v_n^2$, and Liapunov's condition applies to the X_j. But, with an appeal to (3.35),

$$E(Y_r) = \sum_{j=1}^{r} E(X_j) = \sum_{j=1}^{r} \tfrac{1}{2}\log p_j = \tfrac{1}{2}\log n$$

$$V(Y_r) = \sum_{j=1}^{r} V(X_j) = \sum_{j=1}^{r} (\tfrac{1}{2}(\log p_j)^2 - (\tfrac{1}{2}\log p_j)^2) = v_n^2$$

and, since $\log p_r = o(v_n)$,

$$\sum_{j=1}^{r} E(|X_j - \tfrac{1}{2}\log p_j|^3) = \sum_{j=1}^{r} \tfrac{1}{8}(\log p_j)^3 \le \tfrac{1}{2}(\log p_r)v_n^2 = o(v_n^3)$$

The proof is completed. ∎

Exercises

1. Assume that the distribution function $F_n(x)$ of X_n converges weakly to $F(x)$. Prove that $G_n(x) \to F(x)$ weakly as well, where $G_n(x)$ is the distribution function of Y_n and either (i) $Y_n = a_n X_n + b_n$, $a_n \to 1$ and $b_n \to 0$; or (ii) $P(X_n \ne Y_n) \to 0$; or (iii) $X_n - Y_n \to 0$ in probability.

2. Let X be uniformly distributed on $(0, 1)$ (with respect to Lebesgue measure). Let X_n be the nth decimal digit of X and let $Y_n(j) = 1$ if $X_n = j$ and $Y_n(j) = 0$ otherwise. Compute $E(X_n)$, $V(X_n)$, $E(Y_n(j))$, and $V(Y_n(j))$, and apply the classical central limit theorem both to $X_1 + X_2 + \cdots + X_n$ and to $Y_1(j) + Y_2(j) + \cdots + Y_n(j)$, $0 \le j \le 9$.

3. Let X and Y be independent and identically distributed random variables. Let $\varphi(t)$ be their common characteristic function. Assume that $E(X) = 0$ and $V(X) = 1$ and that $X + Y$ and $X - Y$ are independent. Deduce from the classical central limit theorem that X is standard normal by proving the following intermediate steps: show that $\varphi(t)$ satisfies the functional equations $\varphi(2t) = \varphi^3(t)\varphi(-t) = \varphi^2(t)|\varphi(t)|^2$ and $|\varphi(2t)|^2 = |\varphi(t)|^8$, i.e., with $\psi(t) = |\varphi(t)|^2$, $\psi(2t) = \psi^4(t)$.

4. Let X be a random variable with density function $f(x) = |x|^{-3}$ if $|x| \ge 1$ and $f(x) = 0$ for $-1 < x < 1$. Let X_j, $j \ge 1$, be independent random variables, each distributed as X. Set $S_n = X_1 + X_2 + \cdots + X_n$. Show that, even though $V(X) = +\infty$, $S_n/(n \log n)^{1/2}$ is asymptotically normal.

3.4 CHARACTERISTIC FUNCTIONS: INEQUALITIES

The inequalities established in the present section will be used in proofs in subsequent chapters. Inequality (iii) in Theorem 16 below also shows that characteristic functions are not only continuous but also uniformly continuous on the whole real line.

We use the notation $\operatorname{Re} z = x$ and $\operatorname{Im} z = y$ for the complex number $z = x + iy$.

Theorem 16 If $\varphi(t)$ is the characteristic function of a (proper) distribution function $F(x)$ then, for all real numbers t,

(i) $\quad 1 - \operatorname{Re} \varphi(2t) \le 4(1 - \operatorname{Re} \varphi(t))$

and

(ii) $\quad 1 - |\varphi(2t)|^2 \le 4(1 - |\varphi(t)|^2)$

Furthermore, if h also is real,

(iii) $\quad |\varphi(t + h) - \varphi(t)|^2 \le 2(1 - \operatorname{Re} \varphi(h))$

Proof. Starting with the elementary inequality

$$0 \le 2(1 - \cos u)^2 = 3 - 4 \cos u + \cos 2u$$

we get

$$1 - \cos 2u \le 4(1 - \cos u)$$

where u is an arbitrary real number. Hence,

$$1 - \operatorname{Re} \varphi(2t) = \int_{-\infty}^{+\infty} (1 - \cos 2tx)\, dF(x) \le 4 \int_{-\infty}^{+\infty} (1 - \cos tx)\, dF(x)$$

$$= 4(1 - \operatorname{Re} \varphi(t))$$

which is inequality (i). Inequality (ii) is a special case of inequality (i), as $|\varphi(t)|^2$ is a real-valued characteristic function.

To prove (iii), we apply the Cauchy–Schwarz inequality (see Exercise 6 of Section 2.2) in

$$|\varphi(t + h) - \varphi(t)|^2 = \left| \int_{-\infty}^{+\infty} e^{itx}(e^{ihx} - 1)\, dF(x) \right|^2$$

$$\le \left\{ \int_{-\infty}^{+\infty} dF(x) \right\}\left\{ \int_{-\infty}^{+\infty} |e^{ihx} - 1|^2\, dF(x) \right\}$$

$$= 2 \int_{-\infty}^{+\infty} (1 - \cos hx)\, dF(x) = 2(1 - \operatorname{Re} \varphi(h))$$

which is the asserted inequality in part (iii). The proof is complete. ∎

For the next inequality, we need the concept of a median of a distribution. The number m is called a median of the distribution function $F(x)$ if, for each real number $a \geq 0$, $F(m - a) \leq 1/2$ and $F(m + a) \geq 1/2$. An important property of the median is given below.

Lemma **(Symmetrization Inequalities)** Let X and Y be independent random variables with a common distribution function $F(x)$. Let $g(x) \geq 0$ be a function of the real variable x such that $g(x) = g(-x)$ and $g(x)$ is non-decreasing for $x \geq 0$. Then, for any median m of F.

$$2E(g(X - Y)) \geq E(g(X - m))$$

Remark Since the distribution function of $X - Y$ is symmetric about zero, the procedure of expressing a distributional property of X by means of the distribution of $X - Y$ is called symmetrization. Typical functions for $g(x)$ in the Lemma are (i) for a fixed $u \geq 0$, let $g(x) = 1$ if $|x| \geq u$ and zero otherwise—then we get $2P(|X - Y| \geq u) \geq P(|X - m| \geq u)$; (ii) $g(x) = x^2$ if $|x| \leq 1$ and equals 1 for $|x| \geq 1$; and (iii) $g(x) = x^2/(1 + x^2)$.

Proof of Lemma. It may be assumed that $m = 0$ (just replace X and Y by $X - m$ and $Y - m$). Now, since

$$E(g(X - Y)) = \int_{-\infty}^{+\infty} \int_{-\infty}^{+\infty} g(u - v) \, dF(u) \, dF(v)$$

by $g(x) \geq 0$, we decrease if we integrate only over a subset of the plane. Let $A = \{(u, v): u \geq 0, v \leq 0\}$ and $B = \{(u, v): u < 0, v \geq 0\}$. Note that, on both A and B, $g(u - v) \geq g(u)$. Hence,

$$E(g(X - Y)) \geq \iint_A g(u) \, dF(u) \, dF(v) + \iint_B g(u) \, dF(u) \, dF(v)$$

Now, since zero is a median, the integrals with respect to $F(v)$ are at least $1/2$; the right-hand side is further decreased if we write

$$\frac{1}{2} \int_{u \geq 0} g(u) \, dF(u) + \frac{1}{2} \int_{u < 0} g(u) \, dF(u) = \frac{1}{2} E(g(X))$$

This establishes the desired inequality when $m = 0$; thus, the proof is complete. ∎

We now return to inequalities involving characteristic functions.

Theorem 17 Let $F(x)$ be a distribution function whose characteristic function is $\varphi(t)$. Let $r > 0$ be a fixed finite number and set

$$a = a(r) = \int_{|x|<r} x \, dF(x) \qquad F^*(x) = F(x + a) \qquad \varphi^*(t) = e^{-ira}\varphi(t)$$

Then, for each real number $b > 0$, (i) there is a constant $c_1 = c_1(r, a, b) > 0$ such that

$$c_1 \max_{|t|\le b} |\varphi^*(t) - 1| \le \int_{-\infty}^{+\infty} \frac{x^2}{1 + x^2} \, dF^*(x)$$

and (ii) if m is a median of F and if $r > |m|$, then there exists a constant $c_2 = c_2(r, m, b) > 0$ such that

$$\int_{-\infty}^{+\infty} \frac{x^2}{1 + x^2} \, dF^*(x) \le c_2 \int_0^b (1 - |\varphi(t)|^2) \, dt$$

Proof. Since $|\exp[it(x - a) - 1]| \le 2$ and, for $|t| \le b$,

$$e^{it(x-a)} - 1 = it(x - a) + \vartheta \frac{b^2}{2}(x - a)^2 \qquad |\vartheta| \le 1$$

we get from the definition of $\varphi^*(t)$ and a,

$$|\varphi^*(t) - 1| = \left| \int_{-\infty}^{+\infty} (e^{it(x-a)} - 1) \, dF(x) \right| \le 2 \int_{|x|\ge r} dF(x)$$

$$+ b \left| \int_{|x|<r} (x - a) \, dF(x) \right| + \frac{b^2}{2} \int_{|x|<r} (x - a)^2 \, dF(x)$$

$$= (2 + |a|b) \int_{|x|\ge r} dF(x) + \frac{b^2}{2} \int_{|x|<r} (x - a)^2 \, dF(x)$$

Now, since, for $|x| \ne |a|$,

$$1 = \frac{1 + (x - a)^2}{(x - a)^2} \frac{(x - a)^2}{1 + (x - a)^2} \le \frac{1 + (|x| + |a|)^2}{(|x| - |a|)^2} \frac{(x - a)^2}{1 + (x - a)^2}$$

and since the first ratio on the extreme right-hand side is decreasing in $|x|$ for $|x| > |a|$, the inequality $|a| < r$, ensured by the definition of a, implies

$$\int_{|x|\ge r} dF(x) \le \frac{1 + (r + |a|)^2}{(r - a)^2} \int_{|x|\ge r} \frac{(x - a)^2}{1 + (x - a)^2} \, dF(x)$$

Furthermore,

$$\int_{|x|<r} (x-a)^2 \, dF(x) \le (1 + (r + |a|)^2) \int_{|x|<r} \frac{(x-a)^2}{1 + (x-a)^2} \, dF(x)$$

The combination of the appropriate inequalities thus yields

$$|\varphi^*(t) - 1| \le \frac{1}{c_1} \int_{-\infty}^{+\infty} \frac{x^2}{1 + x^2} \, dF^*(x)$$

where $c_1 > 0$ is easily made specific from the previous inequalities. This establishes the inequality asserted in part (i).

To prove the inequality in (ii), first notice that if $\varphi(t)$ is the characteristic function of X, then $|\varphi(t)|^2$ is the characteristic function of the symmetrized variable $X_1 - X_2$, where X_1 and X_2 are independent, each with distribution $F(x)$. Hence, upon denoting the distribution function of $X_1 - X_2$ by $F^{**}(x)$ and interchanging the orders of integration, we get

$$\int_0^b (1 - |\varphi(t)|^2) \, dt = \int_{-\infty}^{+\infty} \left[\int_0^b (1 - \cos tx) \, dt \right] dF^{**}(x)$$

$$= b \int_{-\infty}^{+\infty} \left(1 - \frac{\sin bx}{bx} \right) dF^{**}(x)$$

$$\ge \frac{b}{10} \int_{-\infty}^{+\infty} g(bx) \, dF^{**}(x)$$

where $g(z) = z^2$ if $|z| \le 1$ and equals one otherwise. Now, since $g(bx) \ge \min(1, b^2) g(x)$ and $g(x)$ satisfies the conditions in the Lemma, we have

$$\int_0^b (1 - |\varphi(t)|^2) \, dt \ge \frac{b}{20} \min(1, b^2) \int_{-\infty}^{+\infty} g(x) \, dF(x + m)$$

A further reduction is obtained by observing that $g(x) \ge x^2/(1 + x^2)$. Finally, we have to show that if we change $F(x + m)$ to $F^*(x) = F(x + a)$, no increase is affected if the coefficient of the integral is appropriately modified. To see this, we reverse the procedure of estimation; thus, we begin by integrating with respect to $F(x + a)$ and estimate from above. First, we have

$$\int_{-\infty}^{+\infty} \frac{x^2}{1 + x^2} \, dF(x + a) = \int_{-\infty}^{+\infty} \frac{(x-a)^2}{1 + (x-a)^2} \, dF(x)$$

$$\le \int_{|x|<r} (x-a)^2 \, dF(x) + \int_{|x|\ge r} dF(x)$$

Next, we use the elementary (trivial) inequality

$$(u + v)^2 \leq u^2 + 2uv + 2v^2 = u^2 + 2v(u + v)$$

with $u = x - m$ and $v = m - a$. We have (recall that $|m| < r$)

$$\int_{|x|<r} (x - a)^2 \, dF(x) \leq \int_{|x|<r} (x - m)^2 \, dF(x) + 4r \left| \int_{|x|<r} (x - a) \, dF(x) \right|$$

$$\leq \int_{|x|<r} (x - m)^2 \, dF(x) + 4r^2 \int_{|x| \geq r} dF(x)$$

To complete the proof, one can now proceed in the same manner as at the end of the proof of part (i); one gets that these last two integrals are further increased if their coefficients are appropriately modified and the integrands are changed to $(x - m)^2/[1 + (x - m)^2]$. This repetition of our argument is possible because $|m| < r$. We thus obtain

$$\int_{-\infty}^{+\infty} \frac{x^2}{1 + x^2} \, dF(x + a) \leq c^* \int_{-\infty}^{+\infty} \frac{x^2}{1 + x^2} \, dF(x + m)$$

where $c^* > 0$ and $c^* = c^*(m, r)$. The combination of our chain of estimates leads to the desired inequality. The theorem is established. ∎

Once again, before proceeding to the next inequality involving characteristic functions, we shall establish a widely applicable tool, known as the *Riemann–Lebesgue lemma*: if $g(x)$ is an integrable function on the real line, then

$$\lim \int_{-\infty}^{+\infty} e^{itx} g(x) \, dx = 0 \tag{3.36}$$

as t goes to plus or minus infinity.

The proof follows immediately. If $g(x)$ is a finite step function (i.e., a simple random variable on a finite interval, where probability is Lebesgue measure), then integration "brings t into the denominator" and the desired limit follows. On the other hand, if $g(x)$ is integrable, then it can be approximated by finite step functions and, thus, the previous special case would imply (3.36) for a general integrable function.

We shall now prove an important inequality due to Berry (1941) and Esseen (1945).

Theorem 18 Let F_1 and F_2 be distribution functions with characteristic functions φ_1 and φ_2, respectively. Assume that F_2 is differentiable for all

x and set $f_2(x) = F'_2(x)$. Then, for arbitrary real numbers v and $U > 0$,

$$2\left|\int_{-\infty}^{+\infty} [F_1(x+v) - F_2(x+v)] \frac{1 - \cos xU}{x^2 U} dx\right| \le \int_{-U}^{U} \left|\frac{\varphi_1(t) - \varphi_2(t)}{t}\right| dt$$

which implies

$$\sup|F_1(x) - F_2(x)| \le \frac{1}{\pi} \int_{-U}^{U} \left|\frac{\varphi_1(t) - \varphi_2(t)}{t}\right| dt + \frac{c}{U} \sup f_2(x)$$

where $c > 0$ is a constant and the suprema are taken for all x.

Proof. We go back to the proof of Theorem 8, in particular to formula (3.19). With the characteristic function $\psi(t/U)$, where $\psi(t)$ is defined in (3.16), we form the characteristic functions $\eta_{j,U}(t) = \varphi_j(t)\psi(t/U)$, $j = 1$ and 2. Then, as established in (3.19), the distributions corresponding to $\eta_{j,U}(t)$ have density functions $h_j(x; U), j = 1, 2$, and (3.19) applies. In addition, we know that $h_j(x; U)$ is the density of the convolution of F_j and the distribution whose density is $Ug(Ux)$, $g(x)$ having been determined in (3.17). Hence, we get from (3.19) that

$$\int_{-\infty}^{+\infty} [D(x-v) - D(y-v)] \frac{1 - \cos vU}{v^2 U} dv$$

$$= \frac{1}{2} \int_y^x \int_{-U}^{U} e^{-itz} \Delta(t)\psi(t/U) \, dt \, dz$$

where $D(w) = F_1(w) - F_2(w)$ and $\Delta(t) = \varphi_1(t) - \varphi_2(t)$. Now, if $\Delta(t)/t$ is not integrable, the first inequality of the theorem becomes trivially true. On the other hand, if it is integrable, we can interchange integrations on the right hand side and let $y \to -\infty$. We get, on the left-hand side by the dominated convergence theorem and on the right-hand side by the Riemann–Lebesgue lemma (formula (3.36)),

$$\int_{-\infty}^{+\infty} D(x-v) \frac{1 - \cos vU}{v^2 U} dv = \frac{1}{2} \int_{-U}^{U} e^{-itx} \frac{\Delta(t)}{-it} \psi(t/U) \, dt$$

which establishes the main inequality of the theorem.

In order to prove its stated implication, first note that if $d = \sup|D(x)| = 0$, then there is nothing to prove. The assertion is also evident if $f^* = \sup f_2(x) = +\infty$. So let $d > 0$ and $f^* < +\infty$. Then, since $D(x)$ is bounded and $D(\pm\infty) = 0$, there is a finite x_0 such that either $|D(x_0)| = d$ or $|D(x_0 - 0)| = d$. Also note that there is no change in the main inequality if $D(x)$ is replaced by $-D(x)$; thus, we may as well assume that $D(x_0) = d$

(the reader is invited to check that in all arguments that follow $D(x_0 - 0) = d$ would not require any change). Now, with arbitrary $r > 0$,

$$\left| \int_{-\infty}^{+\infty} \cdots \right| \geq \left| \int_{|x|<r} \cdots \right| - \left| \int_{|x|\geq r} \cdots \right| \qquad (3.37)$$

and

$$\left| \int_{|x|\geq r} D(x+v) \frac{1 - \cos xU}{x^2 U} \, dx \right| \leq \frac{2d}{U} \int_{|x|\geq r} \frac{dx}{x^2} = \frac{4d}{Ur} \qquad (3.38)$$

When integrating for $|x| < r$, we use the fact that v is arbitrary. We even specify r as $r = d/2f^*$ at this stage. We then choose $v = r + x_0$, which yields (since $|x| < r$)

$$x_0 < x + v < x_0 + 2r \qquad x + r > 0, \qquad x - r < 0$$

Next, by the mean value theorem of calculus,

$$F_2(x + v) = F_2(x_0) + \vartheta(x + v - x_0)f_2(x^*) \qquad |\vartheta| \leq 1, \quad x_0 < x^* < x + v$$

and, thus,

$$D(x + v) = F_1(x + v) - F_2(x + v) \geq F_1(x_0) - F_2(x_0) - f^*(x + r)$$
$$= d - f^*(x + r) = 2f^*r - f^*(x + r) = (r - x)f^* > 0$$

Hence

$$\int_{|x|<r} D(x+v) \frac{1 - \cos xU}{x^2 U} \, dx \geq f^* \int_{|x|<r} (r - x) \frac{1 - \cos xU}{x^2 U} \, dx$$

Note that the integral corresponding to x out of $r - x$ is zero by symmetry; thus, the right-hand side is further decreased by

$$f^* r \pi \left(1 - \int_{|x|\geq r} \frac{1 - \cos xU}{x^2 U} \, dx \right) \geq f^* r \pi \left(1 - \frac{4d}{Ur} \right)$$

where the last estimate of (3.38) was used. Finally, upon substituting $r = d/2f^*$ and combining this last result with (3.37) and (3.38), the second inequality of the theorem follows from the first one, which completes the proof. ∎

Exercises

1. Prove that if $|\varphi(t)| = 1$ in some neighborhood of the origin, where $\varphi(t)$ is a characteristic function, then it is the characteristic function of a degenerated distribution.

2. Assume that the characteristic function $\varphi(t)$ satisfies the tail estimate $|\varphi(t)| \le a < 1$ for $|t| \ge b$. Then, upon choosing n such that $b/2^n < |t| \le b/2^{n-1}$, prove, by repeated applications of Theorem 16, that $|\varphi(t)|^2 \le 1 - t^2(1 - a^2)/4b^2$ for $0 < |t| < b$. Hence, conclude that if $\varphi(t)$ is the characteristic function of a lattice variable then, as $t \to +\infty$ or $-\infty$, $\limsup|\varphi(t)| = 1$.

3. Show that if $F(x)$ has a density function then its characteristic function $\varphi(t) \to 0$ as $t \to +\infty$ or $-\infty$.

4. Estimate the error in the normal approximation to the binomial distribution by using the second inequality of Theorem 18.

3.5 MULTIVARIATE CHARACTERISTIC FUNCTIONS

Let $(X_1, X_2, \ldots, X_d)$ be a d-dimensional random vector. Then the function

$$\varphi(t_1, t_2, \ldots, t_d) = E\left[\exp\left(i \sum_{j=1}^{d} t_j X_j \right) \right]$$

defined for all real t_j, $1 \le j \le d$, is called the characteristic function of the vector above.

With no change in the one-dimensional arguments, we get all the basic properties of characteristic functions in higher dimensions. In particular, we have the uniqueness and continuity theorems, inversion formulas, the multiplication property for component by component summation of independent vectors, and others. In addition to these properties, we should mention that multivariate characteristic functions, just like the univariate ones, are bounded by one, equal to one at the origin, and are continuous for all points of the d-dimensional real space R^d. All these properties will be used freely below and will be referred to by the general term "basic properties."

Notice that, by definition, a characteristic function on R^d is uniquely determined by the collection of the univariate characteristic functions of all linear combinations of the components. Thus, from the uniqueness theorem we obtain

Theorem 19 The (multivariate) distribution of the random vector $(X_1, X_2, \ldots, X_d)$ is uniquely determined by the collection of the univariate distributions of the linear forms $a_1 X_1 + a_2 X_2 + \cdots + a_d X_d, a_j$ real, $1 \le j \le d$.

Theorem 19 gives us a convenient tool with which to define multivariate distributions. For example, one may define multivariate normality by the property that each linear form of the components is normally distributed. This definition, in fact, is equivalent to the more commonly used definition in terms of density functions (recall Section 2.9).

Another important consequence of the definition and the basic properties is a new characterization of independence.

Theorem 20 Let $(X_1, X_2, \ldots, X_d)$ and $(Y_1, Y_2, \ldots, Y_s)$ be two random vectors with characteristic functions φ_1 and φ_2. Let the characteristic function of the combined vector $(X_1, X_2, \ldots, X_d, Y_1, Y_2, \ldots, Y_s)$ be ψ. Then the two original vectors are independent if and only if

$$\psi(t_1, \ldots, t_d, t_{d+1}, \ldots, t_{d+s}) = \varphi_1(t_1, \ldots, t_d)\varphi_2(t_{d+1}, \ldots, t_{d+s})$$

Proof. The fact that, for independent vectors, the joint characteristic function splits as stated follows immediately from the definition of multivariate characteristic function and the product rule for expectations of independent variables. The converse, on the other hand, follows from the uniqueness theorem for characteristic functions. ∎

Example Let X and Y be independent standard normal variables. Let the constants a, b, c and d satisfy $ac + bd = 0$. Then $aX + bY$ and $cX + dY$ are independent.

Indeed, if we put $\varphi(t_1, t_2)$ for the characteristic function of the vector $(aX + bY, cX + dY)$, then, upon replacing "variable" by "vector" in (2.36),

$$\varphi(t_1, t_2) = \iint_{R^2} e^{it_1(ax+by) + it_2(cx+dy)} \frac{1}{2\pi} e^{-x^2/2 - y^2/2} \, dx \, dy$$

Now, if we separate the terms containing x and y in the exponent, i.e., if we write the exponent as

$$[ix(at_1 + ct_2) - x^2/2] + [iy(bt_1 + dt_2) - y^2/2]$$

we recognize that the double integral above splits into the product of two normal characteristic functions yielding

$$\varphi(t_1, t_2) = \{\exp[-(at_1 + ct_2)^2/2]\}\{\exp[-(bt_1 + dt_2)^2/2]\}$$
$$= \{\exp[-t_1^2(a^2 + b^2)/2]\}\{\exp[-t_2^2(c^2 + d^2)/2]\}$$

where the mixed terms $t_1 t_2$ canceled out. Hence, Theorem 20 applies and the desired independence follows. ∎

Exercises

1. Use bivariate characteristic functions to show that $\min(X, Y)$ and $d = \max(X, Y) - \min(X, Y)$ are independent, where X and Y are independent exponential variables with unit expectation.
2. Let $X_1, X_2, \ldots, X_n$ be independent and identically distributed normal variables. Show that the mean $M_n = (X_1 + X_2 + \cdots + X_n)/n$ and the sample variance $S_n^2 = (n - 1)^{-1} \sum_{j=1}^{n} (X_j - M_n)^2$ are independent.

3.6 LAPLACE TRANSFORMS

Laplace transforms are usually associated with nonnegative random variables, and we shall limit our discussion to this case. Thus, let X be a random variable with $P(X \geq 0) = 1$. Then

$$\text{Lap}(z) = E(e^{-zX}) = \int_0^{+\infty} e^{-zx} \, dF(x) \qquad z \geq 0 \text{ real}$$

is called the Laplace transform either of X or of its distribution function $F(x)$.

Most of the nice properties of characteristic functions are shared by Laplace transforms. Some of them are given in the following theorems.

Theorem 21 For every distribution function $F(x)$ with $F(0-) = 0$, $0 \leq \text{Lap}(z) \leq 1$, $\text{Lap}(0) = 1$ and continuous, where $z \geq 0$.

Proof. The first two properties are obvious from the definition, while the last one follows from the continuity of e^{-zx} and the dominated convergence theorem. ∎

Theorem 22 If X has Laplace transform $\text{Lap}(z)$, then the Laplace transform of $aX + b$ is $e^{-bz} \text{Lap}(az)$, where a and b are constants. Furthermore, if X and Y are independent, then the Laplace transform of $X + Y$ is the product of the transforms of X and Y.

Proof. This follows easily from the corresponding properties of the integral. ∎

Theorem 23 Let $X \geq 0$ be random variable with distribution function $F(x)$ and Laplace transform $\text{Lap}(z)$. Then, for $z > 0$, $\text{Lap}(z)$ is differentiable any number of times and

$$(-1)^k \text{Lap}^{(k)}(z) = \int_0^{+\infty} e^{-zx} x^k \, dF(x) \tag{3.39}$$

In addition, if the moment $m_k = E(X^k)$ is finite, then (3.39) also applies at $z = 0$, yielding

$$(-1)^k \, \mathrm{Lap}^{(k)}(0) = m_k \tag{3.40}$$

and the inequalities

$$\sum_{j=0}^{2n-1} \frac{(-1)^j m_j z^j}{j!} \le \mathrm{Lap}(z) \le \sum_{j=0}^{2n} \frac{(-1)^j m_j z^j}{j!} \tag{3.41}$$

hold for $2n \le k$.

Proof. First we prove (3.39) for $k = 1$. For $z > 0$, let h be a real number such that $z + h > 0$. Then

$$\frac{\mathrm{Lap}(z + h) - \mathrm{Lap}(z)}{h} = \int_0^{+\infty} \frac{e^{-(z+h)x} - e^{-zx}}{h} \, dF(x)$$

$$= -\int_0^{+\infty} x e^{-z'x} \, dF(x)$$

where the last equation follows from the mean value theorem of calculus and z' lies between z and $z + h$ (hence, $z' > 0$). By letting $h \to 0$, the dominated convergence theorem yields (3.39) with $k = 1$. Assuming that (3.39) is valid for $k - 1$ and repeating the preceding argument, we find that (3.39) obtains for general k by the method of induction. If m_k is finite, the asserted formula for m_k follows from the dominated convergence theorem if we let $z \to 0$ in (3.39). Finally, since for every $n > 0$ and for $xz > 0$,

$$\sum_{j=0}^{2n-1} \frac{(-1)^j x^j z^j}{j!} < e^{-xz} < \sum_{j=0}^{2n} \frac{(-1)^j x^j z^j}{j!}$$

integration with respect to $F(x)$ leads to (3.41). The theorem is established. ∎

The inequalities in (3.41) imply that if the moments m_k are finite for all k, then

$$\mathrm{Lap}(z) = \sum_{j=0}^{+\infty} (-1)^j m_j \frac{z^j}{j!} \tag{3.42}$$

whenever the right-hand side converges. This, of course, also follows from Taylor's expansion because of (3.40).

In Theorem 22, the Laplace transform of $X + Y$ is expressed by means of $\mathrm{Lap}_X(z)$ and $\mathrm{Lap}_Y(z)$, the Laplace transforms of X and Y. An easy calculation shows that the same product rule applies if $X + Y$ is replaced by convolution. That is, if $\text{Lap}_c(z)$ is the Laplace transform of the convolution of X and Y, then

$$\text{Lap}_c(z) = \text{Lap}_X(z)\text{Lap}_Y(z) \tag{3.43}$$

We shall now list the Laplace transforms of some well-known distributions. Instead of repeating earlier computations, we observe that, formally, the Laplace transform of F is its characteristic function at $t = iz$. We thus have

Exponential distribution: $F(x) = 1 - e^{-ax}$, $x \geq 0$.

$$\text{Lap}(z) = \frac{1}{1 + \dfrac{z}{a}}$$

Binomial distribution: $F(x)$ binomial with parameters n and p

$$\text{Lap}(z) = [1 + p(e^{-z} - 1)]^n$$

Poisson distribution: $F(x)$ Poisson with parameter a

$$\text{Lap}(z) = \exp[a(e^{-z} - 1)]$$

Gamma distribution: $F(x) = [a^n/(n-1)!] \int_0^x u^{n-1} e^{-au}\, du, \ x \geq 0$

$$\text{Lap}(z) = \left(\frac{1}{1 + \dfrac{z}{a}}\right)^n \quad n \geq 1$$

This can be obtained directly from the case of the exponential distribution via the convolution formula (3.43).

Uniform distribution on $(0, 1)$: $F(x) = x$, $0 \leq x \leq 1$

$$\text{Lap}(z) = \frac{1 - e^{-z}}{z}$$

Degenerate distribution at $a \geq 0$: $F(x) = 0$ if $x < a$ and $= 1$ if $x \geq a$

$$\text{Lap}(z) = e^{-az}$$

The power of Laplace transforms as a tool in distribution theory is the same as that of characteristic functions: the uniqueness and continuity theorems apply.

Theorem 24 The distribution function $F(x)$ of a random variable $X \geq 0$ is uniquely determined by its Laplace transform $\text{Lap}(z)$. In fact, the sequence

Lap(n), $n \geq 0$ integer, uniquely determines $F(x)$. An explicit method of computing $F(x)$ from Lap(z) is the inversion formula: for continuity points x of $F(x)$,

$$F(x) = \lim_{a = +\infty} \sum_{n \leq ax} (-1)^n \operatorname{Lap}^{(n)}(a) \frac{a^n}{n!} \qquad (3.44)$$

Proof. Introduce the random variable $Y = e^{-X}$. Then $0 \leq Y \leq 1$, $Y^n = e^{-nX}$; thus, $E(Y^n) = \operatorname{Lap}(n)$. Since the moments $m_n = E(Y^n)$, $n \geq 0$, uniquely determine the distribution function $G(y)$ of Y (Corollary 5 of Section 3.2), the relation $F(x) = 1 - G(e^{-x})$, which is valid for continuity points x of $F(x)$, thus ensures that the sequence $\operatorname{Lap}(n)$, $n \geq 0$ integer, uniquely determines $F(x)$.

In order to prove (3.44), we observe that, because of (3.39),

$$\sum_{n \leq ax} (-1)^n \operatorname{Lap}^{(n)}(a) \frac{a^n}{n!} = \int_0^{+\infty} \left(\sum_{n \leq ax} e^{-ay} \frac{(ay)^n}{n!} \right) dF(y)$$

where the sum under the integral sign can also be written as

$$P(U \leq ax) \qquad U \text{ Poisson with } E(U) = ay$$

However, by Chebyshev's inequality, as $a \to +\infty$, $P(U \leq ax)$ converges to 0 or 1 depending on whether $x < y$ or $x > y$. Hence, (3.44) follows by the dominated convergence theorem, which completes the proof. ∎

Theorem 25 Let F_n with $F_n(0-) = 0$ be a distribution function whose Laplace transform is $\operatorname{Lap}_n(z)$. If $F_n \to F$ weakly, then $\operatorname{Lap}_n(z) \to \operatorname{Lap}(z)$, where $\operatorname{Lap}(z)$ is the Laplace transform of F. Conversely, if $\operatorname{Lap}_n(z) \to h(z)$ for $z > 0$ and $h(z) \to 1$ as $z \to 0$, then $h(z)$ is the Laplace transform of a proper distribution function F and $F_n \to F$ completely.

Proof. The first part of the theorem is an immediate consequence of the Helly–Bray theorem (Theorem 2.26). To prove the converse part, we thus assume that $\operatorname{Lap}_n(z) \to h(z)$. By the weak compactness theorem (Theorem 2.24), there is a subsequence F_{n_k} of F_n that converges weakly to some distribution F^*. By the Helly–Bray theorem again, $\operatorname{Lap}_{n_k}(z)$ converges to the Laplace transform $\operatorname{Lap}^*(z)$ of F^*, while, by assumption, the limit is $h(z)$. That is,

$$h(z) = \int_0^{+\infty} e^{-xz} \, dF^*(x), \qquad z > 0$$

Let $z \to 0$. By assumption, $\lim h(z) = 1$; thus, the dominated convergence theorem ensures that F^* is a proper distribution function whose Laplace transform is $h(z)$. By the uniqueness theorem of Laplace transforms, $h(z)$ determines F^*; thus, every weakly convergent subsequence of F_n converges to the same proper distribution function F^*. Consequently, $F_n \to F^*$ completely. The theorem is established.   ■

Exercises

1. Show that if Lap(z) is the Laplace transform of a distribution function F, then, for each $a > 0$, the sequence $p_n = (-1)^n \text{Lap}^{(n)}(a)a^n/n!$, $n \geq 0$, is a discrete distribution on the nonnegative integers and, consequently, Lap($a - ae^{-z/a}$) is another Laplace transform (as a function of $z \geq 0$). Let $a \to +\infty$, and obtain (3.44) again, via Theorem 25.
2. Let X_1, X_2,... be independent and identically distributed exponential variables. Let $N \geq 1$ be a geometric variable, i.e., $P(N = k) = (1 - p)^{k-1}p$, $0 < p < 1$, $k \geq 1$. Assume that N is independent of the X_j. Show that $X_1 + X_2 + \cdots + X_N$ is exponential also.
3. Let X_1, X_2,... be independent and identically distributed nonnegative random variables with $E(X_1)$ finite. Show, by using Laplace transforms, that $(X_1 + X_2 + \cdots + X_n)/n \to E(X_1)$ in probability.

3.7 GENERATING FUNCTIONS

Let X be a discrete random variable whose values are nonnegative integers with distribution $p_k = P(X = k)$. The function of the complex variable z

$$G(z) = p_0 + p_1 z + p_2 z^2 + \cdots + p_n z^n + \cdots$$

is called the generating function of the distribution $\{p_k\}$ or of X. It is well defined for at least $|z| \leq 1$, and $G(1) = 1$ for every distribution. The function $G(z)$ is differentiable any number of times in the unit circle and

$$G^{(n)}(0) = n!p_n \qquad n \geq 1$$

We therefore have that $G(z)$ uniquely determines the distribution $\{p_k\}$.

Note that $G(e^{it})$ is the characteristic function $\varphi(t)$ of X and $G(e^{-u})$ is the Laplace transform Lap(u) of $\{p_k\}$. From previous continuity theorems, we have, therefore, that, if $G_n(z)$ is a sequence of generating functions, then the convergence of $G_n(z)$ to $G(z)$ such that $G(z) \to 1$ as $z \to 1$ through values $|z| \leq 1$ ensures that the respective distributions $\{p_{kn}: k \geq 0\}$, $n \geq 1$, converge to a probability distribution. We refer to this statement as the continuity theorem for generating functions.

The usual product rule for convolutions also applies to generating functions. As a matter of fact, if X and Y are random variables taking the nonnegative integers with distributions $\{p_k\}$ and $\{r_k\}$, respectively, and if their generating functions are $G_X(z)$ and $G_Y(z)$, then the generating function $G_c(z)$ of the convolution

$$s_n = p_0 r_n + p_1 r_{n-1} + \cdots + p_{n-1} r_1 + p_n r_0 \tag{3.45}$$

satisfies

$$G_c(z) = G_X(z) G_Y(z) \tag{3.46}$$

The equivalence of (3.45) and (3.46) is simply the Cauchy rule of multiplying two (infinite) series.

We record two additional simple, but significant, formulas for generating functions. First, if the generating function of X is $G_X(z)$ and if $E(X^k)$ is finite, then

$$E[X(X-1)(X-2)\cdots(X-k+1)] = G_X^{(k)}(1) \tag{3.47}$$

where the right-hand side is the limit of the kth derivative as $z \to 1$ with $z < 1$ (z may be chosen to be real for this purpose).

The proof of (3.47) is simple and left to the reader.

The second formula is about sums of random size. Let $X_1, X_2, \ldots$ be independent nonnegative integer valued random variables, each with the same generating function $G(z)$. Let $N \geq 0$ be an integer-valued random variable with generating function $H(z)$ which is distributed independently of the X_j. Then the generating function $R(z)$ of $X_1 + X_2 + \cdots + X_N$ satisfies (the sum is 0 if $N = 0$)

$$R(z) = H(G(z)) \tag{3.48}$$

For the proof of (3.48) it is convenient to interpret the generating function $G^*(z)$ of a random variable Y as $G^*(z) = E(z^Y)$ with $0^0 = 1$. Then, as a consequence of the total probability rule

$$R(z) = E(z^{X_1 + X_2 + \cdots + X_N}) = \sum_{k=0}^{+\infty} E(z^{X_1 + \cdots + X_N} | N = k) P(N = k)$$

$$= \sum_{k=0}^{+\infty} [E(z^{X_1})]^k P(N = k) = \sum_{k=0}^{+\infty} G^k(z) P(N = k)$$

in which we recognize H at the variable $G(z)$.

The following examples are either representative applications of, or supplements to, the material discussed so far.

Example 1 If X is Poisson with parameter $a > 0$, then its generating function $G(z) = e^{a(z-1)}$.

By the uniqueness property of generating functions and the convolution formula (3.46), we have that sums of independent Poisson variables are Poisson. By (3.48), random sums of Poisson variables, which are called *compound Poisson distributions*, have generating functions of the form $H(\exp(a(z - 1)))$, where $H(\cdot)$ is a generating function. Finally, from (3.47),

$$E[X(X - 1)(X - 2)\cdots(X - k + 1)] = a^k \qquad k \geq 1 \qquad \blacksquare$$

Example 2 If $X = 0$ or 1 with $P(X = 1) = p$, i.e., if X is an indicator variable, then its generating function $G(z) = 1 - p + pz$. If $X_1, X_2, \ldots$ are independent indicator variables with $p_j = P(X_j = 1)$, then the generating function of $X_1 + X_2 + \cdots + X_n$,

$$U_n(z) = \prod_{j=1}^{n} [1 + p_j(z - 1)]$$

If $\max_{1 \leq j \leq n} p_j \to 0$ and $p_1 + p_2 + \cdots + p_n \to a\ (0 < a < +\infty)$ as $n \to +\infty$, then the distribution of $X_1 + X_2 + \cdots + X_n$ is asymptotically Poisson. Indeed, since the p_j are uniformly close to 0,

$$\log U_n(z) = \sum_{j=1}^{n} \log[1 + p_j(z - 1)] = (1 + o(1)) \sum_{j=1}^{n} p_j(z - 1)$$

which converges to $a(z - 1)$. Thus, $U_n(z)$ converges to the Poisson generating function $\exp[a(z - 1)]$, which ensures, by the continuity theorem, the asserted limit.

In particular, if $p_j = p$ for all j, then $U_n(z) = [1 + p(z - 1)]^n$ becomes the generating function of the binomial distribution and the stated limit theorem reduces to the Poisson approximation to the binomial distribution. $\blacksquare$

Example 3 If X_n is binomial with parameters n and p and N is a Poisson variable independent of X_n, then X_N is Poisson.

Indeed, X_n is the sum of independent indicator variables whose generating function $G(z) = 1 + p(z - 1)$. Hence, by (3.48), the generating function of X_N is $\exp[a(G(z) - 1)] = e^{ap(z - 1)}$, which is Poisson, as asserted. $\blacksquare$

Example 4 Let X_n be uniformly distributed on the integers $0, 1, \ldots, n$; i.e., $P(X_n = k) = 1/(n + 1), 0 \leq k \leq n$. Then, for every $k \geq 0$ fixed, $P(X_n = k) \to 0$ as $n \to +\infty$; thus, the limit of the distribution of X_n is not a probability distribution. This is reflected in the generating functions, in that, while $G_n(z) = (1 + z + \cdots + z^n)/(n + 1)$, the generating function of X_n, converges to a function $G(z)$ (which equals 0 for $|z| < 1$ and 1 for $z = 1$), $G(z)$ fails to converge to 1 as $z \to 1$ with $z < 1$. Hence, our requirement on $G(z)$ around $z = 1$ in the continuity theorem is not superfluous. ∎

In order to use easily established recursive formulas to evaluate certain probabilities, or even to evaluate some generating functions, it proves useful to associate with an arbitrary sequence q_n of probabilities the power series

$$p(s) = q_0 + q_1 s + q_2 s^2 + \cdots + q_n s^n + \cdots$$

where s is complex and the series $p(s)$ converges for at least $|s| < 1$, although it might diverge at $s = 1$. It has no relation to characteristic functions or Laplace transforms.

A typical example in which $p(s)$ can be employed with success is the problem of the return to the origin of a random walk. We say that $S_n, n \geq 1$, is a *random walk* if $S_n = X_1 + X_2 + \cdots + X_n$, where the X_j are independent and identically distributed random variables and X_j takes the values $+1$ and -1 only, with $P(X_j = 1) = p$. The random walk $S_n, n \geq 1$, is called symmetric if $p = 1/2$ (the literature uses the term "random walk" more freely, but in this book we shall use our definition). Let $q_n = P(S_n = 0)$ and let f_n be the probability that n is the first time that $S_n = 0$. Then $q_{2k+1} = f_{2k+1} = 0$ and

$$q_{2n} = f_2 q_{2n-2} + f_4 q_{2n-4} + \cdots + f_{2n} q_0 \qquad n \geq 1$$

where we set $q_0 = 1$. We also add the convention $f_0 = 0$. Thus, if $Q(s)$ and $F(s)$ are the power series associated with the sequences $\{q_{2n}\}$ and $\{f_{2n}\}$, respectively, then the above relation yields, for $|s| < 1$,

$$Q(s) = 1 + F(s)Q(s) \qquad \text{i.e.,} \qquad Q(s) = 1/[1 - F(s)]$$

This yields a kind of "if and only if" version of the Borel–Cantelli lemma for the events $\{S_n = 0\}$. That is, if $Q(1)$ is finite, then the Borel–Cantelli lemma (Theorem 1.10) entails that, with probability one, there are only finitely many n such that $S_n = 0$. On the other hand, if $Q(1) = +\infty$, the relation just established implies that $F(1) = 1$, i.e., with probability one, $S_n = 0$ infinitely many times. Now, since

$$q_{2n} = \binom{2n}{n} p^n (1-p)^n,$$

one easily gets from Stirling's formula $n! \sim C(n/e)^n n^{1/2}$ (see the introductory volume, pp. 70–73, for the actual computation) that q_{2n} is of the magnitude

$$\frac{[4p(1-p)]^n}{n^{1/2}}$$

Hence, the inequality $4p(1-p) < 1$, if $0 \leq p \leq 1$ but $p \neq 1/2$, yields that $Q(1) = +\infty$ if and only if the random walk $S_n, n \geq 1$, is symmetric.

Exercises

1. Determine the generating function of the geometric distribution $(1-p)^{k-1}p$, $0 < p < 1$, $k \geq 1$. Let $X_1, X_2, \ldots$ be independent geometric variables with the same parameter p. Find the distribution of $X_1 + X_2 + \cdots + X_n$ (which is called the negative binomial distribution) by means of generating functions. Give conditions under which the negative binomial distribution converges to a Poisson distribution.

2. In a random walk, define t_n and $t_n^{(2)}$ as the probabilities that n is the first time that the random walk reaches 1 and 2, respectively. Establish the relations

$$t_n^{(2)} = t_1 t_{n-1} + t_3 t_{n-3} + \cdots + t_{n-1} t_1$$

and

$$t_1 = p \quad \text{and} \quad t_n = (1-p)t_{n-1}^{(2)} \quad (n \geq 3)$$

Deduce from these relations that the power series $T(s)$ associated with $\{t_{2k+1}\}$ has the explicit form

$$T(s) = \frac{1 - [1 - 4p(1-p)s^2]^{1/2}}{2(1-p)s}$$

From $T(s)$ compute t_{2k+1}. Analyze the result given by $T(1)$.

3. Let f_{2n} be defined as in the text for a random walk. Then, with the notation of the previous exercise, establish $f_{2n} = (1-p)t_{2n-1} + pt_{2n-1}^*$, where t_k^* is as t_k when p and $1-p$ are changed into each other (thus representing the probability of first reaching -1 in k time units). Hence, deduce, from the form for $T(s)$,

$$F(s) = 1 - [1 - 4p(1-p)s^2]^{1/2}$$

Conclude from the form of $F(s)$ that, for a symmetric random walk, $Q(s) = (1 - s^2)^{-1/2}$.

3.8 SPECIAL DISTRIBUTIONS: PART II

We continue our study of special distributions that we started in Section 2.9. We concentrate on two kinds of distributions: (i) those of functions of random variables introduced in Section 2.9 and (ii) convolutions of known distributions.

Lognormal distribution The lognormal distributions have come up in examples and exercises of the previous sections. We say that X has a lognormal distribution if $P(X > 0) = 1$ and $Y = \log X$ is a normal variable. That is, with some $Y \in N(m, \sigma)$, $X = e^Y$. Hence, for $x > 0$,

$$F(x) = P(X \leq x) = P(Y \leq \log x) = \frac{1}{\sqrt{2\pi}\sigma} \int_{-\infty}^{\log x} e^{-(y-m)^2/2\sigma^2}\, dy$$

and its density function

$$f(x) = F'(x) = (\sqrt{2\pi}\sigma x)^{-1} \exp[-(\log x - m)^2/2\sigma^2], \qquad x > 0,$$

and both $f(x)$ and $F(x)$ equal zero for $x \leq 0$.

From the mathematical point of view, the most striking property of a lognormal distribution is what we have observed in Section 3.2: the moments of X do not determine $F(x)$. Therefore, limit theorems in which the limiting distribution is lognormal cannot be proved by the method of moments. In spite of this drawback, the lognormal distributions play a very important role in limit theorems. The following result is widely applicable: Let U_1, $U_2, \ldots$ be independent and identically distributed positive random variables with $m = E(\log U_1)$ and $\sigma^2 = V(\log U_1)$ finite. Then, with any constant $c > 0$,

$$Y_n = cA_n(U_1 U_2 \cdots U_n)^{1/B_n}$$

with $A_n = \exp(-m\sqrt{n}/\sigma)$ and $B_n = \sigma\sqrt{n}$ has a lognormal limiting distribution. This result is a direct consequence of the central limit theorem by turning to logarithm.

Note that we wrote a lognormal variable in the standard form. Upon replacing X by $X - a$ in the definition, where a is an arbitrary constant, a third parameter, a, enters the distribution which gives a large variety to its density. This variation of the density, just as for the Weibull distribution (Section 2.9), contributed to the wide acceptance of lognormality in applied models (whenever the decision on choosing or accepting an underlying distribution for a model is based on statistical observations). In particular, the lognormal distributions are frequently used in engineering and in insurance.

From the properties of the normal distribution, many properties of the lognormal distribution can be deduced with little effort. For example, for a standard lognormal distributed variable X, X^u is lognormal for any constant $u > 0$. Furthermore, for independent standard lognormal variables X_1 and X_2, $X_1^u X_2^v$, $u > 0$, $v > 0$, also is lognormal.

Pareto distribution Let Y be an exponential variable with $E(Y) = 1/a$. Then $X = e^Y$ has the distribution function

$$F(x) = P(X \leq x) = P(Y \leq \log x) = 1 - e^{-a(\log x)} = 1 - 1/x^a$$

for $x \geq 1$ and $F(x) = 0$ otherwise. We call X and $F(x)$ a standard Pareto variable and distribution function, respectively. When we want to emphasize the parameter a, we write $X_{(a)}$. Two additional parameters can be introduced by transforming x in $F(x)$ to $(x - A)/B$, $B > 0$.

In spite of the close relation between the Pareto and exponential variables, it was not the mentioned transformation that induced its initial application in statistics, in particular, in econometrical models. Pareto offered this distribution to explain income distribution in a society: he believed that the proportion of the population with income exceeding x is decreasing as x^{-a}, i.e., that the income size distribution is "Pareto" (the name came later). Pareto decleared it a universal law and it is being debated ever since. In favor of the law two properties, deduced from the exponential, are offered here: (i) if X_1 and X_2 are independent Pareto variables, then $\min(X_1, X_2)$ also is Pareto. Indeed, upon writing $X_j = \exp(Y_j)$, Y_j exponential, the claim follows from the observation that $\min(X_1, X_2) = \exp[\min(Y_1, Y_2)]$. One can interpret this property as saying that even the poorer in an income group is in that income group. (ii) $P(X > uv \mid X > u) = P(X > v)$, $u, v \geq 1$, which is the transformation of the lack of memory (this, in fact, is an "if and only if" statement). The interpretation of (ii) for income levels is that within an income group the growth of income level is the same as the growth in the whole population.

Note that not all moments of X exist. In fact, for $0 < a \leq 1$, $E(X_{(a)}) = +\infty$. Any difficulty arising from the nonfiniteness of some moments can be overcome by turning to X^u, $u > 0$, which is Pareto, too. By choosing u properly, any moment can be made finite. One can also utilize that X^u, $u < 0$, although no longer Pareto distributed, is a bounded random variable. In particular, $X_{(a)}^{-a}$ is uniform on $(0, 1)$.

Gamma distributions   Let $Y_1, Y_2, \ldots, Y_n$ be independent and identically distributed exponential variables with $E(Y_j) = 1/a$. Then, from the convolution formula (3.7) (see also Corollary 1) we get by induction that the density function $f(x; n, a)$ of $X = X_n = Y_1 + Y_2 + \cdots + Y_n$ has the form

$$f(x; n, a) = \frac{a^n x^{n-1} e^{-ax}}{\Gamma(n)} \qquad x \geq 0$$

where

$$\Gamma(n) = \int_0^{+\infty} x^{n-1} e^{-x} \, dx$$

known as the gamma function. Because of the presence of this function in the density above, $f(x; n, a)$ is called a gamma density. Since it is a density function for every positive real number n, not just integers, in the general definition of a gamma density n is not assumed to be an integer. In order to make a distinction, a gamma distribution is frequently called an *Erlang distribution* when n is an integer.

Note that if Y is a standard normal variable, whose distribution function is denoted by $N(x)$, then $X = Y^2$ is a gamma variable. Indeed, by symmetry, $P(X \le x) = P(Y \le \sqrt{x}) - P(Y \le -\sqrt{x}) = 2N(\sqrt{x}) - 1$, and thus the density of X, $f(x) = N'(\sqrt{x})/\sqrt{x} = (2\pi x)^{-1/2}e^{-x/2} = f(x; 1/2, 1/2)$.

Upon appealing once more to convolutions, or using characteristic functions, we find that if X_j are independent gamma variables with the same a-parameter but not necessarily identical n-parameters, then the sum of X_j is gamma. It thus follows that if $Y_1, Y_2, \ldots$ are independent standard normal variables then $T_n = Y_1^2 + Y_2^2 + \cdots + Y_n^2$ is a gamma variable. Since in the just mentioned convolution the n-parameters add up in a summation of gamma variables, T_n has the density $f(x; n/2, 1/2)$. Hence, for $n = 2$, T_2 is an exponential variable and, for even n, T_n is Erlangian. Because of the significance of T_n in statistics, $f(x; n/2, 1/2)$ with integer n has the additional name of *chi-squared distribution* with n degrees of freedom.

We conclude the present section with a method for generating multivariate distributions from univariate marginals. Let $F(x_1, x_2, \ldots, x_t)$ be a t-dimensional distribution function. We know from Section 1.8 that the univariate marginals $F_1(x_1), F_2(x_2), \ldots, F_t(x_t)$ are uniquely determined by F by letting $t - 1$ variables x_j tend to infinity. Write

$$F(x_1, x_2, \ldots, x_t) = D(F_1(x_1), F_2(x_2), \ldots, F_t(x_t)) \tag{3.49}$$

where $D(u_1, u_2, \ldots, u_t)$, $0 \le u_j \le 1$, $1 \le j \le t$, is an appropriate function. In order to see that it is a well defined function in most cases, first assume that F is such that each $F_j(x_j) = u_j$ is a strictly increasing, continuous function in x_j. Then there is a well defined inverse function $x_j = F_j^{-1}(u_j)$, and $F_j(F_j^{-1}(u_j)) = u_j$. We thus have that $D(u_1, u_2, \ldots, u_t)$ is a t-dimensional distribution function whose univariate marginals are uniformly distributed on $(0, 1)$. We shall call this D the dependence function of F. The dependence function D is unique for a given F in our particular case. If we keep continuity of each F_j as a requirement but drop the strict monotonicity, the uniqueness of D still follows, which proof is left to the reader. When some F_j have points of discontinuity, we cannot expect the uniqueness of D since $F_j(x_j)$ does not "assign" any value to $(u_j - h, u_j)$ if $F_j(x_j - 0) = u_j - h < u_j = F_j(x_j)$. In such cases, D can be somewhat arbitrary. Since we restrict our discussion to the case of strictly increasing continuous marginals F_j of F, we invite the reader to analyze (3.49) for various discontinuous marginals. We add that if D is a t-dimensional distribution function with marginals uniformly distributed on

(0, 1), then with arbitrary univariate distribution functions $F_j(x_j)$, $1 \leq j \leq t$, (3.49) defines a t-dimensional distribution function F.

Let us look at some particular examples. If $D(u_1, u_2, \ldots, u_2) = u_1 u_2 \cdots u_t$, then we get the case of independence, whatever the marginals $F_j(x_j)$. Next, let us go back to the bivariate exponential distribution of Section 2.9. For simplicity, we choose each $\lambda_j = 1$, obtaining (we set $x_1 = x$, $x_2 = y$)

$$F(x, y) = 1 - e^{-2x} - e^{-2y} + e^{-x-y-\max(x,y)}.$$

Hence, $F_1(x) = 1 - e^{-2x}$, $F_2(y) = 1 - e^{-2y}$, $x, y \geq 0$, entailing

$$F(x, y) = F_1(x) + F_2(y) - 1$$
$$+ \{[1 - F_1(x)][1 - F_2(y)]\min([1 - F_1(x)][1 - F_2(y)])\}^{1/2},$$

that is,

$$D(u_1, u_2) = u_1 + u_2 - 1 + [(1 - u_1)(1 - u_2)\min(1 - u_1, 1 - u_2)]^{1/2}.$$

We can now use this D function as a dependence function with standard normal marginal distributions, yielding a bivariate distribution function whose marginals are normal but its dependence structure is different from the bivariate normal distribution of Section 2.9. Actually the dependence function of the bivariate normal distribution function of Section 2.9 cannot be given in an explicit form using elementary functions.

A widely used dependence function

$$D(u_1, u_2) = u_1 u_2 [1 + \alpha(1 - u_1)(1 - u_2)] \qquad -1 \leq \alpha \leq 1,$$

which is known as the Morgenstern system. Any pair of marginals will generate a bivariate distribution function with desired marginal distributions.

Our final example is a dependence function which appears in the literature with a variety of marginals:

$$D(u_1, u_2, \ldots, u_t) = \left(\frac{1}{u_1} + \frac{1}{u_2} + \cdots + \frac{1}{u_t} - t + 1\right)^{-1}$$

Its appeal is its symmetry and simplicity. It also has the property that its lower dimensional marginals have the same structure as the one in t-dimensions.

While the method presented here is very effective for generating multi-dimensional distributions, it is not a typical method of probability theory. In probability theory one prefers the introduction of new distributions via probabilistic arguments such as limit theorems, characterizations, or by simply computing the distribution of some specific vectors. However, the introduction of dependence functions is a significant contribution to our understanding of multivariate distributions.

Exercises

1. Let X be a lognormal variable. Show that $(X^u - 1)/u$ is asymptotically normally distributed as $u \to 0$, $u > 0$.

2. Show that for independent and identically distributed lognormal variables X and Y, XY and X/Y are independent. On the other hand, if the common distribution of X and Y is gamma with identical a-parameters, show that $X + Y$ and X/Y are independent.

3. Let Y be a unit exponential variable and let $X = \log(e^Y - 1)$. Show that X has the *logistic distribution function* $F(x) = 1/(1 + e^{-x})$. Prove the following properties of the logistic variables: (i) let $X_1, X_2, \ldots, X_n$ be independent and identically distributed random variables and let x be an arbitrary real number. Define the indicator variables $I_j(x) = 1$ if $X_j \leq x$ and $I_j(x) = 0$ otherwise. Set $S_n(x) = I_1(x) + I_2(x) + \cdots + I_n(x)$. Show that the X_j are logistic variables if, and only if, $(dE(S_n(x))/dx)/V(S_n(x))$ depends neither on n nor on x; and (ii) the distribution function $F(x)$, assumed continuous and symmetric about the origin, of a random variable X is logistic if, and only if, $P(X \geq -x \mid X \leq x)$, $x > 0$, is an exponential distribution function.

4. Let $F(x, y)$ be defined as $\exp(-e^{-x} - e^{-y}/2)$ if $y > x$ and $\exp(-e^{-y} - e^{y-2x}/2)$ if $y \leq x$. Determine the univariate marginals and the dependence function $D(u_1, u_2)$ of $F(x, y)$. Find $D^k(u_1^{1/k}, u_2^{1/k})$.

REFERENCES

Berry, A. C. (1941). The accuracy of the Gaussian approximation to the sum of independent variates. *Trans. Amer. Math. Soc. 49*, 122–136.

Chan, L. K. (1967). On a characterization of distributions by expected values of extreme order satistics. *Amer. Math. Monthly 74*, 950–951.

Elliott, P. D. T. A. (1970). On the mean value of $f(p)$. *Proc. London Math. Soc. (3) 21*, 28–96.

Esseen, C. G. (1945). Fourier analysis of distribution functions. A mathematical study of the Laplace–Gaussian law. *Acta Math. 77*, 1–125.

Feller, W. (1950). *An Introduction to Probability Theory and Its Applications*, 2nd ed. Wiley, New York.

Galambos, J. (1984). *Introductory Probability Theory*. Marcel Dekker, New York.

Galambos, J. (1992). Characterizations. In: *Handbook of the Logistic Distribution*, Chapter 7. (ed.: N. Balakrishnan), Marcel Dekker, New York, pp. 169–188.

Gnedenko, B. V. and A. N. Kolmogorov (1954). *Limit Distributions for Sums of Independent Random Variables.* Addison-Wesley, Reading, Mass.

Heyde, C. C. (1963). On a property of the lognormal distribution. *J. Royal Statist. Soc. Ser. B 25*, 392–393.

Johnson, N. L. and S. Kotz (1970). *Distributions in Statistics, Vol. 2.* Wiley, New York.

Kátai, I. (1969). Distribution of solutions of some Diophantine equations. *Mat. Lapok 20*, 117–122. (in Hungarian).

Liapunov, A. M. (1901). Nouvelle forme du théorème sur la limite des probabilités. *Mem. Acad. Imp. Sci. St. Pétersbourg 12*, 1–24.

Lindeberg, J. W. (1922). Eine neue Herleitung des Exponentialgesetzes in der Wahrscheinlichkeitsrechnung. *Math. Z. 15*, 211–225.

Lukacs, E. (1970). *Characteristic Functions*, 2nd ed. Griffin, London.

Zygmund, A. (1947). A remark on characteristic functions. *Ann. Math. Statist. 18*, 272–276.

4

Infinite Sequences of Independent Random Variables: Weak Convergence

Throughout this chapter, $X_j, j \geq 1$, will be an infinite sequence of independent random variables with distribution function $F_j(x)$ and characteristic function $\varphi_j(t), j \geq 1$. We set $S_n = X_1 + X_2 + \cdots + X_n$, whose distribution function is denoted by $G_n(x)$ and characteristic function by $\psi_n(t)$. The first section is devoted to the weak convergence of S_n and of $S_n - a_n$, where a_n is a suitable sequence of constants. The remarkable result for independent variables is that weak and strong convergence coincide in these cases and, thus, the limiting distribution is always that of an infinite series. The rest of the chapter treats the weak convergence of the normalized sum $(S_n - a_n)/b_n$, where $b_n > 0$ and $b_n \neq 1$ for infinitely many values of n. In fact, most results automatically imply that $b_n \to +\infty$ with n. Emphasis is on the asymptotic normality of the normalized sums. Speed of convergence estimates are also included. A separate chapter will deal with refinements of the results when the X_j are assumed to be identically distributed as well.

4.1 COMPLETE CONVERGENCE OF SUMS WITHOUT NORMALIZATION, INFINITE SERIES

The following four theorems contain a complete solution of the problem of the complete convergence of S_n and of $S_n - a_n$ with some a_n. It should be noted that, upon setting $a_j^* = a_j - a_{j-1}$, where $a_0 = 0$ and $X_j^* = X_j - a_j^*$,

151

$S_n - a_n = S_n^*$, the sum of the X_j^*, $1 \le j \le n$; thus the two problems are closely related.

Theorem 1   (Kawata and Udagawa)   If

$$\psi_n(t) = \varphi_1(t)\varphi_2(t)\cdots\varphi_n(t) \to \psi(t)$$

as $n \to +\infty$ for all t, except perhaps on a set of Lebesgue measure zero, and if $\psi(t) \ne 0$ on a set A of positive Lebesgue measure, then $\psi(t)$ can be extended to the characteristic function of a proper distribution function $G(x)$ and $G_n(x) \to G(x)$ completely.

Theorem 2   (Kolmogorov's Three-Series Theorem)   The infinite series $\sum X_j$ converges a.s. if and only if each of the three series

(i) $\sum P(|X_j| \ge M)$ (ii) $\sum E(X_{j,M})$ (iii) $\sum V(X_{j,M})$

converges, where $0 < M < +\infty$ is arbitrary and $X_{j,M}$ is X_j truncated at M.

Recall that a random variable Y truncated at M is Y if $|Y| \le M$ and is zero otherwise. The symbols $E(\cdot)$ and $V(\cdot)$ signify, as usual, expectation and variance, respectively.

Theorem 3   The infinite series $\sum X_j$ converges a.s. to a random variable X whose distribution function is $G(x)$ if and only if $G_n(x) \to G(x)$ completely. Furthermore, there are constants a_j^*, $j \ge 1$, such that the infinite series $\sum(X_j - a_j^*)$ converges a.s. if and only if there are constants a_n, $n \ge 1$, such that the distribution function of $S_n - a_n$ converges completely to a proper distribution function. When such constants exist, one can always choose $a_j^* = m_j + E[(X_j - m_j)_M]$ and $a_n = a_1^* + a_2^* + \cdots + a_n^*$, where m_j is a median of the distribution of X_j and $(X_j - m_j)_M$ is $X_j - m_j$ truncated at $M > 0$.

Theorem 4   (Two-Series Theorem)   There exist constants a_j^* such that $\sum(X_j - a_j^*)$ converges a.s. if and only if, with arbitrary constant $M > 0$, the two series $\sum P(|X_j - m_j| \ge M)$ and $\sum V[(X_j - m_j)_M]$ converge, where m_j is a median of the distribution of X_j and $(X_j - m_j)_M$ is $X_j - m_j$ truncated at M. When such a_j^* exist, they can be chosen as in Theorem 3.

The proofs require a number of preparatory lemmas, which are of interest in their own right.

Lemma 1   (Steinhaus)   Let A be a set of positive Lebesgue measure μ on the real line. Then the set $D(A)$, which contains all differences $a - b$ such that both a and b belong to A, contains an interval of the form $(-U, U)$, with $U > 0$.

Proof. We assume $\mu < +\infty$. By Remarks 2 and 3 of Section 1.3, there is a denumerable collection of intervals (a_j, b_j) such that

$$A \subseteq \bigcup_{j=1}^{+\infty} (a_j, b_j) \quad \text{and} \quad \sum_{j=1}^{+\infty} (b_j - a_j) < 1.1\mu$$

Let $\lambda(B)$ denote the Lebesgue measure of the (measurable) set B. We can now easily conclude that among our intervals there is one—(a_k, b_k) say— such that $\lambda(A \cap (a_k, b_k)) \geq 0.9(b_k - a_k)$. Indeed, otherwise we would have

$$0 < \mu = \lambda(A) \leq \sum_{j=1}^{+\infty} \lambda(A \cap (a_j, b_j)) < 0.9 \sum_{j=1}^{+\infty} (b_j - a_j) < 0.99\mu$$

which is impossible. We claim that the interval $I_U = (-U, U)$, with $U = 0.8(b_k - a_k)$, is contained in $D(A)$. In other words, if $x \notin D(A)$, then we must show that $|x| \geq 0.8(b_k - a_k)$. We fix $x \notin D(A)$. Note that if A_1 is an arbitrary subset of A and if $A_1(x)$ denotes the set of all numbers of the form $a_1 - x$ with $a_1 \in A_1$, then A_1 and $A_1(x)$ are disjoint. Hence, with $A_1 = A \cap (a_k, b_k)$,

$$b_k - a_k \geq \lambda(A_1) + \lambda[A_1(x) \cap (a_k, b_k)] \geq 2\lambda(A_1) - |x|$$

which, by the choice of k, yields

$$b_k - a_k \geq 1.8(b_k - a_k) - |x|$$

It follows that $|x| \geq 0.8(b_k - a_k)$, which we earlier set out to prove. The lemma is established. ■

Lemma 2 If a sequence $\tau_n(t)$ of characteristic functions converges to one on a set $\{t\} = A$ of positive Lebesgue measure, then it converges to one on the whole real line.

Proof. First we show that if both t and h belong to A then so does $t - h$. We appeal to the inequality in (iii) of Theorem 3.16. Since Re $\tau_n(h) =$ Re $\tau_n(-h)$, we have

$$|\tau_n(t - h) - \tau_n(t)| \leq 2(1 - \text{Re } \tau_n(h)) \to 0 \quad \text{as } n \to +\infty$$

which, because of $t \in A$, also implies $\tau_n(t - h) \to 1$. Thus, Lemma 1 ensures the existence of an interval $I_U = (-U, U)$, with $U > 0$, such that $\tau_n(t) \to 1$ as $n \to +\infty$ for all $t \in I_U$. With one more appeal to Theorem 3.16 (iii), we obtain by induction that, for every integer $j \geq 1$, $\tau_n(t) \to 1$ as $n \to +\infty$ for $-jU < t < jU$. This completes the proof. ■

The following statement is already contained in the proof of Theorem 3.5. We reformulate it here for the sake of easier reference.

Lemma 3 If $\tau_n(t)$ is the characteristic function of a distribution function $R_n(x)$ and if $\tau_n(t)$ converges as $n \to +\infty$ for almost all t with respect to Lebesgue measure, then there is a distribution function $R(x)$ such that, for all continuity points $a < b$ of $R(x)$, $R_n(b) - R_n(a) \to R(b) - R(a)$.

One comment only is needed in regard to the proof of the lemma. In the proof of Theorem 3.5 we proceeded under the assumption that $\tau_n(t)$ converges for all t. However, the proof itself is based on the integrals of the characteristic functions involved; indeed, the conclusion of the theorem (and of the current lemma as well) is based on the integral equation (3.23), which is not affected if the integrand (and thus the limit in the lemma) is not defined on a set of Lebesgue measure zero.

Proof of Theorem 1. By Lemma 3 we know that there is a distribution function $G(x)$ such that $G_n(b) - G_n(a) \to G(b) - G(a)$ for all continuity points $a < b$ of $G(x)$. Therefore, it remains to be shown that $G(x)$ is a proper distribution function. Since $\psi_n(t)$ is known to converge for almost all t (Lebesgue measure), we have

$$\lim_{n = +\infty} \psi_n(t) = \prod_{j=1}^{+\infty} \varphi_j(t) = \psi(t) \qquad \text{(almost all } t) \tag{4.1}$$

(Here, contrary to custom, we consider an infinite product to be convergent when its value is zero.) For those values of t for which (4.1) applies

$$\tau_n(t) = \prod_{j=n+1}^{+\infty} \varphi_j(t) \tag{4.2}$$

is also convergent, and a characteristic function as well (the limit of the finite product of $\varphi_j(t)$ over $n < j \le n + m$ as $m \to +\infty$). Now,

$$\psi_n(t)\tau_n(t) = \psi(t) \qquad \text{(almost all } t) \tag{4.3}$$

Hence, in view of (4.1), on the set A on which $\psi(t) \ne 0$, $|\tau_n(t)|^2 \to 1$ as $n \to +\infty$. Since A is of positive Lebesgue measure, Lemma 2 entails that $|\tau_n(t)|^2 \to 1$ for all t, implying that the distribution function $R_n(x)$ corresponding to $|\tau_n(t)|^2$ converges weakly to the distribution function $R(x)$ degenerate at zero, i.e., $R_n(x) \to 0$ or 1 depending on whether $x < 0$ or $x > 0$. From this fact it is easy to deduce that

$$|\tau_n(t)|^2 \to 1 \text{ uniformly for } |t| \le T < +\infty \qquad \text{as } n \to +\infty \tag{4.4}$$

Indeed, let $a > 0$ and $2a < \varepsilon/T$, where $\varepsilon > 0$ is arbitrary. Then

$$||\tau_n(t)|^2 - 1| \leq R_n(-a) + 1 - R_n(a) + \left| \int_{-a}^{a} e^{itx} \, dR_n(x) - 1 \right|$$

The right-hand side can now be made arbitrarily small for large n and independent of t when $|t| \leq T$ via the estimate

$$\left| \int_{-a}^{a} e^{itx} \, dR_n(x) - 1 \right| \leq \left| \int_{-a}^{a} (e^{itx} - e^{ita}) \, dR_n(x) \right| + \left| e^{ita} \int_{-a}^{a} dR_n(x) - 1 \right|$$

$$\leq 3Ta(R_n(a) - R_n(-a)) + |R_n(a) - R_n(-a) - 1|$$

where we wrote $e^{ita} = (e^{ita} - 1) + 1$ and used the inequality $|e^{itu} - e^{itv}| \leq |t||u - v|$ for real t, u, and v (including $v = 0$). This establishes (4.4). Next we apply the triangular inequality in (4.3), obtaining

$$|\psi(t)| = |\psi_n(t)| \, |\tau_n(t)| - 1 + 1| \geq |\psi_n(t)| - |\psi_n(t)| \, |\tau_n(t)| - 1|$$

Since $\psi_n(t)$ is the characteristic function of a proper distribution function, we get from (4.4) that the absolute value of the extension of $\psi(t)$ to a characteristic function approaches one as $t \to 0$. This suffices for us to conclude that $G(x)$ is a proper distribution function. The theorem is established.   ■

For the proof of Theorem 2 we need one of the inequalities contained in the following lemma, which is due to Doob (1953).

Lemma 4   Assume that the random variable X is bounded, $|X| \leq M < +\infty$. If V is the variance of X and $\varphi(t)$ is its characteristic function, then, for $|t| \leq 1/4M$,

$$e^{-Vt^2} \leq |\varphi(t)| \leq e^{-Vt^2/3}$$

Proof.   Let Y be an independent copy of X and set $Z = X - Y$. Then the characteristic function of Z is $\varphi(t)\varphi(-t) = |\varphi(t)|^2$ and, since the distribution of Z is symmetric, $E(Z^{2k+1}) = 0$ for $k \geq 0$. Furthermore, $E(Z^2) = 2V$ and $E(Z^{2k}) \leq 2V(2M)^{2k-2}$, $k \geq 1$. Thus, from the Taylor expansion (3.5)

$$|\varphi(t)|^2 = 1 - Vt^2 + \sum_{k=2}^{+\infty} (-1)^k E(Z^{2k}) \frac{t^{2k}}{(2k)!}$$

we get the estimates, valid for $|t| \leq 1/4M$,

$$|\varphi(t)|^2 \leq 1 - Vt^2 + 2Vt^2 \sum_{k=2}^{+\infty} \frac{(2Mt)^{2k-2}}{(2k-2)!}$$

$$\leq 1 - Vt^2 + Vt^2 \sum_{k=2}^{+\infty} \frac{(1/2)^{2k-2}}{(k-1)!} = 1 - Vt^2 + Vt^2(e^{1/4} - 1)$$

$$\leq 1 - 2Vt^2/3 \leq e^{-2Vt^2/3}$$

and

$$|\varphi(t)|^2 \geq 1 - Vt^2 - Vt^2(e^{1/4} - 1) \geq 1 - 4Vt^2/3 \geq e^{-2Vt^2}$$

where we used the elementary inequalities

$$1 - x \leq e^{-x} \leq 1 - \tfrac{2}{3}x$$

which are certainly valid for $0 \leq x \leq 1/2$. The desired inequalities now follow from the estimates above.   ∎

It is clear from the proof that one can easily improve upon the inequalities of Lemma 4. See Exercise 3.

Proof of Theorem 2.   First we prove the sufficiency of the conditions. We proved in Theorem 2.13 that the convergence of (iii) implies the a.s. convergence of $\sum [X_{j,M} - E(X_{j,M})]$ and, thus, the convergence of each of (ii) and (iii) yields that $\sum X_{j,M}$ converges a.s. But, by the Borel–Cantelli lemma (Theorem 1.10), the convergence of (i) and the a.s. convergence of $\sum X_{j,M}$ are sufficient for the a.s. convergence of $\sum X_j$. This completes the proof of the sufficiency part of the theorem. For the necessity part, we now assume that $\sum X_j$ converges a.s. This implies that $X_n \to 0$ a.s. as $n \to +\infty$. Consequently, $|X_j| \geq M$ can occur only for a finite number of the subscript j and, thus, in view of the independence of the X_j, the Borel–Cantelli lemma and its converse (Theorems 1.10 and 1.11) imply that the series in (i) must converge. With one more appeal to the Borel–Cantelli lemma, we can conclude that $\sum X_{j,M}$ converges a.s. to a random variable X^*. Let its characteristic function be $\psi^*(t)$ and let the characteristic function of $X_{j,M}$ be $\varphi_{j,M}(t)$. Then, since a.s. convergence implies weak convergence,

$$\lim_{n = +\infty} \prod_{j=1}^{n} \varphi_{j,M}(t) = \psi^*(t) \tag{4.5}$$

Because X^* is a proper random variable, $\psi^*(0) = 1$ and, thus, for some $T > 0, \psi^*(t) \neq 0$ if $|t| \leq T$. Let $0 < t \leq \min(T, 1/4M)$. We have from Lemma 4 and (4.5)

$$\sum_{j=1}^{n} V(X_{j,M}) \leq -\frac{3}{t^2} \sum_{j=1}^{n} \log |\varphi_{j,M}(t)| \leq -\frac{3}{t^2} \log |\psi^*(t)|$$

Thus, the series in (iii) converges. But then, by what has been proved in the first part, this implies that $\sum [X_{j,M} - E(X_{j,M})]$ converges a.s., which, together with the earlier established a.s. convergence of $\sum X_{j,M}$, yields that the series in (ii) also converges. This concludes the proof.   ∎

Note that Theorem 2 does not imply absolute convergence, as the series in (ii) can be sensitive to the signs of the X_j. Yet the situation is not the same as in calculus: there are conditionally convergent series of independent random variables which remain (a.s.) convergent for all rearrangements of its terms. One has only to eliminate the effect of the series in (ii). For example, if each X_j is bounded by M and has a symmetric distribution, then each term in both (i) and (ii) is zero and (iii) is not affected by rearrangements. Thus, if X_j is either $1/j$ or $-1/j$, each with probability $1/2$, then $\sum X_j$, as well as all series obtained by the rearrangement of the X_j, are a.s. convergent, but $\sum |X_j|$ is clearly divergent. We shall return to more examples, and more surprises, after the completion of the proofs.

Both Theorems 3 and 4 follow easily from the following statement.

Lemma 5   Let Y_j be an independent copy of X_j. There are constants a_j^* such that $\sum (X_j - a_j^*)$ converges a.s. if and only if $\sum (X_j - Y_j)$ converges a.s.

Proof.   If, for some a_j^*, $\sum (X_j - a_j^*)$ converges a.s., then so does $\sum (Y_j - a_j^*)$ and thus $\sum (X_j - Y_j)$, as well. Therefore, only the converse requires nontrivial reasoning. Assume that $\sum (X_j - Y_j)$ converges a.s. Then, by the three-series theorem, with M finite, $\sum P(|X_j - Y_j| \geq M)$ and $\sum V[(X_j - Y_j)_M]$ both converge, where $(X_j - Y_j)_M$ is $X_j - Y_j$ truncated at M. But then, if m_j is a median of X_j, we get from the symmetrization inequalities (Lemma of Section 3.4)

$$\sum P(|X_j - m_j| \geq M) \leq 2 \sum P(|X_j - Y_j| \geq M) < +\infty$$

Thus, upon integrating the simple inequality

$$\{(X_j - m_j)_M - E_{j,M} - [(Y_j - m_j)_M - E_{j,M}]\}^2$$
$$\leq (X_j - Y_j)_M^2 + 4M^2 I_j(M) + M^2 I_{X(j)}(M) + M^2 I_{Y(j)}(M)$$

where $I_j(M)$, $I_{X(j)}(M)$, and $I_{Y(j)}(M)$ are the indicators of the events $|X_j - Y_j| \geq M$, $|X_j - m_j| \geq M$, and $|Y_j - m_j| \geq M$, respectively, and $E_{j,M} = E[(X_j - m_j)_M]$, we get $\sum V[(X_j - m_j)_M] < +\infty$. Therefore, by the three-series theorem, $\sum (X_j - m_j - E_{j,M})$ converges a.s. As a side result, we also obtain that we can choose $a_j^* = m_j + E_{j,M}$. The proof is complete.   ∎

Proof of Theorem 4.   The three-series theorem implies that the convergence of each of the series $\sum P(|X_j - m_j| \geq M)$ and $\sum V[(X_j - m_j)_M]$ is sufficient for the a.s. convergence of $\sum (X_j - m_j - E_{j,M})$; thus, the sufficiency part of the theorem is established. For the converse we appeal to Lemma 5. We have that if $\sum (X_j - a_j^*)$ converges a.s. with some constants a_j^*, then $\sum (X_j - Y_j)$ converges a.s. as well, which, as demonstrated in the proof of Lemma 5, implies the convergence of the two series of the theorem. Finally, we obtained at the end of the proof of Lemma 5 that we can choose $a_j^* = m_j + E_{j,M}$. The proof of Theorem 4 is completed. ∎

Proof of Theorem 3. Clearly, we have to prove only that complete convergence implies a.s. convergence in both parts of the theorem. Now, if $G_n(x) \to G(x)$ completely, then

$$\psi_n(t) = \prod_{j=1}^{n} \varphi_j(t) \to \prod_{j=1}^{+\infty} \varphi_j(t) = \psi(t) \tag{4.6}$$

where $\psi(t)$ is the characteristic function of $G(x)$. Since $G(x)$ is a proper distribution function, $\psi(t)$ is continuous and $\psi(0) = 1$. Therefore, there is a real number $b > 0$ such that $|\psi(t)|^2 \geq 1/2$, say, for $0 \leq t \leq b$ and, thus, for each j, $|\varphi_j(t)|^2 \geq 1/2$ for these values of t, as well. Hence, we can take their logarithm for $0 \leq t \leq b$. Now, in view of the elementary inequality $1 - x \leq -\log x$, we can apply the inequality established in the course of proving part (ii) of Theorem 3.17 in the following form: let $F_j^{**}(x)$ be the distribution function of $X_j - Y_j$, where X_j and Y_j are independent and identically distributed; then

$$\frac{b}{10} \int_{-\infty}^{+\infty} g(bx) \, dF_j^{**}(x) \leq - \int_0^b \log |\varphi_j(t)|^2 \, dt$$

where $g(z) = z^2$ if $|z| \leq 1$ and $g(z) = 1$ for $|z| > 1$. Let us sum the above inequality over $1 \leq j \leq n$. On the right-hand side, upon interchanging integration and summation, we get $-\log |\psi_n(t)|^2$, which is further increased if we replace $\psi_n(t)$ by $\psi(t)$. That is,

$$\frac{b}{10} \sum_{j=1}^{n} \int_{-\infty}^{+\infty} g(bx) \, dF_j^{**}(x) \leq - \int_0^b \log |\psi(t)|^2 < +\infty$$

Let $n \to +\infty$. We have that both $\sum P(|X_j - Y_j| \geq M)$ and $\sum V[(X_j - Y_j)_M]$ are convergent (with $M = 1/b$, but its exact value is irrelevant). By Theorem 2, $\sum (X_j - Y_j)$ converges a.s.; thus, Lemma 5 ensures the existence of constants a_j^* such that $\sum (X_j - a_j^*)$ converges a.s., implying

$$\prod_{j=1}^{n} e^{-ita_j^*} \varphi_j(t) \to \psi^*(t) \tag{4.7}$$

where $\psi^*(t)$ is a proper characteristic function. Hence, if t is sufficiently close to zero, $\psi(t)\psi^*(t) \neq 0$, which permits us to divide (4.6) and (4.7), and we see that $\sum a_j^*$ is convergent. This now yields that $\sum X_j$ converges a.s., which completes the proof of the first part of the theorem.

For the second part, our starting point is (4.7). Since, for real a and t, $|e^{ita}| = 1$, the proof of the first part works without change. We thus have that (4.7) implies the a.s. convergence of $\sum (X_j - a_j^{**})$ with some constants a_j^{**}, concluding the proof of the second part of the theorem. The claim about the possible choice of the constants involved has been established at the end of the proof of Lemma 5. Hence, the proof of the theorem is complete. ∎

There is a large number of implications of Theorems 1–4. Some of these, together with some supplements to these theorems, are collected as remarks below.

Remark 1   Recall that an important special case of the sufficiency part of Theorem 4 had been established earlier: if $V(X_j) < +\infty$ and $\sum V(X_j) < +\infty$ then $\sum (X_j - E(X_j))$ converges a.s. That is, in this special case, the constants a_j^* can be chosen as $E(X_j)$. Even though the series in question is not absolutely convergent, it remains convergent for all rearrangements of its terms, since rearrangements do not affect the convergence of $\sum V(X_j)$.

Remark 2   Let $X_j = 0$ or $(-1)^j/j$, each with probability $1/2$. Then, by Theorem 2, $\sum X_j$ converges a.s., but its terms can be rearranged in such a way that the new series diverges a.s. Indeed, if one takes large blocks of even-indexed terms, followed by just one term of odd index, one can get $\sum E(X_{j_k}) = +\infty$. By Theorem 2 and Kolmogorov's zero-one law (Theorem 2.18), $\sum X_{j_k}$ diverges a.s. However, if we apply Theorem 4 rather than Theorem 2 to these particular X_j, we have that there are constants a_j^* such that $\sum (X_j - a_j^*)$ converges a.s. and the suggested values of a_j^* include $(-1)^j/2j$, since it is a median m_j of X_j, $E(X_j - m_j) = 0$, and $M = 1$ is a uniform bound of $(X_j - m_j), j \geq 1$. With this normalization, $\sum (X_j - m_j)$ not only converges a.s., but also remains a.s. convergent after an arbitrary rearrangement of its terms, since its a.s. convergence is guaranteed by the convergence of $\sum V(X_j - m_j) = \sum V(X_j)$, which is not affected by rearrangements. We can derive two facts from this example: first, the normalizing constants a_j^* are not unique ($a_j^* = 0$ was chosen in the first part and $a_j^* = m_j$ in the second) and, second, the choice of a_j^* has a major influence on $\sum (X_j - a_j^*)$.

Remark 3   It follows immediately from Theorem 4 that if there are centering constants a_j^* such that $\sum (X_j - a_j^*)$ converges a.s., then one can always choose these constants such that the a.s. convergence is not affected by the rearrangement of the terms of the series in question (in fact, the constants suggested in Theorem 4 meet this condition). Simply observe that the convergence of the two series that figure in the conditions of Theorem 4 is not affected by the rearrangement of their terms.

Remark 4 One of the most significant parts of Theorems 1–4 is the establishment of the equivalence of weak and strong convergence for sums of independent random variables. Hence, the combination of these theorems permits one to use characteristic functions to determine a.s. convergence and conversely, Theorem 2 and Theorem 4 or its equivalent stated as Lemma 5, can be used in determining whether an infinite product of characteristic functions converges or not. The following statements are examples of these possibilities.

Let $\varphi_j(t)$ be the characteristic function of the random variable X_j. The infinite product

$$\psi(t) = \prod_{j=1}^{+\infty} \varphi_j(t)$$

is the characteristic function of a random variable if and only if each of the series (i)–(iii) of Theorem 2 converges. On the other hand, $\sum X_j$ converges a.s. if and only if $\psi(t)$ is well defined for almost all t (with respect to Lebesgue measure) and $\psi(t) \neq 0$ on a set of positive Lebesgue measure.

The first part of the preceding statement is the combination of Theorems 2 and 3 (and the continuity theorem of characteristic functions), while the second part combines Theorems 1–3. Theorems 3 and 4 and Lemma 5 entail the next statement.

There are constants a_n such that $S_n - a_n$ converges completely if and only if there are constants a_j^* such that $\sum (X_j - a_j^*)$ converges a.s. Both the preceding statements hold if and only if

$$|\psi(t)|^2 = \prod_{j=1}^{+\infty} |\varphi_j(t)|^2$$

converges for almost all t (Lebesgue measure) and $\psi(t) \neq 0$ on a set of positive Lebesgue measure (recall our convention: we call an infinite product of characteristic functions convergent if its value is zero). The constants a_n and a_j^* can always be chosen as in Theorem 3.

Remark 5 We have discussed the extent to which the centering constants a_j^* affect the a.s. convergence of $\sum (X_j - a_j^*)$ after rearrangement of its terms. The following result of Kawata and Udagawa (1949) shows another aspect of rearrangement.

Let $n_1, n_2, \ldots$ be a rearrangement of successive integers. Assume that the series $\sum X_j$ and $\sum X_{n_k}$ converge a.s. to the random variables X and Y, respectively, whose distribution functions are $F(x)$ and $G(x)$. Then there is a constant a such that $F(x) = G(x + a)$.

The proof is simple at this stage. By the last statement of the previous remark and since the convergence or divergence of an infinite product of

nonnegative real numbers not exceeding one is not affected by the rearrangement of its factors,

$$|\psi(t)|^2 = \prod_{j=1}^{+\infty} |\varphi_j(t)|^2 = \prod_{k=1}^{+\infty} |\varphi_{n_k}(t)|^2 = |\psi^*(t)|^2$$

where $\psi(t)$ and $\psi^*(t)$ are the characteristic functions of $F(x)$ and $G(x)$,
respectively. Now, let $m(s) = \max(n_1, n_2, \ldots, n_s)$. Then

$$\prod_{j=1}^{m(s)} \varphi_j(t) = h_s(t) \prod_{k=1}^{s} \varphi_{n_k}(t)$$

where $h_s(t)$ is a (proper) characteristic function (the product of some $\varphi_j(t)$).
Let $s \to +\infty$. Since the products on the left- and right-hand sides converge
to $\psi(t)$ and $\psi^*(t)$, respectively, there is a $b > 0$ such that, for $|t| \leq b$,
$\psi(t)\psi^*(t) \neq 0$; thus, as $s \to +\infty$, $\lim h_s(t) = h(t)$ exists, $|t| \leq b$, and

$$\psi(t) = h(t)\psi^*(t) \qquad |h(t)| = 1 \quad |t| \leq b$$

The above equation immediately yields that $h(0) = 1$ and $h(t)$ is continuous
at $t = 0$. But then, with an appeal to the compactness of distribution
functions (Theorem 2.24) and the continuity theorem of characteristic
functions (Theorem 3.5), we get that $h(t)$ is, in fact, a characteristic function,
and the last set of equations continues to hold without the restriction $|t| \leq b$
(since $h_s(t)$ is a characteristic function—of, say, $H_s(x)$—$H_s(x)$ has a weakly
convergent subsequence, $H_{s(k)}(x) \to H(x)$; in addition, $h_{s(k)}(t) \to h^*(t)$, the
characteristic function of $H(x)$, which coincides with $h(t)$ for $|t| \leq b$—since
$h(t)$ is continuous at $t = 0$, $h^*(t)$ and $H(x)$ are both proper). Hence, there is
a number $-a$ such that $h(t) = e^{-ita}$ (see, e.g., Exercise 3 of Section 3.1),
that is,

$$\psi(t) = e^{-ita}\psi^*(t)$$

which is equivalent to the assertion of the present remark.

Exercises

1. Let $Y_1, Y_2, \ldots, Y_n$ be independent random variables with the common
   exponential distribution function $F(x) = 1 - e^{-x}$, $x \geq 0$. Let $Y_{1:n} \leq$
   $Y_{2:n} \leq \cdots \leq Y_{n:n}$ be the order statistics of the Y_j. Show that the differences
   $Y_{j+1:n} - Y_{j:n}$ are independent exponential variables and find their expectation. From this result, determine the characteristic function of $F^n(x) =$
   $P(Y_{n:n} \leq x)$. Finally, give an infinite product representation of the characteristic function of $H(x) = \exp(-e^{-x})$ by investigating the infinite series
   $\sum (Y_j - 1)/j$.

2. Show that if $\sum |1 - \varphi_n(t)|$ converges on a t-set of positive Lebesgue measure, then $\sum X_j$ converges a.s. Show that $\sum E(X_j)$ is absolutely convergent if, in addition to the previous assumption, $\sum V(X_j) < +\infty$.

3. Get sharper estimates on $|\varphi(t)|$ than those in Lemma 4 in the following two ways. First, for the same range of t as in Lemma 4, follow the proof of Lemma 4 but ignore the negative terms of the infinite sum in the upper estimate and the positive terms in the lower estimate. Second, bring the upper and lower estimates even closer by narrowing the range of t, to $|t| \le b$, say. What happens as $b \to 0$?

4. Let X_j take the values $-a_j$ and $+a_j$, each with probability $1/2$. What conditions must be satisfied in order that the series (i) $\sum X_j$ and (ii) $\sum (X_j + a_j)$, converges a.s.? Find the distribution function of $\sum X_j$ if $a_j = 1/2^j$.

4.2 DECOMPOSITION OF THE NORMAL DISTRIBUTION

We already know that if X and Y are both normal and if X and Y are independent, then $Z = X + Y$ is normal also. This section is devoted to showing the converse of this simple result, which was first discovered by Cramér (1936).

Theorem 5 Let the independent random variables X and Y have finite expectation and variance. If $Z = X + Y$ is a normal variable, then the distribution functions $F(x)$ and $G(x)$ of X and Y, respectively, are either normal or degenerate.

Remark Although the main purpose of Theorem 5 is to provide a result needed in Section 4.4, it also has a direct implication for the central limit problem. In Section 3.3, we saw that (linearly) normalized sums of independent random variables having finite variances are asymptotically normal (under some additional conditions). The question arises, can such sums, with large numbers of summands, in fact become exactly normal? The answer to this, via Theorem 5, is that this is not possible unless the summands themselves are normal.

Proof of Theorem 5. Without loss of generality, we may assume that $E(X) = E(Y) = E(Z) = 0$ and $V(Z) = 1$. Then, if $\varphi_1(t)$ and $\varphi_2(t)$ denote the characteristic functions of X and Y, respectively, we have

$$\varphi_1(t)\varphi_2(t) = \exp(-t^2/2) \tag{4.8}$$

and

$$\varphi_1(0) = \varphi_2(0) = 1 \qquad \varphi_1'(0) = \varphi_2'(0) = 0 \tag{4.9}$$

Next we show that, for complex values z, the functions

$$\varphi_1(z) = \int_{-\infty}^{+\infty} e^{izx}\, dF(x) \qquad \text{and} \qquad \varphi_2(z) = \int_{-\infty}^{+\infty} e^{izx}\, dG(x)$$

are regular (differentiable any number of times) and (4.8) and (4.9) continue to hold. Let $z = t + iu$. Since

$$|\varphi_1(z)| \le \int_{-\infty}^{+\infty} |e^{izx}|\, dF(x) = \int_{-\infty}^{+\infty} e^{-ux}\, dF(x) = \varphi_1(iu)$$

it suffices to show that $\varphi_j(iu)$, $j = 1$ and 2, are finite for all real u. Indeed, once it has been established that $\varphi_j(z)$ is finite for all z, then differentiation under the integral sign and the assumption of finite expectations yield, via the dominated convergence theorem, that $\varphi_j'(z)$ exists for all z. But if a function of a complex variable is differentiable once, then it is differentiable any number of times, that is, it is regular. We thus have (4.9). From another well-known fact of complex analysis, (4.8) follows, also. Namely, the function $\varphi_1(z)\varphi_2(z) - \exp(-z^2/2)$ is now known to be regular and zero on the whole real line. It follows that it is identically zero on the whole complex plane.

In order to establish that $\varphi_j(iu)$, $j = 1, 2$, are finite, we prove the inequality

$$\varphi_1(iu)\varphi_2(iu) \le \exp(u^2/2) \tag{4.10}$$

Note first that (4.8) is equivalent to

$$\int_{-\infty}^{+\infty} F(x - y)\, dG(y) = N(x)$$

where $N(x)$ is the standard normal distribution function. Thus, for $x_j < x_{j+1}$ and arbitrary finite $a < b$,

$$\int_a^b [F(x_{j+1} - y) - F(x_j - y)]\, dG(y) \le N(x_{j+1}) - N(x_j) \tag{4.11}$$

Let $A < B$ be another pair of finite numbers and let us divide the interval $[A, B]$ by the points $A = x_0 < x_1 < \cdots < x_n = B$. If we multiply (4.11) by $\exp(-ux_j)$ and sum over all j, we recognize Riemann–Stieltjes sums on both sides. Since these Riemann–Stieltjes sums are bounded on the finite interval $[A, B]$ and all the integrals involved are finite, we have, by letting $\max(x_{j+1} - x_j) \to 0$,

$$\int_a^b e^{-uy} \left[\int_A^B e^{-u(x-y)}\, dF(x - y) \right] dG(y) \le \int_A^B e^{-ux}\, dN(x)$$

where the dominated convergence theorem has been used. The right-hand side is further increased if integration is extended over the whole real line, which integral is easily seen to be $\exp(u^2/2)$. But then the right-hand side is dependent neither on $[a, b]$ nor on $[A, B]$. Hence, after having substituted $w = x - y$ in the inner integral, if we let $A \to -\infty$ and $B \to +\infty$, and then $a \to -\infty$ and $b \to +\infty$, we obtain (4.10). This, as was established earlier, implies the validity of (4.8) for all complex values of the argument. In particular, since the right-hand side is never zero, we have $\varphi_1(z) \neq 0$ and $\varphi_2(z) \neq 0$, and thus we can take the logarithm (we take that branch of the logarithm for which $\log 1 = 0$). We shall now establish that $|\log \varphi_j(z)|/|z^2|$, $j = 1, 2$ are bounded on the complex plane, from which, by Liouville's theorem, we conclude that $\log \varphi_j(z) = c_j z^2$, c_j constant, $j = 1, 2$, i.e., $\varphi_j(z) = \exp(c_j z^2)$. Since, for $z = t$ real, φ_j is a characteristic function, c_j must be a negative real number (use $|\varphi_j(t)| \leq 1$ and that $\varphi_j(t)\varphi_j(-t)$ is real valued) and, thus, the normality of F and G follows.

It remains, therefore, to be shown that $\log \varphi_j(z)|/|z^2|$ is bounded. Let $d > 0$ be such that $F(d) - F(-d) \geq 1/2$. Then

$$\varphi_1(iu) = \int_{-\infty}^{+\infty} e^{-ux}\, dF(x) \geq \int_{-d}^{d} e^{-|u||x|}\, dF(x) \geq e^{-d|u|}/2$$

thus, from (4.10), for $z = t + iu$,

$$|\varphi_2(z)| \leq \varphi_2(iu) \leq [\exp(u^2/2)]/\varphi_1(iu) \leq 2\exp(u^2/2 + d|u|)$$
$$\leq 2\exp(|z^2| + d|z|)$$

and a similar inequality obtains for $\varphi_1(z)$ with a suitable constant d^*. Thus

$$|\mathrm{Re}[\log \varphi_j(z)]| = |\log|\varphi_j(z)|| \leq |z^2| + D|z| + \log 2$$

where $D = \max(d, d^*)$, implying that $\mathrm{Re}[\log \varphi_j(z)]/|z^2|$ is bounded. Now, if either $V(X) = 0$ or $V(Y) = 0$, then the theorem is trivially true, so we may assume $V(X) > 0$ and $V(Y) > 0$. Then, because of (4.9), $[\log \varphi_j(z)]/z^2$ is regular and, by the preceding estimate, $\mathrm{Re}[\log \varphi_j(z)]/z^2$ is bounded. With a final appeal to complex analysis, the integral formula of Schwarz

$$f(z) = \mathrm{Im}[f(0)] + \frac{1}{2\pi}\int_0^{2\pi} \mathrm{Re}[f(Ce^{iy})]\frac{Ce^{iy} + z}{Ce^{iy} - z}\, dy$$

which is valid for regular functions $f(z)$ in the circle $|z| < C$, yields that $f(z) = [\log \varphi_j(z)]/z^2$ is bounded, since, for every z, we can choose C arbitrarily large. The proof is completed.   ∎

This proof makes several references to elementary complex analysis. Although all quotations are contained in most elementary textbooks on the subject, we refer to Markushevich (1965).

Exercises

1. Let $\varphi(t)$ and $\psi(t)$ be characteristic functions and assume that both $|\varphi(t)| \not\equiv 1$ and $|\psi(t)| \not\equiv 1$. Furthermore,

$$\varphi^{1/2}(t)\psi^{2/3}(t) = e^{-t^2/2}$$

Deduce from Theorem 5 that, if the corresponding expectations and variances are finite, both $\varphi(t)$ and $\psi(t)$ are normal characteristic functions.

2. Use generating functions rather than characteristic functions to prove the following theorem of Raikov (1938): if X and Y are independent and $Z = X + Y$ is a Poisson variable, then the distribution of each of X and Y is Poisson, provided $P(X \geq 0) = P(Y \geq 0) = 1$ and neither X nor Y equals zero a.s.

4.3 LÉVY'S METRIC

Let $F(x)$ and $G(x)$ be two distribution functions. Extensive use will be made of the distance $L(F, G)$ between F and G, which is defined as the infimum of all h such that, for all x,

$$F(x - h) - h \leq G(x) \leq F(x + h) + h \tag{4.12}$$

The distance $L(F, G)$ was introduced by P. Lévy, and we shall refer to $L(F, G)$ as Lévy's distance or Lévy's metric (both names will be justified soon). Zolotarev's theorem—to be presented in the next section—makes this metric very distinctive.

We first establish that $L(F, G)$ is a distance on the set of distribution functions. It is well defined for every F and G and $L(F, G) = 0$ if and only if $F = G$. Upon replacing x once by $x + h$, and, in the upper inequality, by $x - h$, we get from (4.12) that $L(F, G) = L(G, F)$. Finally, the triangular inequality

$$L(F, H) \leq L(F, G) + L(G, H) \tag{4.13}$$

follows immediately from (4.12) if we write down (4.12) with h^* for $L(G, H)$ and with h^{**} for $L(F, H)$. We then have that $h^{**} \leq h + h^*$, from which, if we first take the infimum of h^{**}, (4.13) follows.

Next, we prove the association of the distance $L(F, G)$ with weak convergence.

Theorem 6   Let F and F_n, $n \geq 1$, be distribution functions. Then $F_n \to F$ weakly if and only if $L(F_n, F) \to 0$.

Proof. First we assume $F_n \to F$ weakly; we show that, for all x and for arbitrary $\varepsilon > 0$, if $n \geq N$,

$$F_n(x - 2\varepsilon) - 2\varepsilon \leq F(x) \leq F_n(x + 2\varepsilon) + 2\varepsilon \qquad (4.14)$$

Consequently, $L(F_n, F) \leq 2\varepsilon$ if $n \geq N$, which is the desired assertion. To prove (4.14), choose the finite points $a = x_0 < x_1 < \cdots < x_r = b$ as continuity points of $F(x)$ and such that $F(a) \leq \varepsilon$, $1 - F(b) \leq \varepsilon$, and $|x_{j+1} - x_j| \leq \varepsilon$ and, for $n \geq N$, $|F(x_j) - F_n(x_j)| \leq \varepsilon$ for all j. Now, if $a > x$, then $F_n(x) \leq F_n(a) \leq F(a) + \varepsilon \leq 2\varepsilon \leq F(x) + 2\varepsilon$ and $F(x) \leq F(a) \leq \varepsilon \leq F_n(x) + \varepsilon$, which are stronger statements than (4.14). The establishment of (4.14) is similar if $b < x$. Finally, if, for some j, $x_j \leq x \leq x_{j+1}$, we have

$$F_n(x - \varepsilon) \leq F_n(x_j) \leq F(x_j) + \varepsilon \leq F(x) + \varepsilon.$$

and

$$F(x) \leq F(x_{j+1}) \leq F_n(x_{j+1}) + \varepsilon \leq F_n(x + \varepsilon) + \varepsilon$$

(4.14) thus obtains for all x, as desired.

Conversely, if $L(F_n, F) \to 0$ as $n \to +\infty$, we want to conclude that $F_n \to F$ weakly. We thus assume that (4.14) holds for all x and for arbitrary $\varepsilon > 0$ whenever $n \geq N$. Let x^* be a continuity point of $F(x)$. That is, $|F(x) - F(x^*)| \leq \delta$ if $|x - x^*| \leq 2\varepsilon$. Now, from (4.14) with $x = x^* + 2\varepsilon$, we get

$$F_n(x^*) - 2\varepsilon \leq F(x) \leq F(x^*) + \delta$$

and with $x = x^* - 2\varepsilon$,

$$F_n(x^*) + 2\varepsilon \geq F(x) \geq F(x^*) - \delta$$

obtains. Since both $\varepsilon > 0$ and $\delta > 0$ can be made arbitrarily small, $F_n(x^*) - F(x^*) \to 0$ as $n \to +\infty$. The theorem is established. ∎

The following lemmas are due to Zolotarev (1967).

Lemma 1   If an asterisk between distribution functions signifies convolution, then, for any distribution functions $R_j(x)$ and $Q_j(x)$, $1 \leq j \leq k$,

$$L(R_1 * \cdots * R_k, Q_1 * \cdots * Q_k) \leq \sum_{j=1}^{k} L(R_j, Q_j)$$

Proof. It suffices to prove the lemma for $k = 2$, since, without further argument, it then extends to arbitrary k by induction. Let $L(R_j, Q_j) = h_j$, $j = 1, 2$. Then, by definition, for any values of x and y,

$$\int_{-\infty}^{+\infty} R_1(x - y - h_1)\, dQ_2(y) - h_1 \leq \int_{-\infty}^{+\infty} Q_1(x - y)\, dQ_2(y)$$

and

$$\int_{-\infty}^{+\infty} R_2(x - y - h_2) \, dR_1(y - h_1) - h_2 \leq \int_{-\infty}^{+\infty} Q_2(x - y) \, dR_1(y - h_1)$$

If we integrate by parts on both the left- and right-hand sides of the last inequality and then make a linear substitution, we get

$$\int_{-\infty}^{+\infty} R_1(x - y - h_1 - h_2) \, dR_2(y) - h_2 \leq \int_{-\infty}^{+\infty} R_1(x - y - h_1) \, dQ_2(y)$$

When combined with the first inequality, this becomes

$$(R_1 * R_2)(x - h_1 - h_2) \leq (Q_1 * Q_2)(x)$$

In a similar manner, we can get

$$(Q_1 * Q_2)(x) \leq (R_1 * R_2)(x + h_1 + h_2) + h_1 + h_2$$

The last two inequalities imply

$$L(R_1 * R_2, Q_1 * Q_2) \leq h_1 + h_2 = L(R_1, Q_1) + L(R_2, Q_2)$$

which completes the proof. ∎

The following statements establish relations between $L(R, Q)$ and the distance

$$\rho(R, Q) = \sup_x |R(x) - Q(x)|$$

Lemma 2 For arbitrary distribution functions $R(x)$ and $Q(x)$,

$$L(R, Q) \leq \rho(R, Q)$$

Proof. Let $\rho(R, Q) = r$. Then, for any x,

$$R(x - r) - r \leq R(x) - r \leq Q(x) \leq R(x) + r \leq R(x + r) + r$$

from which, by the definition of $L(R, Q)$, the desired inequality follows. ∎

Lemma 3 Let $R(x)$ and $Q(x)$ be distribution functions. If $Q'(x) = q(x)$ exists and $q(x) \leq s < +\infty$, then

$$\rho(R, Q) \leq (1 + s) L(R, Q)$$

Proof. Set $h = L(R, Q)$. By definition, for any x

$$Q(x - h) - h \leq R(x) \leq Q(x + h) + h$$

and, thus,

$$Q(x - h) - Q(x) - h \leq R(x) - Q(x) \leq Q(x + h) - Q(x) + h$$

By the mean value theorem of calculus, the latter inequalities can also be written as

$$-hq(x_h) - h \leq R(x) - Q(x) \leq hq(x_h^*) + h$$

where $x - h \leq x_h \leq x \leq x_h^* \leq x + h$. In view of the boundedness of $q(x)$, we thus have

$$|R(x) - Q(x)| \leq hs + h$$

which is valid for all x. The proof is completed.   ■

Corollary   If R_n and Q are distribution functions and if $Q'(x) = q(x)$ exists and is bounded, then $L(R_n, Q) \to 0$ as $n \to +\infty$ if and only if $\rho(R_n, Q) \to 0$.

Proof.   The proof follows immediately from the inequalities of Lemmas 2 and 3.   ■

Lemma 4   Let X and Y be random variables with distribution functions $R(x)$ and $Q(x)$, respectively. Assume that $E(X) = E(Y) = E$ and $V(X)$ and $V(Y)$ are finite. Then

$$L(R, Q) \leq \{4 \max[V(X), V(Y)]\}^{1/3}$$

Proof.   Let $\varepsilon > 0$ be a fixed number. If x satisfies $|x - E| \geq \varepsilon$, then, by Chebyshev's inequality,

$$P(X \leq x) = P(X \leq x, Y \leq x) + P(X \leq x, Y > x)$$
$$\leq P(Y \leq x) + V(Y)/\varepsilon^2$$

Upon interchanging the roles of X and Y and combining the two inequalities, we get

$$R(x) - h \leq Q(x) \leq R(x) + h$$

where $h = \max[V(X), V(Y)]/\varepsilon^2$ and $|x - E| \geq \varepsilon$. On the other hand, with $\varepsilon = h/2$ for $|x - E| < \varepsilon = h/2$, one more appeal to Chebyshev's inequality yields

$$1 - R(x + h) = P(X > x + h) \leq 4V(X)/h^2 \leq h$$

and

$$R(x - h) = P(X \leq x - h) \leq 4V(X)/h^2 \leq h$$

Hence, if $\varepsilon = h/2$ and $|x - E| < \varepsilon$,

$$Q(x) \le 1 \le 1 + \{h - [1 - R(x + h)]\} = R(x + h) + h$$

and

$$Q(x) \ge 0 \ge R(x - h) - h$$

Upon collecting all estimates we find that, for any x with $h = \max[V(X), V(Y)]/\varepsilon^2$ and $\varepsilon = h/2$, i.e., $h^3/4 = \max[V(X), V(Y)]$,

$$R(x - h) - h \le Q(x) \le R(x + h) + h$$

which entails $L(R, Q) \le h$, as claimed. ∎

Lemma 5 Let U_n and W be random variables with distribution functions $R_n(x)$ and $Q(x)$, respectively, and assume that the expectations $E(U_n)$ and $E(W)$ and the variances $V_n = V(U_n)$ and $V = V(W)$ are finite. If $U_n \to W$ weakly as $n \to +\infty$, then

$$\liminf V_n \ge V$$

Proof. Set $h = h_n = L(R_n, Q)$. Let $N > 0$ be a large number that ultimately goes to $+\infty$, which will be made specific at a later point. Let I_N and I_N^c be the indicators of the events $|U_n + h| \le N$ and $|U_n + h| > N$, respectively. We then have

$$V_n = E[(U_n + h - h - E(U_n))^2] = E[(U_n + h)^2] - E^2(U_n + h)$$
$$= E[(U_n + h)^2 I_N] + E[(U_n + h)^2 I_N^c]$$
$$- \{E[(U_n + h)I_N] + E[(U_n + h)I_N^c]\}^2 \qquad (4.15)$$

Now, by the Cauchy–Schwarz inequality,

$$\{E[(U_n + h)I_N^c]\}^2 \le E[(U_n + h)^2 I_N^c]E(I_N^c)$$
$$= [1 - P(|U_n + h| \le N)]E[(U_n + h)^2 I_N^c]$$

and, by the definition of I_N^c,

$$|E[(U_n + h)I_N]E[(U_n + h)I_N^c]|$$
$$\le (1/N)|E[(U_n + h)I_N]|E[(U_n + h)^2 I_N^c]$$

Hence, upon setting

$$\alpha = P(|U_n + h| \le N) = R_n(N - h) - R_n(-N - h)$$

$$\beta = E[(U_n + h)I_N] = \int_{-N}^{N} x \, dR_n(x - h)$$

$$= xR_n(x - h) \Big]_{-N}^{N} - \int_{-N}^{N} R_n(x - h) \, dx$$

$$= N[R_n(N - h) + R_n(-N - h)] - \int_{-N}^{N} R_n(x - h) \, dx$$

$$\gamma = E[(U_n + h)^2 I_N] = \int_{-N}^{N} x^2 \, dR_n(x - h)$$

$$= N^2[R_n(N - h) - R_n(-N - h)] - 2\int_{-N}^{N} R_n(x - h)x \, dx$$

and

$$\gamma_c = E[(U_n + h)^2 I_N^c]$$

where the second formulas for β and γ were obtained by integration by parts, we get from (4.15)

$$V_n \ge \gamma - \beta^2 + (\alpha - (2/N)|\beta|)\gamma_c \tag{4.16}$$

Next, we compare α, β, and γ with

$$\alpha' = P(|W| \le N) = Q(N) - Q(-N)$$

$$\beta' = \int_{-N}^{N} x \, dQ(x) = N[Q(N) + Q(-N)] - \int_{-N}^{N} Q(x) \, dx$$

and

$$\gamma' = \int_{-N}^{N} x^2 \, dQ(x) = N^2[Q(N) - Q(-N)] - 2\int_{-N}^{N} Q(x)x \, dx$$

By the definition of the Lévy metric, we immediately have

$$|\alpha - \alpha'| \le 2h \qquad |\beta - \beta'| \le 4Nh \qquad |\gamma - \gamma'| \le 6N^2h \tag{4.17}$$

We return to (4.16) and use (4.17). First, we show that, for large N and small h, the last term on the right-hand side is positive. Indeed, upon writing

$$\alpha - (2/N)|\beta| = \alpha' - 1 + (\alpha - \alpha') - (2/N)|\beta| + 1$$

$$\ge -(1 - \alpha') - |\alpha - \alpha'| - (2/N)(|\beta'| + |\beta - \beta'|) + 1$$

$$\ge -E(W^2)/N^2 - 2h - (2/N)(|\beta'| + 4Nh) + 1$$

where $1 - \alpha'$ is estimated by Markov's inequality, we see that the right-hand side converges to one as $N \to +\infty$ and $h \to 0$ in an arbitrary manner. Now, since $R_n(x) \to Q(x)$ weakly, Theorem 6 ensures that $h = h_n \to 0$ as $n \to +\infty$. Therefore, for sufficiently large n and N, (4.16) becomes

$$V_n \geq \gamma - \beta^2 \geq \gamma' - |\gamma - \gamma'| - (|\beta - \beta'| + |\beta'|)^2$$

that is, by (4.17),

$$V_n \geq \gamma' - 6N^2h - (4Nh + |\beta'|)^2$$

Finally, we choose $N = h^{-1/4}$, say. This guarantees that, as n, $N \to +\infty$, both Nh and N^2h tend to zero. Thus, by letting $n \to +\infty$, the last inequality reduces to the assertion of the lemma. ∎

The next lemma is the last one in the sequence of Zolotarev's lemmas that we need for the main theorem of the next section.

Lemma 6 With the notation of Lemma 5, we again assume that $U_n \to W$ weakly. Furthermore, let $V(U_n)$ be bounded. Then $E(U_n) \to E(W)$ as $n \to +\infty$.

Proof. We use the notation introduced in the preceding proof. In particular, we again set $h = L(R_n, Q)$ and $N = h^{-1/4}$. Hence, in view of Theorem 6, n being large implies that h is small and N is large. Therefore, from the estimate immediately following (4.17), for large n,

$$\alpha - (2/N)|\beta| > 1/2$$

Thus, (4.16) yields

$$\gamma_c = E[(U_n + h)I_N^c] \leq 2(V_n + \gamma - \beta^2)$$

Furthermore, upon writing

$$\gamma - \beta^2 = \gamma' - (\beta' + (\beta - \beta'))^2 + (\gamma - \gamma')$$

the definitions of γ' and β' and the estimates of (4.17) yield that, for large n, say,

$$\gamma - \beta^2 \leq V(W) + 1$$

Since, by assumption, there exists a constant $K < +\infty$ such that $V(U_n) \leq K$, we now get from (4.18) that $\gamma_c \leq 2(K + V(W) + 1)$, i.e., γ_c is bounded as $n \to +\infty$. This, in turn, by the estimate

$$|E[(U_n + h)I_N^c]| \leq (1/N)E[(U_n + h)^2I_N^c]$$

ensures that $E[(U_n + h)I_N^c]$ converges to zero as $n \to +\infty$. This fact and the estimates in (4.17) are sufficient to conclude from the identity

$$E^2(U_n) = \{E[(U_n + h)I_N] + E[(U_n + h)I_N^c] - h\}^2$$
$$= \{\beta' + (\beta - \beta') + E[(U_n + h)I_N^c] - h\}^2$$

that

$$\lim E^2(U_n) = \lim(\beta')^2 = E^2(W)$$

as $n \to +\infty$, which is the assertion of the lemma.   ■

Exercises

1. Prove that every sequence $R_1(x), R_2(x), \ldots$ of distribution functions that is a Cauchy sequence in Lévy's metric, i.e., every $L(R_n, R_m) \to 0$ as both n and m tend to $+\infty$, is weakly convergent. In other words, the space of one-dimensional distribution functions with the distance $L(F, G)$ is complete.

2. Let $F(x) = x$ and $G(x) = x^2$, both for $0 \le x \le 1$, and let $F(x)$ and $G(x)$ be proper distribution functions. Find $L(F, G)$. Furthermore, show by direct computation (and thus with no reference to Theorem 6) that $L(F_n, F) \to 0$ as $n \to +\infty$ if $F_n(x) = [1 + (\log x)/n]^n$, $e_n < x \le 1$, $e_n = e^{-n}$, and $F_n(x) = 0$ or 1 for $x < e_n$ and $x > 1$, respectively.

4.4 ZOLOTAREV'S THEOREM ON ASYMPTOTIC NORMALITY, THE THEOREM OF LINDEBERG AND FELLER

This section is the first one in which the weak convergence of the normalized sums $(S_n - a_n)/b_n$ will be treated. In particular, the main theorem will cover the asymptotic normality of S_n/s_n, where, as in Lindeberg's theorem (Theorem 3.12), $E(X_j) = 0$, $V(X_j) = \sigma_j^2$ is finite, and $s_n^2 = \sigma_1^2 + \sigma_2^2 + \cdots + \sigma_n^2$. In one of the theorems, we shall establish that Lindeberg's condition (3.27) is both necessary and sufficient for the combination of the validity of (3.28) and the asymptotic normality of S_n/s_n. However, when (3.28) is not insisted upon, another set of necessary and sufficient conditions guarantees the asymptotic normality of S_n/s_n. Such a theorem, due to Zolotarev (1967), will be proved in a more general framework. Note that if we denote $X_{j,n} = X_j/s_n$ and $Y_n = X_{1,n} + X_{2,n} + \cdots + X_{n,n} = S_n/s_n$, then, for each j, $E(X_{j,n}) = 0$ and, for all n, $V(Y_n) = 1$. Guided by this observation, we introduce the following model. Let J be either a finite or a denumerably infinite set of positive integers. Let $X_{j,n}$, $j \in J$, be independent random variables with $E(X_{j,n}) = 0$

and $V(X_{j,n}) = \sigma_{j,n}^2$ finite that satisfy $\sum_{j\in J} \sigma_{j,n}^2 = 1$. Set

$$Y_n = \sum_{j\in J} X_{j,n}$$

Then, by the results of Section 4.1, Y_n is finite a.s. and $V(Y_n) = 1$. In addition, Y_n has the same basic properties regardless of whether J is finite or infinite.

As in the past, we reserve the notation $N(x)$ for the standard normal distribution function. When the variable x is suppressed, we write $N_{j,n} = N(x/\sigma_{j,n})$. The distribution function of $X_{j,n}$ is denoted by $F_{j,n}(x)$ and that of Y_n by $G_n^*(x)$.

We can now formulate the results of the present section.

Theorem 7 **(Zolotarev)** If $G_n^*(x) \to N(x)$ weakly as $n \to +\infty$, then

$$\alpha_n = \sup_{j\in J} L(F_{j,n}, N_{j,n}) \to 0 \tag{4.19}$$

Theorem 8 **(Zolotarev)** $G_n^*(x) \to N(x)$ weakly as $n \to +\infty$ if and only if (4.19) holds and, for every $\varepsilon > 0$,

$$\Delta_n(\varepsilon) = \sum_{j\in J_n} \int_{|x|\geq\varepsilon} x^2 \, dF_{j,n}(x) \to 0 \tag{4.20}$$

where $J_n = \{j: j \in J, \sigma_{j,n} < \alpha_n^{1/4}\}$.

Theorem 9 **(Lindeberg and Feller)** The distribution function of S_n/s_n is asymptotically normal and

$$\beta_n = \frac{1}{s_n} \max_{1\leq j\leq n} \sigma_j \to 0 \tag{4.21}$$

as $n \to +\infty$ if and only if, for every $\varepsilon > 0$,

$$\delta_n(\varepsilon) = \frac{1}{s_n^2} \sum_{j=1}^{n} \int_{|x|\geq\varepsilon s_n} x^2 \, dF_j(x) \to 0 \tag{4.22}$$

Before proceeding to the proofs, we shall give, as a set of remarks, the interrelations of the theorems and a clarification of the meaning of some of the conditions.

Remark 1 Clearly, Theorem 7 is a part of the necessity of the conditions of Theorem 8. The separation of the two theorems is intended to point out that if the asymptotic normality of the sums of independent random variables is guaranteed by any conditions, then (4.19) is always implicit in such conditions.

Remark 2 The sufficiency of Theorem 9 has been proved in Section 3.3, which is Lindeberg's theorem. The necessity part of this theorem is due to Feller (1935). Since the property in (4.21) is made a part of the conclusion of the theorem, we in fact seek the asymptotic normality of S_n/s_n under the additional condition that each term X_j/s_n is uniformly small in probability. Indeed, by Chebyshev's inequality (since $E(X_j) = 0$),

$$P(|X_j/s_n| \geq \varepsilon) \leq \sigma_j^2/\varepsilon^2 s_n^2 \leq \beta_n^2/\varepsilon^2 \to 0$$

uniformly in j as $n \to +\infty$. This provides the basic difference between Theorems 8 and 9, when Theorem 8 is restated for $X_{j,n} = X_j/s_n$: in Theorem 8 the summands are not required to be small. See also the following remark.

Remark 3 It is somewhat misleading to replace Theorem 8 by any statement whose conditions are symmetric in the variables $X_{j,n}$. Assume that the asymptotic normality of Y_n (or S_n/s_n) has been established under some conditions. Let us now add some independent normal variables, the aggregate variance of which also equals one, to Y_n, resulting in Y_n^*. It can then be concluded, without any further conditions to be checked, that Y_n^*, also, is asymptotically normal. Evidently, (4.19) is not affected (if $F_{j,n}$ itself is normal, then $L(F_{j,n}, N_{j,n}) = 0$), but other conditions used to deduce the asymptotic normality of Y_n could be. For example, using the notation of Theorem 8, if the variance of each new normal variable is $2\alpha_n^{1/4}$, no condition of Theorem 8 changes when we apply it to either Y_n or Y_n^*; consequently, no condition of Theorem 9 should change in these cases if $X_{j,n} = X_j/s_n$, $J = \{1, 2, \ldots, n\}$, and (4.21) holds in the first place. In other words, the symmetric condition (4.22) is for convenience only; some of its terms (according to Theorem 8, those which are not in J_n) can automatically be neglected when estimating $\delta_n(\varepsilon)$. The reader's attention should be drawn at this point to the fact that the exponent $1/4$ in the definition of J_n is not a characteristic value of the problem; it will be seen in the proof to what extent this exponent can be changed.

Remark 4 Note that the theorems are about weak convergence to the normal distribution, which has bounded density. Hence, both Theorem 6 and the Corollary of the previous section are applicable, implying that, in all cases, convergence is uniform on the whole real line (i.e., both metrics, ρ and L, can be used to express the convergence to normality).

We now turn to the proofs.

Proof of Theorem 7. Assume that $G_n^*(x) \to N(x)$ weakly, but (4.19) fails. That is,

$$L(G_n^*, N) \to 0 \qquad \text{as } n \to +\infty \tag{4.23}$$

and there exists a subsequence n_1 of the integers n and, for every n_1, an integer $j_1 \in J$ such that, as $n_1 \to +\infty$,

$$L(F_{j_1, n_1}, N_{j_1, n_1}) \geq \gamma \tag{4.24}$$

where γ is some positive constant. In order to enable us to compare (4.23) and (4.24), we rewrite (4.23) as follows. By the definition of Y_n, $Y_{n_1}^* = Y_{n_1} - X_{j_1, n_1}$ is independent of X_{j_1, n_1}. Hence, upon denoting the distribution function of $Y_{n_1}^*$ by $R_{n_1}(x)$, we have, from (4.23),

$$L(F_{j_1, n_1} * R_{n_1}, N) \to 0 \qquad \text{as } n_1 \to +\infty \tag{4.25}$$

Here some caution is advised. Since the removal of X_{j_1, n_1} to the first position in the sum representing Y_{n_1} amounts to a rearrangement of the terms of Y_{n_1}, the question arises when J is infinite: Can the terms of Y_n be rearranged without affecting its distribution function? The results of Section 4.1 and Lemma 6 of the previous section ensure that this can be done. We also have $E(Y_n) = E(Y_{n_1}^*) = 0$ (see Exercise 1). Now, we select sequentially two further subsequences of n_1. First we appeal to the compactness of distributions (Theorem 2.24) and select a subsequence n_2 of n_1 such that both $F_{j_2, n_2}(x) \to U(x)$ and $R_{n_2}(x) \to W(x)$ weakly as $n_2 \to +\infty$. Since, by assumption, $V(X_{j_2, n_2}) + V(Y_{n_2} - X_{j_2, n_2}) = 1$, the variances of U and W are finite by virtue of Lemma 5 of the previous section (although the inequality of Lemma 5 is established under the assumption of finite variances, from the last inequality of its proof one can see that the inequality of the lemma holds; thus, both sides are infinite when $V = +\infty$). We, in fact, have

$$\liminf V(X_{j_2, n_2}) \geq V_U \qquad \text{as } n_2 \to +\infty \tag{4.26}$$

where V_U is the variance associated with U (in short, the variance of U). Finally, we select a subsequence n_3 of n_2 such that $V(X_{j_3, n_3}) \to \sigma^2$ and $V(Y_{n_3} - X_{j_3, n_3}) \to \eta^2$ as $n_3 \to +\infty$. Since our selection of subsequences is such that every previous property continues to hold, we have from (4.23) and (4.25) that,

$$\lim L(F_{j_3, n_3} * R_{n_3}, N) = \lim L(F_{j_3, n_3} * R_{n_3}, U * W) = 0$$

as $n_3 \to +\infty$. This, by the triangular inequality of L, implies $N(x) = U(x) * W(x)$, which in turn yields that the sum $V_U + V_W$ of the variances of $U(x)$ and $W(x)$ equals one. Hence, with one more appeal to Lemma 5 of the previous section, we have $V_U = \sigma^2$ and $V_W = \eta^2$. Now, by Lemma 4 of the previous section, the definition of σ^2 and (4.24) imply

$$0 < \gamma \leq (4\sigma^2)^{1/3} \qquad \text{i.e., } V_U = \sigma^2 > 0$$

Thus, U is nondegenerate. Consequently, in view of Cramér's decomposition theorem of $N(x)$ (Theorem 5), $U(x) = N(x/\sigma)$. But since $V(X_{j_3, n_3}) \to \sigma^2$, $N_{j_3, n_3} \to N(x/\sigma) = U(x)$ as well; thus, by the triangular inequality,

$$L(X_{j_3, n_3}, N_{j_3, n_3}) \leq L(X_{j_3, n_3}, U) + L(N_{j_3, n_3}, U) \to 0$$

as $n_3 \to +\infty$, contradicting (4.24). So the initial assumptions of the proof are not possible, which establishes the theorem. ∎

Proof of Theorem 8. Let $G_{n,1}$ and $N_{n,1}$, respectively, be the convolutions of $F_{j,n}$ and $N_{j,n}$, $j \in J_n$. The corresponding convolutions over $j \in J_n^c$ are denoted by $G_{n,2}$ and $N_{n,2}$. Let the characteristic functions of $G_{n,s}$ and $N_{n,s}$, $s = 1, 2$, be $\psi_{n,s}(t)$ and $\tau_{n,s}(t)$, respectively. Finally,

$$V_{n,1} = \sigma_{n,1}^2 = \sum_{j \in J_n} \sigma_{j,n}^2 \qquad V_{n,2} = \sigma_{n,2}^2 = \sum_{j \in J_n^c} \sigma_{j,n}^2$$

First, we show that, for every finite $T > 0$,

$$\lim |\psi_{n,2}(t) - \tau_{n,2}(t)| = 0 \qquad (n \to +\infty) \tag{4.27}$$

uniformly in $|t| \leq T$, under the sole assumption of (4.19). There is, of course, nothing to prove if J_n^c is empty or when $\alpha_n = 0$, in which case both the theorem and (4.27) trivially hold. Therefore, we assume in the sequel that J_n^c is not empty and $\alpha_n > 0$. Now, since $\tau_{n,2}(t) = \tau(t/\sigma_{n,2})$, where $\tau(t)$ is the characteristic function of a standard normal variable and $\sigma_{n,2} \leq 1$, (4.27) will be proved if we show that (4.19) implies

$$\lim |\psi_{n,2}(t/\sigma_{n,2}) - \tau(t)| = 0 \qquad (n \to +\infty) \tag{4.28}$$

uniformly in $|t| \leq T$. However, by Lemmas 3 and 1 of the previous section,

$$\rho(G_{n,2}, N_{n,2}) \leq \left(1 + \frac{1}{\sigma_{n,2}\sqrt{2\pi}}\right) \sum_{j \in J_n^c} L(F_{j,n}, N_{j,n})$$

which, upon observing that $\alpha_n^{1/2} \leq \sigma_{j,n}^2 \leq V_{n,2} \leq 1$ for $j \in J_n^c$ implies $1/\sigma_{n,2} \leq \alpha_n^{-1/4}$ and that the number of elements of J_n^c does not exceed $\alpha_n^{-1/2}$, yields (since $L(F_{j,n}, N_{j,n}) \leq \alpha_n$)

$$\rho(G_{n,2}, N_{n,2}) \leq [1 + (2\pi\alpha_n^{1/2})^{-1/2}]\alpha_n \alpha_n^{-1/2} \leq c\alpha_n^{1/4}$$

where $0 < c(<2)$ is a constant. Hence, as $n \to +\infty$,

$$\rho(G_{n,2}, N_{n,2}) = \sup_x |G_{n,2}(x) - N_{n,2}(x)| = \sup_x |G_{n,2}(x\sigma_{n,2}) - N(x)|$$

converges to zero, which by the continuity theorem of characteristic functions, becomes (4.28).

We now proceed to prove the sufficiency part of the theorem. If we denote the characteristic function of Y_n by $\psi_n^*(t)$, we have to prove that $\psi_n^*(t) \to \tau(t)$

as $n \to +\infty$. But, because $\psi_n^* = \psi_{n,1}\psi_{n,2}$ and $\tau = \tau_{n,1}\tau_{n,2}$, where the variable t is suppressed, the inequality

$$|\psi_n^* - \tau| = |(\psi_{n,1} - \tau_{n,1})\psi_{n,2} + (\psi_{n,2} - \tau_{n,2})\tau_{n,1}|$$

$$\leq |\psi_{n,1} - \tau_{n,1}| + |\psi_{n,2} - \tau_{n,2}|$$

and (4.27) entail that only $\psi_{n,1} - \tau_{n,1} \to 0$ remains to be shown. However, arguing as in the above inequality, we get by induction

$$|\psi_{n,1} - \tau_{n,1}| \leq \sum_{j \in J_n} |\varphi_{j,n} - \tau_{j,n}| \qquad (4.29)$$

where $\varphi_{j,n} = \varphi_{j,n}(t)$ is the characteristic function of $X_{j,n}$ and $\tau_{j,n} = \tau_{j,n}(t) = \tau(t\sigma_{j,n})$. Hence, it suffices to show that the right-hand side of (4.29) tends to zero. Note that, since the expectations and the variances corresponding to $\varphi_{j,n}$ and $\tau_{j,n}$ are equal,

$$\varphi_{j,n}(t) - \tau_{j,n}(t) = \int_{-\infty}^{+\infty} \left(e^{itx} - 1 - itx + \frac{t^2 x^2}{2} \right) dM_{j,n}(x)$$

where $M_{j,n}(x) = F_{j,n}(x) - N(x/\sigma_{j,n})$ (recall Section 2.7). Next, we split up the above integral for $|x| < \varepsilon$ and $|x| \geq \varepsilon$. In the first integral, we use the Taylor expansion of e^{itx} up to the cubic term, while, in the second, we stop at the quadratic term. Thus, for $|t| \leq T$,

$$|\varphi_{j,n}(t) - \tau_{j,n}(t)| \leq (T^3/6)S_{j,n}^{(1)}(\varepsilon) + T^2 S_{j,n}^{(2)}(\varepsilon)$$

where

$$S_{j,n}^{(1)}(\varepsilon) = \int_{|x| < \varepsilon} |x|^3 \, dF_{j,n}(x) + \int_{|x| < \varepsilon} |x|^3 \, dN(x/\sigma_{j,n}) \leq 2\varepsilon\sigma_{j,n}^2$$

and

$$S_{j,n}^{(2)}(\varepsilon) = \int_{|x| \geq \varepsilon} x^2 \, dF_{j,n}(x) + \int_{|x| \geq \varepsilon} x^2 \, dN(x/\sigma_{j,n})$$

$$= \int_{|x| \geq \varepsilon} x^2 \, dF_{j,n}(x) + \sigma_{j,n}^2 \int_{|x| \geq \varepsilon/\sigma_{j,n}} x^2 \, dN(x)$$

If we sum these terms over $j \in J_n$ and observe that $\sigma_{j,n} < \alpha_n^{1/4}$ and $\sigma_{n,1} \leq 1$, then we get from (4.29)

$$|\psi_{n,1}(t) - \tau_{n,1}(t)| \leq (T^3/3)\varepsilon + T^2\Delta_n(\varepsilon) + 2T^2[1 - N(x\alpha_n^{-1/4})]$$

where $|t| \leq T$ and $\varepsilon > 0$ is arbitrary. We thus proved that $\psi_{n,1} - \tau_{n,1} \to 0$ as $n \to +\infty$, which completes the proof of the sufficiency part of the theorem.

We turn to the necessity of the conditions. Because of Theorem 7, (4.19) must hold, which, in turn, as was shown in the first part of the present proof, implies the validity of (4.27). Hence, it remains to show that (4.19), (4.27), and

$$\psi_n^*(t) - \tau(t) \to 0 \qquad \text{as } n \to +\infty \tag{4.30}$$

imply (4.20). The actual meaning of α_n will not be used in this part of the proof, only its convergence to zero and the fact that it is a bound of $\sigma_{j,n}$ for $j \in J_n$, thus guaranteeing the uniform convergence of these $\sigma_{j,n}$ to zero. In particular, by the assumption of zero expectations, Taylor's expansion (see Theorem 3.2) yields

$$\sup_{j \in J_n} |\varphi_{j,n}(t) - 1| \le \tfrac{1}{2} t^2 \alpha_n^{1/2} \tag{4.31}$$

and

$$
\begin{aligned}
\sum_{j \in J_n} |\varphi_{j,n}(t) - 1|^2 &\le \left(\sup_{j \in J_n} |\varphi_{j,n}(t) - 1| \right) \sum_{j \in J_n} |\varphi_{j,n}(t) - 1| \\
&\le \tfrac{1}{2} t^2 \alpha_n^{1/2} \sum_{j \in J_n} \tfrac{1}{2} t^2 \sigma_{j,n}^2 \le \tfrac{1}{4} t^4 \alpha_n^{1/2}
\end{aligned}
\tag{4.32}
$$

One of the consequences of (4.31) is that, for fixed t, however large, $\varphi_{j,n}(t)$, $j \in J_n$, are uniformly close to one, ensuring that they do not vanish. Therefore, we can take their logarithm, which, with Taylor's expansion in the form

$$\log z = \log[1 - (1 - z)] = z - 1 + \vartheta |z - 1|^2 \qquad |\vartheta| \le 1$$

valid for $|z - 1| < 1/2$, yields

$$\log \psi_{n,1}(t) = \sum_{j \in J_n} \log \varphi_{j,n}(t) = \sum_{j \in J_n} (\varphi_{j,n}(t) - 1) + \vartheta^* t^4 \alpha_n^{1/2} \qquad |\vartheta^*| \le 1/4 \tag{4.33}$$

where (4.32) is used for the error term. Now, in order to exploit (4.30), we go back to the formula preceding (4.29). We have

$$|\psi_{n,1} - \tau_{n,1}| \le |\psi_n^* - \tau| + |\psi_{n,2} - \tau_{n,2}| \tag{4.34}$$

Thus, because of (4.27) and (4.30), $\psi_{n,1} - \tau_{n,1} \to 0$ for t fixed as $n \to +\infty$, which we write in the form

$$e^{-V_{n,1} t^2/2} \left[\exp(\log \psi_{n,1}(t) + V_{n,1} t^2/2) - 1 \right] \to 0$$

Since $V_{n,1} \le 1$, this implies that, for t fixed,

$$\log \psi_{n,1}(t) + V_{n,1} t^2/2 \to 0$$

or, when combined with (4.33).

$$\sum_{j \in J_n} (\varphi_{j,n}(t) - 1) + V_{n,1}t^2/2 \to 0 \qquad (4.35)$$

for fixed t as $n \to +\infty$. Next, we write

$$\frac{t^2}{2} \Delta_n(\varepsilon) = \tfrac{1}{2}t^2\sigma_{n,1}^2 - \sum_{j \in J_n} \int_{|x| < \varepsilon} \frac{x^2 t^2}{2} dF_{j,n}(x)$$

$$\leq \tfrac{1}{2}t^2\sigma_{n,1}^2 - \sum_{j \in J_n} \int_{|x| < \varepsilon} (1 - \cos tx) \, dF_{j,n}(x)$$

$$= \tfrac{1}{2}t^2\sigma_{n,1}^2 + \sum_{j \in J_n} \left\{ \int_{|x| \geq \varepsilon} (1 - \cos tx) \, dF_{j,n}(x) + \mathrm{Re}[\varphi_{j,n}(t)] - 1 \right\}$$

Now, by Chebyshev's inequality,

$$\int_{|x| \geq \varepsilon} (1 - \cos tx) \, dF_{j,n}(x) \leq 2P(|X_{j,n}| \geq \varepsilon) \leq 2\sigma_{j,n}^2/\varepsilon^2$$

which, when summed over $j \in J_n$, gives the estimate $2\sigma_{n,1}^2/\varepsilon^2 \leq 2/\varepsilon^2$. Hence, we have proved

$$\Delta_n(\varepsilon) \leq 4/\varepsilon^2 t^2 + (1/t^2) \left\{ \sum_{j \in J_n} (\mathrm{Re}[\varphi_{j,n}(t)] - 1) + (\sigma_{n,1}^2 t^2)/2 \right\}$$

where t is arbitrary (but fixed). We thus have, by (4.35) and upon letting $n \to +\infty$, that, for arbitrary t,

$$0 \leq \limsup \Delta_n(\varepsilon) \leq 4/\varepsilon^2 t^2$$

which, by choosing t large, yields (4.20). The theorem is established. ∎

Proof of Theorem 9. The sufficiency part of the theorem has been established twice. Once directly in Theorem 3.12 and once as a special case in Theorem 8. For the latter, one has only to observe that a simple estimate in the proof of Theorem 3.12 has shown that (4.22) implies (4.21) and, in view of Lemma 4 of the previous section, $\alpha_n \leq \beta_n$. Finally, it is obvious that $\Delta_n(\varepsilon) \leq \delta_n(\varepsilon)$. Conversely, if we assume asymptotic normality and (4.21), then (4.19) becomes a part of the assumption, so (4.27) holds. In addition, Theorem 8 ensures that (4.20) holds as well. The fact that $\delta_n(\varepsilon) - \Delta_n(\varepsilon) \to 0$ can now be seen as follows (although this is not really necessary, but rather an automatic consequence of (4.21) via (4.19) and (4.27); recall Remark 3). It was emphasized in the proof of Theorem 8, immediately following (4.30), that in the necessity part of the proof of (4.20), the sole role of α_n is as a uniform bound on $\sigma_{j,n}$, $j \in J_n$, and that $\alpha_n \to 0$ as $n \to +\infty$. In Theorem 9, this role can be taken by β_n for all j; the restriction $j \in J_n$ can thus be dropped. That is, not only $\Delta_n(\varepsilon)$, but also $\delta_n(\varepsilon)$, tends to zero. It is worth noting that, while (4.30), via (4.34), is fully utilized for $\Delta_n(\varepsilon) \to 0$, it is irrelevant for $\delta_n(\varepsilon) - \Delta_n(\varepsilon) \to 0$. That is, the role of (4.34) is now taken over by (4.27), which is ensured by (4.19) (and thus by (4.21)) alone. This reinforces Remark 3 and completes the proof. ∎

Exercises

1. Let $X_1, X_2, \ldots$ be independent random variables with $E(X_j) = 0$ and with finite variances $V(X_j)$ that satisfy $\sum V(X_j) = 1$. Let $Y = \sum X_j$ and $Z = \sum X_{n_j}$, where $\{n_j\}$ is a rearrangement of the consecutive integers. Show that Y and Z are identically distributed and that $E(Y) = E(Z) = 0$.
2. Upon integrating by parts in the formula following (4.29), prove that, for Y_n, defined prior to Theorem 7, to be asymptotically normal, it is sufficient that, for any $\varepsilon > 0$ as $n \to +\infty$,

$$\sum_{j \in J_n} \int_{|x| > \varepsilon} |x| \, |F_{j,n}(x) - N(x/\sigma_{j,n})| \, dx \to 0$$

V. I. Rotar (1975)

3. Let the random variables $X_1, X_2, \ldots$ be independent, where X_j takes the values $-\sqrt{j}$ and $\sqrt{j}$ each with probability $1/2$. Show that the arithmetical mean $(X_1 + X_2 + \cdots + X_n)/n$ is asymptotically normal. Comment on this result in the light of the weak law of large numbers (Section 2.3).

4.5 SPEED OF CONVERGENCE: THE BERRY–ESSEEN THEOREM

This section is devoted to an estimate of the speed of convergence to normality in Liapunov's theorem (Section 3.3). That is, we have independent random variables $X_1, X_2, \ldots$ with $E(X_j) = 0$, $V(X_j) = \sigma_j^2 > 0$, and finite third moments. We set $E_{3,j} = E(|X_j|^3)$

$$s_n^2 = \sum_{j=1}^n \sigma_j^2 \qquad \beta_n^3 = \sum_{j=1}^n E_{3,j} \qquad \text{and} \qquad r_n = \beta_n^3/s_n^3$$

Theorem 10 (Berry and Esseen) As throughout this chapter, $G_n(x)$ denotes the distribution function of S_n. For $n \to +\infty$,

$$\sup_x |G_n(xs_n) - N(x)| \le 0.8 r_n \qquad (4.36)$$

In particular, if the X_j are identically distributed,

$$\sup_x |G_n(x\sigma\sqrt{n}) - N(x)| \leq 0.8 E_{(3)}/\sigma^3\sqrt{n} \tag{4.37}$$

where $E_{(3)} = E_{3,1}$ and $\sigma = \sigma_1$.

Remark 1 We prove a somewhat weaker version of this theorem as we shall replace the number 0.8 in the estimates with a universal constant $C > 0$ for which no specific value will be given. The reason for this is that we base our proof on the original Berry–Esseen inequality (Theorem 3.18), which itself contains an undetermined constant. In fact, in order to obtain the value $C = 0.8$, Theorem 3.18 should be strengthened (Zolotarev (1965)); a fine optimization technique is also needed (P. van Beek (1972)), who actually obtained $C \leq 0.7975$; the previous best estimate had been Zolotarev's (1967a) result of $C \leq 0.9051$ in (4.36) and $C \leq 0.82$ in (4.37). These have further been improved by I. S. Shiganov (1987): $C \leq 0.7915$ in (4.36) and $C \leq 0.7655$ in (4.37).

Remark 2 The estimates (4.36) and (4.37) cannot be improved (apart from the value 0.8) without further assumptions on the distribution functions of the X_j. Indeed, if the X_j are identically distributed, taking the values 1 and -1, each with probability 1/2, then

$$P\left(\sum_{j=1}^{2n} X_j = 0\right) = \binom{2n}{n}\left(\frac{1}{2}\right)^{2n} \sim \frac{1}{\sqrt{\pi n}}$$

(use Stirling's formula $n! \sim (n/e)^n\sqrt{2\pi n}$—see the introductory volume, Chapter 2, for details) and, thus, $G_{2n}(x(2n)^{1/2})$ has a jump of the magnitude of $(\pi n)^{-1/2}(1 + o(1))$ at $x = 0$. Consequently, $G_{2n}(x(2n)^{1/2}) -- N(x)$, where $N(x)$ is continuous, cannot be smaller (at $x = 0$) than $(\pi n)^{-1/2}(1 + o(1))/2$. From this remark it also follows that $C \geq (2\pi)^{-1/2} = 0.3989$.

Proof of Theorem 10. We apply Theorem 3.18, which, with the standard notation of this chapter, states that there is a constant $c > 0$ such that, for all $U > 0$,

$$\sup_x |G_n(xs_n) - N(x)| \leq \frac{1}{\pi}\int_{-U}^{U}\left|\frac{\psi_n(t/s_n) - e^{-t^2/2}}{t}\right| dt + \frac{c}{U\sqrt{2\pi}}$$

Hence, by choosing $U = c_1/r_n$ and establishing that, for $|t| \leq U$,

$$|\psi_n(t/s_n) - e^{-t^2/2}| \leq c_2 r_n|t|^3|e^{-t^2/3} \tag{4.38}$$

where $c_1 > 0$ and $c_2 > 0$ are suitable constants, we get (4.36) with some $C > 0$ replacing 0.8 (see Remark 1) by noting that

$$\int_{-U}^{U} t^2 e^{-t^2/3} \, dt \le \int_{-\infty}^{+\infty} t^2 e^{-t^2/3} \, dt = c_3$$

On the other hand, (4.37) is a special case of (4.36); thus, it does not require a separate proof. Therefore, it remains to prove (4.38). We shall, in fact, prove it with $c_1 = 1/4$ and $c_2 = 16$. We repeat the Taylor expansions used in Liapunov's theorem, except that we take slightly more care in some of the estimates. Thus, we recall from the proof of Theorem 3.11 that, for every $j \ge 1$,

$$(\sigma_j/s_n)^3 \le E_{3,j}/s_n^3 \le r_n \tag{4.39}$$

and

$$\varphi_j(t/s_n) = 1 - \frac{\sigma_j^2}{s_n^2} \frac{t^2}{2} + \vartheta \frac{E_{3,j}}{s_n^3} \frac{t^3}{6} \qquad |\vartheta| \le 1 \tag{4.40}$$

We establish (4.38) in two steps. First, we assume that $1/2r_n^{1/3} \le |t| \le 1/4r_n$. Then $16 r_n |t|^3 \ge 2$; thus, it suffices to prove that

$$|\psi_n(t/s_n) - e^{-t^2/2}| \le 2 e^{-t^2/3}$$

which, in turn, follows from

$$|\psi_n(t/s_n)|^2 \le e^{-2t^2/3} \tag{4.41}$$

because $e^{-t^2/2} \le e^{-t^2/3}$ and

$$|\psi_n(t/s_n) - e^{-t^2/2}| \le |\psi_n(t/s_n)| + e^{-t^2/2}$$

Now, since

$$|\psi_n(t/s_n)|^2 = |\varphi_1(t/s_n)|^2 |\varphi_2(t/s_n)|^2 \cdots |\varphi_n(t/s_n)|^2$$

and $|\varphi_j(t)|^2$ is the characteristic function of $X_j - Y_j$—where X_j and Y_j are independent and identically distributed, and thus $E(X_j - Y_j) = 0$, $V(X_j - Y_j) = 2\sigma_j^2$, and

$$E(|X_j - Y_j|^3) \le E[(|X_j| + |Y_j|)^3] = 2E_{3,j} + 6E(|X_j|)E(Y_j^2) \le 8E_{3,j}$$

(we used independence and Exercise 5 of Section 2.2)—the expansion similar to (4.40)

$$|\varphi_j(t/s_n)|^2 \le 1 - \sigma_j^2 t^2/s_n^2 + 4E_{3,j}|t|^3/3s_n^3$$

and the elementary inequality $1 - x \le e^{-x}$ yield

$$|\psi_n(t/s)|^2 \le \exp(-t^2 + 4r_n|t|^3/3) \le \exp(-2t^2/3)$$

where the last inequality is obtained from $|t| \leq 1/4r_n$. For the remainder of the proof, we can assume that both $|t| \leq 1/4r_n$ and $|t| < 1/2r_n^{1/3}$. In particular, $|t|r_n^{1/3} < 1/2$ and, thus, by (4.39) and (4.40),

$$|1 - \varphi_j(t/s_n)| < 1/8 + 1/48 = 7/48$$

and

$$|1 - \varphi_j(t/s_n)|^2 \leq 2\left(\frac{\sigma_j^2 t^2}{2s_n^2}\right)^2 + 2\left(\frac{E_{3,j}|t|^3}{6s_n^3}\right)^2 < 0.3 \frac{E_{3,j}|t|^3}{s_n^3}$$

The first estimate ensures that $\varphi_j(t/s_n)$ does not vanish in the given range of t. Therefore, its logarithm can be taken, which we expand by the formula $(1/2 \leq z < 3/2)$

$$\log z = \log[1 - (1 - z)] = z - 1 + \gamma|z - 1|^2 \qquad |\gamma| \leq 1 \qquad (4.42)$$

The second estimate above and (4.40) thus yield

$$\log \varphi_j(t/s_n) = -\sigma_j^2 t^2/2s_n^2 + \eta_j E_{3,j}|t|^3/2s_n^3 \qquad |\eta_j| \leq 1$$

from which we get, by summation over j,

$$\log \psi_n(t/s_n) = -t^2/2 + \eta r_n|t|^3/2 \qquad |\eta| \leq 1$$

Consequently,

$$|\psi_n(t/s_n) - e^{-t^2/2}| = e^{-t^2/2} |\exp(\eta r_n|t|^3/2) - 1|$$

Now, since $|e^z - 1| \leq |z|e^{|z|}$ (which follows immediately from the Taylor expansions of the two sides) and $r_n|t|^3/2 < 1/16$, we have proved

$$|\psi_n(t/s_n) - e^{-t^2/2}| \leq [r|t|^3/2] e^{-t^2/2} \exp(r_n|t|^3/2)$$
$$\leq c_4 r_n|t|^3 e^{-t^2/2}$$

where $c_4 < (1/2)e^{1/16} < 1$. Thus, (4.38) is established, which concludes the proof.   ∎

In the uniform estimates of Theorem 10, one seems to lose when the difference

$$G_n(xs_n) - N(x) = [1 - N(x)] - [1 - G_n(xs_n)]$$

is small due to the value of x rather than the value of n. The following result, due to Esseen (1945), shows that this is not the case: it is implicit in the estimates of Theorem 10 that its approximations are even better when $|x|$ is large.

Theorem 11   Under the conditions of Theorem 10, there is a constant $c_5 > 0$ such that, for all x,

$$|G_n(xs_n) - N(x)| \leq \frac{c_5 r_n |\log r_n|}{1 + x^2}$$

whenever $r_n \leq (1/e)^{1/2}$.

Proof.   We set $G_n^*(x) = G_n(xs_n)$. First, note that

$$\int_{-\infty}^{+\infty} x^2 \, dG_n^*(x) = \int_{-\infty}^{+\infty} x^2 \, dN(x) = 1 \qquad (4.43)$$

Now let $a \geq 1$ be such that both a and $-a$ are continuity points of $G_n^*(x)$. Then, by integrating by parts (recall Exercise 4 of Section 2.7),

$$\int_{-a}^{a} x^2 \, dG_n^*(x) = \int_{-a}^{a} x^2 \, d(G_n^*(x) - N(x)) + \int_{-a}^{a} x^2 \, dN(x)$$

$$= a^2(G_n^*(a) - N(a)) - a^2(G_n^*(-a) - N(-a))$$

$$- 2 \int_{-a}^{a} x(G_n^*(x) - N(x)) \, dx + \int_{-a}^{a} x^2 \, dN(x)$$

Hence, by virtue of (4.36) and (4.43),

$$\int_{|x| \geq a} x^2 \, dG_n^*(x) \leq 3.2a^2 r_n + \int_{|x| \geq a} x^2 \, dN(x)$$

But obviously, for $y \geq a$,

$$\int_{|x| \geq a} x^2 \, dG_n^*(x) \geq \int_{|x| \geq y} x^2 \, dG_n^*(x) \geq y^2(1 - G_n^*(y)) \geq y^2(N(y) - G_n^*(y))$$

and, for $y \leq -a$,

$$\int_{|x| \geq a} x^2 \, dG_n^*(x) \geq y^2 \int_{|x| \geq |y|} dG_n^*(x) \geq y^2 G_n^*(y) \geq y^2(G_n^*(y) - N(y))$$

which estimates also work if we interchange the roles of $G_n^*(x)$ and $N(x)$. Thus, for $y \geq a$,

$$y^2(N(y) - G_n^*(y)) \leq 3.2a^2 r_n + \int_{|x| \geq a} x^2 \, dN(x)$$

and

$$y^2(G_n^*(y) - N(y)) \leq \int_{|x| \geq a} x^2 \, dN(x)$$

That is,

$$y^2|G_n^*(y) - N(y)| \le 3.2a^2r_n + \int_{|x|\ge a} x^2 \, dN(x) \tag{4.44}$$

The same inequality is obtained from the previous sequence of inequalities for $y \le -a$, a yielding (4.44) for $|y| \ge a$. But (4.44) is trivially true for $|y| \le a$; thus, (4.44) is valid for all y and all $a \ge 1$ (the continuity requirements imposed earlier in relation to a are clearly not needed any more). A more convenient inequality, entailed by (4.44), is

$$(1 + y^2)|G_n^*(y) - N(y)| \le 4.2a^2r_n + \int_{|x|\ge a} x^2 \, dN(x)$$

Finally, since

$$a^{-1}e^{a^2/2}\int_{|x|\ge a} x^2 \, dN(x) \qquad a \ge 1$$

is continuous and has a finite positive limit as $a \to +\infty$ (which is immediately clear after integrating by parts), there exists a constant $c_6 > 0$ such that, for all $a \ge 1$,

$$\int_{|x|\ge a} x^2 \, dN(x) \le c_6 ae^{-a^2/2} \le c_6 a^2 e^{-a^2/2}$$

The choice $a = [2\log(1/r_n)]^{1/2}$, which satisfies $a \ge 1$ whenever $0 < r_n \le 1/\sqrt{e}$, now gives the desired inequality. Note that the reader is asked in Exercise 2 to find the value of c_6 and to conclude that $c_5 < 15$. The proof is completed. ∎

One immediate consequence of Theorems 10 and 11 is the following corollary.

Corollary (Strong Global Central Limit Theorem) Under the conditions of Theorem 10, for arbitrary $p > 1/2$,

$$\lim_{n=+\infty} \int_{-\infty}^{+\infty} |G_n(xs_n) - N(x)|^p \, dx = 0$$

and, in the identically distributed case,

$$\sum_{n=1}^{+\infty} \frac{1}{n} \int_{-\infty}^{+\infty} |G_n(x\sigma\sqrt{n}) - N(x)|^p \, dx < +\infty$$

Exercises

1. Let the X_j take the values 1 and -1, each with probability 1/2. Prove that, for all x and for sufficiently large n,

$$|G_n(x\sqrt{n}) - N(x)| \le c_7 n^{-1/2}(1 + |x|^3)^{-1}$$

2. In the proof of Theorem 11, determine the value of c_6 and deduce that $c_5 < 15$.

4.6 THE CLASS L OF LIMITING DISTRIBUTIONS

We now extend the results on the asymptotic normality of sums S_n of independent random variables X_j, $j \ge 1$, and seek conditions under which there exist constants a_n and $b_n > 0$ such that $(S_n - a_n)/b_n$ has a limiting distribution function $G(x)$ in the sense of complete convergence. No assumptions are to be made on expectations or any other moments. We would also like to determine the class of possible limiting distributions $G(x)$. However, without some assumptions on the X_j and on the sequences a_n and b_n, the answer could be too general to be of any value. For example, if X_1 is a random variable with distribution function $F(x)$ and X_j is degenerate at zero for each $j \ge 2$, then, with $a_n = 0$ and $b_n = 1$, $(S_n - a_n)/b_n$ has the (limiting) distribution function $F(x)$, so $G(x)$ could be any $F(x)$. A much less trivial example that also leads to a much too large class of possible limits is when one again takes X_1 with an arbitrary distribution function and $X_j, j \ge 2$, are such that $E(X_j) = 0$ and $\sum_{j \ge 2} V(X_j) = 1$. Then S_n has the limiting distribution function $F(x) * U(x)$, where $F(x)$ is arbitrary and $U(x)$ is the distribution function of $\sum_{j \ge 2} X_j$ (see Section 4.1). In order to avoid such situations, we want to get rid of the possibility of the domination of S_n/b_n by a single term (X_1/b_n). Hence, we assume that $b_n > 0$ is such that, for arbitrary $\varepsilon > 0$,

$$\max_{1 \le j \le n} P(|X_j| > \varepsilon b_n) \to 0 \qquad \text{as } n \to +\infty \qquad (4.45)$$

In other words, uniformly in j, $|X_j|/b_n$ is asymptotically small in probability. We shall drop the label "in probability" and shall refer to (4.45) by saying that the summands X_j/b_n are uniformly asymptotically negligible (UAN). Recall Remark 2 of Section 4.4 in connection with the Lindeberg–Feller theorem, where we first faced the concept of uniformly small summands. Indeed, the model of the present section is a direct extension of the Lindeberg–Feller theorem.

We are now able to introduce the class L of limiting distributions. We say that the distribution function $G(x)$ belongs to *class L* (of limiting distributions) if there exist independent random variables $X_1, X_2, \ldots$ and constants a_n and $b_n > 0$ such that

(i) the terms X_j/b_n, $1 \leq j \leq n$, are UAN
(ii) the limiting distribution function of $(S_n - a_n)/b_n$ exists and equals $G(x)$.

In this section, we shall give a characterization of the characteristic functions of the members of class L and prove a limit theorem that generates explicit members $G(x)$ of L. More general necessary and sufficient conditions for members of L will be obtained at the end of the next chapter. The notation of the introduction to this current chapter is used.

We begin with some lemmas.

Lemma 1 Condition (4.45) is equivalent to

$$\max_{1 \leq j \leq n} |\varphi_j(t/b_n) - 1| \to 0 \qquad \text{as } n \to +\infty \tag{4.46}$$

uniformly in $|t| < b$, $b > 0$ finite.

Proof. Since $|e^{iu} - 1| \leq |u|$ for real u,

$$|\varphi_j(t/b_n) - 1| \leq \left| \int_{|x| \leq \varepsilon} (e^{itx} - 1)\, dF_j(xb_n) \right| + \left| \int_{|x| > \varepsilon} (e^{itx} - 1)\, dF_j(xb_n) \right|$$

$$\leq b\varepsilon + 2 \int_{|x| > \varepsilon} dF_j(xb_n) = b\varepsilon + 2P(|X_j| > \varepsilon b_n)$$

which entails that (4.46) follows from (4.45).

For the converse, we begin with a classroom example from calculus. Integrating by parts twice, we get

$$\int_0^{+\infty} (1 - \cos tx) e^{-t}\, dt = \frac{x^2}{1 + x^2}$$

We now integrate the two sides with respect to $F_j(xb_n)$ and interchange the orders of integration on the left-hand side:

$$\int_0^{+\infty} e^{-t} \operatorname{Re}(1 - \varphi_j(t/b_n))\, dt = \int_{-\infty}^{+\infty} \frac{x^2}{1 + x^2}\, dF_j(xb_n)$$

Hence, since $|1 - \operatorname{Re} z| \leq |z - 1|$ and

$$1 = \frac{1 + x^2}{x^2}\, \frac{x^2}{1 + x^2} \leq \frac{1 + \varepsilon^2}{\varepsilon^2}\, \frac{x^2}{1 + x^2} \qquad |x| > \varepsilon \tag{4.47}$$

we get from the estimates

$$\max \int_{|x| > \varepsilon} dF_j(xb_n) \leq \frac{1 + \varepsilon^2}{\varepsilon^2} \max \int_{|x| > \varepsilon} \frac{x^2}{1 + x^2}\, dF_j(xb_n)$$

and

$$\int_{|x|>\varepsilon} \frac{x^2}{1+x^2}\, dF_j(xb_n) \leq \int_{-\infty}^{+\infty} \frac{x^2}{1+x^2}\, dF_j(xb_n)$$

$$\leq \int_0^{+\infty} e^{-t}|\varphi_j(t/b_n) - 1|\, dt$$

$$\leq \int_0^M \max |\varphi_j(t/b_n) - 1|\, dt + 2 \int_M^{+\infty} e^{-t}\, dt$$

where max is always taken over $1 \leq j \leq n$, that (4.46) implies (4.45). That is, M can be chosen to be arbitrarily large in the last expressions and (4.46) is uniform on the finite interval $[0, M]$. The proof is completed. ∎

Next, we prove a result whose special cases have been established in two previous exercises.

Lemma 2   Let $U_n(x)$ be a sequence of distribution functions. Assume that there are two sets (a_n, b_n) and (c_n, d_n) of sequences of constants such that $b_n > 0$ and $d_n > 0$, and

(i) $U_n(a_n + b_n x) \to V(x)$ $U_n(c_n + d_n x) \to W(x)$

weakly, where $V(x)$ and $W(x)$ are nondegenerate. Then there are finite numbers A and $B > 0$ such that

(ii) $b_n/d_n \to B$ and $(a_n - c_n)/d_n \to A$

and

(iii) $V(x) = W(A + Bx)$

Conversely, (ii) and one of the limits of (i) imply the other limit in (i); thus, (iii) holds.

Proof.   Let the characteristic functions of U_n, V, and W be $\tau_n(t)$, $\tau(t)$, and $\tau^*(t)$, respectively. Then (i) is equivalent to

$$e^{-ita_n/b_n}\tau_n(t/b_n) \to \tau(t) \qquad e^{-itc_n/d_n}\tau_n(t/d_n) \to \tau^*(t) \qquad (4.48)$$

uniformly for $|t| \leq T$, $T > 0$ finite. Now, one of the ratios b_n/d_n and d_n/b_n must have a subsequence that converges to a finite limit. Assume that $b_{n_k}/d_{n_k} \to B < +\infty$. We shall show that $B = 0$ is not possible. That is, if $B = 0$, then by (4.48),

$$|\tau_{n_k}(t/d_{n_k})|^2 \to |\tau^*(t)|^2$$

for every t and, since (4.48) is uniform on finite intervals,

$$|\tau_{n_k}(t/d_{n_k})|^2 = |\tau_{n_k}(tb_{n_k}/d_{n_k}b_{n_k})|^2 \to |\tau(0)|^2 = 1$$

Consequently, $|\tau^*(t)| = 1$ for all t, which contradicts the assumption that $W(x)$ is nondegenerate (see Exercise 1 of Section 3.4). Hence, $0 < B < +\infty$. Then, by the same argument as above, we have

$$|\tau^*(t)|^2 = |\tau(tB)|^2$$

This same form must obtain if we operate with another subsequence of b_n/d_n, whose limit is B^*, say, implying $|\tau(tB)|^2 = |\tau(tB^*)|^2$. Now, either $B = B^*$, or one of the ratios B/B^* or B^*/B is smaller than one. By symmetry in B and B^*, we may assume that $r = B^*/B < 1$. Then, with t/B in our last equation, we have $|\tau(t)|^2 = |\tau(rt)|^2$ and, by iteration, $|\tau(t)|^2 = |\tau(r^n t)|^2$ for every $n \geq 1$. This would only be possible if $|\tau(t)| = 1$, which would imply that $V(x)$ is degenerate. This contradiction ensures that $B = B^*$; thus, the first limit in (ii) is proved.

Next, we replace t by tb_n/d_n in the first limit of (4.48) and take the ratio of the two limits there. We get

$$\exp[-it(a_n - c_n)/d_n] \to \tau(Bt)/\tau^*(t) \qquad \text{as } n \to +\infty$$

which limit is well defined if t is sufficiently close to 0. This now yields that the second limit of (ii) is also valid and $\tau(Bt) = e^{-itA}\tau^*(t)$, which is equivalent to (iii).

The converse statement is more straightforward. If we use the equivalence of (i) and (4.48), we immediately see that (ii) and one of the limits in (4.48) imply the validity of the second limit of (4.48) together with $\tau(Bt) = \tau^*(t)\exp(-itA)$, i.e., (iii) holds. The lemma is established. ∎

Lemma 3 For a nondegenerate member G of class L, the normalizing constant b_n, which appears in the definition of class L, is such that $b_n \to +\infty$ with n and $b_{n+1}/b_n \to 1$.

Proof. Assume first that b_n fails to go to infinity as $n \to +\infty$. Then there is a subsequence n_k of the integers such that $b_{n_k} \to B < +\infty$ and (4.45) ensures that $B > 0$. Since G is nondegenerate, there exists a j such that X_j is nondegenerate at zero. Then, there is a $\delta > 0$ for such an X_j such that $P(|X_j| \geq \delta) \geq 1/2$. Choose $\varepsilon > 0$ so that $\varepsilon B < \delta$. Hence, (4.45) fails on the subsequence n_k; that is, such a subsequence cannot exist. We thus have $b_n \to +\infty$ with n.

Next, we observe that, because of (4.45),

$$(S_n - a_{n+1})/b_{n+1} = (S_{n+1} - a_{n+1})/b_{n+1} - X_{n+1}/b_{n+1}$$

has the same limiting distribution $G(x)$ as $(S_n - a_n)/b_n$ does (see Exercise 1 of Section 3.3). We thus have from Lemma 2 (ii) that $b_{n+1}/b_n \to 1$, which concludes the proof. ∎

The following theorem gives a characterization of the characteristic function of $G(x)$ belonging to class L.

Theorem 12 A distribution function $G(x)$ with characteristic function $\psi(t)$ belongs to class L if and only if there is a characteristic function $\tau_a(t)$ for every $0 \leq a < 1$ such that

$$\psi(t) = \psi(at)\tau_a(t) \tag{4.49}$$

Proof. First note that $\psi(t)$ satisfying (4.49) never vanishes. Indeed, if $\psi(t_0) = 0$ and $\psi(t) \neq 0$ for $0 \leq t < t_0$, then (4.49) would imply $\tau_a(t_0) = 0$. But, by the inequality (ii) of Theorem 3.16,

$$1 = 1 - |\tau_a(t_0)|^2 \leq 4(1 - |\tau_a(t_0/2)|^2)$$

that is,

$$|\tau_a(t_0/2)|^2 \leq 3/4 \qquad \text{for all } 0 < a < 1$$

while, by (4.49),

$$\tau_a(t_0/2) = \psi(t_0/2)/\psi(at_0/2) \to 1 \qquad \text{as } a \to 1$$

It was therefore wrong to assume that $\psi(t_0) = 0$.

Now, let (4.49) hold. Let X_k be a random variable with the characteristic function $\tau_{a(k)}(kt)$, where $a(k) = (k - 1)/k$, $k \geq 1$. Let the X_k be independent. Then the characteristic function of $S_n/n = (X_1 + X_2 + \cdots + X_n)/n$ equals

$$\prod_{k=1}^{n} \tau_{a(k)}(kt/n) = \prod_{k=1}^{n} \frac{\psi(kt/n)}{\psi[(k-1)t/n]} = \psi(t)$$

In addition, writing $t_{k,n} = (k - 1)t/n$,

$$|\tau_{a(k)}(kt/n) - 1| = \left| \frac{\psi(t_{k,n} + t/n)}{\psi(t_{k,n})} - 1 \right|$$

converges to zero uniformly in k and t, $1 \leq k \leq n$, $|t| \leq b$, $b > 0$ finite, because characteristic functions are uniformly continuous, $|t_{k,n}| \leq |t| \leq b$, and we know from calculus that a continuous function that does not vanish is bounded away from zero on a closed interval. Hence, Lemma 1 entails the validity of (4.45); that is $G(x)$ determined by $\psi(t)$ of (4.49) is in class L.

Conversely, if $G(x)$ is the limiting distribution function of $(S_n - a_n)/b_n$ for some independent random variables X_j satisfying (4.45), then we must show that the characteristic function $\psi(t)$ of $G(x)$ satisfies (4.49) with some characteristic function $\tau_a(t)$. Note that if $G(x)$ is degenerate then (4.49) holds with an appropriate degenerate characteristic function $\tau_a(t)$, $0 \leq a < 1$. Hence, we can assume that $G(x)$ is nondegenerate. We again show that $\psi(t) \neq 0$ for all t. We assume the opposite, i.e., that there is some t for which $\psi(t) = 0$, and show that it leads to a contradiction. Let $\psi(t_0) = 0$ but $\psi(t) \neq 0$ for all $0 \leq t < t_0$. Since (4.45) and Lemma 1 guarantee that, for all $1 \leq j \leq n$, $\varphi_j(t_0/b_n) \neq 0$, the elementary inequalities

$$e^{-2x} \leq 1 - x \leq e^{-x} \qquad 0 \leq x \leq 1/2 \tag{4.50}$$

yield that

$$0 = |\psi(t_0)|^2 = \lim_{n = +\infty} \prod_{j=1}^{n} |\varphi_j(t_0/b_n)|^2$$

$$= \lim_{n = +\infty} \prod_{j=1}^{n} \{1 - [1 - |\varphi_j(t_0/b_n)|^2]\}$$

is only possible if

$$\lim_{n = +\infty} \sum_{j=1}^{n} [1 - |\varphi_j(t_0/b_n)|^2] = +\infty$$

But then, because of the inequality (ii) of Theorem 3.16, a similar sum also diverges at $t_0/2$, which, with one more appeal to (4.50), yields that $\psi(t_0/2) = 0$. This contradicts the choice of t_0 and, thus, the existence of any t with $\psi(t) = 0$.

Next, we observe that an implication of Lemma 3 is that, as $n \to +\infty$, there is an integer $m(n) < n$ such that both $m(n)$ and $n - m(n)$ tend to infinity and $b_{m(n)}/b_n \to a$, where $0 < a < 1$ is a fixed number. That is, if we form all the ratios b_m/b_n, $1 \leq m \leq n$, some fall below a and some stay above it, because $b_m/b_n \to 0$ with m bounded, while $b_m/b_n \to 1$ if $n - m$ is bounded. Let $m(n)$ be that integer for which

$$b_{m(n)}/b_n \leq a < b_{m(n)+1}/b_n = (b_{m(n)}/b_n)(b_{m(n)+1}/b_{m(n)})$$

It is now clear that this $m(n)$ has the desired property.

We are now able to prove (4.49). Consider the following three characteristic functions:

$$\psi_n^*(t) = \exp(-ita_n/b_n) \prod_{j=1}^{n} \varphi_j(t/b_n)$$

$$\psi_{n,1}(t) = \exp(-ita_{m(n)}/b_{m(n)}) \prod_{j=1}^{m(n)} \varphi_j\left[\frac{b_{m(n)}}{b_n} \frac{t}{b_{m(n)}}\right]$$

and

$$\psi_{n,2}(t) = \exp\left[-it\frac{a_n}{b_n} - a\frac{a_{m(n)}}{b_{m(n)}}\right] \prod_{j=m(n)+1}^{n} \varphi_j\left(\frac{t}{b_n}\right) \tag{4.51}$$

Clearly, $\psi_n^*(t) = \psi_{n,1}(t)\psi_{n,2}(t)$, $\psi_n^*(t) \to \psi(t)$, and $\psi_{n,1}(t) \to \psi(at)$. Furthermore, $\psi_{n,2}(t)$ is a characteristic function because it is the product of characteristic functions and has a finite limit

$$\psi_{n,2}(t) = \psi_n^*(t)/\psi_{n,1}(t) \to \psi(t)/\psi(at) = \tau_a(t) \tag{4.52}$$

that is continuous at $t = 0$. Consequently, the limit $\tau_a(t)$ is a characteristic function (Theorem 3.5). We have also established in (4.52) that (4.49) applies. This completes the proof. ∎

In the following limit theorem, the additional assumption of finite variances is made. A general theorem of the next chapter will get rid of this assumption.

Theorem 13 (Kolmogorov) Let the random variables X_j, $j \geq 1$, be independent with $E(X_j) = 0$ and assume that their finite variances $V(X_j) = \sigma_j^2$ with $s_n^2 = \sigma_1^2 + \sigma_2^2 + \cdots + \sigma_n^2$ satisfy

(i) $(1/s_n^2)\max\{\sigma_j^2 : 1 \leq j \leq n\} \to 0$ as $n \to +\infty$

Let

$$K_n(x) = \frac{1}{s_n^2} \sum_{j=1}^{n} \int_{-\infty}^{xs_n} y^2 \, dF_j(y)$$

and assume

(ii) $K_n(x) \to K(x)$ completely

where $K(x)$ is a distribution function; S_n/s_n then has a limiting distribution $G(x)$ whose characteristic function is

(iii) $\psi(t) = \exp\left\{\int_{-\infty}^{+\infty} (e^{itx} - 1 - itx)\frac{1}{x^2} \, dK(x)\right\}$

where the integrand at $x = 0$ is defined by continuity as $-t^2/2$. Furthermore,

(iv) $\int_{-\infty}^{+\infty} x \, dG(x) = 0$ and $\int_{-\infty}^{+\infty} x^2 \, dG(x) = 1$

Remark 1   Since the integrand in the exponent of $\psi(t)$ of (iii) and its first and second derivatives with respect to t are bounded in x, we have that

$\log \psi(t)$ is well defined and both $\psi(t)$ and $\log \psi(t)$ are twice differentiable (ensured by the dominated convergence theorem). We have

$$\psi'(0) = 0 \quad \text{and} \quad \psi''(0) = -1$$

thus, (iv) follows from (iii) (recall (3.2)). Furthermore,

$$-[\log \psi(t)]'' = \int_{-\infty}^{+\infty} e^{itx} \, dK(x)$$

Hence, $\psi(t)$, via the uniqueness theorem of characteristic functions, uniquely determines $K(x)$. In other words, different $K(x)$ generate different $\psi(t)$ and vice versa.

Remark 2 If $K(x)$ is degenerate at zero, i.e., $K(x) = 1$ or 0 depending on whether $x > 0$ or $x < 0$, we get $\psi(t) = \exp(-t^2/2)$; thus, the limit law $G(x)$ is normal. In this particular case, condition (ii) reduces to Lindeberg's condition (Theorem 3.12) and (i) follows from (ii). For all other $K(x)$, the limit distribution $G(x)$ is different from normal. Of course, not every distribution function can serve as $K(x)$ in this theorem. Theorem 12, by means of (4.49), characterizes the family of those $K(x)$ which occur in (ii) and (iii) (see also Exercise 3 and Section 5.2).

Proof. First, note that Chebyshev's inequality and (i) imply (4.45) with $b_n = s_n$ (i.e., the X_j/s_n, $1 \le j \le n$, are UAN), from which, by Lemma 1, (4.46) follows. Next, we follow a method of proof that has been used before.

It follows from (4.46) that $\log \varphi_j(t/s_n)$ is finite for all $1 \le j \le n$, $n \ge n_0$; thus, so is $\log \psi_n(t/s_n)$, where, as before, $\psi_n(t)$ is the characteristic function of S_n. Hence, the Taylor expansions (see (4.42) and (3.2))

$$\log \psi_n(t/s_n) = \sum_{j=1}^{n} \log \varphi_j(t/s_n) = \varphi_j(t/s_n) - 1 + \vartheta |\varphi_j(t/s_n) - 1|^2$$

and

$$\varphi_j(t/s_n) = 1 - \vartheta^* t^2 \sigma_j^2 / 2 s_n^2$$

where both $|\vartheta|$ and $|\vartheta^*|$ are bounded by 1, and the estimate

$$\sum_{j=1}^{n} |\varphi_j(t/s_n) - 1|^2 \le (t^2/2) \sum_{j=1}^{n} \sigma_j^4 / s_n^4$$

$$\le (t^2/2) \left[\max_{1 \le j \le n} (\sigma_j^2 / s_n^2) \right] \sum_{j=1}^{n} \sigma_j^2 / s_n^2$$

$$= (t^2/2) \max_{1 \le j \le n} \sigma_j^2 / s_n^2$$

together with (4.46), yield that

$$\log \psi_n(t/s_n) = \sum_{j=1}^{n} [\varphi_j(t/s_n) - 1] + o(1)$$

$$= \sum_{j=1}^{n} \int_{-\infty}^{+\infty} (e^{itx} - 1) \, dF_j(xs_n) + o(1)$$

for $|t| \le b, b > 0$ finite. Now, because $E(X_j) = 0$ and since, with a substitution,

$$K_n(x) = \sum_{j=1}^{n} \int_{-\infty}^{x} y^2 \, dF_j(ys_n)$$

we have

$$\sum_{j=1}^{n} \int_{-\infty}^{+\infty} (e^{itx} - 1) \, dF_j(xs_n) = \sum_{j=1}^{n} \int_{-\infty}^{+\infty} (e^{itx} - 1 - itx) \frac{x^2}{x^2} \, dF_j(xs_n)$$

$$= \int_{-\infty}^{+\infty} (e^{itx} - 1 - itx) \frac{1}{x^2} \, dK_n(x)$$

that is,

$$\psi_n(t/s_n) = \exp\left\{ \int_{-\infty}^{+\infty} (e^{itx} - 1 - itx) \frac{1}{x^2} \, dK_n(x) + o(1) \right\}$$

Here, the integrand is continuous in x and converges to zero as x goes to $\pm\infty$. Therefore, the Helly–Bray theorem (Theorem 2.26) is applicable; this, with an appeal to the continuity theorem of characteristic functions, yields the desired limit. Finally, we refer to Remark 1 concerning (iv), which completes the proof.   ∎

It is left to the reader to check (in Exercise 3) that if $K(x) = 0$ for $x < 0$ and $M_a(x) = K(x) - a^2 K(x/a)$ is a continuous distribution function in the extended sense for each $0 < a < 1$, then $G(x)$, whose characteristic function is given by formula (iii) of Theorem 13, belongs to class L. Hence, in particular, if, for some $A > 0$,

$$K(x) = \begin{cases} 1 & \text{if } x \ge A \\ A^{-2} x^2 & \text{if } 0 \le x < A \\ 0 & \text{if } x < 0 \end{cases} \tag{4.53}$$

then the corresponding $G(x)$ belongs to class L. In the following example, we shall give an explicit sequence X_j for which the limiting distribution of S_n/s_n leads to $K(x)$ of (4.53). The example, although it looks very special, is not actually far from what is possible for random variables that take only two values (see Kubik (1959)).

Example Let Y_j take the two values j^a, $a > 0$ and 0 only, with probabilities $1/j$ and $1 - 1/j$, respectively, and let $X_j = Y_j - E(Y_j)$. Then, with the notation of Theorem 13, S_n/s_n has the limiting distribution $G(x)$ determined by $\psi(t)$ of the formula (iii) of Theorem 13, in which $K(x)$ is the function (4.53) with $A = (2a)^{1/2}$.

First, note that condition (i) of Theorem 13 is satisfied. Since $E(X_j) = 0$,

$$\sigma_j^2 = E(X_j^2) = j^{2a-1} - j^{2a-2} = j^{2a-1}(1 - 1/j) \tag{4.54}$$

the maximum of which is at $j = n$, i.e., it asymptotically equals n^{2a-1} and

$$s_n^2 = \sum_{j=1}^{n} \sigma_j^2 = \sum_{j=1}^{n} j^{2a-1}(1 - 1/j) \sim n^{2a}/2a \tag{4.55}$$

from which (i) of Theorem 13 follows. Next, we compute $K_n(x)$. Since $F_j(y)$ is discrete with discontinuities at $-j^{a-1}$ and $j^a - j^{a-1}$, where the increments of $F_j(y)$ are $1 - 1/j$ and $1/j$, respectively, the jth integral occurring in the definition of $K_n(x)$ is either zero, $j^{2a-2}(1 - 1/j)$, or σ_j^2, depending on the value of x. But the aggregate contribution of the second types to $K_n(x)$ equals

$$\frac{1}{s_n^2} \sum_{j=1}^{n} j^{2a-2}(1 - 1/j) \sim \frac{n^{2a-1}/(2a - 1)}{n^{2a}/2a} \to 0 \qquad \text{as } n \to +\infty$$

where (4.55) is used, and, thus, nonzero contributions to the asymptotic value of $K_n(x)$ can come only from the terms σ_j^2. This immediately implies that $K_n(x) \to 0$ for $x \le 0$. On the other hand, it is evident that if $xs_n > j^a - j^{a-1}$ for all j, i.e., if $xn^a(2a)^{-1/2} > n^a - n^{a-1}$, where we appeal again to (4.55), then $K_n(x) = 1$. We thus have that $K_n(x) \to 1$ if $x \ge (2a)^{1/2} = A$. Finally, if $0 < x < A$, then, counting only those terms whose contribution is σ_j^2,

$$K_n(x) = s_n^{-2} \sum_{n,x} j^{2a-1}(1 - 1/j) + o(1)$$

where $\sum_{n,x}$ signifies summation over j, for which $j^a - j^{a-1} \le xs_n$. With (4.55), since a finite number of terms do not matter under UAN, this becomes $j \le (x/A)^{1/a}n$. Hence, by (4.54) and (4.55), $K_n(x)$ is asymptotically equal to

$$\frac{[(x/A)^{1/a}n]^{2a}/2a}{n^{2a}/2a} = \left(\frac{x}{A}\right)^2$$

This completes the proof that $K_n(x)$ converges completely to $K(x)$ of (4.53). ∎

Exercises

1. Let $X_j, j \geq 1$, be independent random variables with a common Pareto distribution function $F(x) = 1 - 1/x$, $x \geq 1$. Show that X_j/n are UAN, but that, for $\varepsilon \leq 1/3$.

$$P\left(\max_{1 \leq j \leq n} X_j \geq \varepsilon n \right) \geq 0.95 \qquad \text{as } n \to +\infty$$

2. Deduce from Theorem 12 that class L is closed under (finite) convolutions.
3. Let $M(x)$ be a distribution function in the extended sense and define

$$\psi^*(t) = \exp\left\{ \int_{-\infty}^{+\infty} (e^{itx} - 1 - itx) \frac{1}{x^2} \, dM(x) \right\}$$

Use Riemann–Stieltjes sums and the passage to the limit to show that $\psi^*(t)$ is a characteristic function. Then deduce from Theorems 12 and 13 that if $K(x)$ is a distribution function and if, for every $0 < a < 1$, $M_a(x) = K(x) - a^2K(x/a)$, $x > 0$, and $N_a(x) = K(x) - a^2K(x/a)$, $x < 0$, are continuous distribution functions in the extended sense, then the $G(x)$ determined by $\psi(t)$ of (iii) of Theorem 13 belongs to class L.

REFERENCES

Beek, van P. (1972). An application of Fourier methods to the problem of sharpening the Berry–Esseen inequality. *Zeitschrift für Wahrscheinlichkeits. verw Geb. 23*, 187–196.

Berry, A. C. (1941). (See the reference in Chapter 3.)

Cramér, H. (1936). Über eine Eigenschaft der normalen Verteilungsfunktion. *Math. Z. 41*, 405–414.

Doob, J. L. (1953). *Stochastic Processes*. Wiley, New York.

Esseen, C. G. (1945). (See the reference in Chapter 3).

Feller, W. (1935). Über den zentralen Grenzwertsatz der Wahrscheinlichkeitsrechnung. *Math. Z. 40*, 521–559.

Gnedenko, B. V. (1939). On the theory of limit theorems for sums of independent random variables. *Izv. Akad. Nauk SSSR, Ser. Mat. 4*, 181–232 and 643–647. (In Russian.)

Gnedenko, B. V. and A. N. Kolmogorov (1954). *Limit Distributions for Sums of Independent Random Variables*. Addison–Wesley, Cambridge, Mass.

Hengartner, W. and R. Theodorescu (1973). *Concentration Functions.* Academic Press, New York.

Kawata, T. and M. Udagawa (1949). On infinite convolutions. *Kodai Math. Seminar Reports 3*, 15–22.

Kolmogorov, A. N. (1933). *Grundbegriffe der Wahrscheinlichkeitsrechnung.* Springer, Berlin. (English translation: (1956). *Foundations of the Theory of Probability.* Chelsea, New York.)

Kubik, L. (1959). The limiting distributions of cumulative sums of independent two-valued random variables. *Studia Math. 18*, 295–309.

Lévy, P. (1937). *Théorie de l'Addition des Variables Aléatoires.* Gauthier-Villars, Paris.

Liapunov, A. M. (1901). (See the reference in Chapter 3.)

Lindeberg, J. W. (1922). (See the reference in Chapter 3.)

Markushevich, A. I. (1965–67). *Theory of Functions of a Complete Variable,* Volumes I–III. Prentice Hall, Englewood Cliffs, New Jersey.

Raikov, D. A. (1938). On the decomposition of Gauss's and Poisson's laws. *Izv. Akad. Nauk SSSR, Ser. Mat., 2*, 91–124. (In Russian.)

Rotar, V. I. (1975). An extension of the Lindeberg–Feller theorem. *Math. Notes 18*, 660–663.

Shiganov, I. S. (1987). On an improvement of the constant in the upper estimate of the remainder term in the central limit theorem. In: Problems of Stability of Stochastic Models. (Ed.: V. M. Zolotarev), *Izd. VNIISI,* Moscow, pp. 109–115.

Steinhaus, H. (1920). Sur les distances des points des ensembles de mesure positive. *Fund. Math. 1*, 93–104.

Zolotarev, V. M. (1965). On the closeness of the distributions of two sums of independent random variables. *Theory Prob. Appl. 10*, 472–479.

Zolotarev, V. M. (1967). A generalization of the Lindeberg–Feller theorem. *Theory Prob. Appl. 12*, 608–618.

Zolotarev, V. M. (1967a). Some inequalities in probability theory and their application in sharpening the Lyapunov theorem. *Soviet Math. Dokl. 8*, 1427–1430.

5

Triangular Arrays of Independent Random Variables, Infinitely Divisible Distributions

5.1 INTRODUCTION

The standard assumptions of this chapter are as follows. We deal with double indexed sequences X_{jn} of random variables, where $1 \leq j \leq k(n)$, $n \geq 1$, and $k(n) \to +\infty$ with n. For fixed n, the finite number $k(n)$ of variables X_{jn} are independent. The collection $\{X_{jn}\}$ is called a triangular array. The distribution function and the characteristic function of X_{jn} are denoted by $F_{jn}(x)$ and $\varphi_{jn}(t)$, respectively. No assumptions are made on moments. The variables X_{jn} are assumed to satisfy the *uniform asymptotic negligibility* (UAN) assumption: for every $\varepsilon > 0$,

$$\max_{1 \leq j \leq k(n)} P(|X_{jn}| \geq \varepsilon) \to 0 \qquad \text{as } n \to +\infty \tag{5.1}$$

We set

$$S_{(n)} = \sum_{j=1}^{k(n)} X_{jn} \qquad \text{and} \qquad G_n^*(x) = P(S_{(n)} \leq x)$$

The characteristic function of $S_{(n)}$ is denoted by $\psi_n^*(t)$. The aim of this chapter is to present the theory of complete convergence of $G_n^*(x)$ under UAN.

As was pointed out in the preceding chapter in the special case $X_{jn} = X_j/b_n$, where $X_j, j \geq 1$, is an infinite sequence of independent random

198

variables and $b_n > 0$ is a sequence of constants, the limiting theory of sums without the UAN assumption is quite meaningless when one of the aims is to determine the set of all possible limit distributions ("limiting distribution functions" and "limit laws" are alternative terms). However, when convergence to a particular distribution, such as asymptotic normality, is the sole interest, then (5.1) can be dropped (recall Zolotarev's theorem from Section 4.4, which has already been formulated for double-indexed sequences). We shall not discuss this aspect any further.

The motivation for turning to triangular arrays from normalized sums is best demonstrated by the binomial distribution. Let $Y_j, j \geq 1$, be independent random variables that take only the values 1 and 0 with respective probabilities p and $q = 1 - p$. Then $S_n = \sum_{j=1}^{n} Y_j$ is binomial and, in the normal approximation to the binomial, we sum $X_{jn} = (Y_j - p)/(npq)^{1/2}$, which satisfies (5.1). Another limit theorem for S_n is the Poisson approximation when $X_{jn} = Y_j$ with $p = p_n = a/n$. Once again, (5.1) holds. These two limit theorems are significantly different in nature: the approach we take here will unify them.

One of the remarkable results of this chapter is that the family of possible limit laws for $G_n^*(x)$ consists of those distribution functions $G(x)$ whose characteristic functions

$$\psi(t) = \tau_n^n(t) \qquad \tau_n(t) \text{ characteristic function,} \qquad n \geq 2 \qquad (5.2)$$

In other words, one of the nth roots of $\psi(t)$ must be a characteristic function for every $n \geq 2$. An equivalent formulation of (5.2) is that there are independent random variables X_{jn} with common distribution function $F_n(x)$ such that

$$S_{(n)} = \sum_{j=1}^{n} X_{jn} \qquad E(e^{itX_{jn}}) = \tau_n(t) \qquad P(S_{(n)} \leq x) = G(x) \qquad (5.2a)$$

for every $n \geq 2$. A distribution function $G(x)$, or its characteristic function $\psi(t)$, is called *infinitely divisible* (to be abbreviated i.d.) if (5.2)—or, equivalently, (5.2a)—holds. Immediate examples of i.d. distributions are the normal, Poisson, and Cauchy distributions, whose characteristic functions obviously satisfy (5.2) when $\tau_n(t)$ is of the same type as $\psi(t)$ for the normal and Cauchy distributions, and $\tau_n(t)$ is Poisson for Poisson $\psi(t)$ (see Sections 3.1 and 3.2). Other i.d. distributions are the degenerate ones and the family of gamma distribution functions $G(x; a, \gamma)$ given by the densities

$$g(x) = g(x; a, \gamma) = \frac{a^\gamma}{\Gamma(\gamma)} x^{\gamma-1} e^{-ax} \qquad x \geq 0 \qquad (5.3)$$

where $a > 0$, $\gamma > 0$,

$$\Gamma(u) = \int_0^{+\infty} x^{u-1} e^{-x} \, dx \qquad u > 0 \tag{5.4}$$

and whose characteristic function

$$\psi(t) = \psi(t; a, \gamma) = (1 - it/a)^{-\gamma} \tag{5.5}$$

In order to establish (5.5), substitute $y = x(a - it)$ in the integral

$$\psi(t) = \int_0^{+\infty} e^{itx} \frac{a^\gamma}{\Gamma(\gamma)} x^{\gamma-1} e^{-ax} \, dx$$

which transforms the path of integration from the positive real line to the line $y = x(a - it)$. Then an easy estimate shows that the complex integral $\psi(z)$ on the line segment $z = N - iu$, u varying between 0 and Nt, tends to zero as $N \to +\infty$ and, thus, via the residue theorem of complex integrals (the integrand is regular, so its integral on closed curves is zero), integration can be moved back to the positive real line, yielding

$$\psi(t) = (a - it)^{-\gamma} \int_0^{+\infty} \frac{a^\gamma}{\Gamma(\gamma)} y^{\gamma-1} e^{-y} \, dy$$

which is (5.5) in view of (5.4). Now, by (5.5),

$$\psi^{1/n}(t; a, \gamma) = \psi(t; a, \gamma/n) = \tau_n(t)$$

which is a characteristic function; thus, (5.2) holds. Consequently, every gamma distribution is i.d.

Let X be a random variable with standard normal distribution function $F(x)$. Then the distribution function of X^2 is $F^*(x) = 2F(\sqrt{x}) - 1$, whose density function is of the form (5.3), that is, gamma. Consequently, $F^*(x)$ is i.d. Next, let

$$f_2(X) = X^2 + 2aX + c = (X + a)^2 + c - a^2$$

be a quadratic polynomial in X. Since $X + a$ is normal with parameters a and 1, by direct computation, the characteristic function $\rho(t)$ of $Y = (X + a)^2$ is

$$\rho(t) = \frac{1}{1 - 2it} \exp\left(\frac{ita^2}{1 - 2it}\right) = \frac{1}{1 - 2it} \exp\left[\frac{a^2}{2}\left(\frac{1}{1 - 2it} - 1\right)\right]$$

where, on the extreme right, we recognize the product of two i.d. characteristic functions: the first one is a (special) gamma (i.e., exponential), while the second one is of the form of Exercise 5 shown there to be i.d. Theorem 3, established in the forthcoming section, therefore, implies that Y, and thus $f_2(X)$, is i.d. On the other hand, if we take polynomials $f_n(X)$ of degree $n \geq 3$, one can choose the coefficients so that $f_n(X)$ is no longer i.d. Rohatgi and Székely (1992) give such examples by the following very clever approach. Keeping the preceding notation and assumptions, consider $Z = F(X) - 1/2$. Then Z is uniformly distributed over the interval $(-1/2, 1/2)$, and thus its characteristic function $u(t) = \sin(t/2)/(t/2)$, whose minimum, taken at the smallest positive t satisfying $t = 2\tan(t/2)$, must be negative. The approximate solution of this last equation, by numerical methods, comes close to $t = 9$. A convenient choice of t for trigonometric computations, whose value is close to nine, is $t = 3\pi$. It can then be suspected that, if $f_n(x)$ is the Taylor polynomial approximation of degree n to $F(x) - 1/2$, the characteristic function $u_n(t)$ of $f_n(X)$ is negative at $t = 3\pi$ for each $n \geq 3$. If indeed $u_n(3\pi) < 0$, then the distribution of $f_n(X)$ cannot be i.d. because the facts that $u_n(0) = 1$ and $u_n(t)$ is continuous would entail $u_n(t) = 0$ for some $0 < t < 3\pi$, which, by Theorem 1, is impossible for an i.d. characteristic function. Now, to show that $u_n(3\pi) < 0$, one can proceed by setting up the integral expressing $u_n(t)$ and then using numerical integration to compute $u_n(3\pi)$. We limit the details to $n = 3$. We have

$$f_3(X) = (2\pi)^{-1/2}(X - X^3/6)$$

and thus

$$u_3(t) = \frac{1}{\sqrt{2\pi}} \int_{-\infty}^{+\infty} \left[\cos\left(x - \frac{x^3}{6} \right) \frac{t}{\sqrt{2\pi}} \right] e^{-x^2/2} \, dx$$

Numerical computations yield $u_3(3\pi) = -0.2734$ (the authors Rohatgi and Székely used Mathematica).

    An example of a distribution that is not i.d. is given by the characteristic function $\varphi(t) = |1 + 3e^{it}|^2/16$. Its square roots are not characteristic functions; this is the result of Exercise 4 of Section 3.2. This is a special case of a large class of distributions (those of bounded random variables) that are not i.d. (see Exercise 1).

    The following theorem can be used to make τ_n of (5.2) specific.

Theorem 1   If (5.2) holds, then $\psi(t) \neq 0$ and $\tau_n(t) \to 1$ as $n \to +\infty$ for all t.

Proof. We can write $|\psi(t)|^2 = [|\tau_n(t)|^2]^n$, where $|\psi(t)|^2$ is a real-valued characteristic function. Hence, $|\tau_n(t)|^2 = |\psi(t)|^{2/n}$, where the real-valued positive root is taken for every n. First, we show that $|\tau_n(t)|^2$ has a limit $\tau(t)$ for every t. As a matter of fact, if $\psi(t) = 0$, then $|\tau_n(t)|^2 = 0$ for all n, while

$$|\tau_n(t)|^2 = \exp[(2/n)\log|\psi(t)|] \to 1$$

for $\psi(t) \neq 0$. Hence, $\tau(t) = 0$ or 1 depending on whether $\psi(t) = 0$ or $\neq 0$. The latter case applies in an interval containing zero because $\psi(0) = 1$ and $\psi(t)$ is continuous. Consequently, $\tau(t) = 1$ in this interval, implying that $\tau(t)$ is continuous at $t = 0$. Thus, by the continuity theorem, $\tau(t)$ is a characteristic function that must be continuous everywhere, so it cannot "jump" from one to zero. That is, $\tau(t) = 1$ and $\psi(t) \neq 0$ for all t. But then $\log \psi(t)$ is finite and well defined if we require that $\log \psi(0) = 0$. Thus the unique nth root

$$\tau_n(t) = \exp[(1/n)\log \psi(t)] = \psi^{1/n}(t) \tag{5.6}$$

defines $\tau_n(t)$, which converges to one as $n \to +\infty$. The theorem is established. ∎

Remark In the sequel, if $\psi(t)$ is an i.d. characteristic function, then $\psi^{1/n}(t)$ signifies the root given in (5.6) (for which $\tau_n(0) = 1$).

As a corollary to the following theorem, we get a close relation between the UAN condition and the class of i.d. distributions.

Theorem 2 The UAN condition (5.1) is equivalent to either of the following two conditions:

(i) $\displaystyle \max_j \int_{-\infty}^{+\infty} \frac{x^2}{1 + x^2} \, dF_{jn}(x) \to 0$

(ii) $\displaystyle \max_j |\varphi_{jn}(t) - 1| \to 0$ uniformly in $|t| \le T < +\infty$

Proof. A substantial part of this theorem was proved in Lemma 1 of Section 4.6. Indeed, if we make the notational changes $\varphi_j(t/b_n) = \varphi_{jn}(t)$ and $F_j(xb_n) = F_{jn}(x)$ in the proof of Lemma 1 of Section 4.6, then, without any further changes, we have that (5.1) and (ii) are equivalent and

$$\int_0^{+\infty} e^{-t} \operatorname{Re}(1 - \varphi_{jn}(t)) \, dt = \int_{-\infty}^{+\infty} \frac{x^2}{1 + x^2} \, dF_{jn}(x)$$

The left-hand side is increased if $\operatorname{Re}(1 - \varphi_{jn}(t))$ is replaced by $|1 - \varphi_{jn}(t)|$. Hence, upon taking the maximum on both sides with respect to j and then letting $n \to +\infty$, the dominated convergence theorem shows that (ii) implies (i). Finally, if (i) holds, then the maximum over j of the same integral—when integration is limited to $|x| \ge \varepsilon$—also tends to zero, which, as seen from the inequality following (4.47), implies (5.1). This completes the proof. ∎

Corollary The family of complete limits of $G_n^*(x)$ contains all i.d. distribution functions.

Proof. By (5.2a), we need only demonstrate that the special summands X_{jn} in (5.2a) satisfy (5.1) or, equivalently, that Theorem 2 (ii) applies with $\varphi_{jn}(t) = \tau_n$. This, however, follows from Theorem 1, since convergence of characteristic functions is uniform on finite intervals.   ∎

The converse statement to the Corollary is more difficult, which will be proved in Section 5.3.

Exercises

1. Let the random variable X be bounded, i.e., with some finite constants $a < b$, $P(a \le X \le b) = 1$. Show that the distribution function $F(x)$ of X is not i.d., provided that $F(x)$ is nondegenerate.
2. Clearly, every member of the class L is a possible limit of $G_n^*(x)$. The converse is not true. Show that the Poisson distribution, which is i.d., is not in class L.
3. From Exercise 7 of Section 3.2, reconstruct the characteristic function of $G(x)$ whose density $g(x) = (1 - \cos x)/\pi x^2$, and give a reason why $G(x)$ is not i.d.

5.2   MORE ON INFINITELY DIVISIBLE DISTRIBUTIONS

We shall now analyze in detail the family of i.d. distributions introduced in (5.2) and (5.2a).

Theorem 3   The family of i.d. distributions is closed under convolution and complete convergence.

Proof. If $\psi_1(t)$ and $\psi_2(t)$ are characteristic functions that satisfy (5.2) with $\tau_{n,1}(t)$ and $\tau_{n,2}(t)$, respectively, then the convolution of their respective distributions has the characteristic function $\psi_1(t)\psi_2(t)$, which satisfies $\psi_1(t)\psi_2(t) = [\tau_{n,1}(t)\tau_{n,2}(t)]^n$, where $\tau_{n,1}(t)\tau_{n,2}(t)$ is a characteristic function. Consequently, (5.2) holds and the convolution is i.d.

Next, if ψ_n is a sequence of i.d. characteristic functions such that $\psi_n \to \psi$, where ψ is a characteristic function, we will show that ψ is i.d. Now, for every integer $m \ge 1$, $|\psi_n|^{2/m}$ is a characteristic function and converges to $|\psi|^{2/m}$. Since ψ is continuous at zero, so is $|\psi|^{2/m}$, i.e., $|\psi|^{2/m}$ is a characteristic function for every $m \ge 1$. But then $|\psi|^2$ is i.d.; thus, by Theorem 1, $|\psi(t)|^2 \ne 0$ for all t. Hence, by the convention made at (5.6), both $\log \psi_n(t)$ and $\log \psi(t)$ are well defined and finite and, for $m \ge 1$ fixed,

$$\psi_n^{1/m} = \exp[(1/m)\log \psi_n] \to \exp[(1/m)\log \psi] = \psi^{1/m}$$

which, because it is continuous at zero, is a characteristic function. We thus have shown that ψ is i.d. The proof is completed.   ∎

For the following theorem, the reader is reminded that if X is a Poisson variable with parameter $a > 0$, then its characteristic function $\varphi(t) = \exp[a(e^{it} - 1)]$. Now, if A and $B \neq 0$ are constants, then $Y = A + BX$ is a Poisson-type variable (or has a Poisson-type distribution), whose characteristic function

$$\varphi_Y(t) = e^{itA} + a(e^{itB} - 1) \qquad \text{or} \qquad \log \varphi_Y(t) = itA + a(e^{itB} - 1)$$

In the sequel, these forms will routinely be used.

The following theorem, which shows the significance of Poisson-type random variables, is quite surprising.

Theorem 4   A distribution function $G(x)$ is i.d. if and only if $G(x)$ is the complete limit of a finite convolution of Poisson-type distributions.

Proof.   It has been remarked that a Poisson-type distribution is i.d. Hence, by Theorem 3, their convolutions and complete limits remain i.d.

Conversely, if some $G(x)$ with a characteristic function $\psi(t)$ is i.d., then $\psi^{1/n}$ is the characteristic function of a distribution $F_n(x)$ and, by Theorem 1, $\psi^{1/n} \to 1$ as $n \to +\infty$. Hence,

$$\psi^{1/n}(t) - 1 = \int_{-\infty}^{+\infty} (e^{itx} - 1) \, dF_n(x) \tag{5.7}$$

and

$$n[\psi^{1/n}(t) - 1] = n\{\exp[(1/n)\log \psi(t)] - 1\} \to \log \psi(t) \tag{5.8}$$

Next, we replace the right-hand side of (5.7) by Riemann–Stieltjes sums. Let $x_1 < x_2 < \cdots < x_s$ be continuity points of $F_n(x)$ such that

$$\left| \sum_{j=0}^{s} (e^{itz_j} - 1)[F_n(x_{j+1}) - F_n(x_j)] - \int_{-\infty}^{+\infty} (e^{itx} - 1) \, dF_n(x) \right| \leq \frac{1}{n^2}$$

where $x_0 = -\infty$, $x_{s+1} = +\infty$, and z_j is an arbitrary finite point in $[x_j, x_{j+1}]$, $0 \leq j \leq s$. Then (5.7) and (5.8) can be combined into

$$n \sum_{j=0}^{s} (e^{itz_j} - 1)[F_n(x_{j+1}) - F_n(x_j)] \to \log \psi(t)$$

The left-hand side can be recognized as the sum of a finite number of the logarithm of Poisson-type characteristic functions by setting $a_{jn} = n[F_n(x_{j+1}) - F_n(x_j)]$ and $B_{jn} = z_j$. This concludes the proof.   ∎

Formulas (5.7) and (5.8) suggest that, for an i.d. characteristic function $\psi(t)$, $\log \psi(t)$ can be represented by an integral of the type appearing in (5.7). This is not quite true, as $nF_n(x)$ might not remain finite as $n \to +\infty$. However, with one more correction term, we indeed get an integral representation of $\log \psi(t)$.

Theorem 5 The characteristic function $\psi(t)$ is i.d. if and only if there are a constant L and a nondecreasing, right-continuous function $K^*(x)$ with $K^*(-\infty) = 0$ and $K^*(+\infty)$ finite such that

$$\log \psi(t) = itL + \int_{-\infty}^{+\infty} \left(e^{itx} - 1 - \frac{itx}{1 + x^2} \right) \frac{1 + x^2}{x^2} \, dK^*(x) \tag{5.9}$$

where the integrand at $x = 0$ is defined by continuity as $-t^2/2$.

This representation (5.9) will be referred to as the Lévy–Khintchine formula (for $\log \psi(t)$) and the function $K^*(x)$ will be called the Lévy–Khintchine spectral function. The reader is asked in Exercise 1 to establish a formula equivalent to (5.9) and in Exercise 2 to show that (5.9) reduces to Kolmogorov's formula when the variance associated with $\psi(t)$ is finite; this contains, as a special case, the formula in Theorem 4.13 (iii).

The proof is essentially given in the next three lemmas.

Lemma 1 $\psi(t)$ of (5.9) is an i.d. characteristic function.

Proof. Since (5.9) reduces to a degenerate i.d. characteristic function when $K^*(x) \equiv 0$, we may assume for the remainder of the proof that $K^*(x)$ is not identically zero. Now, note that the integrand on the right-hand side of (5.9) is bounded in x and continuous in t for fixed x; therefore, the integral itself, and thus $\psi(t)$, is continuous in t. In addition, if we form Riemann–Stieltjes sums in which each division point x_j is different from zero and is a continuity point of $K^*(x)$ such that $K^*(x_{j+1}) - K^*(x_j) > 0$, we again recognize them to be sums of the logarithm of Poisson-type characteristic functions. Hence, their limit, $\log \psi(t)$, for which we have established continuity and $\log \psi(0) = 0$, is necessarily the logarithm of a characteristic function by the continuity theorem and $\psi(t)$ is i.d. by Theorem 4. The lemma is established. ∎

Lemma 2 $\psi(t)$ of (5.9) uniquely determines the pair $(L, K^*(x))$.

Proof. Set $\rho(t) = \log \psi(t)$ and define

$$\varphi(t) = \rho(t) - \int_0^1 \frac{\rho(t + u) + \rho(t - u)}{2} \, du \tag{5.10}$$

Hence, from (5.9) and (5.10), we get

$$\varphi(t) = \int_0^1 \left[\int_{-\infty}^{+\infty} e^{itx}(1 - \cos ux) \frac{1 + x^2}{x^2} \, dK^*(x) \right] du$$

which, after interchanging the order of integration and substituting

$$U(x) = \int_{-\infty}^x \left(1 - \frac{\sin y}{y} \right) \frac{1 + y^2}{y^2} \, dK^*(y) \tag{5.11}$$

becomes

$$\varphi(t) = \int_{-\infty}^{+\infty} e^{itx} \, dU(x) \tag{5.12}$$

Clearly, $U(x)$ is nondecreasing, continuous from the right, $U(+\infty) < +\infty$, and $U(-\infty) = 0$. Thus, with an appeal to the uniqueness theorem of characteristic functions (Theorem 3.4; see also the remark immediately following Definition 3 of Section 2.7), $U(x)$ is uniquely determined by $\varphi(t)$, which, in turn, is determined by $\psi(t)$ via $\rho(t)$ in (5.10). Finally, $U(x)$ uniquely determines $K^*(x)$ by the formula

$$K^*(x) = \int_{-\infty}^x \left(1 - \frac{\sin y}{y} \right)^{-1} \left(\frac{1 + y^2}{y^2} \right)^{-1} dU(y) \tag{5.13}$$

because, with universal constants $0 < c_1 < c_2 < +\infty$,

$$c_1 \le \left(1 - \frac{\sin y}{y} \right) \frac{1 + y^2}{y^2} \le c_2$$

Clearly, the functions $\psi(t)$ and $K^*(x)$ assign a unique value to L in (5.9), which completes the proof. ∎

Lemma 3 Let $\psi_n(t)$, L_n, and $K_n^*(x)$ be defined as at (5.9) and assume that they satisfy (5.9). If $K_n^*(x) \to K^*(x)$ completely and $L_n \to L$ as $n \to +\infty$, then $\psi_n(t) \to \psi(t)$, as determined by L and $K^*(x)$ via (5.9). Conversely, if $\psi_n(t)$ converges to a function $\eta(t)$ that is continuous at zero, then $L_n \to L$ and $K_n^*(x) \to K^*(x)$ completely and $\eta(t) = \psi(t)$, as determined by L, $K^*(x)$, and (5.9).

Proof. The first part of the lemma is an immediate consequence of the Helly–Bray theorem (Theorem 2.26). Turning to the converse, we note that, by the continuity theorem, $\eta(t)$ is a characteristic function and the convergence $\psi_n(t) \to \eta(t)$ is uniform on finite intervals. Furthermore, Theorem 3 ensures that $\eta(t)$ is i.d.; thus, by Theorem 1, $\eta(t) \ne 0$ for all t. Hence, $\rho^*(t) = \log \eta(t)$ is finite and continuous and $\rho_n(t) = \log \psi_n(t)$ converges to

$\rho^*(t)$ uniformly on finite intervals. It thus follows that, if $\varphi_n(t)$ and $\varphi^*(t)$ denote the transformation (5.10) of $\rho_n(t)$ and $\rho^*(t)$, respectively, then

$$\varphi_n(t) \to \varphi^*(t) \tag{5.14}$$

In particular, the case $t = 0$, via (5.12), implies $U_n(+\infty) = \varphi_n(0) \to \varphi^*(0)$ which is finite. Thus, there is a constant $c > 0$ such that, for each n, $U_n(x)/c$ is a distribution function in the extended sense, where both (5.11) and (5.12) apply to $U_n(x)$. Hence, the continuity theorem, by (5.14), yields $U_n(x) \to U^*(x)$ completely, where $U^*(x)$ is the function determined by $\varphi^*(t)$ in (5.12). Therefore, using the Helly–Bray theorem in (5.13), we get that $K_n^*(x) \to K^*(x)$ completely, where $K^*(x)$ is given by (5.13) when $U(y)$ is replaced by $U^*(y)$. Finally, appealing once more to the Helly–Bray theorem, (5.9) entails $L_n \to L$. We also obtain that $\eta(t) = \psi(t)$, as determined by L, $K^*(x)$, and (5.9), and the proof is concluded. ∎

Completion of the proof of Theorem 5. Given Lemma 1, we need only prove that if $\psi(t)$ is i.d., then, with suitable L and $K^*(x)$, (5.9) applies. To this effect, we rewrite (5.7) and use (5.8) and Lemma 3. That is, if we set

$$L_n = \int_{-\infty}^{+\infty} \frac{nx}{1 + x^2} dF_n(x) \qquad K_n^*(x) = n \int_{-\infty}^{x} \frac{y^2}{1 + y^2} dF_n(y) \tag{5.15}$$

(5.7), when multiplied by n, becomes

$$n(\psi^{1/n}(t) - 1) = itL_n + \int_{-\infty}^{+\infty} \left(e^{itx} - 1 - \frac{itx}{1 + x^2} \right) \frac{1 + x^2}{x^2} dK_n^*(x)$$

where the right-hand side has the form as (5.9). Now, (5.8) ensures that the left-hand side converges to $\log \psi(t)$; thus, the right-hand side converges as well. But then Lemma 3 implies that L_n converges to some L and $K_n^*(x) \to K^*(x)$ completely, so (5.9) obtains. This terminates the proof. ∎

The following two particular cases of (5.9) deserve special attention.

Example 1 If V is a positive number and

$$K^*(x) = \begin{cases} V & \text{if } x \geq 0 \\ 0 & \text{if } x < 0 \end{cases}$$

then (5.9) becomes

$$\log \psi(t) = itL - t^2 V/2$$

i.e., $\psi(t)$ is the characteristic function of a normal variable with expectation L and variance V. By the uniqueness expressed in Lemma 2, no other spectral function can lead to the normal distribution.   ■

Example 2   Let $C \neq 0$ and $D > 0$. Then $\psi(t)$ is Poisson type, i.e.,

$$\log \psi(t) = itA + a(e^{itB} - 1)$$

with some A, $B \neq 0$, and $a > 0$, if the spectral function in (5.9)

$$K^*(x) = \begin{cases} D & \text{if } x \geq C \\ 0 & \text{if } x < C \end{cases}$$

It is easily computed that $A = L - D/C$, $a = D(1 + C^2)/C^2$, and $B = C$. Again, Lemma 2 ensures that no other kind of Lévy–Khintchine spectral function can provide a Poisson-type characteristic function. From the numbers we can see that we get an exactly Poisson distribution from (5.9) if $C = 1$ and $L = D = a/2$.   ■

Other interesting consequences of the results of the present section are contained in the following examples.

Example 3   Let X be a geometric variable (discrete waiting time) with parameter $0 < p < 1$. That is, X takes the positive integers $1, 2, \ldots$ with distribution $P(X = k) = pq^{k-1}$, where $q = 1 - p$. Then the distribution of X is i.d.

Indeed, the characteristic function of X

$$\psi(t) = p \sum_{k=1}^{+\infty} e^{itk} q^{k-1} = (1 - q)e^{it}/(1 - qe^{it})$$

Thus, Taylor's expansion of $\log(1 - z)$ yields

$$\log \psi(t) = it + \sum_{k=1}^{+\infty} (e^{itk} - 1)q^k/k$$

Now, each term in the infinite sum is the logarithm of a Poisson-type characteristic function, which is i.d. Hence, the whole sum, in view of Theorem 3, generates an i.d. distribution, whose convolution with the degenerate distribution corresponding to the characteristic function e^{it} remains i.d., i.e., the geometric distribution is i.d.   ■

Example 4   If $\psi(t)$ is an i.d. characteristic function, then so is $|\psi(t)|^\alpha$ for arbitrary real number $\alpha > 0$. In particular, $|\psi(t)|$ is always a characteristic function (compare this with Exercise 4 of Section 3.2).

Indeed, since, in view of Theorem 5, the infinite divisibility of $\psi(t)$ implies that of $\psi(-t)$, their convolution $|\psi(t)|^2 = \psi(t)\psi(-t)$ is i.d. (Theorem 3). Let L and $K^*(x)$ be the constant and the spectral function, respectively, in the representation (5.9) of $|\psi(t)|^2$. Then, with $\alpha L/2$ and $\alpha K^*(x)/2$, (5.9) applies to $|\psi(t)|^\alpha$, i.e., $|\psi(t)|^\alpha$ is an i.d. characteristic function.   ∎

Exercises

1. Let $\psi(t)$ be an i.d. characteristic function with representation (5.9). Set $\sigma^2 = K^*(0) - K^*(0-)$ and define

$$M(x) = -\int_x^{+\infty} \frac{1 + y^2}{y^2}\, dK^*(y) \qquad \text{if } x > 0$$

and

$$m(x) = \int_{-\infty}^x \frac{1 + y^2}{y^2}\, dK^*(y) \qquad \text{if } x < 0$$

The representation (5.9) then becomes (Lévy's formula)

$$\log \psi(t) = itL - \frac{\sigma^2 t^2}{2} + \int_{-\infty}^{0-} \left(e^{itx} - 1 - \frac{itx}{1 + x^2} \right) dm(x)$$
$$+ \int_{0+}^{+\infty} \left(e^{itx} - 1 - \frac{itx}{1 + x^2} \right) dM(x)$$

State those properties of $m(x)$ and $M(x)$ for which this new representation becomes equivalent to (5.9).

2. Let X be a random variable with i.d. characteristic function $\psi(t)$. Assume that $0 < V(X) < +\infty$. Show that $\log \psi(t)$ admits the representation (Kolmogorov's formula)

$$\log \psi(t) = itL_1 + \int_{-\infty}^{+\infty} (e^{itx} - 1 - itx) \frac{1}{x^2}\, dK(x)$$

where L_1 is a constant and, with a suitable constant $c > 0$, $cK(x)$ is a distribution function. Conversely, $\psi(t)$ given by Kolmogorov's formula is an i.d. characteristic function with finite variance. Establish relations between $(L, K^*(x))$ of (5.9) and $(L_1, K(x))$.

3. The Riemann zeta function of number theory is defined as

$$\zeta(s) = \sum_{n=1}^{+\infty} \frac{1}{n^s} = \prod_p (1 - p^{-s})^{-1}$$

where $s = \sigma + it$ and the product is over all prime numbers p. Show that $\psi(t) = \zeta(\sigma + it)/\zeta(\sigma)$ is an i.d. characteristic function ($\sigma > 1$).

4. Let $X \geq 0$ be a random variable with distribution function $F(x)$ and characteristic function $\psi(t)$. Assume $F(0) = 0$ and $F(x) > 0$ for all $x > 0$. Upon using Riemann–Stieltjes sums and Theorem 3 with gamma variables, show that if

$$\log \psi(t) = -\int_0^{+\infty} \log(1 - it/y) \, dU(y)$$

where $U(y)$ is a nondecreasing, right-continuous function with $U(0) = 0$ and

$$\int_0^1 |\log y| \, dU(y) < +\infty, \qquad \int_1^{+\infty} (1/y) \, dU(y) < +\infty$$

then $\psi(t)$ is i.d.

5. Let $\varphi(t)$ be an arbitrary characteristic function. Then, for every $0 < a < n$, $1 - a/n + (a/n)\varphi(t)$ is a characteristic function also (Exercise 2 of Section 3.1), as is $[1 - a/n + (a/n)\varphi(t)]^n$ if n is an integer. By letting $n \to +\infty$, show that $\psi(t) = \exp[a(\varphi(t) - 1)]$ is an i.d. characteristic function.

6. Let $F(x)$ be a distribution function with characteristic function $\varphi(t)$. Show, by interchanging the orders of integration, that

$$u(t) = \int_0^t \int_0^v \varphi(z) \, dz \, dv = -\int_{-\infty}^{+\infty} (e^{itx} - 1 - itx) \frac{dF(x)}{x^2}$$

and thus that $\psi(t) = \exp[-u(t)]$ is the characteristic function of an i.d. distribution function with finite variance. Apply this result and the result of Exercise 5 with $\varphi(t) = \exp(-|t|)$ and conclude that if X is a standard Cauchy variable, then there are independent random variables Y and Z with nondegenerate i.d. distributions such that $X = Y + Z$ and $V(Y) < +\infty$. Does Kolmogorov's formula of Exercise 2 apply to the characteristic function of Z as well?

7. Deduce from the proof of Theorem 5, starting with (5.15), that if $\psi(t)$ is the characteristic function of the gamma density (5.3), then, in the Lévy–Khintchine representation (5.9) of $\log \psi(t)$,

$$L = \gamma \int_0^{+\infty} \frac{e^{-ax}}{1 + x^2} \, dx, \qquad K^*(x) = \gamma \int_0^x \frac{y}{1 + y^2} e^{-ay} \, dy$$

for $x > 0$ and $K^*(x) = 0$ for $x < 0$. Compute L_1 and $K(x)$ in the Kolmogorov formula for $\log \psi(t)$.

5.3 CONVERGENCE UNDER UAN

As throughout the present chapter, we assume that the random variables X_{jn}, $1 \le j \le k(n)$, for fixed n, are independent. The notation of the introduction applies.

An important tool of proof will be the inequalities of Theorem 3.17. In order to apply these inequalities, we make the following transformation. Let $r > 0$ be a fixed number and

$$a_{jn} = a_{jn}(r) = \int_{|x| < r} x \, dF_{jn}(x)$$

We set $F_{jn}^*(x) = F_{jn}(x + a_{jn})$ and $\varphi_{jn}^*(t) = \varphi_{jn}(t)\exp(-ita_{jn})$. Note that, under UAN, for every $\varepsilon > 0$,

$$A_n = \max_j |a_{jn}| \le \int_{|x| < \varepsilon} |x| \, dF_{jn}(x) + \int_{\varepsilon \le |x| < r} |x| \, dF_{jn}(x)$$

$$\le \varepsilon + r \int_{|x| \ge \varepsilon} dF_{jn}(x) \to \varepsilon \qquad \text{as } n \to +\infty$$

Consequently, $A_n \to 0$ as $n \to +\infty$, implying that, together with X_{jn}, $X_{jn} - a_{jn}$ satisfies the UAN condition (5.1). We thus have from Theorem 2 that, under UAN,

$$\max_j |\varphi_{jn}^*(t) - 1| \to 0 \qquad \text{uniformly in } |t| \le T < +\infty \qquad (5.16)$$

as $n \to +\infty$.

Theorem 6 A characteristic function $\psi(t)$ is i.d. if and only if there is a triangular array $\{X_{jn}\}$ of random variables that satisfy the UAN condition (5.1) and $\psi(t)$ is the characteristic function of the limiting distribution of $S_{(n)}$.

Proof. We have established in the Corollary of Section 5.1 that the above-mentioned family of limiting characteristic functions contains all i.d. characteristic functions. Therefore, it remains to be proven that if $\psi(t) = \lim \psi_n^*(t)$, where $\psi_n^*(t)$ is the characteristic function of $S_{(n)}$, then $\psi(t)$ is i.d. First, we prove that if $\psi_n^*(t) \to \psi(t)$, then for all t,

$$\sum_j \{\log \varphi_{jn}^*(t) - [\varphi_{jn}^*(t) - 1]\} \to 0 \qquad \text{as } n \to +\infty \qquad (5.17)$$

We start with the observation that $|\varphi_{jn}(t)| = |\varphi_{jn}^*(t)|$ and, thus,

$$|\psi_n^*(t)| = \prod_j |\varphi_{jn}^*(t)| \to |\psi(t)| \qquad (5.18)$$

Since $\psi(t)$ is a characteristic function, there exists a $T > 0$ such that $\psi(t) \neq 0$ for $|t| \leq T$. In addition, by (5.16), we can choose T so that $\varphi_{jn}^*(t) \neq 0$ for $|t| \leq T$, all j, and $n \geq n_0$. Hence, we can take the logarithm of (5.18) and get

$$\sum_j \log|\varphi_{jn}^*(t)| \to \log|\psi(t)| \qquad \text{uniformly in } |t| \leq T \tag{5.19}$$

This implies

$$-\sum_j \int_0^T \log|\varphi_{jn}^*(t)| \, dt \to -\int_0^T \log|\psi(t)| \, dt < +\infty \tag{5.20}$$

Now, since

$$1 - |\varphi_{jn}^*(t)|^2 \leq -\log|\varphi_{jn}^*(t)|^2 = -2\log|\varphi_{jn}^*(t)|$$

(5.20) entails that, for $n \geq n_0$ and some $T > 0$, there is a constant $c > 0$ such that

$$\sum_j \int_0^T (1 - |\varphi_{jn}^*(t)|^2) \, dt \leq c < +\infty \tag{5.21}$$

Hence, with an appeal to Theorem 3.17 (see also Exercise 1),

$$\sum_j \int_{-\infty}^{+\infty} \frac{x^2}{1+x^2} \, dF_{jn}^*(x) \leq c' \qquad n \geq n_0 \tag{5.22}$$

which, in turn, yields that, for arbitrary $b > 0$ (upon applying the lower inequality of Theorem 3.17),

$$\sum_j |\varphi_{jn}^*(t) - 1| \leq c^* \qquad n \geq n_0, \quad |t| \leq b \tag{5.22a}$$

where c' and c^* are (finite) positive constants. Finally, from the frequently applied Taylor expansion (see (4.42))

$$\log \varphi_{jn}^*(t) = \varphi_{jn}^*(t) - 1 + \vartheta_{jn}|\varphi_{jn}^*(t) - 1|^2 \qquad |\vartheta_{jn}| \leq 1$$

by the estimate

$$\sum_j |\varphi_{jn}^*(t) - 1|^2 \leq \left[\max_j |\varphi_{jn}^*(t) - 1| \right] \sum_j |\varphi_{jn}^*(t) - 1|$$

(5.17) follows from (5.16) and (5.22a).

Next, observe that

$$\log \varphi_{jn}^*(t) - [\varphi_{jn}^*(t) - 1] = \log \varphi_{jn}(t) - \left[it a_{jn} + \int_{-\infty}^{+\infty} (e^{itx} - 1) \, dF_{jn}^*(x) \right]$$

We can thus rewrite (5.17) as

$$\log \psi_n^*(t) - \sum_j \left[ita_{jn} + \int_{-\infty}^{+\infty} (e^{itx} - 1)\, dF_{jn}^*(x) \right] \to 0$$

as $n \to +\infty$, where we know that $\psi_n^*(t) \to \psi(t)$. With the further transformations (recall the transformation of (5.7) in the paragraph containing (5.15))

$$L_n = \sum_j \left[a_{jn} + \int_{-\infty}^{+\infty} \frac{x}{1 + x^2}\, dF_{jn}^*(x) \right] \tag{5.23}$$

and

$$K_n^*(x) = \sum_j \int_{-\infty}^{x} \frac{y^2}{1 + y^2}\, dF_{jn}^*(y) \tag{5.24}$$

the preceding limits become

$$\psi(t) = \lim \psi_n^*(t) = \lim \psi_n^{(id)}(t) \tag{5.25}$$

where $\psi_n^{(id)}(t) = \exp[\eta_n^{(id)}(t)]$ and

$$\eta_n^{(id)}(t) = itL_n + \int_{-\infty}^{+\infty} \left(e^{itx} - 1 - \frac{itx}{1 + x^2} \right) \frac{1 + x^2}{x^2}\, dK_n^*(x)$$

Because of Theorem 5, $\psi_n^{(id)}(t)$ is an i.d. characteristic function; thus, Theorem 3 ensures that its limit $\psi(t)$ is i.d. also. The proof is completed. ∎

The closing lines of the preceding proof and Lemma 3 of Section 5.2 immediately yield the following limit theorem.

Theorem 7 The distribution function $G_n^*(x)$ of $S_{(n)}$ in a triangular array $\{X_{jn}\}$ of independent random variables that satisfy the UAN condition (5.1) converges completely to a distribution function $G(x)$ if and only if there are constants L and $c > 0$ and a function $K^*(x)$ such that $cK^*(x)$ is a distribution function in the extended sense and, with L_n and $K_n^*(x)$ of (5.23) and (5.24),

$$L_n \to L \quad \text{and} \quad K_n^*(x) \to K^*(x) \text{ completely} \quad (n \to +\infty)$$

Proof. If $G_n^*(x) \to G(x)$ completely, that is, if $\psi_n^*(t) \to \psi(t)$, where $\psi_n^*(t)$ and $\psi(t)$ are the respective characteristic functions, then, as demonstrated in (5.25), $\psi_n^{(id)}(t) \to \psi(t)$ as well, where $\psi_n^{(id)}(t)$ is an i.d. characteristic function whose Lévy–Khintchine representation is characterized by L_n and $K_n^*(x)$ of (5.23) and (5.24). Lemma 3 of Section 5.2 therefore implies that the conditions of the theorem are necessary for the complete convergence of $G_n^*(x)$. Conversely, if $L_n \to L$ and $K_n^*(x) \to K(x)$ completely, then, by the same

Lemma 3, $\psi_n^{(id)}(t)$, introduced in (5.25), converges to the i.d. characteristic function $\psi(t)$, which is characterized by L and $K^*(x)$ in its Lévy–Khintchine representation. But the other limit of (5.25) also applies, because the fact that $K_n^*(+\infty) \to K^*(+\infty)$ ensures the validity of (5.22) and, thus, of (5.22a), which, together with the UAN condition (by (5.16)), yields (5.17). The theorem is established.   ∎

Remark   Notice that the fact that $\psi_n^*(t) \to \psi(t)$ is not fully used in the proofs of Theorems 6 and 7. Instead, the basic assumption is (5.18), where we took absolute values. Therefore, the arguments based on (5.18) are not sensitive to a translation of $S_{(n)}$, implying that the necessity parts of these theorems would continue to hold if we were to consider $S_{(n)} - a_n$ with some sequence a_n of constants rather than just $S_{(n)}$. That is, the limiting distribution of $S_{(n)} - a_n$ is necessarily i.d. and if $S_{(n)} - a_n$ has a limiting distribution, then $a_n - L_n$ must converge to a finite number L and $K_n^*(x) \to K^*(x)$ completely. Conversely, since the characteristic function of $S_{(n)} - a_n$ equals $\psi_n^*(t)\exp(-ita_n)$, it can also be approximated by an i.d. characteristic function that is determined by $L_n - a_n$ and $K_n^*(x)$. Consequently, if these latter converge ($K_n^*(x)$ completely), then $S_{(n)} - a_n$ has a limiting distribution. We thus have that, under UAN, $S_{(n)} - a_n$ can have a limiting distribution only if a_n is essentially L_n (i.e., if $a_n - L_n$ converges to a finite value).

In the next theorem, we show that the extremes of the summands of $S_{(n)}$ have a major influence on the existence of a limit law of $S_{(n)} - a_n$, a_n constant. We set

$$W_n^* = \min_j X_{jn} \qquad Z_n^* = \max_j X_{jn}$$

and

$$L_n^*(x) = P(W_n^* \le x) \qquad H_n^*(x) = P(Z_n^* \le x)$$

Theorem 8   Let $\{X_{jn}\}$ be a triangular array of independent random variables satisfying the UAN condition. Then there are constants a_n such that $S_{(n)} - a_n$ has a limiting distribution $G(x)$ if and only if

(i) $\displaystyle \lim_{\varepsilon = 0} \lim_{n = +\infty} \sum_j V(X_{jn}(\varepsilon)) = K^*(0) - K^*(0-) = \sigma^2$

where $X_{jn}(\varepsilon) = X_{jn}$ if $|X_{jn}| < \varepsilon$ and $X_{jn}(\varepsilon) = 0$ otherwise,

(ii) $\displaystyle \lim_{n = +\infty} H_n^*(x) = H(x) = \begin{cases} \exp[M(x)] & \text{if } x > 0 \\ 0 & \text{if } x < 0 \end{cases}$

and

$$\text{(iii)} \quad \lim_{n=+\infty} L_n^*(x) = L(x) = \begin{cases} 1 & \text{if } x > 0 \\ 1 - \exp[-m(x)] & \text{if } x < 0 \end{cases}$$

where $K^*(x)$ is the Lévy–Khintchine spectral function of $G(x)$, while $M(x)$, $x > 0$, and $m(x)$, $x < 0$, are its Lévy spectral functions, i.e.,

$$M(x) = -\int_x^{+\infty} \frac{1 + y^2}{y^2} \, dK^*(y) \qquad x > 0$$

and

$$m(x) = \int_{-\infty}^x \frac{1 + y^2}{y^2} \, dK^*(y) \qquad x < 0$$

Remark By Theorem 6 and the Remark following the proof of Theorem 7, $G(x)$ is i.d.; thus, its characteristic function can be characterized by either the Lévy–Khintchine formula or Lévy's representation (Exercise 1 of Section 5.2).

Proof. By independence,

$$H_n^*(x) = \prod_j F_{jn}(x) = \prod_j \{1 - [1 - F_{jn}(x)]\}$$

and

$$1 - L_n^*(x) = \prod_j [1 - F_{jn}(x)]$$

Thus, for $x < 0$, the UAN condition implies $F_{jn}(x) \to 0$ and, for $x > 0$, $F_{jn}(x) \to 1$, both uniformly in j. Hence,

$$H_n^*(x) \to 0 \quad \text{if } x < 0 \qquad \text{and} \qquad L_n^*(x) \to 1 \quad \text{if } x > 0 \quad (n \to +\infty)$$

while, in the other cases, we can take logarithms. We get

$$\log H_n^*(x) = -(1 + o(1)) \sum_j [1 - F_{jn}(x)] \qquad x > 0$$

and

$$\log[1 - L_n^*(x)] = -(1 + o(1)) \sum_j F_{jn}(x) \qquad x < 0$$

whenever n is sufficiently large. Next, observe that, once again setting $A_n = \max_j |a_{jn}|$, $A_n \to 0$ as $n \to +\infty$ (see the second paragraph of the current section) and

$$F_{jn}(x - A_n) \le F_{jn}^*(x) \le F_{jn}(x + A_n) \tag{5.26}$$

Therefore, in limit relations, we can replace F_{jn} by F_{jn}^*, obtaining

$$\log H_n^*(x) = -(1 + o(1)) \sum_j [1 - F_{jn}^*(x)] \qquad x > 0 \tag{5.27}$$

and

$$\log[1 - L_n^*(x)] = -(1 + o(1)) \sum_j F_{jn}^*(x) \qquad x < 0 \tag{5.28}$$

Now, by Theorem 7 and the Remark following its proof, there is a sequence a_n such that $S_{(n)} - a_n$ has a limiting distribution if and only if $K_n^*(x) \to K^*(x)$ completely, where $K_n^*(x)$ is defined in (5.24). This, by (5.27), (5.28), and the Helly–Bray theorem (Theorem 2.26), is equivalent to the two limits (ii) and (iii) of the theorem and to

$$\lim_{\varepsilon = 0} \lim_{n \to \infty} \sum_j \int_{|x| < \varepsilon} \frac{x^2}{1 + x^2} \, dF_{jn}^*(x) = K^*(0) - K^*(0-) \tag{5.29}$$

It remains to establish that (5.29), when combined with (ii) and (iii), is equivalent to (i). First, note

$$\frac{1}{1 + \varepsilon^2} \int_{|x| < \varepsilon} x^2 \, dF_{jn}^*(x) \le \int_{|x| < \varepsilon} \frac{x^2}{1 + x^2} \, dF_{jn}^*(x) \le \int_{|x| < \varepsilon} x^2 \, dF_{jn}^*(x)$$

which entails that the integrand $x^2/(1 + x^2)$ can be replaced by x^2 in (5.29). Furthermore,

$$\sum_j V(X_{jn}(\varepsilon)) = \sum_j \left\{ \int_{|x| < \varepsilon} x^2 \, dF_{jn}(x) - \left[\int_{|x| < \varepsilon} x \, dF_{jn}(x) \right]^2 \right\}$$

and

$$\left| \sum_j \left\{ \int_{|x| < \varepsilon} x^2 \, dF_{jn}^*(x) - \int_{|x| < \varepsilon} x^2 \, dF_{jn}(x) + \left[\int_{|x| < \varepsilon} x \, dF_{jn}(x) \right]^2 \right\} \right|$$

$$\le \left| \sum_j \left[\int_{|x| < \varepsilon} x^2 \, dF_{jn}^*(x) - \int_{|x| < \varepsilon} (x - a_{jn})^2 \, dF_{jn}(x) \right] \right|$$

$$+ \left| \sum_j \left\{ \int_{|x| < \varepsilon} [(x - a_{jn})^2 - x^2] \, dF_{jn}(x) + \left[\int_{|x| < \varepsilon} x \, dF_{jn}(x) \right]^2 \right\} \right|$$

$$= |\cdots|_1 + |\cdots|_2, \text{ say} \tag{5.30}$$

Now,

$$|\cdots|_1 \le \sum_j \int_{\varepsilon - |a_{jn}| \le |x| \le \varepsilon + |a_{jn}|} (|x| + |a_{jn}|)^2 \, dF_{jn}(x)$$

$$\le (\varepsilon + 2A_n)^2 \sum_j \int_{\varepsilon - A_n \le |x| \le \varepsilon + A_n} dF_{jn}(x)$$

and, upon using the definition of a_{jn} and choosing $0 < \varepsilon < r$, we have (details to follow)

$$|\cdots|_2 \le \left[r^2 \max_j \int_{|x| \ge \varepsilon} dF_{jn}(x) + A_n^2 \right] \sum_j \int_{|x| \ge \varepsilon} dF_{jn}(x) \tag{5.31}$$

Thus, by letting $n \to +\infty$, the UAN condition and the limits (ii) and (iii) of the theorem (see also (5.26)–(5.28)) yield

$$\lim \sup |\cdots|_1 = 0 \qquad \lim \sup |\cdots|_2 = 0$$

Hence, (5.30) ensures that (5.29) and (i) are equivalent.

To prove (5.31), we write

$$a_{jn}^2 \int_{|x| < \varepsilon} dF_{jn}(x) = a_{jn}^2 - a_{jn}^2 \int_{|x| \ge \varepsilon} dF_{jn}(x)$$

and

$$\left[\int_{|x| < \varepsilon} x \, dF_{jn}(x) \right]^2 = \left[a_{jn} - \int_{\varepsilon \le |x| < r} x \, dF_{jn}(x) \right]^2$$

which yield

$$|\cdots|_2 = \left| \sum_j \left[\left(\int_{\varepsilon \le |x| < r} x \, dF_{jn}(x) \right)^2 - a_{jn}^2 \int_{|x| \ge \varepsilon} dF_{jn}(x) \right] \right|$$

Finally, by the Cauchy–Schwarz inequality

$$\left(\int_{\varepsilon \le |x| < r} x \, dF_{jn}(x) \right)^2 \le \left(\int_{\varepsilon \le |x| < r} x^2 \, dF_{jn}(x) \right) \left(\int_{|x| \ge \varepsilon} dF_{jn}(x) \right)$$

$$\le r^2 \left[\max_j \int_{|x| \ge \varepsilon} dF_{jn}(x) \right] \int_{|x| \ge \varepsilon} dF_{jn}(x)$$

which, with $a_{jn}^2 \le A_n^2$, entails (5.31). The theorem is established. ∎

The following exercises and the next section, which deals with special cases, provide a large number of examples of the results of the theorems.

Exercises

1. Show that if the independent random variables X_{jn} satisfy the UAN condition and if m_{jn} is a median of X_{jn}, then $\max_j |m_{jn}| \to 0$ as $n \to +\infty$.

2. Let $X_j, j \geq 1$, be independent random variables with a common distribution function $F(x)$, where $F'(x) = 1/2x^2$ for $|x| \geq 1$ and $F'(x) = 0$ otherwise. Let $c_{jn}, 1 \leq j \leq n, n \geq 1$, be real numbers such that the random variables $X_{jn} = c_{jn}X_j$ are UAN. Give conditions on $\{c_{jn}\}$ under which $S_{(n)}$ has a limiting distribution $G(x)$. Choose a particular case and find $G(x)$.

3. Let X_{jn} satisfy the UAN condition and assume that $S_{(n)}$ has a limiting distribution function $G(x)$ whose characteristic function $\psi(t)$ is determined by $L = 0$ and $K^*(x)$ in its Lévy–Khintchine representation, where $K^*(x)$ is the distribution function of a simple random variable taking the values 1, 2, and 3, each with probability $1/3$. Find

$$\lim P(S_{(n)} \leq 2) = G(2) \qquad \text{and} \qquad \lim P(Z_n^* \leq 2) = H(2)$$

as $n \to +\infty$.

4. Show that, under the UAN condition, if $S_{(n)}$ has a limiting distribution,

$$\lim \sum_j \int_{-\infty}^{+\infty} \frac{x}{1+x^2} \, dF_{jn}^*(x) \text{ exists (finite)}$$

as $n \to +\infty$.

5.4 CONVERGENCE TO SPECIAL DISTRIBUTIONS

In this section, the general results of the current chapter are made specific for cases of complete convergence to members of class L and to three particular distributions: degenerate, normal, and Poisson. The UAN condition is imposed throughout the section; thus, Zolotarev's theorem (Section 4.4) will not be repeated when we deal with the asymptotic normality. We shall not specialize further within the class L to identically distributed summands: this will be dealt with in the next chapter. Besides its historical significance, the convergence to the degenerate, normal, and Poisson limits is special in that their Lévy–Khintchine spectral function is either identically zero (the degenerate case) or degenerate to a constant c ($c = 0$ for the normal distribution and $c = 1$ for the Poisson; $c \neq 0$ leads to a Poisson-type distribution that differs very little from the Poisson itself).

A. Class L: Convergence of Normalized Sums of Independent Random Variables

Class L was first introduced in Section 4.6. Recall that we have independent random variables $X_1, X_2, \ldots$ and we seek conditions under which there exist some constants a_n and $b_n > 0$ such that $S_n/b_n - a_n$ (here the meaning of a_n conforms to the notation of the current chapter, and thus slightly differs from that of Section 4.6) has an asymptotic distribution $G(x)$. The limits $G(x)$ form the class L. The UAN condition implies that $b_n \to +\infty$ with n and $G(x)$ is i.d. Let $\psi(t)$ be the characteristic function of $G(x)$. It is established in Theorem 4.12 that class L is characterized by the property that, for every real number $0 < a < 1$,

$$\psi(t)/\psi(at) = \tau_a(t) \tag{5.32}$$

is a characteristic function. In fact, we found in (4.51) and (4.52) that $\tau_a(t)$ is the limit of a product of characteristic functions, which product is the characteristic function of a translated sum $S_{(n)} - c_n$, c_n constant, and that the summands of $S_{(n)}$ satisfy the UAN condition (see Lemma 1 of Section 4.6). Hence, $\tau_a(t)$ is i.d. (Theorem 6, combined with the Remark following the proof of Theorem 7). Let the Lévy–Khintchine spectral functions of $\psi(t)$ and $\tau_a(t)$ be $K^*(x)$ and $K_a^*(x)$, respectively, and let the corresponding Lévy spectral functions be $M(x)$, $M(x;a)$, $x > 0$, and $m(x)$, $m(x;a)$, $x < 0$ (Exercise 1 of Section 5.2 or Theorem 8). Finally, the spectral functions in Kolmogorov's formula (the case of finite variances; see Theorem 13 (iii) and Exercise 2 of Section 5.2) are denoted by $K(x)$ and $K_a(x)$, respectively.

Theorem 9   Let $G(x)$ be an i.d. distribution function with characteristic function $\psi(t)$. Then $G(x)$ belongs to the class L if and only if

(i)   both $M(x)$, $x > 0$, and $m(x)$, $x < 0$, are continuous functions with finite left and right derivatives, and

$$xM'(x) \quad x > 0 \quad \text{and} \quad xm'(x) \quad x < 0$$

are nonincreasing, where the prime can denote either kind of derivative, or equivalently,

(ii) $K^*(x)$ is a continuous function with a finite left and right derivative for every $x \neq 0$ and, either derivative being denoted by $K^{*\prime}(x)$, $[(1 + x^2)/x]K^{*\prime}(x)$ is a nonincreasing function on both $(-\infty, 0)$ and $(0, +\infty)$.

Furthermore, if $G(x)$ has finite variance and $\psi(t)$ can thus be represented by Kolmogorov's formula, then (i) and (ii) are both equivalent to

(iii) $K(x)$ is continuous with finite left and right derivatives for every $x \neq 0$, and $K'(x)/x$ is a nonincreasing function both when the domain is $x > 0$ and when it is $x < 0$. Once again, the prime signifies either right or left derivative.

Proof. Because of the relations between any pair of the functions $K^*(x)$, $M(x)$, and $K(x)$ when $x > 0$ or $K^*(x)$, $m(x)$, and $K(x)$ when $x < 0$, the equivalence of the three criteria of the theorem is obvious. Therefore, it suffices to prove any one of them. We prove the theorem in the form of (i). The proof is simply a reformulation of (5.32) and the fact that $\tau_a(t)$ is i.d. This is then combined with well-known properties of convex functions, which, however, we shall prove in detail.

Now, we start with (5.32) and the Lévy representation of an i.d. characteristic function (Exercise 1 of Section 5.2, or one can substitute into (5.9)). If we combine L with the terms

$$-a \int_{-\infty}^{0-} \left(\frac{x}{1+x^2} - \frac{x}{1+a^2x^2} \right) dm(x)$$

$$= a(1 - a^2) \int_{-\infty}^{0-} \frac{x^3}{(1+x^2)(1+a^2x^2)} \, dm(x)$$

and

$$a(1 - a^2) \int_{0+}^{+\infty} \frac{x^3}{(1+x^2)(1+a^2x^2)} \, dM(x)$$

to form the linear part (itL_a) of the Lévy representation of $\tau_a(t)$, we get that $\tau_a(t)$ is an i.d. characteristic function if and only if

$$M(x; a) = M(x) - M(x/a) \qquad x > 0 \tag{5.33a}$$

and

$$m(x; a) = m(x) - m(x/a) \qquad x < 0 \tag{5.33b}$$

are nondecreasing (and right continuous) (the equations above also utilize the uniqueness between an i.d. characteristic function and the appropriate spectral function; see Lemma 2 of Section 5.2). Since the functions $M(x)$, $x > 0$, and $m(x)$, $x < 0$, exhibit similar properties, we proceed with the analysis of $M(x)$ only. Let $0 < x_1 < x_2$ be arbitrary real numbers. Then (5.33a) entails that, for every $0 < a < 1$,

$$M(x_1) - M(x_1/a) \le M(x_2) - M(x_2/a)$$

that is,

$$M(x_2) - M(x_1) \ge M(x_2/a) - M(x_1/a) \tag{5.34}$$

Let $D(y) = M(e^y)$, y real. Then (5.34) becomes

$$D(y_2) - D(y_1) \ge D(y_2 + h) - D(y_1 + h) \tag{5.35}$$

where $y_1 < y_2$ and $h > 0$ are arbitrary and $D(y)$ is nondecreasing.

First note that these properties of $D(y)$ imply that, if $u < v < w$, then none of the points $(v, D(v))$ of the graph of $z = D(y)$ fall below the line connecting the points $(u, D(u))$ and $(w, D(w))$ (i.e., the graph is concave). Indeed, if we divide (5.35) by $y_2 - y_1$, then it becomes an inequality between the slopes of two particular lines. Using this, we choose $y_1 = u$, $y_2 = (u + w)/2 = y_1 + h$, and $y_2 + h = w$, thus obtaining our claim from (5.35) when v is the middle point v_1 of (u, w). A repetition of this argument ensures that the points of the graph $z = D(y)$ at the middle points v_2 and v_3 of the intervals (u, v_1) and (v_1, w), respectively, do not fall below the line connecting $(u, D(u))$ and $(w, D(w))$. Continuing this procedure, we get by induction that no points $(v_k, D(v_k))$ with $v_k = u + k(w - u)/2^n$, $1 \le k < 2^n$, $n \ge 1$, can be below the line in question. The monotonicity of $D(y)$ now easily yields that v can be arbitrary with the desired property.

Another way of expressing the property just established is that, in the triangle formed by $P_1 = (u, D(u))$, $P_2 = (v, D(v))$, and $P_3 = (w, D(w))$, the side $P_1 P_2$ has the largest slope and $P_2 P_3$ the smallest. Furthermore, the points of $z = D(y)$ for y outside of $[u, w]$ cannot be above the line $P_1 P_3$. Now, let w vary in such a way that $w \to v^+$. The slope of $P_2 P_3$ is nondecreasing and bounded by the slope of $P_1 P_2$. Consequently, the slope of $P_2 P_3$ has a finite limit, that is, the right derivative of $D(y)$ is finite. Letting u vary and $u \to v-$, we similarly find that $D(y)$ has a finite left derivative. A particular implication of the existence of these derivatives is that $D(y)$ is continuous. Finally, the inequality (5.35) yields that these tangent slopes are nonincreasing. Therefore, since $D'(y) = e^y M'(e^y)$, whether it is a left or right derivative, the substitution $x = e^y$ establishes the theorem in its form of (i) by recalling that the argument for $m(x)$ is completely identical to the one for $M(x)$. The other forms follow immediately from (i). The proof is completed. ∎

Another consequence of (5.34) is the subject of the next theorem (Fisz (1963)).

Theorem 10   With the notation of Theorem 9, if $G(x)$ belongs to the class L, then either $M(x) \equiv 0$ for all $x > 0$ or $M(0+) = -\infty$. Furthermore, if there exists an $x_0 > 0$ such that $M(x_0) < 0$, then $M(x)$ is strictly increasing for $0 < x \le x_0$. Similarly for $m(x)$, $x < 0$, $m(0-) = +\infty$ unless $m(x) \equiv 0$ for $x < 0$ and $m(x)$ is strictly increasing for $x_0^* \le x < 0$ if $m(x_0^*) > 0$.

Proof.   Because of the similarities of behavior between $m(x)$ and $M(x)$, we again prove the claimed properties for $M(x)$ only. Since $M(x)$ is nondecreasing, $M(0+)$ exists. Assume $M(0+) > -\infty$ and $M(x_0) < 0$, $x_0 > 0$. We proved in the previous theorem that $M(x)$ is continuous; therefore, it is uniformly continuous on $[0, A]$ for every $A > 0$. That is, to every $\varepsilon > 0$ there is a

$\delta = \delta(\varepsilon) \to 0$ with ε such that $M(x_2) - M(x_1) < \varepsilon$ whenever $0 < x_1 < x_2 \leq \delta$. Then, by (5.34), $M(x_2/a) - M(x_1/a) < \varepsilon$ for every $0 < a < 1$ as well. By choosing $a > 0$ small, the length $(x_2 - x_1)/a$ of the interval $[x_1/a, x_2/a]$ can be made arbitrarily large, while the position of this interval is again arbitrary by the arbitrary choice of ε. That is, if we make our choices such that $x_1/a < x_0 < x_2/a$ and $|M(x_2/a)| < \varepsilon$ (we use the fact that $M(+\infty) = 0$), then we get

$$-\varepsilon - M(x_0) \leq M(x_2/a) - M(x_0) \leq M(x_2/a) - M(x_1/a) < \varepsilon$$

i.e., $-M(x_0) = |M(x_0)| < 2\varepsilon$ with arbitrary $\varepsilon > 0$. This contradicts the assumption on x_0, that is, $M(0+)$ cannot be bounded when there is an $x_0 > 0$ with $M(x_0) < 0$.

Next, with $M(x_0) < 0$ again, let $0 < x_3 < x_4 \leq x_0$ be such that $M(x_3) = M(x_4)$. Since $M(x)$ is nondecreasing, we have that $M(x)$ is constant for $x_3 \leq x \leq x_4$. With the transformation $D(y) = M(e^y)$ we thus have $D(y_4) - D(y_3) = 0$, where $y_j = \log x_j$, $j = 3$, 4, and (5.35) yields that $D(y_4 + h) - D(y_3 + h) = 0$ for every $h > 0$ as well. Applying this sequentially, we have $D(y_3) = D(y_4) = D(y_3 + kh)$, where $h = y_4 - y_3$ and $k = 1, 2, \dots$. By the monotonicity of $D(y)$, this entails that $D(y)$ is constant for $y \geq y_3$; that is, $M(x)$ is constant for $x \geq x_3$. But $x_3 < x_0$ and $M(+\infty) = 0$ while $M(x_0) < 0$. Hence, we have arrived at a contradiction; for $0 < x_3 < x_4 \leq x_0$, $M(x_3) < M(x_4)$. The theorem is established. ∎

Finally, we summarize the conditions under which there exist constants a_n and $b_n > 0$ that guarantee that the UAN condition holds and $S_n/b_n - a_n$ has a nondegenerate asymptotic distribution $G(x)$. The degenerate case will be discussed in a separate subsection. We know that $G(x)$ is necessarily an i.d. distribution; thus, its characteristic function $\psi(t)$ admits the Lévy–Khintchine as well as the Lévy representation. The general notation of the previous sections becomes $X_{jn} = X_j/b_n$, $F_{jn}(x) = F_j(xb_n)$, $\varphi_{jn}(t) = \varphi_j(t/b_n)$,

$$a_{jn} = a_{jn}(r) = \frac{1}{b_n} \int_{-b_n r}^{b_n r} x \, dF_j(x) \qquad r > 0 \text{ arbitrary}$$

and

$$K_n^*(x) = \sum_j \int_{-\infty}^{b_n x} \frac{y^2}{b_n^2 + y^2} \, dF_j(y + a_{jn})$$

The following theorem specializes previous results to the case of this subsection.

Theorem 11 There exist constants a_n and $b_n > 0$ such that the triangular array X_j/b_n, $1 \leq j \leq n$, $n \geq 1$, with the X_j, $j \geq 1$, independent, satisfies the

UAN condition and the distribution function of $S_n/b_n - a_n$ converges completely to a nondegenerate distribution function $G(x)$ if and only if there exist a function $K^*(x)$ and a constant $c > 0$ such that $cK^*(x)$ is a distribution function and, with $b_n = B_n$, determined by the equation

$$\sum_{j=1}^{n} \int_{-\infty}^{+\infty} \frac{x^2}{B_n^2 + x^2} \, dF_j^{(s)}(x) = 2K^*(+\infty) \tag{5.36}$$

where $F_j^{(s)}(x)$ is the distribution function obtained from $F_j(x)$ by symmetrization, that is, whose characteristic function is $|\varphi_j(t)|^2$,

$$\max_j \int_{-\infty}^{+\infty} \frac{x^2}{b_n^2 + x^2} \, dF_j(x) \to 0 \tag{5.37}$$

and

$$K_n^*(x) \to K^*(x) \text{ completely} \tag{5.38}$$

The limiting distribution $G(x)$ is i.d. and its Lévy–Khintchine spectral function is $K^*(x)$. The norming constants a_n and $b_n > 0$ can always be chosen as $b_n = B_n(1 + o(1))$ and $a_n = \sum_j a_{jn} + L + o(1)$, where L is an arbitrary constant.

Further properties of the convergence of the normalized sum $S_n/b_n - a_n$, which was asserted in the preceding theorem, are given below.

Theorem 11a The notation of Theorem 11 applies and it is again assumed that $G(x)$ is nondegenerate. Furthermore, set

$$W_n = \min\{X_j : 1 \leq j \leq n\} \qquad Z_n = \max\{X_j : 1 \leq j \leq n\}$$

If (5.37) and (5.38) hold, then the normalized extremes $W_n^* = W_n/b_n$ and $Z_n^* = Z_n/b_n$ have the respective limiting distributions $L(x)$ and $H(x)$, defined in (ii) and (iii) of Theorem 8. The functions $L(x)$ and $H(x)$ are either degenerate at zero or continuous on the whole real line with $L(0) = 1$ and $H(0) = 0$. They are strictly increasing for all x for which their value lies in the open interval $(0, 1)$ and $\log H(e^x)$ and $\log[1 - L(-e^x)]$ are concave for the same values of x. Both $L(x)$ and $H(x)$ are degenerate at zero if and only if $S_n/b_n - a_n$ is asymptotically normal.

Finally, before the short proofs of these theorems, the reader is advised to go through the conditions of Theorem 11 and to establish that its sufficiency part reduces to Kolmogorov's theorem (Theorem 4.13) when the additional assumption of finite variances is made.

Proof of Theorem 11. The sufficiency of the conditions follows from
Theorem 2(i) and Theorem 7 when combined with the Remark following its
proof. In addition, Theorem 7 (with the above-mentioned Remark) and
Exercise 4 of Section 5.3 yield that $G(x)$ is i.d. and a_n can always be chosen
as stated. That the conditions are also necessary will follow if we show that
the existence of the asymptotic distribution of $S_n/b_n - a_n$ implies that
$b_n/B_n \to 1$ as $n \to +\infty$. Since a_n is not involved in this condition, it is only
natural to turn to the symmetrized forms. Let X_j and X_j^* be independent
with a common distribution function $F_j(x)$. Then the distribution function
of $Y_j = X_j - X_j^*$ is $F_j^{(s)}(x)$. Upon turning to characteristic functions, one
immediately sees that the assumed convergence of the distribution of
$S_n/b_n - a_n$ ensures that $(Y_1 + Y_2 + \cdots + Y_n)/b_n$ also has an asymptotic
distribution and that the Y_j/b_n, $1 \leq j \leq n$, satisfy the UAN condition. Thus,
in view of Theorems 2 and 7, (5.37) and (5.38) must hold when $F_j(x)$ is
replaced by $F_j^{(s)}(x)$ (since $F_j^{(s)}(x)$ is symmetric, the value corresponding to a_{jn}
is zero). The limit $K^*(x)$ in (5.38) becomes $K^{**}(x) = K^*(x) + K^*(+\infty) - K^*(-x + 0)$. Hence, $K^{**}(+\infty) = 2K^*(+\infty)$ and, thus, (5.36) and (5.38)
entail

$$\sum_{j=1}^{n} \left[\int_{-\infty}^{+\infty} \frac{x^2}{b_n^2 + x^2} \, dF_j^{(s)}(x) - \int_{-\infty}^{+\infty} \frac{x^2}{B_n^2 + x^2} \, dF_j^{(s)}(x) \right] \to 0$$

that is,

$$|b_n^2 - B_n^2| \sum_{j=1}^{n} \int_{-\infty}^{+\infty} \frac{x^2}{(b_n^2 + x^2)(B_n^2 + x^2)} \, dF_j^{(s)}(x) \to 0$$

Up to this point we have not used the fact that $G(x)$ is nondegenerate, that
is, that $K^*(x)$ in (5.38) is not identically zero (and thus, with a suitable $c > 0$,
$cK^*(x)$ is a distribution function). Therefore, by the definition of $K_n^*(x)$
applied to the sequence $F_j^{(s)}(x)$, (5.38) entails the existence of a number $u > 0$
such that, for all sufficiently large n,

$$\sum_{j=1}^{n} \int_{-ub_n}^{ub_n} \frac{x^2}{b_n^2 + x^2} \, dF_j^{(s)}(x) \geq \eta > 0$$

Now, with the estimate

$$\int_{-\infty}^{+\infty} \frac{x^2}{(b_n^2 + x^2)(B_n^2 + x^2)} \, dF_j^{(s)}(x) \geq \frac{1}{b_n^2 + u^2} \int_{-ub_n}^{ub_n} \frac{x^2}{B_n^2 + x^2} \, dF_j^{(s)}(x)$$

we get from the last limit relation

$$\frac{|b_n^2 - B_n^2|}{b_n^2 + u^2} \eta = \frac{|1 - B_n^2/b_n^2|}{1 + u/b_n^2} \eta \to 0$$

which ensures that $B_n/b_n \to 1$ as $n \to +\infty$. This fact also implies that we can choose b_n as $B_n(1 + o(1))$. The proof is complete. ∎

Proof of Theorem 11a. This theorem is an immediate consequence of Theorems 8, 9, and 10. The claimed concavity is not explicitly stated in Theorem 9, which is the relevant statement in this connection, but it is made clear in the course of its proof. Hence, the theorem is established. ∎

B. Asymptotic Normality

Recall Example 1 of Section 5.2, in which it is demonstrated that the limiting distribution of sums in a triangular array of independent random variables satisfying the UAN condition is normal with parameters L and $V = \sigma^2$ if and only if the Lévy–Khintchine spectral function of the limit is $K^*(x) = V$ if $x \geq 0$ and 0 if $x < 0$, implying that the Lévy spectral functions $M(x)$, $x > 0$, and $m(x)$, $x < 0$ (see Theorem 8) are identically zero. Hence, the general convergence theorems of the present chapter take very simple forms in the case of asymptotic normality. The reader is also reminded that weak and complete convergence can be replaced by convergence in the supremum metric because the normal distribution has bounded density (see Theorem 4.6 and the Corollary of Section 4.3). We thus have the following results.

Theorem 12 Let the random variables X_{jn}, $1 \leq j \leq k(n)$, $n \geq 1$, be independent for fixed n. Then the X_{jn} satisfy the UAN condition and there are constants a_n such that $S_{(n)} - a_n$ is asymptotically normal if and only if for every $\varepsilon > 0$ and for some fixed number $r > 0$,

(a) $\sum_j P(|X_{jn}| \geq \varepsilon) \to 0$

and

(b) $\sum_j \left[\int_{|x| < r} x^2 \, dF_{jn}(x) - a_{jn}^2(r) \right] \to \sigma^2 > 0$

as $n \to +\infty$, where

$$a_{jn}(r) = \int_{|x| < r} x \, dF_{jn}(x)$$

The constants a_n can always be chosen as $a_n = \sum_j a_{jn}(r)$, in which case the parameters of the limiting normal distribution are zero and σ^2.

Proof. First note that (a) is a stronger requirement than the UAN condition since

$$\max_j P(|X_{jn}| \geq \varepsilon) \leq \sum_j P(|X_{jn}| \geq \varepsilon)$$

Next, we observe that $|a_{jn}(r)| < r$ and, for $r_1 < r_2$,

$$a_{jn}(r_2) - a_{jn}(r_1) = \int_{r_1 \leq |x| < r_2} x \, dF_{jn}(x)$$

Hence, upon setting

$$V(X_{jn}(r)) = \int_{|x| < r} x^2 \, dF_{jn}(x) - a_{jn}^2(r)$$

we have

$$\left| \sum_j V(X_{jn}(r_2)) - \sum_j V(X_{jn}(r_1)) \right|$$

$$\leq \sum_j \int_{r_1 \leq |x| < r_2} x^2 \, dF_{jn}(x) + 2r_2 \sum_j \int_{r_1 \leq |x| < r_2} |x| \, dF_{jn}(x)$$

$$\leq 3r_2^2 \sum_j \int_{r_1 \leq |x|} dF_{jn}(x)$$

which tends to zero as $n \to +\infty$, whenever (a) holds. That is, both the lim sup and lim inf of the sum of $V(X_{jn}(r))$ over j, as $n \to +\infty$, are independent of r when (a) is valid. Therefore, apart from the claimed form of a_n, Theorem 8 concludes this proof, since (a) is equivalent to (ii) and (iii) in Theorem 8 (see also (5.26)–(5.28)) and, thus, (b) becomes equivalent to (i) of Theorem 8. That a_n can always be chosen as claimed follows from the Remark following the proof of Theorem 7 and from Exercise 4 of Section 5.3. This completes the proof. ∎

From Theorem 8 we also get the following result.

Theorem 13 Let X_{jn} be as in Theorem 12. Assume that, with some sequence a_n of constants, $S_{(n)} - a_n$ has a nondegenerate limiting distribution $G(x)$. Then the summands X_{jn} satisfy the UAN condition and $G(x)$ is normal if and only if

$$\max_j |X_{jn}| \to 0 \qquad \text{in probability}$$

No proof is required. This is a direct consequence of Theorem 8, and all steps of the deduction are implicitly made in the proof of Theorem 12.

The class L version of Theorem 12 is as follows.

Theorem 14 (**Feller**) Let X_j be a sequence of independent random variables and let $b_n > 0$ be a sequence of constants. Then

$$\max_{1 \le j \le n} P(|X_j| \ge \varepsilon b_n) \to 0 \qquad \text{for every fixed } \varepsilon > 0$$

and there exists a sequence a_n of constants such that

$$\sup_x |P(S_n/b_n - a_n \le x) - N(x)| \to 0$$

where $N(x)$ is the standard normal distribution function, if and only if, for every fixed $\varepsilon > 0$ and for some $r > 0$ (and thus for any $r > 0$) as $n \to +\infty$,

$$\sum_{j=1}^{n} P(|X_j| \ge \varepsilon b_n) \to 0$$

and

$$\frac{1}{b_n^2} \sum_{j=1}^{n} \left\{ \int_{|x| < rb_n} x^2 \, dF_j(x) - \left[\int_{|x| < rb_n} x \, dF_j(x) \right]^2 \right\} \to 1$$

The constant a_n can always be chosen as

$$a_n = \frac{1}{b_n} \sum_{j=1}^{n} \int_{|x| < rb_n} x \, dF_j(x)$$

Since this statement is just a reformulation of Theorem 12, no proof is required. Notice that if the X_j have finite variances and b_n^2 is a priori chosen as the sum of the variances of the X_j, then Theorem 14 reduces to the Lindeberg–Feller theorem discussed in Section 4.4.

C. Weak Laws of Large Numbers

In the setting of the current chapter, we say that the X_{jn} obey the weak law of large numbers if, for some constants a_n, $S_{(n)} - a_n$ has a degenerate asymptotic law at zero. Therefore, this weak law of large numbers is characterized by $K^*(x) \equiv 0$, which can be seen as a degenerate asymptotic normality with $\sigma = 0$. Hence, from the previous subsection, or directly from the relevant general theorems, we have the following criterion.

Theorem 15   The triangular array X_{jn}, the X_{jn} being independent for fixed n, obeys the weak law of large numbers and satisfies the UAN condition if and only if conditions (a) and (b) of Theorem 12 are satisfied, the latter with $\sigma = 0$. The constant a_n can be chosen as in Theorem 12.

It is possible to simplify the conditions of Theorem 15 somewhat in the class L setting. This is reflected in the following statement.

Theorem 16   Let $X_j, j \geq 1$, be independent random variables and let $b_n > 0$ be an increasing sequence of numbers diverging to infinity. Then $S_n/b_n \to 0$ in probability if and only if

(i) $\displaystyle\sum_{j=1}^{n} P(|X_j| \geq b_n) \to 0$

(ii) $\displaystyle\frac{1}{b_n^2} \sum_{j=1}^{n} \left\{ \int_{|x|<b_n} x^2\, dF_j(x) - \left[\int_{|x|<b_n} x\, dF_j(x) \right]^2 \right\} \to 0$

and

(iii) $\displaystyle\frac{1}{b_n} \sum_{j=1}^{n} \int_{|x|<b_n} x\, dF_j(x) \to 0$

Proof.   First note that $S_n/b_n \to 0$ in probability (which is equivalent to saying that the asymptotic distribution of S_n/b_n exists and is degenerate at zero) implies that the UAN condition holds for the summands X_j/b_n. Indeed, upon writing

$$X_n/b_n = S_n/b_n - (b_{n-1}/b_n)S_{n-1}/b_{n-1}$$

we have

$$P\left(\left|\frac{X_n}{b_n}\right| < \varepsilon\right) \geq P\left(\left|\frac{S_n}{b_n}\right| < \frac{\varepsilon}{2}, \left|\frac{S_{n-1}}{b_{n-1}}\right| < \frac{\varepsilon}{2}\right)$$

which approaches one as $n \to +\infty$. By the monotonicity of b_n, this suffices for the validity of the UAN condition. Hence, Theorem 7 is applicable, which yields, using the same type of computations as in the case of asymptotic normality, that the conditions of the theorem must hold if $S_n/b_n \to 0$ in probability. Now, let us assume that (i)–(iii) hold. Define $X_{jn} = X_j/b_n$ if $|X_j| < b_n$ and $X_{jn} = 0$ otherwise. Then

$$P(S_{(n)} \neq S_n/b_n) \leq \sum_{j=1}^{n} P(X_{jn} \neq X_j/b_n) = \sum_{j=1}^{n} P(|X_j| \geq b_n)$$

and, thus, condition (i) and Exercise 1 of Section 3.3 imply that it suffices to show that $S_{(n)} \to 0$ in probability, which, however, follows from Chebyshev's inequality and from conditions (ii) and (iii). The proof is completed. ■

D. Convergence to the Poisson Law

Once again, with the standard assumptions of the present chapter, we have to restate our results when $K^*(x) = a/2$ or 0 depending on whether $x \geq 1$ or $x < 1$ (see Example 2 in Section 5.2). Here, $a > 0$ is the parameter of the Poisson distribution. Thus, in the same manner as in the normal case, we have the following conditions for convergence to the Poisson law.

Theorem 17 Let X_{jn} be a triangular array with independent X_{jn} for fixed n. Then the X_{jn} satisfy the UAN condition and $S_{(n)}$ is asymptotically Poisson if and only if, for every $0 < \varepsilon < 1$ and for some $0 < r < 1$,

$$\sum_{j=1}^{n} \int_{|x-1| < \varepsilon} dF_j(x) \to a, \qquad \sum_{j=1}^{n} \int_{|x| \geq \varepsilon, |x-1| \geq \varepsilon} dF_j(x) \to 0$$

and

$$\sum_{j=1}^{n} a_{jn}(r) \to 0 \qquad \sum_{j=1}^{n} V(X_{jn}(r)) \to 0$$

as $n \to +\infty$, where $a_{jn}(r)$ and $V(X_{jn}(r))$ are defined as in Theorem 12 (including its proof).

The argument is similar to that for Theorem 12; thus, the details are omitted.

Exercises

1. Decide whether or not the geometric distribution belongs to the class L. Give a new reason why the Poisson distribution is not a member of class L.
2. Let $G(x)$ be the i.d. distribution function whose characteristic function is determined by the Lévy–Khintchine spectral function $K^*(x) = \pi/2 + \arctan x$. Find $G(x)$ (you may assume $L = 0$) and show that $G(x)$ belongs to class L. Given that X_{jn} is a triangular array of random variables with X_{jn} independent for fixed n that satisfies the UAN condition and that the limiting distribution of $S_{(n)}$ is $G(x)$, determine the limiting distributions $L(x)$ and $H(x)$ of the extremes W_n^* and Z_n^*.
3. Give conditions for convergence to the gamma distribution within the general model of the current chapter.

5.5 SPECIAL DISTRIBUTIONS: PART III

This section supplements the result obtained in Theorem 11a. This theorem essentially states that in order for a normalized sum of independent random variables to have a limiting distribution one has to guarantee that the extremes of the summands have limiting distributions as well. The theorem also characterizes the structures of these extreme value distributions (the stated concavity) whenever the limiting distribution is not normal. In the present section we want to separate the extremes from the sums, and in particular we want to drop the restriction of the nonnormality of sums. We shall deal with positive random variables only, since our guide of studying the extremes is life distributions in an extended sense: life means the time elapsed up to the occurrence of a certain event. One can think of the breakdown of an instrument, recovery from a disease after some treatment or actual death as the event in question. In any case, we think of something that has a large number of components, and "life ends" with the first breakdown of the components (known as the weakest link principle: a chain breaks at its weakest link). We assume that component lives are independent, so we arrived at the following mathematical model.

Let $X_1, X_2, \ldots, X_n$ be independent positive random variables. We want to study $W_n = \min(X_1, X_2, \ldots, X_n)$ by utilizing properties of the distribution functions $F_j(x) = P(X_j \leq x)$, $j \geq 1$. We assume throughout this section that each $F_j(x) > 0$ if $x > 0$, and a further restriction to be stated below will imply that each $F_j(x)$ is continuous at zero (with $F_j(0) = 0$). In order to make some results comparable with properties of failure rates (see Section 2.9), we sometimes write

$$F_j(x) = 1 - \exp(-u_j(x)) \qquad u_j(x) = -\log[1 - F_j(x)] \qquad x \geq 0 \qquad (5.39)$$

The function $u_j(x)$ is known as the hazard (failure) function of $F_j(x)$ because, when differentiable, its derivative $u_j'(x) = r_j(x)$, the hazard (failure) rate function.

Note that by developing results on W_n, one automatically gets results on $\max(X_1, X_2, \ldots, X_n) = Z_n$ by turning to reciprocals. We do not exploit this possibility any further except in one of the exercises at the end of the section.

One cannot develop a meaningful theory for W_n without making further restrictions. Indeed, if $F(x)$ is the distribution function of a positive random variable with hazard function $u(x)$, and if we define the random variables $X_j, j \geq 1$, as independent with $F_j(x) = 1 - \exp(-u(x)/2^j)$, then

$$P(W_n > x) = \prod_{j=1}^{n} [1 - F_j(x)] = \exp\left[-u(x) \sum_{j=1}^{n} 2^{-j}\right]$$

whose limit, as $n \to +\infty$, is clearly $F(x)$. This way, every distribution function with $F(0) = 0$ would become a limiting distribution of some W_n. This triviality can be avoided by reimposing UAN in the following form: we seek limiting distributions in the form $P(W_n \leq zb_n)$, where the sequence $b_n > 0$ of normalizing constants satisfies

$$\lim_{n \to +\infty} m_n(z) = 0 \quad \text{where} \quad m_n(z) = \max_{1 \leq j \leq n} [F_j(zb_n)] \tag{5.40}$$

Theorem 18 Under the preceding assumptions, including UAN, on the sequence $X_j, j \geq 1$, if, as $n \to +\infty$,

$$\lim \sum_{j=1}^{n} F_j(zb_n) = a(z) \tag{5.41}$$

exists with $0 < a(z) < +\infty$, then

$$\lim P(W_n \leq zb_n) = 1 - e^{-a(z)} = L(z), \text{ say} \tag{5.42}$$

Proof. The proof is simple and it utilizes the same ideas which we learned at Theorem 8. Clearly,

$$P(W_n > zb_n) = \prod_{j=1}^{n} [1 - F_j(zb_n)] = \exp\left[-\sum_{j=1}^{n} u_j(zb_n)\right] \tag{5.43}$$

By Taylor's expansion $\log(1 - x) = -x + \vartheta x^2$ with $|\vartheta| \leq 1$ if $|x| < 1/2$, we get

$$\left| \sum_{j=1}^{n} u_j(ab_n) - \left[-\sum_{j=1}^{n} F_j(zb_n) \right] \right| \leq m_n(z) \sum_{j=1}^{n} F_j(zb_n)$$

where the right hand side converges to zero by (5.40) and (5.41). Hence, with one more appeal to (5.41), (5.43) entails (5.42), which completes the proof. ∎

Note in (5.42) that $a(z)$ of (5.41) is the hazard function of the limiting distribution $L(z)$.

Interesting results can be obtained by further specializing.

Theorem 19 If, in addition to the assumptions of Theorem 18, for every $0 < t \leq 1$, as $n \to +\infty$,

$$\lim \sum_{j=1}^{nt} F_j(zb_n) = a_t(z) \quad 0 < a_t(z) < \infty \tag{5.44}$$

holds, then the relation $b_{nt}/b_n \to t^r$ for some $r \geq 0$ holds as $n \to +\infty$. Here, when nt appears in a subscript we automatically mean its integer part. If $r > 0$, then the hazard function $a(z)$ is convex, implying that, if $a(z)$ is differentiable, the hazard rate $r_L(z)$ of $L(z)$ is monotonic.

Proof. Since all assumptions of Theorem 18 are repeated, $L(z)$ exists and (5.42) applies. But then, for every $0 < t \leq 1$, as $n \to +\infty$,

$$\lim P(W_{nt} \leq zb_{nt}) = 1 - e^{-a(z)} \tag{5.45}$$

as well. On the other hand, by repeating the computation following (5.43), we get

$$\lim P(W_{nt} \leq zb_n) = 1 - \exp(-a_t(z)) \qquad 0 < t \leq 1 \qquad n \to +\infty \tag{5.46}$$

Since (5.45) and (5.46) are limit theorems for the same sequence of random variables, we can apply Lemma 2 of Section 4.6. It entails that

$$\lim \frac{b_{nt}}{b_n} = B_t > 0 \text{ exists} \qquad \text{and} \qquad a_t(z) = a(zB_t) \tag{5.47}$$

where $n \to +\infty$. Now, upon writing

$$\frac{b_{nts}}{b_n} = \frac{b_{nts}}{b_{nt}} \frac{b_{nt}}{b_n} \qquad 0 < t \leq 1 \qquad 0 < s \leq 1$$

(the reader is advised to check that it makes no difference when passing to the limit whether we understand nts as the integer part of nts or, after putting m for the integer part of nt, nts signifies the integer part of ms) and letting $n \to +\infty$, we get the functional equation

$$B_{ts} = B_s B_t \qquad \text{all} \quad 0 < t \leq 1 \qquad \text{and} \qquad 0 < s \leq 1$$

This, however, is just a transformation of Cauchy's functional equation (Exercise 2 of Section 1.8) from which, by the evident monotonicity of B_t in t, either $B_t = 1$ for all t or $B_t = t^r$ with some $r > 0$.

Next, we take the complements of (5.42) and (5.46) and utilize (5.47) in the just established form for B_t. We get

$$\prod_{j=nt+1}^{n} [1 - F_j(zb_n)] \to \exp\{-[a(z) - a(zt^r)]\}$$

as $n \to +\infty$. The left hand side above is decreasing in z, hence so does the right hand side which implies that $a(z) - a(zt^r)$ is an increasing function of z for a fixed t and r, $0 < t < 1$, $r > 0$. By introducing the new variables $z = e^y$ and $t = e^v$ and putting $a^*(y) = a(z)$, the property of $a(z)$ above translates to $a^*(y)$ as $a^*(y) - a^*(y + rv)$, $v < 0$, is an increasing function in y. Upon keeping rv fixed, we have that the slopes of the secants connecting y and $y + rv$ are increasing in y, that is, $a(e^y)$ is a convex function. But then $a(z)$ is a convex function as well which can be seen by a variety of ways. The simplest one is to use the fact that a convex function is characterized by its (one-sided) tangent's being below the curve (recall the proof of Theorem 9 about the existence of one-sided derivatives for convex functions). Another property of convex functions is that the slopes of their tangents are nondecreasing which yields the monotonicity of $r_L(z)$ (when defined). This completes the proof. ∎

In the next theorem we assume that each $F_j(x)$ is differentiable, so we can speak of the hazard rate $r_j(x)$ of $F_j(x)$.

Theorem 20 Assume that the hazard rate $r_j(x)$ of $F_j(x)$ is defined for each j and that $r_j(x)$ is twice differentiable on some interval $0 \le x \le c$. Furthermore, we assume (i) $r_j(0) = r > 0$ for all $j \ge 1$, (ii) there are two constants $-\infty < k_1 < 0$ and $0 < k_2 < +\infty$ such that $k_1 < r_j'(0) < k_2$, $j \ge 1$, and (iii) $r_j''(x)$ is uniformly bounded in j for $0 \le x \le c$. Then, under the assumptions of Theorem 18, the limiting distribution $L(z)$ of W_n/b_n is exponential.

Proof. The hazard functions $u_j(x)$ of (5.39) now take the form

$$u_j(x) = \int_0^x r_j(y) \, dy$$

Because $b_n \to 0$ as $n \to +\infty$ and the $r_j'(0)$ are uniformly bounded by k_1 and k_2, the assumption on the second derivatives of $r_j(x)$ entail (use finite Taylor expansion of $r_j(y)$) that the above integral for $u_j(zb_n)$ equals rzb_n with an error term which in uniformly bounded in j by a multiple of b_n^2. Hence, by Taylor's expansion

$$F_j(zb_n) = u_j(zb_n) + O(u_j^2(zb_n)) = rzb_n + O(b_n^2),$$

where the constant in the last error term is uniformly bounded in j. Hence, by (5.41), $nb_n r$ must have a finite limit $\lambda > 0$, and $a(z) = \lambda z$. The proof is complete. ∎

No change is needed in the preceding proof if we allow $r_j(0) = r_j$ to vary with j but the first derivative of $r_j(x)$ at zero is assumed positive and uniformly bounded in j.

Theorem 21 Using the same notation as in Theorem 20, we assume that $r_j(0) = r_j > 0$ for each j and that the right hand derivatives $r_j'(0)$ satisfy $0 \le r_j'(0) < k_2 < +\infty$. Then, with second derivatives $r_j''(x)$, $0 \le x \le c$, which are uniformly bounded in j, the conditions of Theorem 18 imply that W_n/b_n is asymptotically exponentially distributed.

Just as in the proof of Theorem 20, one will find that (5.41) reduces to

$$b_n \sum_{j=1}^{n} r_j \to \lambda \qquad 0 < \lambda < +\infty \qquad \text{and} \qquad a(z) = \lambda z$$

The details are the same as for Theorem 20 and thus omitted.

We conclude this section by adding that recent investigations show that the linear normalization zb_n can be replaced by monotonic functions $g_n(z)$ is which case the limiting distributions $L(z)$ may have characteristics and forms different from those discussed here. See E. Pancheva (1994). In the volume in which Pancheva's paper appears, a variety of applications can be found. For extensions to dependent cases, see Galambos (1987).

Exercises

1. Restate the results of the present section for the maximum of independent positive random variables by turning to reciprocals in each theorem.
2. With the notation of the present section, determine $L(z)$ and the normalizing constants b_n if $F_j(x) = 1 - \exp(-(x + x^2)/j)$, $j \ge 1$, $x \ge 0$.

REFERENCES

Finetti, B. de (1930). Le funzioni caratteristiche di legge instantanea. *Rend. Accad. Lincei, Ser. 6, 12,* 278–282.

Fisz, M. (1962). Infinitely divisible distributions: recent results and applications. *Ann. Math Statist. 33,* 68–84.

Fisz, M. (1963). On the orthogonality of measures induced by L-processes. *Trans. Amer. Math. Soc. 106,* 185–192.

Galambos, J. (1987). *The Asymptotic Theory of Extreme Order Statistics, 2nd ed.* Krieger, Malabar, Florida.

Gnedenko, B. V. and A. N. Kolmogorov (1954). *Limit Distribution for Sums of Independent Random Variables.* Addison-Wesley, Cambridge, Mass.

Khintchine, A. Yu. (1938). *Limit Laws for Sums of Independent Random Variables.* GONTI, Moscow (in Russian).

Lévy, P. (1937). *Théorie de l'addition des variables aléatoires.* Gauthier–Villas, Paris.

Loeve, M. (1963). *Probability Theory*, 3rd ed. Van Nostrand, New York.

Machis, Yu. Yu. (1971). Non-classical limit theorems. *Theory Prob. Appl.* *16*, 175–182.

Pancheva, E. (1994). Extreme value limit theory with nonlinear normalization. In: *Extreme Value Theory and Applications*. (eds.: J. Galambos, J. Lechner and E. Simiu), Kluwer, Dordrecht, pp. 305–318.

Petrov, V. V. (1975). *Sums of Independent Random Variables*. Springer, Berlin (original in Russian: Nauka, Moscow, 1972).

Rohatgi, V. K. and G. J. Székely (1992). On the infinite divisibility of polynomials in infinitely divisible random variables. In: *Probability Theory and Applications*. (eds.: J. Galambos and I. Kátai), Kluwer, Dordrecht, pp. 103–106.

Steutel, F. W. (1970). Preservation of Infinite Divisibility under Mixing. *Math. Centre Tracts*, *33*, Amsterdam.

Steutel, F. W. (1973). Some recent results in infinite divisibility. *Stochastic Process. Appl. 1*, 125–143.

Thorin, O. (1977). On the infinite divisibility of the lognormal distribution. *Scand. Actuarial J.*, 121–148.

6
Independent and Identically Distributed Random Variables

One of the basic assumptions of statistics is that observations are taken independently on a random variable, i.e., that the data form a sequence of independent and identically distributed (i.i.d.) random variables. For this reason alone we are justified in devoting much attention to such sequences and in studying them in depth. Throughout this chapter, $X_1, X_2, \ldots$, will denote i.i.d. random variables. Their common distribution function and characteristic function will be denoted by $F(x)$ and $\varphi(t)$, respectively. We again set $S_n = X_1 + X_2 + \cdots + X_n$, whose distribution function is denoted by $G_n(x)$ and characteristic function by $\varphi_n(t)$. Several results of the previous chapters can be applied directly to these X_j, and we shall do so. As one would expect, however, most of the previous results can be refined when the additional assumption of i.i.d. is made. Two specific results, the classical central limit theorem (Theorem 3.10) and Kolmogorov's strong law of large numbers (Theorem 2.1) have already been established. Both will come up frequently in the current chapter.

Sections 6.1 and 6.3 are studies on special functions, which are of interest in many branches of mathematics. They play important roles in the current chapter. The other sections cover only i.i.d. random variables.

6.1 CAUCHY'S FUNCTIONAL EQUATION

Let $g(x)$ be a finite, measurable function (with respect to Lebesgue measure) on the positive real line. We call any one of the following four (equivalent) functional equations Cauchy's functional equation

$$g(xy) = g(x) + g(y) \qquad x, y > 0 \tag{6.1}$$

$$u(x + y) = u(x) + u(y) \qquad x, y \text{ real} \tag{6.2}$$

$$W(x + y) = W(x)W(y) \qquad x, y \text{ real } W(x) > 0 \tag{6.3}$$

and

$$U(xy) = U(x)U(y) \qquad x, y > 0 \quad U(x) > 0 \tag{6.4}$$

The functions u, g, and U are the following transformations of g for arbitrary real numbers x and $z > 0$:

$$u(x) = g(e^x) \qquad W(x) = \exp[g(e^x)] \qquad U(z) = e^{g(z)} \tag{6.5}$$

The main aim of the present section is to show that (6.1) (and thus each of (6.2)–(6.4) as well) has a unique solution among the finite, measurable functions. Recall that we have encountered (6.3) in Exercise 2 of Section 1.8 under the additional assumption that $1 - W(x)$ is a nondegenerate distribution function. We begin with a case that is similarly easy to solve by first establishing the following result.

Theorem 1 Let $g(x)$, $x > 0$, be a continuous function satisfying (6.1). Then there exists a constant c such that $g(x) = c \log x$.

Proof. By induction, we get from (6.1) that, for $x_j > 0$, $1 \le j \le n$, and $n \ge 2$,

$$g(x_1 x_2 \cdots x_n) = g(x_1) + g(x_2) + \cdots + g(x_n)$$

and thus, with $x_j = x > 0$, $1 \le j \le n$,

$$g(x^n) = ng(x) \tag{6.6}$$

Upon defining $y = \log x$ and $u(y) = g(e^y)$, (6.6) becomes

$$u(ny) = nu(y) \tag{6.6a}$$

Next, we choose $y = mz/n$ and apply (6.6a) repeatedly. We get

$$nu(y) = u(mz) = mu(z) \qquad \text{i.e.,} \qquad u(rz) = ru(z) \qquad r > 0 \text{ rational}$$

which, with $z = 1$, yields $u(r) = ru(1)$ for all positive rational numbers. Because $g(x)$ is continuous by assumption, so is $u(y)$, and thus $u(y) = yu(1)$, $y > 0$. That is,

$$g(x) = u(\log x) = c \log x \qquad x > 1$$

An appeal to (6.6) ensures that this same form will continue to hold if $x < 1$ (choose $n = -1$), while (6.1) with $y = 1$ yields that $g(1) = 0$ as desired. This completes the proof.   ∎

Notice in the proof that the statement "$u(r) = ru(1)$ for all positive rational numbers" was deduced from (6.1) without using the continuity of $g(x)$. Therefore, since the proof could be concluded from that point in the same manner as above if continuity were replaced by monotonicity, we have that all finite monotonic solutions of (6.1) are of the form $g(x) = c \log x$, c constant. The change to seeking monotonic rather than continuous solutions is significant for generalizations when the domain of $g(x)$ is changed to a discrete set. For example, the fact that, when the argument is restricted to positive integers, the only monotonic solution of (6.1) is again $g(x) = c \log x$ (with an arbitrary constant c) has important applications in information theory.

We want to relax the assumption of continuity to measurability in Theorem 1 and to show that its conclusion continues to hold. It is, in fact, possible to show that measurable functions satisfying Cauchy's equation are continuous. To do so, we need the following general result on measurable functions.

While the proof given for Theorem 1 is equally applicable under the assumption of continuity or monotonicity, the following second proof fully exploits the continuity assumption. The idea of this second proof as well as the completion of the proof of Theorem 2 below is based on an idea of Kac (1937) and was communicated to the author by J. Aczél (1989).

Second Proof of Theorem 1. The assumption of continuity for $g(x)$ extends to $W(x)$ of (6.3) via (6.5). Hence, $W(x)$ is integrable on any finite interval and such integral is clearly positive. Now, let a be a fixed positive number and put

$$0 < C = \int_0^a W(y)\,dy < +\infty$$

Therefore, upon integrating equation (6.3) with respect to y, from 0 to a, we get

$$CW(x) = \int_0^a W(x+y)\,dy = \int_x^{x+a} W(t)\,dt = \int_0^{x+a} W(t)\,dt - \int_0^x W(t)\,dt$$

Note that the extreme right-hand side is differentiable due to the continuity of the integrand $W(t)$, and thus so is the extreme left-hand side; that is, $W(x)$ is differentiable. Let us differentiate (6.3) with respect to y and set $y = 0$. We obtain

$$W'(x) = W'(0)\,W(x)$$

which entails

$$W(x) = A \exp[W'(0)x] = A e^{Bx}, \text{ say}$$

where A is some constant. By substituting into (6.3) we get $A = 1$ and, by (6.5), $g(e^x) = Bx$, which completes the proof. ∎

Note that in the preceding discussion, $B = W'(0) = 0$ also is allowed, which case yields $W(x) = 1$ for all x and $g(x) = 0$ for all $x > 0$.

Theorem 2 The conclusion of Theorem 1, that $y(x) = c \log x$, still holds if the assumption of continuity of $g(x)$ is replaced by the condition that $g(x)$, $x > 0$, is a finite Lebesgue measurable function.

Proof. Once again, we transform away from $g(x)$ by (6.5) and argue with $u(x)$ of (6.2). The measurability of $g(x)$ entails the measurability of $u(x)$, and thus of $W_1(x) = \exp(iu(x))$ as well. By (6.2), $W_1(x)$ satisfies (6.3) with the modification that $W_1(x)$ is complex valued with $|W_1(x)| = 1$. The latter property implies that $W_1(x)$ is Lebesgue integrable on any finite interval and that there is a real number a such that

$$C_1 = \int_0^a W_1(y) \, dy \neq 0$$

(otherwise the above integral would be identically zero in a, entailing that $W_1(y) = 0$ for almost all y. This would contradict $|W_1(y)| = 1$ for all y). But then, by (6.3), just as in the second proof of Theorem 1,

$$C_1 W_1(x) = \int_0^{x+a} W_1(t) \, dt - \int_0^x W_1(t) \, dt \tag{6.7}$$

This time, the right-hand side is continuous in x by the measurability of $W_1(t)$, and thus so is the left-hand side, that is, $W_1(x)$. With one more appeal to (6.7), we now have that the right-hand side is differentiable in x since the integrand is continuous. From this point on, we can follow the last part of the second proof of Theorem 1 and obtain $W_1(x) = \exp(iB^*x)$, B^* and x real. In the last form we utilized that, given that $W_1(x)$ is a complex valued exponential function then its variable is a purely imaginary number by definition. We thus have

$$u(x) = B^*x + 2\pi k(x), k(x) \text{ integer valued} \tag{6.8}$$

However, as remarked at the end of the first proof of Theorem 1, any finite solution $u(x)$ of (6.2) satisfies $u(r) = ru(1)$ for all positive rational numbers r.

Consequently, (6.8) with $x = r$ and $x = 1$ implies $k(r) - rk(1) = 0$ for all rational $r > 0$. For the integer valued $k(x)$ this is possible only if $k(x) = 0$ for all x (note that $k(x) = (u(x) - B^*x)/2\pi$ satisfies (6.2)). The proof is completed by an appeal to (6.5). ∎

In many applications one faces the equations (6.1–6.4) with restrictive domains. If these equations are related to distributional properties of integer valued random variables then their domain is necessarily a subset of integers x and y. See Exercise 1 and the lack of memory property of the geometric distribution at (2.43). Here, the additional monotonicity requirement is essential: Exercise 3 shows that measurability alone does not lead to a unique solution in the Cauchy equations with integer domains. The equation resulting from the problem of Exercise 2 leads to interesting questions concerning (6.3) with restrictive domains which we discuss below.

The domain of (6.3) is the whole xy-plane, which we reduce for the present discussion to the first quadrant $Q = \{(x, y): x > 0, y > 0\}$. Within Q, we investigate (6.3) along straight lines passing through the origin. First we observe that the line $y = x$ (restricted to Q) is not sufficient for $W(x)$ to be characterized by (6.3) even among the monotonic functions. Indeed, when $y = x > 0$, (6.3) becomes

$$W(2x) = W^2(x) \qquad x > 0 \tag{6.3a}$$

which uniquely determines $W(2^n)$, n positive or negative integer, from the single value $W(1)$, but if $W(1) \neq 1$ (recall that $W(x) > 0$ on its domain) then an arbitrary monotonic function $w(x)$, $1/2 \leq x \leq 1$ with $w(1) = W(1)$ and $w(1/2) = W^{1/2}(1)$ will generate a monotonic solution of (6.3a) which coincides with $w(x)$ on $[1/2, 1]$. Quite surprising, however, is the fact that if (6.3) is required to hold on the two lines $y = x$ and $y = 2x$, $x > 0$, then (6.3) has a unique solution among the monotonic functions.

Theorem 3 Let $W(x)$ be monotonic for $x > 0$ and assume that $W(x)$ satisfies (6.3) on the semi-lines $y = x$ and $y = 2x$, $x > 0$. That is,

$$W(2x) = W^2(x) \qquad \text{and} \qquad W(3x) = W^3(x) \qquad \text{for} \qquad x > 0 \tag{6.3b}$$

Then, if there is a single value $c > 0$ such that $W(c) = 0$, $W(x) = 0$ for all $x > 0$. Otherwise, there is a real number a such that $W(x) = e^{ax}$, $x > 0$.

Proof. If $W(c) = 0$ for some $c > 0$ then $W(2^nc) = 0$ for all integers n (positive and negative). Thus, by the monotonicity of $W(x)$, $W(x) = 0$ for all $x > 0$. Next, note that if $W(x) \neq 0$, then upon writing $x = 2x_1$, the first equation of (6.3b) entails that $W(x) > 0$. Hence, for the sequel we can assume that $W(x) > 0$ for all $x > 0$.

By iteration of the equations at (6.3b) we obtain for all integers m and n, both positive and negative, that

$$W(x) = e^{ax} \qquad a = \log W(1) \qquad x = 2^m 3^n.$$

Let us put $x_{m,n} = 2^m 3^n$. We shall show that the numbers $x_{m,n}$ are everywhere dense on the positive real line as m and n go through all positive and negative integers. This will mean that to every $x > 0$ we can find two sequences $x_{m(t),n(t)}$ and $x_{m_1(t),n_1(t)}$ such that $x_{m(t),n(t)} \leq x \leq x_{m_1(t),n_1(t)}$ and both sequences converge to x. Since the claimed exponential form applies both to $W(x_{m(t),n(t)})$ and to $W(x_{m_1(t),n_1(t)})$, the assumed monotonicity will lead to the same exponential form for $W(x)$ at all $x > 0$.

It remains to show that the set $\{x_{m,n}\}$ is dense on the positive real line, which we show in the form that the set

$$z_{m,n} = \log x_{m,n} = m \log 2 + n \log 3$$

where m and n go through the positive and negative integers, is dense on the whole real line. Indeed, assume that z is a real number and that there is a smallest $z^*_{m,n}$ such that $z < z^*_{m,n}$ and $z \neq z^*_{m,n}$. In other words, the interval $(z, z^*_{m,n})$ does not contain any further number of the form $z_{m,n}$. That is, every $z_{u,v} < z^*_{m,n}$ also satisfies $z_{u,v} \leq z$. For example, $z_{u,v} = z^*_{m,n} - \log 2$ should be such a number. We put $z'_{u,v}$ for the largest members of $\{z_{m,n}\}$ not exceeding z. Then no two members of $\{z_{m,n}\}$ can be closer to each other than $\alpha = z^*_{m,n} - z'_{u,v} > 0$ because differences of the members of $\{z_{m,n}\}$ belong to this same set, and thus with $z_{m_1,n_1} > z_{m_2,n_2}$ and $z_{u,v} = z_{m_1,n_1} - z_{m_2,n_2} < \alpha$ the points $z'_{u,v} + z_{u,v}$ and $z^*_{m,n} - z_{u,v}$ would violate the definition of both $z'_{u,v}$ and $z^*_{m,n}$. With the same argument we get that no two consecutive members of $\{z_{m,n}\}$ can have a larger distance than α. Consequently, any difference in the set $\{z_{m,n}\}$ is an integer multiple of α, entailing $(\log 2 + \log 3) - \log 2 = k\alpha$, $(\log 2 + \log 3) - \log 3 = r\alpha$ where k and r are integers. The ratio of these two equations contradicts that $(\log 3)/\log 2$ is irrational. So, our choice of z above is impossible, that is, $\{z_{m,n}\}$ is dense everywhere. This concludes the proof. ■

It is clear from the proof that the only role 2 and 3 had at (6.3b) is that $(\log 3)/\log 2$ is irrational. Therefore, Theorem 3 remains unchanged if we require (6.3) to hold on the two semi-lines $y = n_1 x$ and $y = n_2 x$, $x > 0$, where n_1 and n_2 are positive integers such that $[\log(n_1 + 1)]/\log(n_2 + 1)$ is irrational.

Exercises

1. Note that Exercise 2 of Section 1.8 does not lead to (6.3), in that the domain there is $x, y \geq 0$. Show that if the domain is further restricted to

"x and y are nonnegative integers," the monotonic solution of (6.3) is still $G(x) = e^{cx}$, $x \geq 0$ integer, c constant.

2. Let $X_1, X_2, \ldots, X_n$ be i.i.d. nonnegative random variables and assume $W_n = \min(X_1, X_2, \ldots, X_n)$ has the property that nW_n is distributed as X_1 for all $n \geq 1$. (Compare this assumption with (6.6) and the appropriate transformation of (6.3) when the domain is $x, y > 0$. See also Exercise 6 of Section 1.8.) Show that if the X_j are nondegenerate, then they are exponentially distributed.

3. Consider (6.1) with the domain "$x, y > 0$ integers." Show that the general solution of (6.1) with this modified domain is of the following form: for prime numbers p and positive integers k and x, define $e_{p,k}(x) = 1$ if p^k divides x but p^{k+1} does not and $e_{p,k}(x) = 0$ otherwise. Then, with arbitrary values $g(p)$,

$$g(x) = \sum_p \sum_{k \geq 1} kg(p)e_{p,k}(x)$$

Check that $g(x)$ is indeed $c \log x$ if $g(p) = c \log p$ and give an interpretation of $g(x)$ if $g(p) = 1$ for all primes p. (Functions $g(x)$ satisfying (6.1) for positive integers are called completely additive arithmetical functions. Their probabilistic theory grew into a huge field; see the books by Elliott (1979/80) and the paper by Galambos (1982) for further relations to Cauchy's equation.)

6.2 STABLE DISTRIBUTIONS

In this section, we shall characterize all distribution functions that are limiting distributions of $S_n/b_n - a_n$ for some sequences a_n and $b_n > 0$ and for some population distribution $F(x)$ (a reference to the common distribution function of the summands of S_n). We shall, in fact, show that the above-mentioned family of limiting distributions coincides with the following class of stable distributions. We say that $G(x)$, or its characteristic function $\psi(t)$, is *stable* if, for every pair of positive real numbers B_1 and B_2 there exist real numbers a and $B > 0$ such that

$$\psi(B_1 t)\psi(B_2 t) = e^{-iat}\psi(Bt) \qquad \text{for all real } t \tag{6.9}$$

In other words, $G(x)$ is stable if $G(x)$ is the distribution function of each of the independent random variables U and V and $B_1 U + B_2 V$ has the distribution function $G(a + x/B)$, where the meanings of B_1, B_2, B, and a are as above.

Notice that, with appropriate changes of variable, (6.9) can be written in the form $\psi(t) = \psi(At)\tau_A(t)$, where $\tau_A(t)$ is guaranteed to be a characteristic function and $A \geq 0$ is arbitrary. Hence, Theorem 4.12 implies that every stable distribution belongs to class L. The following theorem shows more than this observation by establishing the claimed equivalence of the classes of stable distributions and limiting distributions of (linearly) normalized sums of i.i.d. random variables.

Theorem 4   The set of distribution functions that are limiting distribution functions of $S_n/b_n - a_n$, where the summands of S_n are i.i.d., coincides with the set of stable distribution functions.

Proof.   Let $G(x)$ be a stable distribution with characteristic function $\psi(t)$. Let $X_1, X_2, \ldots, X_n$ be independent random variables with a common distribution function $G(x)$. Then the characteristic function of S_n is $\psi^n(t)$, which, by induction in (6.9), has the representation

$$\psi^n(t) = [\exp(-ita_n)]\psi(b_n t)$$

with some a_n and $b_n > 0$. Consequently, the distribution function of $(S_n + a_n)/b_n$ is $G(x)$ for all n and, thus, $G(x)$ is a limiting distribution of the desired type.

   Conversely, let $G(x)$ be the limiting distribution of $S_n/b_n - a_n$, where a_n and $b_n > 0$ are suitable constants and the summands of S_n are i.i.d. The theorem is obviously true with $\psi(t)$ degenerate; hence, we assume that $G(x)$ is nondegenerate. Then, with $\psi(t)$ as the characteristic function of $G(x)$ and using the standard notation of this chapter,

$$[\exp(-ita_n)]\varphi^n(t/b_n) \to \psi(t)$$

where $|\psi(t)|$ is not identically one. Consequently, for some $\delta > 0$, $\mathrm{Re}[\varphi(t/b_n)] \to 1$ as $n \to +\infty$ for $|t| \le \delta$. But then, by the inequality in Theorem 3.16 (ii), $\mathrm{Re}[\varphi(t/b_n)] \to 1$ for $|t| \le 2\delta$ and, thus, when this is applied repeatedly, we get that $\varphi(t/b_n) \to 1$ for all t. This, by Theorem 5.2 (ii), yields that the sequence X_j/b_n, $1 \le j \le n$, satisfies the UAN condition. Hence, Lemma 3 of Section 4.6, which states that $b_n \to +\infty$ with n and $b_n/b_{n+1} \to 1$ as $n \to +\infty$, applies. As was demonstrated earlier (see the paragraph preceding (4.51)), the latter limit implies the existence of a sequence $m(n)$ of integers such that $m(n) < n$, both $m(n)$ and $n - m(n)$ tend to infinity, and $b_{m(n)}/b_n \to B_1/B_2$, where $B_1 > 0$ and $B_2 > 0$ with $B_1/B_2 < 1$ are arbitrary. Now, if we write

$$B_2\left(\frac{1}{b_n}S_n - a_n\right) + B_2\frac{b_{m(n)}}{b_n}\left(\frac{1}{b_{m(n)}}\sum_{j=n+1}^{n+m(n)} X_j - a_{m(n)}\right)$$

$$= \frac{B_2}{b_n}S_{n+m(n)} - B_2 a_n - B_2\frac{b_{m(n)}}{b_n}a_{m(n)}$$

we recognize that the asymptotic distribution of the left-hand side exists and, thus, so does that of the right-hand side. The limiting characteristic function of the left-hand side equals $\psi(B_2 t)\psi(B_1 t)$ and that of the right-hand side, in view of the general result expressed in Lemma 2 of Section 4.6, equals $\psi(Bt)\exp(-itA)$ with some constants A and $B > 0$. Consequently, (6.9) holds, which completes the proof. ∎

Because a stable distribution is infinitely divisible, its characteristic function can be represented by its Lévy–Khintchine form. The following theorem gives a complete characterization of the spectral function.

Theorem 5 The distribution function $G(x)$ is stable if and only if it is infinitely divisible and the Lévy–Khintchine spectral function $K^*(x)$ of its characteristic function $\psi(t)$ satisfies one of the following three conditions
(i) $K^*(t)$ is identically zero.
(ii) $K^*(x)$ is degenerate at zero $(K^*(0) - K^*(0-)) = \sigma^2 > 0)$.
(iii) $K^*(t)$ is continuous at zero and is differentiable for all $x \neq 0$ with

$$\frac{1 + x^2}{|x|} \frac{dK^*(x)}{dx} = c_i |x|^{-\alpha} \qquad i = 1, 2$$

where $c_i \geq 0$ are constants, c_1 applies for $x > 0$, and c_2 applies for $x < 0$. Furthermore, $c_1 + c_2 > 0$ and $0 < \alpha < 2$.

Proof. Cases (i) and (ii) apply when $G(x)$ is degenerate and normal, respectively. These are both i.d. and stable. Furthermore, regarding $K^*(x)$ in (iii), an easy substitution shows that the i.d. characteristic function generated by $K^*(x)$ via the Lévy–Khintchine representation satisfies (6.9). Therefore, it remains to be shown that if $G(x)$ is stable—that is, neither degenerate nor normal—then (iii) applies. Let us apply (6.9) to the Lévy–Khintchine representation (5.9) of $\psi(t)$. That is, after taking the logarithm in (6.9) (since $\psi(t)$ is i.d., $\psi(t) \neq 0$), and substituting it into the form

$$\log \psi(t) = iLt + \int_{-\infty}^{+\infty} \left(e^{itx} - 1 - \frac{itx}{1 + x^2} \right) \frac{1 + x^2}{x^2} \, dK^*(x)$$

which is a unique representation, we get that there is a number $B > 0$ for every $B_1 > 0$ and $B_2 > 0$ such that

$$[K^*(0) - K^*(0-)](B_1^2 + B_2^2 - B^2) = 0 \qquad (6.10)$$

and

$$M(x/B_1) + M(x/B_2) = M(x/B) \qquad x > 0 \qquad (6.11a)$$

$$m(x/B_1) + m(x/B_2) = m(x/B) \qquad x < 0 \qquad (6.11b)$$

where $M(x)$, $x > 0$, and $m(x)$, $x < 0$, are the Lévy spectral functions (recall

Theorem 5.8). Our aim is to show that (6.11a, b) imply the validity of one of Cauchy's functional equations and thus that the solution is unique. The current problem is that B is not arbitrary, but a function of B_1 and B_2. The calculations are somewhat more convenient with the transformations $D(y) = M(e^y)$ and $m(-e^y) = d(y)$, where y is an arbitrary real number. Hence, with $1/B_i = \exp(b_i)$, $i = 1, 2$, and $1/B = \exp(b)$, (6.11a, b) become

$$D(y + b_1) + D(y + b_2) = D(y + b) \tag{6.12a}$$

$$d(y + b_1) + d(y + b_2) = d(y + b) \tag{6.12b}$$

where $b = b(b_1, b_2)$ and, for given b_1, b_2, and b, the equations above hold for all real numbers y. Let us choose $b_1 = b_2 = x$ and apply (6.12a) repeatedly with x and its multiples. We get that, for every integer $n \geq 1$ and for any x, there is a number $b(n, x)$ such that

$$nD(y + x) = D(y + b(n, x)) \qquad \text{for all } y$$

Thus, for rational numbers $r = s/t$, s and t positive integers,

$$D(y + b(s, x)) = sD(y + x) = rtD(y + x) = rD(y + b(t, x))$$

which, with the transformation $z = y - b(t, x)$, takes the form

$$rD(z) = D(z + b^*(r, x))$$

where x is an arbitrary fixed number and, thus, $b^*(r, x) = u(r)$, a number depending on r alone. Now, let r_1 and r_2 be two rational numbers. Then $r_1 D(0) = D(u(r_1))$ and $r_2 D(u(r_1)) = D(u(r_1) + u(r_2))$, as well as $r_2 D(0) = D(u(r_2))$. That is, abbreviating $u(r_i) = u_i$,

$$r_1 r_2 D(0) = r_2 D(u_1) = D(u_1 + u_2)$$

and thus

$$D(u_1)D(u_2) = r_1 r_2 D^2(0) = D(0)D(u_1 + u_2)$$

Next, we show either that $D(y)$ is identically zero or that $D(y) < 0$ for all y and, in the latter case, that $D(y)$ is continuous and $u(r)$ is dense on the whole real line as r goes through the rational numbers. Hence, if $D(y) \neq 0$, we can divide our last equation by $D^2(0)$, getting that $W(y) = D(y)/D(0)$ satisfies

$$W(u_1)W(u_2) = W(u_1 + u_2) \qquad u_1, u_2 \text{ arbitrary real numbers}$$

Since $W(y)$ is continuous, Theorem 1, via the transformation (6.5), yields that there exists a constant β such that $W(y) = \exp(\beta y)$, i.e., $D(y) = c_1' \exp(\beta y)$, where $c_1' = D(0) < 0$. If we add that c_1' can also be zero, the above form also covers the case when $D(y) \equiv 0$. Finally, since $D(y)$ is nondecreasing (and

$D(y) < 0$), $\beta = -\alpha < 0$, i.e., $\alpha > 0$. Hence, when our claims on $D(y)$ are established, we have $D(y) = c_1' \exp(-\alpha y)$, $c_1' \leq 0$, and, thus,

$$M(x) = c_1' x^{-\alpha} \qquad \alpha > 0 \quad c_1' \leq 0 \quad x > 0 \tag{6.13}$$

Let us now prove the preceding claims on $D(y)$. Since $G(x)$ is stable, it belongs to class L and, thus, Theorem 5.9 yields that $D(y)$ is continuous. Now, if $D(y) \not\equiv 0$, then there is a y_0 such that $D(y_0) < 0$. We thus have, with an appeal to Theorem 5.10, that $D(y)$ is strictly increasing for $y \leq y_0$ and $D(-\infty) = -\infty$. We show that it is not possible that $D(z) = 0$ for some $z > y_0$. That is, if $D(z_0) = 0$ and $D(y) < 0$ for all $y < z_0$, then (6.12a) first yields, in view of the strict monotonicity of $D(y)$, that $b(0, b_2) < 0$ for small values of $b_2 > 0$. But then, by (6.12a) again, $D(z_0 + b(0, b_2)) = D(z_0) + D(z_0 + b_2) = 0$, which contradicts that, with $y = z_0 + b(0, b_2) < z_0$, $D(y) < 0$. We have thus established that if $D(y) \not\equiv 0$, then $D(y) < 0$ for all y, $D(-\infty) = -\infty$, $D(+\infty) = 0$, and $D(y)$ is a continuous, strictly increasing function on the real line. Thus the values $u(r)$ defined by the equation $rD(0) = D(u(r))$ form a dense set as r goes through the rational numbers. This now completes the argument leading to (6.13).

Because the properties of $m(x)$, $x < 0$ are similar to those of $M(x)$, $x > 0$, we immediately have that, with some constants $c_2' \geq 0$ and $\alpha' > 0$,

$$m(x) = c_2' |x|^{-\alpha'} \qquad \alpha' > 0, \quad c_2' \geq 0, \quad x < 0 \tag{6.14}$$

Substituting the actual forms (6.13) and (6.14) into (6.11a, b), the (not identically) zero solutions yield (with $B_1 = B_2 = 1$)

$$B^\alpha = B^{\alpha'} = 2 \qquad \text{i.e., } \alpha = \alpha'$$

Thus, if we transform our results back to $K^*(x)$, we have that $K^*(x)$ is indeed differentiable for all $x \neq 0$ and the desired form of its derivative follows immediately from (6.13) and (6.14) (since $\alpha = \alpha'$). The condition $c_1 + c_2 > 0$ merely expresses the fact that parts (i) and (ii) are not repeated in part (iii). The limit $\alpha < 2$ is a consequence of the particular form of the derivative of $K^*(x)$ and the fact that $K^*(+\infty) < +\infty$. That is, since $K^*(x) - K^*(0)$ must be finite—and proportional to the integral of $y^{1-\alpha}/(1 + y^2)$ from zero to $x > 0$—$y^{1-\alpha}$ must be integrable on $(0, 1)$, say, and thus $\alpha < 2$. Finally, since we have seen that, when $B_1 = B_2 = 1$, $B^\alpha = 2$, and $\alpha < 2$, (6.10) can hold for $B_1 = B_2 = 1$ only if $K^*(0) - K^*(0-) = 0$. This completes the proof. ∎

By evaluating the integral in the Lévy–Khintchine representation of a stable characteristic function, the following criterion can be obtained from Theorem 5.

Theorem 6 The distribution function $G(x)$ is stable if and only if its characteristic function

$$\psi(t) = \exp\left\{ist - c|t|^\alpha\left[1 + i\gamma\frac{t}{|t|}u(t,a)\right]\right\}$$

where s, $c \geq 0$, $0 < \alpha \leq 2$, and $-1 \leq \gamma \leq 1$ are constants and

$$u(t,\alpha) = \begin{cases} \tan\dfrac{\pi\alpha}{2} & \text{if } \alpha \neq 1 \\[2mm] \dfrac{2}{\pi}\log|t| & \text{if } \alpha = 1 \end{cases}$$

The expressions involved are defined by continuity at $t = 0$, yielding $\psi(0) = 1$.

Remarks 1 The case $c = 0$ leads to the degenerate distribution and the case $\alpha = 2$ (with $c > 0$), to the normal distributions. In the remaining cases, i.e., $0 < \alpha < 2$ and $c > 0$, the form of $\psi(t)$ in Theorem 6 is equivalent to part (iii) of Theorem 5.

2 If $\psi(t)$ is a nondegenerate stable characteristic function, then $|\psi(t)| = \exp[-c|t|^\alpha]$, $c > 0$, $0 < \alpha \leq 2$. Since this is integrable on the whole real line, Theorem 3.7 ensures that every nondegenerate stable distribution function $G(x)$ is differentiable. The density function $g(x) = G'(x)$ can, at least in principle, be recovered by the inversion formula (3.15).

3 Real-valued stable characteristic functions are of the form $\psi(t) = \exp[-c|t|^\alpha]$, $c \geq 0$, $0 < \alpha \leq 2$.

4 The number α in the representation of a stable characteristic function is called its characteristic exponent.

Proof of Theorem 6. We apply Theorem 5. Its parts (i) and (ii) give the degenerate and normal distributions, respectively, their characteristic functions having the desired form. Therefore, it remains to compute $\psi(t)$ from the Lévy–Khintchine representation, when $K^*(x)$ satisfies the conditions of Theorem 5 (iii). In this case,

$$\log\psi(t) = iLt + c_2\int_{-\infty}^{0-}\left(e^{itx} - 1 - \frac{itx}{1 + x^2}\right)\frac{1}{|x|^{1+\alpha}}\,dx$$

$$+ c_1\int_{0+}^{+\infty}\left(e^{itx} - 1 - \frac{itx}{1 + x^2}\right)\frac{1}{x^{1+\alpha}}\,dx \tag{6.15}$$

where $c_1 \geq 0$, $c_2 \geq 0$, and $c_1 + c_2 > 0$. Furthermore, $0 < \alpha < 2$.

First, we compute the integrals in (6.15) for $\alpha = 1$. We further limit $t > 0$, and we evaluate the last integral first. Then

$$\int_{0+}^{+\infty} \left(e^{itx} - 1 - \frac{itx}{1+x^2} \right) \frac{1}{x^2} \, dx = \int_{0+}^{+\infty} \frac{\cos tx - 1}{x^2} \, dx$$

$$+ i \int_{0+}^{+\infty} \left(\sin tx - \frac{tx}{1+x^2} \right) \frac{1}{x^2} \, dx$$

Substituting $tx = y$ in the first term on the right-hand side and integrating by parts, we get

$$\int_{0+}^{+\infty} \frac{\cos tx - 1}{x^2} \, dx = t \int_{0+}^{+\infty} \frac{\cos y - 1}{y^2} \, dy = -t \int_{0}^{+\infty} \frac{\sin y}{y} \, dy$$

which equals $-\pi t/2$ (see the Lemma of Section 3.2). On the other hand,

$$\int_{0+}^{+\infty} \left(\sin tx - \frac{tx}{1+x^2} \right) \frac{1}{x^2} \, dx = \lim_{\varepsilon=0} \int_{\varepsilon}^{+\infty} \left(\frac{\sin tx}{x^2} - \frac{t}{x(1+x^2)} \right) dx$$

which, upon splitting the right-hand side into two integrals and substituting $y = tx$ in the first one, becomes

$$t \lim_{\varepsilon=0} \left[\int_{t\varepsilon}^{+\infty} \frac{\sin y}{y^2} \, dy - \int_{\varepsilon}^{+\infty} \frac{dy}{y(1+y^2)} \right]$$

$$= t \left[\lim_{\varepsilon=0} \int_{\varepsilon}^{+\infty} \left(\frac{\sin y}{y^2} - \frac{1}{y(1+y^2)} \right) dy - \lim_{\varepsilon=0} \int_{\varepsilon}^{\varepsilon t} \frac{\sin y}{y^2} \, dy \right]$$

The first limit on the right-hand side is finite, which can easily be seen by expanding $\sin y - y/(1+y^2)$ by Taylor's formula for small values of y. For the second limit, we write

$$\lim_{\varepsilon=0} \int_{\varepsilon}^{\varepsilon t} \frac{\sin y}{y^2} \, dy = \lim_{\varepsilon=0} \left[\int_{\varepsilon}^{\varepsilon t} \left(\frac{\sin y}{y} - 1 \right) \frac{1}{y} \, dy + \int_{\varepsilon}^{\varepsilon t} \frac{1}{y} \, dy \right] = \log t$$

because the limit of the first integral following the first equation sign is zero (again using the Taylor expansion).

The combination of the results above thus yields that the last term in (6.15) equals

$$itL_1 - c_1 \frac{\pi}{2} t \left(1 + i \frac{2}{\pi} \log t \right) \tag{6.16}$$

for $\alpha = 1$ and $t > 0$, with some finite number L_1 and with $c_1 \geq 0$. This same term of (6.15), for $t < 0$, is the complex conjugate of (6.16), with $(-t)$

replacing t, and the last but one term is the complex conjugate of the last integral. Hence, for $t > 0$,

$$\log \psi(t) = iL_2 t - (c_1 + c_2)\frac{\pi}{2}t - i(c_1 - c_2)t \log t$$

and, for $t < 0$,

$$\log \psi(t) = iL_2 t - (c_1 + c_2)\frac{\pi}{2}|t| - i(c_1 - c_2)t \log |t|$$

which can be combined to the desired form by setting

$$c = (c_1 + c_2)\frac{\pi}{2} > 0 \qquad \gamma = \frac{c_1 - c_2}{c_1 + c_2} \qquad (\text{thus } |\gamma| \le 1)$$

Next, we assume $0 < \alpha < 1$. Then the improper integral

$$\int_{-\infty}^{+\infty} \frac{|x|}{1 + x^2}\frac{1}{|x|^{1+\alpha}}dx = \int_{-\infty}^{+\infty} \frac{dx}{|x|^\alpha(1 + x^2)} < +\infty$$

and, thus, by a change in L, (6.15) reduces to

$$\log \psi(t) = iL_3 t + c_2 \int_{-\infty}^{0} \frac{e^{itx} - 1}{|x|^{1+\alpha}}dx + c_1 \int_{0}^{+\infty} \frac{e^{itx} - 1}{x^{1+\alpha}}dx$$

Again, the two integrals on the right-hand side are complex conjugates of each other, and the change from t to $(-t)$ leads to complex conjugates within the same integral. Thus, we easily get $\log \psi(t)$ if we evaluate the last integral above for $t > 0$. Now, by the substitution $y = tx$,

$$\int_{0}^{+\infty} \frac{e^{itx} - 1}{x^{1+\alpha}}dx = t^\alpha \int_{0}^{+\infty} \frac{e^{iy} - 1}{y^{1+\alpha}}dy$$

To evaluate the right-hand side, we view y as a complex number and use contour integration, choosing as the contour of integration the closed curve consisting of the segment $[0, R]$ of the real line, $i[0, R]$ of the imaginary axis, and the arc $u \ge 0$, $v \ge 0$, of the circle $u^2 + v^2 = R^2$. An easy estimate ensures that the integral on the circular segment of the contour will approach 0 as $R \to +\infty$ and, thus, by the theorem of residues from elementary complex analysis, we get

$$\int_{0}^{+\infty} \frac{e^{iy} - 1}{y^{1+\alpha}}dy = \int_{0}^{+\infty i} \frac{e^{iv} - 1}{v^{1+\alpha}}dv = i^{-\alpha}\int_{0}^{+\infty} \frac{e^{-u} - 1}{u^{1+\alpha}}du$$

Writing

$$i^{-\alpha} = e^{-i\alpha\pi/2} \qquad \text{and} \qquad c^*(\alpha) = \int_0^{+\infty} \frac{e^{-u} - 1}{u^{1+\alpha}} \, du < 0$$

we thus have, for $t > 0$, $0 < \alpha < 1$,

$$\log \psi(t) = itL_3 + t^\alpha c^*(\alpha) \left[(c_1 + c_2)\cos \frac{\pi}{2}\alpha + i(c_2 - c_1)\sin \frac{\pi}{2}\alpha \right]$$

which, with

$$c = -c^*(\alpha)(c_1 + c_2)\cos \frac{\pi}{2}\alpha > 0 \qquad \text{and} \qquad \gamma = \frac{c_2 - c_1}{c_1 + c_2} \qquad \text{(thus } |\gamma| \le 1\text{)}$$

takes the simpler form

$$\log \psi(t) = itL_3 - ct^\alpha \left\{ 1 + i\gamma \tan \frac{\pi}{2}\alpha \right\}$$

The change from t to $(-t)$ and the simultaneous transformation to complex conjugates requires that we change t^α to $|t|^\alpha$ and γ to $\gamma t/|t|$, which completes the proof for $0 < \alpha < 1$.

Finally, for $1 < \alpha < 2$, first observe that the integral

$$\int_{-\infty}^{+\infty} \left(itx - \frac{itx}{1 + x^2} \right) \frac{1}{|x|^{1+\alpha}} \, dx = \int_{-\infty}^{+\infty} \frac{itx^3}{(1 + x^2)|x|^{1+\alpha}} \, dx$$

is finite. Hence, by changing L, the integrals in (6.15) can be modified to (we write the case of $x > 0$ only)

$$c_1 \int_0^{+\infty} (e^{itx} - 1 - itx) \frac{1}{x^{1+\alpha}} \, dx$$

which becomes, by the substitution $y = tx$ $(t > 0)$ and contour integration along the same path as in the preceding case,

$$c_1 t^\alpha i^{-\alpha} \int_0^{+\infty} \frac{e^{-v} - 1 + v}{v^{1+\alpha}} \, dv = t^\alpha e^{-i\alpha\pi/2} c^{**}(\alpha) c_1$$

where $c^{**}(\alpha) > 0$ is the (finite) value of the integral. Thus, if we set

$$c = -c^{**}(\alpha)(c_1 + c_2)\cos \frac{\pi}{2}\alpha > 0$$

and if γ is the same ratio as in the two previous cases, then, by the same calculations as before (taking complex conjugates and changing t to $(-t)$), the desired form of $\log \psi(t)$ is obtained. The theorem is established. ∎

Exercises

1. In a simple random walk $S_n = X_1 + X_2 + \cdots + X_n$, where the X_j are i.i.d. and take the values 1 and -1 with probabilities $1/2$ each, define $Y_k = n_k - n_{k-1}$, where $n_0 = 0$ and n_k, $k \geq 1$, are successive positive integers such that $S_{n_k} = 0$. Show that the Y_k are i.i.d. Assume that it has been proved that the probability that the number of $m \leq 2n$ such that $S_m = 0$ is r equals $\binom{2n-r}{n} \Big/ 2^{2n-r}$. Use Stirling's formula (see the introductory volume) to show that

$$P(Y_1 + Y_2 + \cdots + Y_N \leq N^2 x) \to 2[1 - N(x^{-1/2})]$$

as $N \to +\infty$ $(x > 0)$, where $N(x)$ is the standard normal distribution function. Consequently, $G(x) = 2[1 - N(x^{-1/2})]$ is a stable distribution function.

2. Let X_j, $j \geq 1$, be i.i.d. random variables whose common distribution function $G(x)$ is stable with characteristic exponent α. Describe the normalizing constants a_n and b_n such that $S_n/b_n - a_n$ has an asymptotic distribution. Use this result to give the value of the characteristic exponent of $G(x)$ of the preceding exercise.

3. Let X be a random variable with a stable distribution function $G(x)$. For which values of u is $E(|X|^u)$ finite?

6.3 REGULARLY VARYING FUNCTIONS

An important property of distribution functions $F(x)$ that can serve as the underlying distribution of an i.i.d. sequence such that the asymptotic distribution of $S_n/b_n - a_n$ is a nonnormal, nondegenerate stable distribution is that of regular variation. This same property comes up later in connection with the theory of extremes. The foundations of regular variation were laid down by Karamata in the early 1930s. This section is devoted to those fundamental results of this theory which we need in this book. For the history, and for a more detailed development of the theory itself, the reader is referred to Seneta's monograph (1976).

Throughout this section, "measurable" means measurability with respect to Lebesgue measure and integrals are Lebesgue integrals.

Definition   A positive, measurable function $R(x)$ that is defined (and finite) on $(A, +\infty)$ with some $A \geq 0$, is *regularly varying* at infinity if

$$\lim \frac{R(tx)}{R(x)} = U(t) \tag{6.17}$$

exists as $x \to +\infty$ for every $t > 0$ and $0 < U(t) < +\infty$.

Theorem 7 The function $U(t)$ that appears in the definition of regular variation is of the form $U(t) = t^\beta$, $-\infty < \beta < +\infty$.

Proof. Let $R(x)$ be a regularly varying function. Let $t_1 > 0$ and $t_2 > 0$ be two fixed numbers. Then the identity

$$\frac{R(t_1 t_2 x)}{R(x)} = \frac{R(t_1 t_2 x)}{R(t_2 x)} \frac{R(t_2 x)}{R(x)}$$

entails

$$U(t_1 t_2) = U(t_1)U(t_2)$$

upon letting $x \to +\infty$. Since the limits of measurable functions are measurable, we find $U(t)$ to be a measurable function satisfying (6.4). Hence, (6.5) and Theorem 2 ensure that $U(t)$ is of the desired form, and the proof is concluded. ∎

By Theorem 7, we thus have that there exists a number β for every regularly varying function $R(x)$ such that

$$s(x) = x^{-\beta} R(x) \tag{6.18}$$

satisfies the properties (i) $s(x)$ is defined (finite), positive, and measurable on $(A, +\infty)$ for some $A \geq 0$ and (ii) for every fixed $t > 0$

$$\lim \frac{s(tx)}{s(x)} = 1 \tag{6.19}$$

as $x \to +\infty$. A function $s(x)$ that satisfies (i) and (ii) above is called a *slowly varying* function at infinity. The number β in (6.18) is called the *exponent of regular variation* of $R(x)$.

For monotonic functions $R(x)$, regular variation is ensured by the following simple criterion. (Throughout this section, both regular and slow variations are at infinity.)

Theorem 8 Let $R(x)$ be a positive monotonic function on $(A, +\infty)$ with some $A \geq 0$. Assume that there are two sequences B_n and b_n such that

$$B_{n+1}/B_n \to 1 \qquad b_n \to +\infty \tag{6.20}$$

as $n \to +\infty$ and, for $t > 0$,

$$B_n R(b_n t) \to v(t) \tag{6.21}$$

where $0 < v(t) < +\infty$ on some interval (a, b) with $0 \leq a < 1 < b$ (outside of (a, b), $v(t) = +\infty$ is also permitted). Then $v(t) = ct^\beta$, $c > 0$, and $R(x)$ varies regularly with exponent β.

Proof. Let us assume that $R(x)$ is nondecreasing. Let $x \geq A$ and $x \to +\infty$. Let n be the smallest positive integer such that $b_n \leq x < b_{n+1}$. Then, for $t > 0$ (and when $b_n t \geq A$),

$$R(b_n t) \leq R(xt) \leq R(b_{n+1} t)$$

and thus

$$\frac{R(b_n t)}{R(b_{n+1})} \leq \frac{R(xt)}{R(x)} \leq \frac{R(b_{n+1} t)}{R(b_n)}$$

Now, by rewriting

$$R(b_n t)/R(b_{n+1}) = [B_n R(b_n t)/B_{n+1} R(b_{n+1})](B_{n+1}/B_n)$$

and similarly the extreme right-hand side, we get from (6.20) and (6.21) that, as $x \to +\infty$, both extreme terms, and thus the middle one as well, tend to $v(t)/v(1)$, which is finite and positive on (a, b). From this new relation of $v(t)$ to $R(x)$ it is clear that $v(t)$ is monotonic and, just as in the proof of Theorem 7, $v(t_1 t_2) = v(t_1)v(t_2)$ for all $t_1 > 0$ and $t_2 > 0$. Thus, if $v(t) = +\infty$ for some $t > 0$, then $v(t^n) = +\infty$ for all rational $n > 0$. This, upon choosing n close to zero, would imply that $v(y) = +\infty$ for $y = t^n$ arbitrarily close to one, contradicting that $v(y)$ is finite for all $1 \leq y < b$. Consequently, $v(t) < +\infty$ for all $t > 0$. The assumption $v(t) = 0$ for some $t > 0$ leads to a similar contradiction; thus, $0 < v(t) < +\infty$ for all $t > 0$. Hence, if $R(x)$ is non-decreasing, Theorem 7 applies with $U(x) = v(x)/v(1)$. By reversing all inequalities, a similar result obtains for nonincreasing $R(x)$. The theorem is established. ∎

The theorem can be sharpened to a considerable extent, in that the role of an interval (a, b) on which $v(t)$ is finite and positive can be replaced by a variety of sets. See Seneta (1976) for a discussion. A remarkable related result is formulated in Exercise 1.

The following important properties of slowly varying functions will lead to the celebrated Karamata representation of regularly varying functions.

First, let $s(x)$ be a continuous slowly varying function. Assume that r is such that the improper Riemann integral of $x^r s(x)$ on $(A, +\infty)$ is infinite. It then follows immediately by L'Hospital's rule that the function

$$R_r(x) = \int_A^x y^r s(y)\, dy \tag{6.22}$$

is regularly varying with exponent $r + 1$. The result, however, continues to hold if the continuity assumption is dropped.

Theorem 9 Let $s(x)$ be a slowly varying function defined on $(A, +\infty)$, $A \geq 0$. Then the function $R_r(x)$ of (6.22) tends to infinity with x for $r > -1$ and is regularly varying with exponent $r + 1$ for all $r \geq -1$. Furthermore, $x^{r+1}s(x)/R_r(x) \to r + 1$ as $x \to +\infty$ $(r + 1 \geq 0)$.

Proof. Let $t > 0$ be a fixed number. Then, for every $\varepsilon > 0$, there is a number $B > A$ such that, for all $x \geq B$,

$$(1 - \varepsilon)s(x) \leq s(xt) \leq (1 + \varepsilon)s(x) \tag{6.23}$$

Next, we write, for $A < B < x$ such that $tx > A$,

$$\int_A^{tx} y^r s(y)\, dy = \int_A^{Bt} y^r s(y)\, dy + \int_{Bt}^{tx} y^r s(y)\, dy$$

Let us substitute $y = tu$ in the last integral and use (6.23). We get

$$(1 - \varepsilon)t^{r+1} \leq \frac{R_r(tx) - R_r(Bt)}{R_r(x) - R_r(B)} \leq (1 + \varepsilon)t^{r+1}$$

(note that, because $s(y) > 0$ for all $y \geq A$, $R_r(x)$ is strictly increasing, so the denominator is positive). By letting $x \to +\infty$, we find that $R_r(+\infty) < +\infty$ is not possible for $r + 1 > 0$ and, thus, these same inequalities imply that $R_r(x)$ is indeed regularly varying with exponent $r + 1$ if $r + 1 > 0$. We also reach the same conclusion for $r + 1 = 0$ if, in addition, we know that $R_{-1}(+\infty) = +\infty$. However, if $R_{-1}(+\infty) = c < +\infty$, then $R_{-1}(x)$ is slowly varying as a direct consequence of the definition of slow variation. Hence, the first part of the theorem is proved.

Turning to the second part, we define

$$d_r(x) = x^{r+1}s(x)/R_r(x) \qquad r \geq -1$$

It follows from the first part of the theorem that $d_r(x)$ is slowly varying. Consider now $\log R_r(x)$. It will be demonstrated in the next chapter that functions represented by integrals like $R_r(x)$ are differentiable for almost all x (Lebesgue measure) and that the derivative coincides with the integrand, again for almost all x. Thus

$$\frac{d_r(x)}{x} = \frac{d \log R_r(x)}{dx} \qquad \text{for almost all } x \geq A$$

from which, by integration, we get

$$\log \frac{R_r(tx)}{R_r(x)} = \int_x^{xt} \frac{d_r(y)}{y}\, dy = \int_1^t \frac{d_r(ux)}{u}\, du = d_r(x) \int_1^t \frac{d_r(ux)}{u\, d_r(x)}\, du$$

In view of the result in the first part of the theorem, the left-hand side converges to $(r + 1)\log t$ as $x \to +\infty$. We now establish that $d_r(x)$ converges to a finite constant c as $x \to +\infty$ and, thus, that the middle term converges to $c \log t$, yielding $c = r + 1$, as desired. In fact, if $d_r(x_n) \to c < +\infty$ on a subsequence x_n, it follows that $c = r + 1$; thus, it suffices to prove that there exists no sequence x_n such that $d_r(x_n) \to +\infty$ with n. But this follows immediately from the extreme right-hand side, since $d_r(ux_n)/d_r(x_n) \to 1$ and Fatou's lemma (Theorem 2.15) ensure that the integral on the extreme right-hand side is at least $\log t$ and, thus, when we compare it with the extreme left-hand side, we get

$$\limsup d_r(x) \le r + 1$$

as $x \to +\infty$. This completes the proof. ∎

Notice the following consequence of Theorem 9. If $R(x)$ is regularly varying with exponent $r + 1 \ge 0$, then, by (6.18), $s(x) = x^{-r-1}R(x)$ is slowly varying. With this $s(x)$, we define $R_r(x)$ via (6.22), which is strictly increasing and continuous; thus, Theorem 9 yields that $R(x)/R^*(x) \to 1$ as $x \to +\infty$, where $R^*(x) = (r + 1)R_r(x)$ is regularly varying with the same exponent as $R(x)$ and $R^*(x)$ is strictly increasing and continuous.

Another consequence of Theorem 9 is the following representation theorem of regularly varying functions.

Theorem 10 (Karamata's representation) To every regularly varying function $R(x)$ defined on $(A, +\infty)$, there exist a positive number $B \ge A$ and two functions $u(x)$ and $\varepsilon(x)$, both of which are measurable and bounded and the former of which converges to a finite constant while $\varepsilon(x) \to 0$ as $x \to +\infty$, and

$$R(x) = x^\beta \exp\left[u(x) + \int_B^x \frac{\varepsilon(y)}{y}\, dy \right] \tag{6.24}$$

where β is the exponent of regular variation of $R(x)$.

Proof. By (6.18), it suffices to prove (6.24) for slowly varying functions $s(x)$, i.e., with $\beta = 0$. With the notation of the proof of Theorem 9, every slowly varying function $s(x)$ can be written in the form

$$s(x) = d_0(x)R_0(x)/x \qquad x \ge A > 0$$

where $d_0(x) \to 1$ as $x \to +\infty$ and

$$R_0(x) = R_0(A)\exp\left[\int_A^x \frac{d_0(y)}{y}\, dy \right]$$

The combination of these expressions yields (6.24), where

$$u(x) = \log d_0(x) + \log R_0(A) \qquad \varepsilon(y) = d_0(y) - 1$$

The theorem is established. ∎

A substitution into (6.24) immediately yields the following uniformity criterion.

Corollary (**The uniform convergence theorem**) If $s(x)$ is a slowly varying function, then the convergence in (6.19) is uniform in t on every finite interval $0 < a \leq t \leq b < +\infty$.

It is evident that every function $R(x)$ of the form (6.24) is regularly varying and, thus, that Theorem 10 is a necessary and sufficient condition for regular variation.

Formula (6.24) can be used both to construct regularly varying functions and to set general bounds on their order of magnitude. Thus, the choices $B = e$, $u(x) = 0$, and $\varepsilon(x) = c/\log x$ show that $(\log x)^c$ is slowly varying for every c (if $x \geq e$). Another set of choices shows that every power of $\log \log x$ is slowly varying. The latter, however, does not follow from the former just because its order of magnitude is smaller. The function $z(x) = \log x$ if $e^{2n} < x \leq e^{2n+1}$ with a positive integer n and $z(x) = 1$ otherwise is clearly not slowly varying, even though $z(x)$ is of smaller magnitude than $s(x) = (\log x)^2$, say, which is slowly varying. This justifies the name emphasizing that the variation (the change in it) is slow. Exercises 3 and 4 contain general estimates of the order of magnitude of slowly varying functions.

Exercises

1. Show that if $s(x)$ is a positive, monotonic function on $(A, +\infty)$, $A \geq 0$, and if $s(t_0 x)/s(x) \to 1$ for one $t_0 > 0$, $t_0 \neq 1$, as $x \to +\infty$, then $s(x)$ is slowly varying.
2. Show that if $s_1(x)$ and $s_2(x)$ are slowly varying, then so is $s_1(x) + s_2(x)$.
3. Show that if $s(x)$ is slowly varying, then, for any $c > 0$, $x^c s(x) \to +\infty$ and $x^{-c} s(x) \to 0$ as $x \to +\infty$.
4. Show that, for every slowly varying function $s(x)$,

   $$[\log s(x)]/\log x \to 0 \qquad \text{as } x \to +\infty$$

5. Let $s_1(x)$ and $s_2(x)$ be slowly varying functions and assume that $s_2(x) \to +\infty$ with x. Show that $s_1(s_2(x))$ is slowly varying.

6.4 DOMAINS OF ATTRACTION FOR STABLE DISTRIBUTIONS

It is demonstrated in Theorem 4 that, for every stable distribution function $G(x)$, there exists at least one distribution function $F(x)$ such that, if the random variables $X_1, X_2, \ldots$ are independent with a common distribution function $F(x)$, then, with some constants a_n and $b_n > 0$, the asymptotic distribution of $S_n/b_n - a_n$ is $G(x)$. The set of distribution functions $F(x)$ with this property is called the domain of attraction of $G(x)$, to be abbreviated $D(G)$. With this new concept, we, in fact, found, in the course of the proof of Theorem 4, that $G(x) \in D(G)$. It follows immediately that if $F(x) \in D(G)$ then so do all distribution functions that are of the same type as $F(x)$. Furthermore, Lemma 2 of Section 4.6 implies that if $G_1(x)$ and $G_2(x)$ are not of the same type, then $D(G_1)$ and $D(G_2)$ are disjoint. In this section, we shall give complete characteristics of the domains of attraction of stable distributions.

We start with the normal distribution $N(x)$. The classical central limit theorem can now be restated as saying that if F has a finite variance then $F \in D(N)$. However, Exercise 4 of Section 3.3 shows that $D(N)$ has additional members. The following theorem characterizes $D(N)$.

Theorem 11 $F \in D(N)$ if and only if the function

$$s(x) = \int_{-x}^{x} y^2 \, dF(y)$$

is slowly varying (at infinity).

Proof. First note that if $s(x) \to V < +\infty$ as $x \to +\infty$, i.e., if the variance V associated with $F(x)$ is finite, then both $s(x)$ is slowly varying and, by the classical central limit theorem, $F \in D(N)$. Hence, we need to prove the theorem when $s(x) \to +\infty$ with x.

Assume that $F \in D(N)$. That is, with some constants a_n and $b_n > 0$, $S_n/b_n - a_n$ has an asymptotic normal distribution, where the summands X_j of S_n are i.i.d. with a common distribution function $F(x)$. Now, if $\varphi(t)$ is the characteristic function of F, then $\varphi^n(t/b_n)\exp(-ita_n) \to \exp(-t^2/2)$ implies that $|\varphi(t/b_n)| \to 1$ as $n \to +\infty$; thus, by Lemma 1 of Section 4.6, the variables X_j/b_n, $1 \le j \le n$, $n \ge 1$, satisfy the UAN condition. Hence, $b_n \to +\infty$ and $b_n/b_{n+1} \to 1$ as $n \to +\infty$ (Lemma 3 of Section 4.6). Furthermore, by the criterion for asymptotic normality in class L (Theorem 5.14)

$$\frac{n}{b_n^2} \left\{ s(xb_n) - \left[\int_{|y| \le xb_n} y \, dF(y) \right]^2 \right\} \to 1 \qquad (x > 0)$$

and $nP(|X_j| \ge \varepsilon b_n) \to 0$. This latter condition can be used for a specific construction of a function $w(x) > 0$ such that $w(x) \to +\infty$ at the infinities and

$$\int_{-\infty}^{+\infty} w(x) \, dF(x) = W < +\infty$$

But then, by the Cauchy–Schwarz inequality,

$$\left[\int_{-x}^{x} y\, dF(y)\right]^2 \le \left[\int_{-x}^{x} w^{1/2}(y)\, \frac{|y|}{w^{1/2}(y)}\, dF(y)\right]^2$$

$$\le \left(\int_{-x}^{x} w(y)\, dF(y)\right)\left(\int_{-x}^{x} \frac{y^2}{w(y)}\, dF(y)\right)$$

which has a smaller order of magnitude than $s(x)$. Consequently, the previously quoted criterion reduces to

$$(n/b_n^2)s(xb_n) \to 1 \qquad \text{as } n \to +\infty \qquad \text{all } x > 0$$

Because $B_n = n/b_n^2$ and b_n satisfy (6.20) and $s(x)$ is nondecreasing, Theorem 8 ensures that $s(x)$ is slowly varying.

We now turn to the converse. That is, $F(x)$ is a distribution function such that $s(x) \to +\infty$ with x and $s(x)$ is slowly varying. We explicitly give constants a_n and $b_n > 0$ such that the asymptotic distribution of $S_n/b_n - a_n$ is standard normal. For this, we note that the assumptions on $s(x)$ imply that $E = E(X_j)$ is finite (we use the notation of the first part of this proof). Indeed, by writing

$$s(kx) = s(x) + \int_{x}^{kx} y^2\, dF(y) + \int_{-kx}^{-x} y^2\, dF(y)$$

$$\ge s(x) + x^2[F(kx) - F(x) + F(-x) - F(-kx)]$$

for $k > 1$ and then choosing k large but fixed, such that $F(kx) > 1 - \varepsilon$ and $F(-kx) < \varepsilon$ for all $x \ge x_0$, we get, as a consequence of the slow variation of $s(x)$, that

$$\delta(x) = \frac{x^2[1 - F(x) + F(-x)]}{s(x)} \to 0 \tag{6.25}$$

as $x \to +\infty$. Therefore, there is an $x_0 > 0$ such that $\delta(x) \le 1/2$ for all $x \ge x_0$. Thus, upon choosing $x_0 \le a < b$ and integrating by parts,

$$\int_{a}^{b} y\, dF(y) = a[1 - F(a)] - b[1 - F(b)] + \int_{a}^{b} [1 - F(y)]\, dy$$

and

$$-\int_{-b}^{-a} y\, dF(y) = aF(-a) - bF(-b) + \int_{-b}^{-a} F(y)\, dy$$

That is,

$$\int_{a\leq|y|\leq b} |y|\, dF(y) \leq a[1 - F(a) + F(-a)] + \int_a^b [1 - F(y) + F(-y)]\, dy$$

$$\leq \frac{1}{2}\frac{s(a)}{a} + \frac{1}{2}\int_a^b \frac{s(y)}{y^2}\, dy \leq \frac{1}{2}\frac{s(a)}{a} + \int_a^{+\infty} \frac{s(y)}{y^2}\, dy$$

where the extreme right-hand side is finite because $s(y)$ is slowly varying and, thus, for any $c > 0$,

$$s(y)y^{-c} \to 0 \qquad \text{as } y \to +\infty \tag{6.26}$$

(see Exercise 3 in the previous section). Since the extreme right-hand side does not depend on b, the claimed finiteness of the expectation follows.

From now on we shall assume that $E = E(X_j) = 0$; thus, $a_n = 0$ can be chosen (otherwise, $a_n = nE/b_n$ would do). Next, we choose $b_n > 0$. We note from (6.26) that $s(y)/y^2 \to 0$ as $y \to +\infty$ and, thus, we can choose b_n such that

$$ns(b_n)/b_n^2 \to 1 \qquad \text{as } n \to +\infty$$

Then $b_n \to +\infty$ with n and, for any $\varepsilon > 0$, $ns(\varepsilon b_n)/b_n^2 \to 1$ because of the slow variation of $s(x)$. Hence, (6.25) ensures that $n[1 - F(\varepsilon b_n) + F(-\varepsilon b_n)] \to 0$. Thus, for $\varepsilon > 0$ small, the inequalities

$$\frac{1}{b_n^2(1 + \varepsilon)}\int_{-b_n\varepsilon}^{b_n\varepsilon} y^2\, dF(y) \leq \int_{-b_n\varepsilon}^{b_n\varepsilon} \frac{y^2}{b_n^2 + y^2}\, dF(y)$$

$$\leq \frac{1}{b_n^2}\int_{-b_n\varepsilon}^{b_n\varepsilon} y^2\, dF(y)$$

and

$$\int_{-\infty}^{-b_n\varepsilon} \frac{y^2}{b_n^2 + y^2}\, dF(y) + \int_{b_n\varepsilon}^{+\infty} \frac{y^2}{b_n^2 + y^2}\, dF(y) \leq 1 - F(b_n\varepsilon) + F(-b_n\varepsilon)$$

entail that the function

$$K_n^*(x) = n\int_{-\infty}^{b_n x} \frac{y^2}{b_n^2 + y^2}\, dF(y)$$

satisfies

$$\left| K_n^*(x) - \frac{ns(\varepsilon b_n)}{b_n^2} \right| \leq \varepsilon\frac{ns(\varepsilon b_n)}{b_n^2} + n[1 - F(\varepsilon b_n) + F(-\varepsilon b_n)]$$

for every $0 < \varepsilon < x$, from which we have that $K_n^*(x) \to 1$ as $n \to +\infty$ for every $x > 0$, including $x = +\infty$, while, by a direct estimate, $K_n^*(x) \to 0$ if $x < 0$. The sufficiency part of Theorem 5.11 (recall Example 1 of Section 5.2, which gave the spectral function $K^*(x)$ for the normal distribution) now completes the proof. ∎

Remark Note from the proof that b_n can be written as $b_n = n^{1/2}g(n)$, where $g(n) \to +\infty$ with n if $s(x) \to +\infty$ with x (the case of infinite variance), but $g(n)$ has a smaller order of magnitude than any power of n. On the other hand, by the classical central theorem, we can choose $g(n) = V$ when the variance V associated with $F(x)$ is finite.

While $D(N)$ is very large and the members of $D(N)$ can differ from $N(x)$ in a number of ways regarding shape and characteristics, this is no longer true of $D(G)$ for any other nondegenerate stable distribution $G(x)$: they can attract only those distributions which, to a large extent, are similar to themselves.

Theorem 12 Let $G(x)$ be a nondegenerate stable distribution function whose type is different from the normal distributions. Then $F \in D(G)$ if and only if the following two conditions are satisfied:

(i) $T(x) = 1 - F(x) + F(-x)$ is regularly varying with exponent $-\alpha$, where $0 < \alpha < 2$ is the characteristic exponent of $G(x)$
(ii) there is a number d, $0 \le d \le 1$, such that

$$[1 - F(x)]/T(x) \to d \qquad \text{as } x \to +\infty$$

Remark Clearly, a function satisfying (ii) also satisfies $F(-x)/T(x) \to 1 - d$ as $x \to +\infty$. Furthermore, by (6.18), (i) is equivalent to the existence of a slowly varying function $s_F(x)$ such that $T(x) = x^{-\alpha}s_F(x)$. Hence, (ii), together with the first part of the present remark, becomes

$$1 - F(x) = (d + \varepsilon(x))x^{-\alpha}s_F(x) \qquad F(-x) = (1 - d - \varepsilon(x))x^{-\alpha}s_F(x)$$

where $\varepsilon(x) \to 0$ as $x \to +\infty$. Since $G \in D(G)$, these same properties also hold for $G(x)$, which shows both that $D(G)$ is restrictive and that, to some extent, all members of $D(G)$ "try to be like G."

Proof. With our standard notation for this chapter and the additional notation $X_{jn} = X_j/b_n$ and $F_{jn}(x) = F(xb_n)$ for all $j \ge 1$, Theorem 5 (iii) and Theorem 5.8 (also consult the two displayed formulas preceding (5.26) as well as (6.13) and (6.14), where $\alpha = \alpha'$) state that $F \in D(G)$ if and only if, with some constants $c_1 \geq 0$ and $c_2 \geq 0$ such that $c_1 + c_2 > 0$

$$nF(-xb_n) \to c_2 x^{-\alpha} \qquad n[1 - F(xb_n)] \to c_1 x^{-\alpha} \qquad x > 0 \qquad (6.27)$$

as $n \to +\infty$ and

$$\lim_{\varepsilon=0} \lim_{n=+\infty} \frac{n}{b_n^2} \left\{ \int_{-b_n\varepsilon}^{b_n\varepsilon} y^2 \, dF(y) - \left[\int_{-b_n\varepsilon}^{b_n\varepsilon} y \, dF(y) \right]^2 \right\} = 0 \qquad (6.28)$$

In applying Theorem 5.8 we have, in fact, presupposed that X_j/b_n satisfy the UAN condition, which reduces here to $b_n \to +\infty$ with n. But this is quite obvious if we once more transform the weak convergence of $S_n/b_n - a_n$ to characteristic functions (we did this, as may be recalled, in the previous proof). Now, if we add up the two terms in (6.27), (i) follows from Theorem 8. To establish (ii), we note that, since $b_n \to +\infty$, there is an integer n such that $b_n \leq y \leq b_{n+1}$ for every $y > 0$. Then

$$F(-b_{n+1}) \leq F(-y) \leq F(-b_n) \qquad 1 - F(b_{n+1}) \leq 1 - F(y) \leq 1 - F(b_n)$$

and, thus, if $c_2 > 0$,

$$\frac{1 - F(b_{n+1})}{F(-b_n)} \leq \frac{1 - F(y)}{F(-y)} \leq \frac{1 - F(b_n)}{F(-b_{n+1})}$$

i.e., by (6.27), $[1 - F(y)]/F(-y) \to c_1/c_2$. This entails (ii) with $d = c_1/(c_1 + c_2)$. Should only $c_1 > 0$, then we argue with $F(-y)/[1 - F(y)]$ and (ii) obtains with $d = 1$.

We turn to the converse, assuming that (i) and (ii) hold. We now have to establish the limits (6.27) and (6.28). First, we show that the first integral in (6.28) dominates the second. Let us set

$$R(x) = \int_{-x}^{x} y^2 \, dF(y)$$

(We intentionally changed the notation to $R(x)$ from the $s(x)$ used in the previous proof, in order to avoid the suggestion that it might be slowly varying.) Since $F \notin D(N)$, $R(x) \to +\infty$ with x. Now, setting

$$E(x) = \int_{-x}^{x} y \, dF(y) \qquad x > 0$$

the Cauchy–Schwarz inequality yields

$$[E(x) - E(y)]^2 \leq R(x)[1 - F(y) + F(-y)]$$

for $0 < y < x$. Thus, by choosing y large but fixed and then letting $x \to +\infty$, we get $E^2(x)/R(x) \to 0$ as stated. Next we refer to Exercises 2 and 3, where the reader is asked to show that the regular variation of $T(x)$ with exponent $-\alpha$ implies the regular variation of $R(x)$ with exponent $2 - \alpha$. Thus, if we choose b_n by the rule $(n/b_n^2)R(b_n) \to 1$, then

$$(n/b_n^2)R(\varepsilon b_n) = [R(\varepsilon b_n)/R(b_n)]R(b_n)(n/b_n^2) \to \varepsilon^{2-\alpha}$$

as $n \to +\infty$, which, because $0 < \alpha < 2$, tends to zero as $\varepsilon \to 0$, i.e., (6.28) holds. From Exercises 2 and 3 we also have that, with the same b_n as above, $nT(xb_n) \to cx^{-\alpha}$ with $c > 0$; thus, (ii) (see also the Remark) ensures the validity of (6.27). This completes the proof.   ■

Exercises

1. In Theorem 11 we have shown that the slow variation of $s(x)$ implies (6.25). Prove the converse: if (6.25) holds then $s(x)$ is slowly varying.
2. Let $T(x)$ and $R(x)$ be as in Theorem 12 including its proof. Prove that if $T(x)$ is regularly varying with exponent $-\alpha$, then $R(x)$ is regularly varying with exponent $2 - \alpha$, $0 < \alpha < 2$.
3. With $T(x)$ and $R(x)$ as in the preceding exercise, show that $R(x)$ is regularly varying with exponent $2 - \alpha$, $0 < \alpha < 2$, if $x^2 T(x)/R(x) \to (2 - \alpha)/\alpha$.
4. Show that if $F \in D(G)$, where the characteristic exponent of G is α, then the rth moment associated with $F(x)$ is finite for $0 \le r < \alpha$.
5. Let the common density function of the i.i.d. random variables X_j be

$$f(x) = \begin{cases} c|x|^{-2.5} \log |x| & \text{for } |x| \ge 1 \\ 0 & \text{otherwise} \end{cases}$$

Find normalizing constants a_n and $b_n > 0$ that guarantee that $S_n/b_n - a_n$ has a nondegenerate asymptotic distribution $G(x)$. Give the characteristic function of $G(x)$.

6.5   THE ASYMPTOTIC THEORY OF THE EXTREMES

As in the previous sections of this chapter, $X_1, X_2, \ldots$ are i.i.d. with a common distribution function $F(x)$. We set

$$\alpha(F) = \inf \{x: F(x) > 0\} \quad \text{and} \quad \omega(F) = \sup\{x: F(x) < 1\}$$

Let

$$W_n = \min\{X_j: 1 \le j \le n\} \qquad Z_n = \max\{X_j: 1 \le j \le n\}$$

and

$$L_n(x) = P(W_n \le x) \qquad H_n(x) = P(Z_n \le x)$$

Clearly,

$$L_n(x) = 1 - [1 - F(x)]^n \qquad H_n(x) = F^n(x)$$

Since $Z_n \leq Z_{n+1}$ and $P(Z_n > x) > 0$ for every $x < \omega(F)$, Example 6 of Section 2.6 yields that $Z_n \to \omega(F)$ almost surely (a.s.). We similarly get that $W_n \to \alpha(F)$ a.s. Therefore, in order to keep the extremes W_n and Z_n from escaping to the endpoints of F, we seek constants a_n, $b_n > 0$, c_n, and $d_n > 0$ such that

$$L_n(c_n + d_n x) \to L(x) \qquad H_n(a_n + b_n x) \to H(x)$$

in the sense of complete convergence, where $L(x)$ and $H(x)$ are some non-degenerate distribution functions.

We already have some results of this nature. If $F \in D(G)$, where G is a stable distribution with exponent $0 < \alpha < 2$ and, thus, there exist constants a_n and $b_n > 0$ such that the asymptotic distribution of $S_n/b_n - a_n$ is $G(x)$, then the combination of Theorem 5 and Theorem 5.8 yields that

$$H_n(b_n x) \to \begin{cases} \exp[M(x)] & \text{if } x > 0 \\ 0 & \text{if } x < 0 \end{cases}$$

as $n \to +\infty$, where $M(x) = -c_1 x^{-\alpha}$, $c_1 \geq 0$, and

$$L_n(b_n x) \to \begin{cases} 1 & \text{if } x > 0 \\ 1 - \exp[-m(x)] & \text{if } x < 0 \end{cases}$$

where $m(x) = c_2|x|^{-\alpha}$, $c_2 \geq 0$, and $c_1 + c_2 > 0$. In particular, if $F(x) < 1$ for all $x > 0$, then $c_1 > 0$ and we get the limiting distribution of $H_n(b_n x)$

$$H(x) = \exp(-c_1 x^{-\alpha}) \qquad c_1 > 0, \quad x > 0, \quad 0 < \alpha < 2$$

Because the normalizing factor b_n is the same for both S_n and Z_n, we have that a single term (Z_n) takes a positive percentage of the aggregate (S_n), which is the basis of the assumption in the insurance industry that large claims requiring reinsurance have distribution functions $F(x)$ belonging to $D(G)$. These are characterized in Theorem 12. Hence, for reinsurance, the underlying distribution for large claims is of the form

$$1 - F(x) = x^{-\alpha} s(x) \qquad x > 0 \quad 0 < \alpha < 2$$

where $s(x)$ is some slowly varying function.

Other applications such as the strength of materials when the assumption of the weakest link principle is justified (the material breaks at its weakest point and, thus, its strength is W_n, the smallest of the strengths of its parts), failures of systems with a large number of components and others, however, justify the development of the theory of the extremes independently of that of sums, and we do so now. This remarkably well-developed theory owes much to its founders: Fisher, Tippett, Fréchet, and Gnedenko. The reader can find a large variety of applications and a good account of the history of the field in another book of the author's, Galambos (1987).

Note that if we transform each X_j into $(-X_j)$, Z_n of the X_j will become W_n of the $(-X_j)$ and, thus, the mathematical theory of W_n is readily obtained from that of Z_n. Therefore, we concentrate on Z_n only.

We shall establish the following results.

Theorem 13 Let $\omega(F) = +\infty$. Assume that $1 - F(x)$ is regularly varying with exponent $-\gamma$, $\gamma > 0$. Then there is a sequence $b_n > 0$ such that

$$\lim H_n(b_n x) = H_{1,\gamma}(x)$$

as $n \to +\infty$, where

$$H_{1,\gamma}(x) = \begin{cases} \exp(-x^{-\gamma}) & \text{if } x > 0 \\ 0 & \text{otherwise} \end{cases} \tag{6.29}$$

The normalizing constant b_n can be chosen as

$$b_n = \inf\{x: 1 - F(x) \le 1/n\}$$

Theorem 14 Let $\omega(F)$ be finite. Define $F^*(x) = F(\omega(F) - 1/x)$, $x > 0$. If $1 - F^*(x)$ is regularly varying with exponent $-\gamma$, then there are sequences a_n and $b_n > 0$ such that

$$\lim H_n(a_n + b_n x) = H_{2,\gamma}(x)$$

as $n \to +\infty$, where

$$H_{2,\gamma}(x) = \begin{cases} 1 & \text{if } x > 0 \\ \exp(-(-x)^\gamma) & \text{if } x < 0 \end{cases} \tag{6.30}$$

The normalizing constants a_n and b_n can be chosen as $a_n = \omega(F)$ and

$$b_n = \omega(F) - \inf\{x: 1 - F(x) \le 1/n\}$$

Theorem 15 Assume that, for every finite $t < \omega(F)$,

$$\int_t^{\omega(F)} [1 - F(y)]\, dy < +\infty$$

For $\alpha(F) < t < \omega(F)$, define

$$M(t) = [-F(t)]^{-1} \int_t^{\omega(F)} [1 - F(y)]\, dy$$

Assume that, for all real x as $t \to \omega(F)$ with $t < \omega(F)$,

$$\lim \frac{1 - F(t + xM(t))}{1 - F(t)} = e^{-x}$$

Then there are sequences a_n and $b_n > 0$ such that

$$\lim H_n(a_n + b_n x) = H_{3,0}(x)$$

as $n \to +\infty$, where

$$H_{3,0}(x) = \exp(-e^{-x}) \qquad \text{all real } x \tag{6.31}$$

The normalizing constants a_n and $b_n > 0$ can be chosen as

$$a_n = \inf\{x: 1 - F(x) \le 1/n\} \qquad b_n = r(a_n)$$

Theorem 16 There are only three types of nondegenerate distribution functions that can appear as the (complete) limit of $H_n(a_n + b_n x) = F^n(a_n + b_n x)$, where a_n and $b_n > 0$ are suitable constants. These types are $H_{1,\gamma}(x)$, $H_{2,\gamma}(x)$, and $H_{3,0}(x)$.

Although we shall not prove it in the present book, each of Theorems 13–15 provides both necessary and sufficient conditions for the validity of its conclusion. These theorems are due to Gnedenko (1943), except that the specific form of $M(t)$ was first recommended by de Haan (1970). Note that we obtained $H_{1,\gamma}(x)$ of (6.29) as the limiting distribution of Z_n/b_n in connection with stable distributions, where the value of γ was limited to $0 < \gamma < 2$.

In the proofs, we shall repeatedly apply the elementary limit relation

$$\lim_{n = +\infty} \left[1 + \frac{u(n)}{n} \right]^n = e^u \qquad \text{if } u(n) \to u \text{ (finite)} \tag{6.32}$$

Proof of Theorem 13. First, note that if b_n is chosen by the rule suggested in the theorem, then $b_n \to +\infty$ with n and

$$1 - F(b_n) \le 1/n \le 1 - F(b_n - 0) \le 1 - F(b_n \varepsilon)$$

where $0 < \varepsilon < 1$ is arbitrary. Thus,

$$\frac{1 - F(b_n)}{1 - F(b_n \varepsilon)} \le n[1 - F(b_n)] \le 1$$

from which the regular variation of $1 - F(x)$ ensures that

$$\lim n[1 - F(b_n)] = 1 \tag{6.33}$$

as $n \to +\infty$. Now, writing

$$H_n(b_n x) = F^n(b_n x) = \left\{1 - \frac{1 - F(b_n x)}{1 - F(b_n)} n[1 - F(b_n)] \frac{1}{n}\right\}^n$$

we again appeal to the regular variation of $1 - F(x)$. By (6.33), we get that

$$u(n) = \frac{1 - F(b_n x)}{1 - F(b_n)} n[1 - F(b_n)] \to x^{-\gamma}$$

as $n \to +\infty$ for $x > 0$, which enables us to apply (6.32), from which (6.29) obtains for $x > 0$. Since it is trivial that $H_n(b_n x) \to 0$ for $x < 0$ ($b_n \to +\infty$), the proof is concluded. ∎

Proof of Theorem 14. With

$$b_n^* = \inf\{x: 1 - F^*(x) \le 1/n\} = \inf\{x: 1 - F[\omega(F) - 1/x] \le 1/n\}$$
$$= [\omega(F) - \inf\{x: 1 - F(x) \le 1/n\}]^{-1} = 1/b_n$$

Theorem 13 entails that

$$F^{*n}(b_n^* x) = F^n(\omega(F) - 1/b_n^* x) = F^n(\omega(F) - b_n/x) \to H_{1,\gamma}(x)$$

as $n \to +\infty$ for $x > 0$. In other words, with $a_n = \omega(F)$

$$F^n(a_n + b_n x) \to H_{1,\gamma}(-1/x) = H_{2,\gamma}(x)$$

as $n \to +\infty$ for $x < 0$. On the other hand, since $a_n = \omega(F)$ and $b_n > 0$, $F(a_n + b_n x) \equiv 1$ for $x > 0$. The proof is completed. ∎

Proof of Theorem 15. Let a_n and $b_n > 0$ be the numbers recommended in the theorem. Then $a_n \to \omega(F)$ as $n \to +\infty$ and, thus

$$\frac{1 - F(a_n + b_n x)}{1 - F(a_n)} \to e^{-x} \qquad \text{for all } x \tag{6.34}$$

as $n \to +\infty$. Furthermore,

$$1 - F(a_n) \le 1/n \le 1 - F(a_n - 0) \le 1 - F(a_n - \varepsilon b_n)$$

where $\varepsilon > 0$ is arbitrary. That is,

$$\frac{1 - F(a_n)}{1 - F(a_n - \varepsilon b_n)} \le n[1 - F(a_n)] \le 1$$

from which, via (6.34) and by ultimately letting $\varepsilon \to 0$, we get,

$$\lim n[1 - F(a_n)] = 1 \tag{6.35}$$

as $n \to +\infty$. From this point, the completion of the proof is similar to that of Theorem 13. By writing

$$H_n(a_n + b_n x) = F^n(a_n + b_n x) = \left[1 - \frac{u(n)}{n} \right]^n$$

where

$$u(n) = \frac{1 - F(a_n + b_n x)}{1 - F(a_n)} \, n[1 - F(a_n)]$$

we have from (6.34) and (6.35) that $u(n) \to e^{-x}$ as $n \to +\infty$; thus, (6.32) implies (6.31). The theorem is established. ■

The fact that the conditions of Theorem 13 for $H_n(b_n x)$ converging to $H_{1,\gamma}(x)$ are also necessary is not difficult to prove in view of the results developed for regularly varying functions (see Exercise 1). However, the converse of Theorem 15 is very difficult.

As a preparation for establishing Theorem 16, we prove the following lemmas.

Lemma 1 A nondegenerate limiting distribution function $H(x)$ of $F^n(a_n + b_n x)$ satisfies the functional equation

$$H^m(A_m + B_m x) = H(x) \qquad \text{all real } x \quad (m \ge 2) \tag{6.36}$$

with some constants A_m and $B_m > 0$. Furthermore, the normalizing constants a_n and $b_n > 0$ are such that, for fixed integers $m > 1$

$$\frac{a_{nm} - a_n}{b_n} \to A_m \qquad \frac{b_{nm}}{b_n} \to B_m$$

as $n \to +\infty$.

Proof. Let $m > 1$ be fixed. Then

$$F^{nm}(a_{nm} + b_{nm} x) \to H(x)$$

and

$$F^{nm}(a_n + b_n x) \to H^m(x)$$

as $n \to +\infty$. The lemma is therefore just a restatement of Lemma 2 of Section 4.6 in light of the present situation. ■

For easier reference, we introduce the terminology applied earlier in connection with sums. We say that a nondegenerate distribution function $H(x)$ is *max-stable* if (6.36) holds. From Lemma 1 it is clear that $H(x)$ is max-stable if and only if it is the limiting distribution of the (linearly) normalized maximum of some i.i.d. sequence. In addition, if $F^n(a_n + b_n x) \to H(x)$, which is nondegenerate, then we say that $F(x)$ is in the domain of attraction of $H(x)$; in our notation, $F \in D(H)$. In order to avoid confusion, we shall always add (outside this section) the stipulation that H must be max-stable to the notation $D(H)$. Just as in the case of stable distributions for sums, $D(H_1)$ and $D(H_2)$ are either identical (and then H_1 and H_2 are of the same type) or disjoint. Furthermore, distribution functions of the same type always belong to the same domain of attraction.

Lemma 2   Let $H(x)$ be a nondegenerate distribution function satisfying (6.36). Then there exist functions $A(s)$ and $B(s) > 0$, defined for all $s > 0$, such that

$$H^s(A(s) + B(s)x) = H(x) \qquad s > 0, \quad \alpha(H) < x < \omega(H)$$

Proof.   Let $s > 0$ be fixed. For an integer m, let $n(s) = [ms]$ denote the integer part of ms. Now, if $m > 1/s$, then (6.36) yields

$$H^{n(s)}(A_{n(s)} + B_{n(s)}x) = H(x) \qquad \text{for all } x$$

Let $\alpha(H) < x < \omega(H)$ and $m \to +\infty$. Then, of course, $n(s) \to +\infty$ and $A_{n(s)} + B_{n(s)}x \to \omega(H)$ for all $x < \omega(H)$. Furthermore, $H(A_{n(s)} + B_{n(s)}x) \to 1$ and, since

$$H^{ms}(\cdot) \le H^{n(s)}(\cdot) \le H^{ms-1}(\cdot) \qquad 0 < H(\cdot) < 1$$

we have that

$$H^m(A_{n(s)} + B_{n(s)}x) \to H^{1/s}(x)$$

as $m \to +\infty$ for $\alpha(H) < x < \omega(H)$. This, together with (6.36), once again brings us back to Lemma 2 of Section 4.6, which completes the proof. ∎

We now turn to the proof of Theorem 16, in which we follow the basic ideas of de Haan (1976).

Proof of Theorem 16.   In view of Lemma 1, we have to solve the functional equation (6.36), in which, by Lemma 2, m can be replaced by a continuous variable $s > 0$.

Let $\alpha(H) < x < \omega(H)$ and define $g(x) = -\log\log(1/H(x))$. Then, $g(x) \to -\infty$ as $x \to \alpha(H)$ and $g(x) \to +\infty$ as $x \to \omega(H)$, since, by (6.36), the left-hand limit of $H(x)$ at $\omega(H)$ must be one and the right-hand limit of $H(x)$ at $\alpha(H)$ must be zero. Hence, we can define, for all real y,

$$g^{-1}(y) = \inf\{x: g(x) \ge y\}$$

Clearly, both $g(x)$ and $g^{-1}(y)$ are monotonic on their domains, and, in view of Lemma 2, with some functions $A(s)$ and $B(s) > 0$, $s > 0$,

$$g(A(s) + B(s)x) - \log s = g(x)$$

or, equivalently,

$$\frac{g^{-1}(y + \log s) - A(s)}{B(s)} = g^{-1}(y)$$

We can eliminate $A(s)$ by writing this same equation for $y = 0$ and subtracting the two equations. That is,

$$\frac{g^{-1}(y + \log s) - g^{-1}(\log s)}{B(s)} = g^{-1}(y) - g^{-1}(0)$$

for all $s > 0$ and all real y. This, with the new variable $t = \log s$ and the transformed functions $B^*(t) = B(e^t)$ and $g_0^{-1}(y) = g^{-1}(y) - g^{-1}(0)$, becomes

$$g_0^{-1}(y + t) - g_0^{-1}(t) = B^*(t)g_0^{-1}(y) \tag{6.37}$$

for all real y and t. Now, if $B^*(t) = 1$ for all t, then (6.37) reduces to the Cauchy equation

$$g_0^{-1}(y + t) = g_0^{-1}(y) + g_0^{-1}(t) \qquad \text{all } y, t$$

with the additional limitation that $g_0^{-1}(y)$ be monotonic. Hence, Theorem 2 and (6.5) entail that $g_0^{-1}(y) = cy$ with some constant c, i.e., $g^{-1}(y) = cy + g^{-1}(0)$. We thus obtain from the definition of $g^{-1}(y)$ that $g(x) = a + bx$, $b > 0$, for all $\alpha(H) < x < \omega(H)$, which yields $H(x) = \exp[-\exp(-a - bx)]$ for all x, which is of the same type as $H_{3,0}(x)$ (recall that no jumps are possible at $\alpha(H)$ and $\omega(H)$, so x cannot be limited).

Next, we consider the case when $B^*(t) \neq 1$ for some t. Writing (6.37) in the form

$$g_0^{-1}(t + y) - g_0^{-1}(y) = B^*(y)g_0^{-1}(t)$$

and subtracting it from (6.37), we get

$$g_0^{-1}(y) - g_0^{-1}(t) = B^*(t)g_0^{-1}(y) - B^*(y)g_0^{-1}(t)$$

which can be rearranged to

$$g_0^{-1}(y) = \frac{g_0^{-1}(t)}{1 - B^*(t)}[1 - B^*(y)] \tag{6.38}$$

We set $v(t) = g_0^{-1}(t)/(1 - B^*(t))$ and combine (6.37) and (6.38). We get

$$v(t)[B^*(t)B^*(y) - B^*(y + t)] = 0 \tag{6.39}$$

which is valid for all y and those t for which $B^*(t) \neq 1$. Note that $v(t) \neq 0$ for any permissible t, because (6.38) would yield $g_0^{-1}(y) = 0$ for all y, which is impossible for our $g(x)$. Furthermore, since $g_0^{-1}(y)$ is monotonic, (6.38) ensures that $B^*(y)$ is also. Therefore, the set of permissible t is at least a semiline. But since we are free to change the values of y in (6.39), we have that the only possibility for $B^*(y) = 1$ is $y = 0$. Thus, by simplifying by $v(t)$ in (6.39), we have established that the Cauchy equation

$$B^*(t)B^*(y) = B^*(y + t)$$

holds for all t and y and that $B^*(y) > 0$ with $B^*(y) \neq 1$ with the exception of $y = 0$. Thus, by one more appeal to Theorem 2, we get $B^*(t) = e^{at}$ with some constant $a \neq 0$. By noting that, by (6.38), $g_0^{-1}(y)/(1 - B^*(y))$ is constant, we have, in fact, determined that $g^{-1}(y) = g^{-1}(0) + b(1 - e^{at})$ with some constants a and b. Now, depending on whether $a > 0$ or $a < 0$, we get two solutions for $g(x)$ or $H(x)$, leading to the types $H_{1,\gamma}(x)$ and $H_{2,\gamma}(x)$. This completes the proof. ∎

The following examples are typical for Theorems 13–15.

Example 1 Let $F(x) = 1 - 1/x$, $x \geq 1$. Then $\omega(F) = +\infty$ and $1 - F(x)$ is regularly varying with exponent -1. Hence, Theorem 13 applies and, with some $b_n > 0$, the asymptotic distribution of Z_n/b_n is $H_{1,1}(x) = \exp(-1/x)$, $x > 0$. The suggested value of $b_n = \inf\{x: 1/x \leq 1/n\}$, i.e., $b_n = n$. ∎

Example 2 Let $F(x) = x$, $0 \leq x \leq 1$. Then $\omega(F) = 1$ and $F^*(x) = F(1 - 1/x)$, $x \geq 1$, is regularly varying. In fact, given the result from the preceding example and the suggested normalizing constants, $n(Z_n - 1)$ has the asymptotic distribution $H_{2,1}(x) = e^x$, $x < 0$. ∎

Example 3 Let $F(x) = 1 - e^{-x}$, $x \geq 0$. Then $\omega(F) = +\infty$ and $1 - F(x)$ is integrable on $(0, +\infty)$. For $t > 0$, $M(t) = 1$ and, thus, $[1 - F(t + xM(t))]/[1 - F(t)] = e^{-x}$ for all $t \geq 0$. Hence, Theorem 15 applies, $a_n = \log n$, and $b_n = 1$, which yield that the asymptotic distribution of $Z_n - \log n$ is $H_{3,0}(x)$. ∎

In Examples 1–3 the recommended normalizing constants are easy to calculate, which is not always the case. For example, for $F(x) = N(x)$, the standard normal distribution function, the exact solution of $1/n = 1 - N(x)$ is impossible in closed form. However, for applying Theorems 13–15 one is free to deviate from the recommended values of a_n and b_n, and, in particular, one can replace recommended exact solutions by approximate ones. Upon applying Lemma 2 of Section 4.6 we have that if we replace a_n and $b_n > 0$

by a_n^* and $b_n^* > 0$ then b_n^*/b_n and $(a_n^* - a_n)/b_n$ must have finite limits and the first limit is required to be positive. This is the guide on the extent to which the solution of $1/n = 1 - N(x)$ can be approximate only (see Exercise 3).

Another way of achieving simplifications in the computation of normalizing constants is as follows. Assume that $F(x)$ and $G(x)$ satisfy $\omega(F) = \omega(G) = \omega$ and, as $x \to \omega$ ($x < \omega$),

(i) $\quad \lim \dfrac{1 - F(x)}{1 - G(x)} = D \qquad 0 < D < +\infty$

Then F and G belong to the same domain of attraction, and the normalizing constants for F can be computed from those for G. To see this, first note that, for $x > \alpha(H)$, as $n \to +\infty$,

(ii) $\quad G^n(a_n + b_n x) \to H(x) \Leftrightarrow n[1 - G(a_n + b_n x)] \to -\log H(x)$

One way in this equivalence comes from (6.32) by writing

$$G^n(a_n + b_n x) = \left[1 - \frac{n(1 - G(a_n + b_n x))}{n} \right]^n$$

For the converse we use Taylor's expansion $\log(1 - z) = -z + O(z^2)$, as $z \to 0$, to yield

$$
\begin{aligned}
G^n(a_n + b_n x) &= \exp[n \log G(a_n + b_n x)] \\
&= \exp(n \log\{1 - [1 - G(a_n + b_n x)]\}) \\
&= \exp\{-n[1 - G(a_n + b_n x)](1 + o(1))\}
\end{aligned}
$$

since, whenever the extreme left-hand side converges to a finite value, $G(a_n + b_n x) \to 1$. Therefore, by one more appeal to (6.32) with

$$u(n) = \frac{1 - F(a_n + b_n x)}{1 - G(a_n + b_n x)} n[1 - G(a_n + b_n x)]$$

we get that whenever either limit of (ii) holds

$$F^n(a_n + b_n x) \to \exp(-D \log H(x)) = H^D(x) = H(A + Bx)$$

where the last equation is by (6.36) (and so $A, B > 0$ depend on $1/D$).

Example 4 Let $F(x) = 1/2 + (1/\pi) \arctan x$ and $G(x) = 1 - 1/x$, $x \geq 1$. Then $\omega(F) = \omega(G) = +\infty$, and by L'Hospital's rule, as $x \to +\infty$,

$$\lim \frac{1 - F(x)}{1 - G(x)} = \lim \frac{1}{\pi} \frac{x^2}{1 + x^2} = \frac{1}{\pi} = D$$

By Example 1, $G \in D(H_{1,1})$ and $b_n = n$. Hence, $F^n(nx) \to H_{1,1}^D(x) = \exp(-1/\pi x)$, that is, $F^n(nx/\pi) \to \exp(-1/x)$.

Example 5   Let $F(x) = (1 + e^{-x})^{-1}$, the logistic distribution, and $G(x) = 1 - e^{-x}$, $x > 0$. By Example 3, $G \in D(H_{3,0})$ and $a_n = \log n$ and $b_n = 1$ for G. Furthermore, $\omega(F) = \omega(G) = +\infty$ and, as $x \to +\infty$,

$$\lim \frac{1 - F(x)}{1 - G(x)} = \lim e^x [1 - (1 + e^{-x})^{-1}] = \lim \frac{1}{1 + e^{-x}} = 1$$

Therefore, $F^n(\log n + x) \to \exp(-e^{-x})$.

In order to see that most of the well-known discrete distributions do not belong to any one of the three domains of attraction for the maximum, we establish the following result.

Theorem 17   Assume that, for the distribution function $F(x)$, there are sequences a_n and b_n such that

$$F^n(a_n + b_n x) \to H(x) \qquad n \to +\infty$$

where $H(x)$ is one of the three possible nondegenerate distribution functions of Theorem 16. Then, as $x \to \omega(F)$ with $x < \omega(F)$,

$$\lim \frac{F(x) - F(x - 0)}{1 - F(x)} = 0$$

Proof.   Since $H_{2,\gamma}(x)$ can be handled via the case of $H_{1,\gamma}(x)$, it suffices to give details for $F \in D(H_{1,\gamma})$ and $F \in D(H_{3,0})$. In either case, we write

$$F(x) - F(x - 0) = 1 - F(x - 0) - [1 - F(x)],$$

and thus we prove that $[1 - F(x - 0)]/[1 - F(x)]$ converges to one under the conditions of the theorem. First, let $F \in D(H_{1,\gamma})$. Then $\omega(F) = +\infty$ and $1 - F(x)$ is regularly varying at infinity as it follows from the remark made after the statement of Theorem 16 and from Theorem 13. Hence, for every $0 < \varepsilon < 1$, as $t \to +\infty$,

$$(1 - \varepsilon)^{-\gamma} = \lim \frac{1 - F((1 - \varepsilon)t)}{1 - F(t)} \geq \limsup \frac{1 - F(t - \varepsilon)}{1 - F(t)} \geq 1$$

The claimed limit follows upon letting $\varepsilon \to 0$. The case $F \in D(H_{3,0})$ is proved similarly by arguing with the relation

$$e^\varepsilon = \lim \frac{1 - F(t - \varepsilon M(t))}{1 - F(t)} \geq 1 \qquad (t \to \omega(F))$$

The theorem is established.   ∎

Note that for a discrete random variable X taking the nonnegative integers with distribution $P(X = k) = p_k$,

$$F(k) - F(k - 0) = p_k \quad \text{and} \quad 1 - F(k) = \sum_{t > k} p_t$$

so their ratio should converge to zero in Theorem 17.

Example 6 If X is a geometric variable, $p_k = p(1 - p)^{k-1}$, $k \geq 1$. Hence,

$$\sum_{t > k} p(1 - p)^{t-1} = p(1 - p)^k \frac{1}{1 - (1 - p)} = (1 - p)^k$$

and thus Theorem 17 fails, since the required ratio equals $p/(1 - p)$ rather than converging to zero. Consequently, the distribution function of X is not in any of the domains of attraction of the three limiting types for the maximum.

We conclude this section with a discussion of Theorem 15. Let X be a random variable with distribution function $F(x)$ and assume that $M(t)$ defined in Theorem 15 is meaningful. First, note that, for $\alpha(F) < t < \omega(F)$,

$$P(X \leq y | X > t) = \frac{P(t < X \leq y)}{1 - F(t)} = \frac{F(y) - F(t)}{1 - F(t)} = 1 - \frac{1 - F(y)}{1 - F(t)}$$

and thus, the major condition of Theorem 15 is that, for $x > 0$, $(X - t)/M(t)$ should be conditionally asymptotically unit exponential, given $X > t$. This is a limit theorem in which t and $M(t)$ are normalizing constants which can be modified without violating the asymptotic exponentiality requirement. In particular, $M(t)$ can be replaced by another positive function $a(t)$ such that $M(t)/a(t) \to 1$ as $t \to \omega(F)$. It turns out that, for several differentiable $F(x)$, the reciprocal of the failure rate function $r(t)$ (recall Section 2.9) can serve as $a(t)$. This is the case whenever $M'(t) \to 0$ as $t \to \omega(F)$, since, by simple differentiation, we get $M'(t) = M(t)r(t) - 1$.

Another close relation between $r(t)$ and $M(t)$ can be established in terms of the conditions of Theorem 15. Since from the mathematical form of $r(t)$ it immediately follows that

$$\exp\left\{ -\int_A^B r(t)\, dt \right\} = \frac{1 - F(B)}{1 - F(A)}$$

for all $A < B < \omega(F)$, assuming that $F'(x)$ exists for all $A \leq x$, the previously discussed conditional distribution can also be written as

$$P(X > t + xM(t) | X > t) = \exp\left\{ -\int_t^{t + xM(t)} r(u)\, du \right\}$$

The reader is advised to restate the condition of Theorem 15 in this new form, and deduce from it, by substituting $x = 1$ and then $x = -1$ into this formula, that if $(X - t)/M(t)$ is conditionally unit exponential, given $X > t$, then $M(t)r(t) \to 1$ as $t \to \omega(F)$, whenever $r(t)$ is defined and monotonic (Galambos and Obretenov (1987)).

The conditional distribution $F_t(y) = P(X \le y | X > t)$ is closely related to $M(t)$. Indeed, by calculating integrals with respect to $F_t(y)$, we have $M(t) = E(X - t | X > t)$. Due to this formula, $M(t)$ is called the expected residual life at age t (even when X does not represent age).

For an important subset of population distributions $F(x)$ Theorem 15 can be simplified considerably. Let $F(x)$ be such that $\omega(F) = +\infty$ and $M(t)$ is finite. Then the asymptotic exponentiality

(a) $\quad \lim \dfrac{1 - F(t + xM(t))}{1 - F(t)} = e^{-x} \qquad (t \to +\infty, x \text{ arbitrary})$

assumed to hold in Theorem 15, implies that

(b) $\quad \lim M(t)/t = 0 \qquad (t \to +\infty)$

Indeed, (a) implies that, for every x, $t + xM(t) \to +\infty$ with t, entailing that $t + xM(t) > 0$ for large t. With $x < 0$ this is possible only if (b) holds. Now, Galambos and Xu (1990) observed that if $M(t)$ is regularly varying, which is the case for most of the popular continuous distributions belonging to the domain of attraction of $H_{3,0}(x)$, then (b) implies (a). *Consequently, if F is such that $\omega(F) = +\infty$ and $M(t)$ is regularly varying then* (b) *alone implies that F is in the domain of attraction of $H_{3,0}(x)$.*

In order to prove this last claim, first note that $1/M(t)$ is the derivative of $\log[\int_t^{+\infty} (1 - F(y)) \, dy]$ at every continuity point t of $F(t)$. Hence, using Lebesgue integrals, we have the uniqueness formula

$$1 - F(u) = \dfrac{C}{M(u)} \exp\left[- \int_B^u \dfrac{1}{M(v)} \, dv \right]$$

where $C > 0$ and $B > \alpha(F)$ are suitable constants. Next, we use the assumption of regular variation. By (6.18),

(c) $\quad M(t) = t^\beta s(t)$

where $\beta \le 1$ is a constant (under (b)), and $s(t)$ is a slowly varying function defined by the relation (6.19). We want to prove (a) assuming that (b) and

(c) hold. When we utilize the uniqueness formula in (a), we face the limits, as $t \to +\infty$, of

$$\frac{M(t + xM(t))}{M(t)} = \left[1 + x\,\frac{M(t)}{t}\right]^{\beta} \frac{s(t + xM(t))}{s(t)} \to 1$$

and

$$\int_{t}^{t+xM(t)} \frac{1}{M(v)}\,dv = \frac{xM(t)}{M(w)} = x\left(\frac{t}{w}\right)^{\beta} \frac{s(t)}{s(w)} \to x$$

where in the first case (c) while in the second the mean value theorem for integrals was applied. The mean value theorem, of course, entails that w is some value between t and $t + xM(t)$. In both limits the ratio involving $s(t)$ is of the form $s[\lambda(t)t)]/s(t)$ where $\lambda(t)$ is bounded when (b) applies. Therefore, the uniform convergence version of (6.19) (recall the Corollary following Karamata's representation in Theorem 10) allows us to use (6.19). This completes the proof of our claim that (b) and (c) imply (a).

Note that, in view of Exercise 3 of Section 6.3, (b) is automatically satisfied if (c) holds with $\beta < 1$ (and (b) fails if $\beta > 1$ in (c)). So, the exponential distribution $F(x) = 1 - e^{-x}$, $x \geq 0$, is in $D(H_{3,0})$ by the simple fact that $M(t) = 1 = t^0$. Similarly, the fact alone that, for the standard normal distribution function $F(x)$, $M(t)t \to 1$, as $t \to +\infty$, suffices to conclude that $F \in D(H_{3,0})$, since the mentioned limit implies (c) with $\beta = -1$ (see Exercise 3 below). When $\beta = 1$ in (c) then (b) reduces to $s(t) \to 0$ as $t \to +\infty$. For demonstrating that this is indeed necessary, take $F(x) = 1 - 1/x^2$, $x \geq 1$. We have $M(t) = t$, but $F \in D(H_{1,\gamma})$. On the other hand, for

$$F(x) = 1 - C[(\log x)/x]\exp[-(\log x)^2/2] \qquad x \geq e$$

where $C > 0$ is a constant, $M(t) = t/(\log t)$, that is, both (b) and (c) apply entailing $F \in D(H_{3,0})$. Here, once again, $\beta = 1$.

The combination of Theorem 15 and its simplified version can lead to significant saving in computations. As an example, take $F(x) = 1 - \exp(-\sqrt{x})$, $x > 0$. In order to avoid the difficulty in evaluating $M(t)$, we introduce $F_{1/2}(x) = 1 - (1/2\sqrt{x})\exp(-\sqrt{x})$, $x \geq 1$. Let $M_{1/2}(t)$ be the expected residual life function of $F_{1/2}(x)$. Clearly, $M_{1/2}(t) = t^{1/2}$ yielding the validity of both (b) and (c), so (a) holds as well for $F_{1/2}(x)$. However, by l'Hospital's rule, $M_{1/2}(t)/M(t) \to 1$ as $t \to +\infty$, and thus (a) holds with $F(x)$, too, which, by Theorem 15, suffices for concluding $F \in D(H_{3,0})$.

This example can be generalized as follows. For a continuous distribution function $F(x)$, introduce the family of distributions $F_A(x) = 1 - x^{-A}[1 - F(x)]$, $x \geq 1$, $A \geq 0$. Let $M_A(t)$ be the expected residual life function of $F_A(t)$. Now, if $M_A(t)$ for one $A \geq 0$ satisfies both (b) and (c), then (a) holds with this A, but then (a) holds with all $A \geq 0$ by an appeal to l'Hospital's rule just as above. Consequently, all $F_A(x)$, including $F_0(x) = F(x)$, belong to $D(H_{3,0})$. Note that the technique just described allows us to divide as well as to multiply by a power of x when introducing $F_A(x)$ by simply redefining $F_0(x)$. For example, if $F(x) = 1 - \exp(-x^2)$, $x \geq 0$, then we can call $F(x) = F_1^*(x)$, $x \geq 1$, in the family whose initial distribution is $F^*(x) = 1 - x\exp(-x^2)$, $x \geq 1$. This time, the initial distribution $F^*(x)$ is the "easy one" to handle, but then the technique above shows that all $F_A^*(x)$ are in $D(H_{3,0})$ which conclusion we were aiming at with $A = 1$.

Exercises

1. By applying the equivalence relation (ii) following Example 3 conclude that if $\omega(F) = +\infty$ and $F \in D(H_{1,\gamma})$, then $1 - F(x)$ is regularly varying with exponent $-\gamma$.

2. Show that if the function $g(x)$ is nondecreasing and right continuous and if $g^{-1}(y) = \inf\{x: g(x) \geq y\}$ is continuous, then $g^{-1}(g(x)) = x$.

3. By writing

$$\int_x^{+\infty} e^{-u^2/2} \, du = \int_x^{+\infty} (ue^{-u^2/2})u^{-1} \, du$$

and integrating by parts, show that, for the standard normal distribution function $N(x)$, $M(t) = 1/t + O(t^{-3})$ and

$$[1 - N(t + x/t)]/[1 - N(t)] \to e^{-x} \qquad \text{as } t \to +\infty$$

Hence, Theorem 15 applies. Follow the rules of Theorem 15 to compute the normalizing constants and obtain

$$a_n = (2 \log n)^{1/2} - \frac{(1/2)(\log \log n + \log 4\pi)}{(2 \log n)^{1/2}}$$

and $b_n = (2 \log n)^{-1/2}$.

4. Show that if $F(x)$ is the distribution function of a Poisson variable then $F(x)$ is not in any of the domain of attraction of an asymptotic distribution function for the maximum.

5. Restate Theorems 13–15 for the case of the minimum.

6. Let $F(x)$ be a distribution function with $\omega(F) = +\infty$ whose density $f(x) = F'(x)$ exists for $x_0 < x < \omega(F)$. Show that, if $xf(x)/[1 - F(x)] \to \gamma$, $0 < \gamma < +\infty$ as $x \to +\infty$, then $F \in D(H_{1,\gamma})$.

7. Assume that $F(x)$ is a distribution function with $\omega(F) = +\infty$ and that the conditions of Theorem 15 hold. Show that the normalizing constants a_n and b_n satisfy $b_n/a_n \to 0$ as $n \to +\infty$.

6.6 THE LAW OF THE ITERATED LOGARITHM

For i.i.d. random variables X_j with $E(X_j) = 0$ and finite variance $V > 0$, the classical central limit theorem ensures that the probabilities

$$P\left(\lim \inf \frac{S_n}{n^{1/2}} < -c\right) = P\left(\lim \sup \frac{S_n}{n^{1/2}} > c\right) \geq \lim \sup P\left(\frac{S_n}{n^{1/2}} > c\right)$$

are positive and, thus, by the zero-one law (Section 2.6), both probabilities on the left-hand side equal one. By letting $c \to +\infty$, we obtain that $\lim \inf S_n/n^{1/2} = -\infty$ and $\lim \sup S_n/n^{1/2} = +\infty$ as $n \to +\infty$, both a.s. On the other hand, the strong law of large numbers (Theorem 2.1) yields that $S_n/n \to 0$ a.s. It is, therefore, of interest to find a function $\rho(n)$ of n such that $\lim \sup S_n/\rho(n) = 1$ a.s. as $n \to +\infty$. Clearly, $n^{1/2} < \rho(n) < n$. We shall show that the proper value of $\rho(n)$ is $(2Vn \log \log n)^{1/2}$. In order to avoid lengthy computations, we shall prove the above claim under the additional assumption that the X_j are bounded. In such a form, the theorem goes back to Khintchine; it was later put into its final form for i.i.d. variables by Hartman and Wintner (1941).

Theorem 18   Let X_1, $X_2, \ldots$ be bounded and i.i.d. with $E(X_j) = 0$ and $V = V(X_j) > 0$. Then

$$P\left(\lim_{n = +\infty} \sup \frac{S_n}{(2Vn \log \log n)^{1/2}} = 1\right) = 1$$

Remark   By changing each variable X_j to $(-X_j)$, the theorem also states that the lim inf of the ratio in the theorem is -1 a.s.

For the proof, we need two inequalities, which we shall state as lemmas.

Lemma 1   Let Y_1, $Y_2, \ldots$ be independent random variables with $E(Y_j) = 0$, $|Y_j| \leq K$, and $V(Y_j) = V_j > 0$. Then, for arbitrary $0 < a < 2s_n/K$, where $s_n^2 = V_1 + V_2 + \cdots + V_n$,

$$P(|Y_1 + Y_2 + \cdots + Y_n| \geq as_n) \leq 2 \exp\left[-\frac{a^2}{2(1 + aK/2s_n)^2}\right]$$

Proof.   Let us consider the moment-generating functions

$$g_j(z) = E(\exp(zY_j)) \qquad z > 0 \text{ real}$$

and $G_n(z) = E\{\exp[z(Y_1 + Y_2 + \cdots + Y_n)]\}$. By the assumption of independence,

$$G_n(z) = g_1(z)g_2(z)\cdots g_n(z)$$

and, because the Y_j are bounded, each $g_j(z)$ is finite for all $z > 0$. Furthermore, we can expand $\exp(zY_j)$ into its Taylor series and integrate term by term. That is,

$$g_j(z) = 1 + \frac{z^2 V_j}{2} + \sum_{n=3}^{+\infty} \frac{z^n E(Y_j^n)}{n!}$$

where the linear term is missing because $E(Y_j) = 0$. By the estimates $|E(Y_j^n)| \leq K^{n-2} V_j$, $n \geq 3$ and $1/n! \leq (1/6)(1/(n-3)!)$, $n \geq 3$,

$$\sum_{n=3}^{+\infty} \frac{z^n E(Y_j^n)}{n!} \leq \frac{z^2 V_j}{6} \sum_{n=3}^{+\infty} \frac{(zK)^{n-2}}{(n-3)!} = \frac{z^2 V_j}{6} (zKe^{zK})$$

and, thus,

$$G_n(z) \leq \prod_{j=1}^{n} \left[1 + \frac{z^2 V_j}{2} (1 + \tfrac{1}{3} zKe^{zK}) \right] \leq \exp\left[\frac{z^2 s_n^2}{2} \left(1 + \frac{zK}{3} e^{zK} \right) \right]$$

where the last inequality is obtained from $1 + x < e^x$, $x > 0$. Next, we apply Markov's inequality (Theorem 2.7) to the variable $\exp(zU_n)$, where $U_n = Y_1 + Y_2 + \cdots + Y_n$. We get

$$P\left[U_n \geq \frac{1}{z} \log(e^u G_n(z)) \right] \leq e^{-u} \qquad u > 0$$

and, thus,

$$P\left[U_n \geq \frac{u}{z} + \frac{z s_n^2}{2} \left(1 + \frac{zK}{3} e^{zK} \right) \right] \leq e^{-u} \qquad u > 0$$

Let us first eliminate z by setting $z = (2u)^{1/2}/s_n$ and then change the variable u to $b = (2u)^{1/2}$. The above inequality becomes

$$P\left[U_n \geq bs_n \left(1 + \frac{bK}{6s_n} e^{bK/s_n} \right) \right] \leq e^{-b^2/2}$$

which is also applicable to $(-U_n)$, yielding

$$P\left[|U_n| \geq bs_n \left(1 + \frac{bK}{6s_n} e^{bK/s_n} \right) \right] \leq 2e^{-b^2/2}$$

We change the variables further by setting $a = b(1 + bK/2s_n)$. Then $b \leq a \leq b(1 + aK/2s_n)$ and, thus, for $b \leq s_n/K$, $\exp(bK/s_n) < 3$ and $a \leq b + a/2$, i.e., $a \leq 2s_n/K$, by which the last displayed inequality simplifies to the desired one. The proof is completed. ∎

Remark The inequality of Lemma 1, or its version without the absolute value signs established in the course of the proof, is known as a large deviation inequality. Note that, if $s_n \to +\infty$ with n, by classical versions of the central limit theorem, for fixed a as $n \to +\infty$,

$$\lim P(|Y_1 + Y_2 + \cdots + Y_n| > as_n) = 2(1 - N(a)),$$

where $N(x)$ is the standard normal distribution function. Therefore, for large but still fixed a, the left-hand side of the inequality of Lemma 1 is asymptotic to $(c/a)\exp(-a^2/2)$ where $c = (2/\pi)^{1/2}$ (use partial integration as in Exercise 3 of the preceding section), but for $a = a(n) \to +\infty$ with n we cannot claim this asymptotic relation from the central limit theorem (except that the limit above is zero). The significance of Lemma 1 is that a is allowed to vary with n, and in fact the magnitude of a can be as large as s_n. Although the boundedness of the Y_j for a large deviation inequality like our Lemma 1 is not essential, more than the typical assumptions for the central limit theorem are needed. For example, the finiteness of the moment generating function $G_n(z)$ is unavoidable for the kind of proof we have for Lemma 1.

   Assume that the Y_j are not only independent but identically distributed as well. We drop the boundedness requirement but assume that moment generating functions are finite and we choose units so that $V_j = V = 1$. Then $s_n^2 = n$. Then it is known that the left hand side in Lemma 1 is asymptotic to $2(1 - N(a))$, which in turn is asymptotic to $(c/a)\exp(-a^2/2)$, if $a = a(n) \le n^r$ with $r < 1/2$. Should the previous asymptotic results be valid with $a = zn^{1/2}$, then, upon taking logarithms, we would get

$$\frac{1}{n}\log P(|Y_1 + \cdots + Y_n| > zn) \to -z^2/2$$

For such logarithmic limit results we do not need the preceding asymptotic formula: inequalities like in Lemma 1 coupled with a lower bound whose main term is similar to the one in the upper bound would lead to such limits. This led to investigations of two-sided large deviation inequalities. We do not go into details here; see a book by Varadhan (1984) for a variety of results.

Lemma 2 For i.i.d. variables X_j with $E(X_j) = 0$ and $0 < V = V(X_j) < +\infty$,

$$P\left(\max_{1 \le k \le n} S_k \ge x\right) \le \tfrac{4}{3}P(S_n \ge x - 2Vn^{1/2})$$

Proof. Set $A_1 = \{S_1 \ge x\}$ and, for $k \ge 2$,

$$A_k = \{S_1 < x, S_2 < x, \ldots, S_{k-1} < x, S_k \ge x\}$$

Furthermore, let $B_k = \{S_n - S_k > -2Vn^{1/2}\}$, $k \geq 1$, where B_n is the sure event. Then the A_k, and, thus, $A_k \cap B_k$ as well, are mutually exclusive and the events A_k and B_k, for fixed k, are independent since they depend on mutually exclusive sets of the X_j. Now

$$\bigcup_{k=1}^{n} A_k = \left\{ \max_{1 \leq k \leq n} S_k \geq x \right\} \quad \text{and} \quad \bigcup_{k=1}^{n} (A_k \cap B_k) \subset C_n(x)$$

where $C_n(x) = \{S_n \geq x - 2Vn^{1/2}\}$. Thus, from the observed properties of the A_k and B_k,

$$\sum_{k=1}^{n} P(A_k) = P\left(\max_{1 \leq k \leq n} S_k \geq x \right) \tag{6.40}$$

and

$$\sum_{k=1}^{n} P(A_k)P(B_k) \leq P(S_n \geq x - 2Vn^{1/2}) \tag{6.41}$$

Finally, since, by Chebyshev's inequality,

$$1 - P(B_k) \leq P(|S_n - S_k| \geq 2Vn^{1/2}) \leq \frac{n-k}{4n} < \frac{1}{4}$$

i.e., $P(B_k) > 3/4$, (6.40) and (6.41) imply the stated inequality, which completes the proof. ∎

Proof of Theorem 18.   First, we prove that, with probability one, there exists a finite number n_0 such that, for all $n \geq n_0$,

$$S_n < (1 + \varepsilon)(2Vn \log \log n)^{1/2} \quad \text{a.s.} \tag{6.42}$$

where $\varepsilon > 0$ is arbitrary. To show this, we choose the subsequence N_k of the integers, defined as the smallest integer exceeding $(1 + \varepsilon)^{k/2}$, and show that, with probability one,

$$\max_{N_k \leq n < N_{k+1}} S_n < (1 + \varepsilon)(2VN_k \log \log N_k)^{1/2} \tag{6.43}$$

for $k \geq k_0$. In order to see that (6.43) indeed suffices for (6.42), note that

$$|(2N_k \log \log N_k)^{1/2} - (2n \log \log n)^{1/2}| \leq 2\varepsilon(N_k \log \log N_{k+1})^{1/2}$$
$$\leq 2\varepsilon(2n \log \log n)^{1/2}$$

for $N_k \leq n \leq N_{k+1}$. Let D_k^c be the event that (6.43) holds for the value k. Then

$$D_k \subset \left\{ \max_{1 \leq n \leq N_{k+1}} S_n \geq (1 + \varepsilon)(2N_k \log \log N_k)^{1/2} \right\}$$

and, thus, by Lemma 2,

$$P(D_k) \le 4/3 P[S_{N_{k+1}} \ge (1 + \varepsilon)(2N_k \log \log N_k)^{1/2} - 2VN_{k+1}^{1/2}]$$

Hence, with the choice $a = a_k = (1 + \varepsilon)^\delta \{[(2N_k \log \log N_k)/N_{k+1}]^{1/2}\}$, $\delta > 1$ and an appeal to Lemma 1, we obtain

$$P(D_k) \le \frac{8}{3} \frac{c(\varepsilon)}{k^{1+\varepsilon}} \qquad k \ge k_1 \tag{6.44}$$

where $c(\varepsilon)$ is a constant depending on $\varepsilon > 0$ alone. Here, we used the fact that $N_k/N_{k+1} \to (1 + \varepsilon)^{-1/2}$ and $a_k/N_{k+1}^{1/2} \to 0$ as $k \to +\infty$ and, thus, for large k,

$$\frac{a^2}{2[1 + a_k K/2(VN_{k+1})^{1/2}]^2} \ge (1 + \varepsilon) \log \log N_k \ge (1 + \varepsilon) \log k + c$$

where the constant c depends on ε only. We thus have ((6.44))

$$\sum_{k=1}^{+\infty} P(D_k) < +\infty$$

from which the Borel–Cantelli lemma (Theorem 1.10) ensures that, with perhaps a finite number of exceptions, all D_k^c occur with probability one. That is, (6.42) has been established via (6.43).

It remains to show that, for every $\varepsilon > 0$,

$$P[S_n \ge (1 - \varepsilon)(2nV \log \log n)^{1/2} \text{ i.o.}] = 1 \tag{6.45}$$

Since we want to establish the inequality in (6.45) for infinitely many values of n (a.s.) it suffices if we do so on a subsequence of the subscripts. We now choose $N_k' \sim s^k$ with some $s > 1$. For brevity, we set

$$v(n) = (2nV \log \log n)^{1/2} \qquad b(k) = [2(N_k' - N_{k-1}')V \log \log N_k']^{1/2}$$

Now, upon applying (6.42) to the sequence $(-X_j)$, we know that, almost surely, $(-S_n) \le 2v(n)$ for all but a finite number of n (we choose $\varepsilon = 1$). That is, $S_n \ge -2v(n)$ and, thus,

$$S_{Nk} \ge S_{Nk} - S_{Nk-1} - 2v(N_{k-1}') \quad \text{a.s.}$$

Hence, if we show that

$$P(S_{Nk} - S_{Nk-1} \ge (1 - \varepsilon)v(N_k') + 2v(N_{k-1}') \text{ i.o.}) = 1 \tag{6.46}$$

then (6.45) follows. Note that, there is an $\varepsilon_1 > 0$ that goes to zero with ε for every $\varepsilon > 0$ such that

$$(1 - \varepsilon)v(N_k') + 2v(N_{k-1}') < (1 - \varepsilon_1)b(k)$$

which entails that, for (6.46), it suffices to prove that

$$P(S_{N_k} - S_{N_{k-1}} \geq (1 - \varepsilon_1)b(k) \text{ i.o.}) = 1 \tag{6.47}$$

for every $\varepsilon_1 > 0$. The advantage of (6.47) over (6.45) is that the differences $S_{N_k} - S_{N_{k-1}}$ are independent (they are defined in terms of exclusive subsets of the X_j) and, thus, for (6.47), we can readily apply the converse to the Borel–Cantelli lemma (Theorem 1.11), which states that (6.47) is ensured by

$$\sum_{k=1}^{+\infty} P(S_{N_k} - S_{N_{k-1}} \geq (1 - \varepsilon_1)b(k)) = +\infty \tag{6.48}$$

Therefore, the proof will be completed if we show that, for some $0 < \gamma < 1$,

$$P(S_{N_k} - S_{N_{k-1}} \geq (1 - \varepsilon_1)b(k)) \geq \frac{1}{k^\gamma}$$

Now, note that

$$P(S_{N_k} - S_{N_{k-1}} \geq (1 - \varepsilon_1)b(k))$$
$$\geq P(S_{N_k} \geq (1 - \tfrac{1}{2}\varepsilon_1)b(k) \text{ and } S_{N_{k-1}} < \tfrac{1}{2}\varepsilon_1 b(k))$$
$$\geq P(S_{N_k} \geq (1 - \tfrac{1}{2}\varepsilon_1)b(k)) - P(S_{N_{k-1}} \geq \tfrac{1}{2}\varepsilon_1 b(k))$$

where the last inequality is justified by the elementary inequality

$$P(A \cap B) = P(A) - P(A \cap B^c) \geq P(A) - P(B^c)$$

where A and B are arbitrary events. These inequalities reduce our goal to establishing

$$P(S_{N_k} \geq (1 - \tfrac{1}{2}\varepsilon_1)b(k)) \geq \frac{c}{k^\gamma} \tag{6.49}$$

and

$$P(S_{N_{k-1}} \geq \tfrac{1}{2}\varepsilon_1 b(k)) \leq \tfrac{1}{2} \frac{c}{k^\gamma} \tag{6.50}$$

where $c > 0$ does not depend on k and $0 < \gamma < 1$.
Lemma 1 applies to the left-hand side of (6.50), yielding

$$P(S_{N_{k-1}} \geq \tfrac{1}{2}\varepsilon_1 b(k)) \leq c \exp\left[-\frac{(s-1)\varepsilon_1^2}{4} \log k \right] \tag{6.51}$$

where c does not depend on k. Hence, (6.51) entails (6.50) when s is sufficiently large. To establish (6.49), we go back to the proof of Lemma 1.

Denoting the common bound of the X_j by K and setting $g(z) = E[\exp(zX_j)]$, $z \geq 0$, we get, as in Lemma 1, but now estimating from below,

$$g(z) \geq 1 + \frac{z^2 V}{2}\left(1 - \frac{zK}{3} - \frac{z^2 K^2}{12} - \cdots\right)$$

$$\geq 1 + \frac{z^2 V}{2}\left(1 - \frac{zK}{2}\right) \qquad \text{if } zK < 1$$

from which the elementary inequality $1 + x \geq e^{x(1-x)}$, $x \geq 0$, yields

$$g(z) \geq \exp\left[\frac{z^2 V}{2}\left(1 - \frac{zK}{2}\right)\left(1 - \frac{z^2 V}{2} + \frac{z^3 KV}{4}\right)\right] \qquad zK < 1$$

From now on, we restrict the value of z to $z = z_n \to 0$ as $n \to +\infty$. Thus, for large n, with arbitrary $\varepsilon > 0$,

$$g(z) \geq \exp\left[\frac{z^2 V}{2}(1 - \varepsilon)\right]$$

from which we have

$$E(e^{zS_n}) = g^n(z) \geq \exp[\tfrac{1}{2}nz^2 V(1 - \varepsilon)] \tag{6.52}$$

Now, we have to squeeze $P(S_n > y)$ with $y = y(n)$ suitable for (6.49) out of the left-hand side. We write

$$E(e^{zS_n}) = \int_{-nK}^{nK} e^{zy}\, dP(S_n \leq y) = -\int_{-nK}^{nK} e^{zy}\, dP(S_n > y)$$

$$= z \int_{-nK}^{nK} e^{zy} P(S_n > y)\, dy$$

$$= a_1 + z \int_0^{nK} e^{zy} P(S_n > y)\, dy \tag{6.53}$$

where the last but one equation is obtained by integrating by parts (both limits are included in all integrals.) The estimate $0 < a_1 < 1$ follows immediately by replacing $P(S_n > y)$ for $y < 0$ by the value one. We can further reduce the path of integration by applying Lemma 1 to $P(S_n > y)$ for "large values of y," while, for "small y," the coefficient z in the last integral makes the contribution small. It is convenient to specify z at this stage. We choose $z = y(n)/(1 - \eta)nV$, where $0 < \eta < 1$ is arbitrary and $y(n)$ is the value needed for (6.49). Thus, since $y(n)$ is of the magnitude of $(n \log \log n)^{1/2}$, $z \to 0$ as $n \to +\infty$, which was required eaarlier, and $n^{1/2}z \to +\infty$ with n, which expression appears in (6.52). Hence, the two sides of (6.52) are large for large n and when we say that a contribution to (6.53) is small, we mean it in the light of (6.52). Clearly, a_1 is small in (6.53). So are

$$a_2 = z \int_0^{y(n)} e^{zy} P(S_n > y) \, dy \qquad a_3 = z \int_{z(1+\eta)nV}^{nK} e^{zy} P(S_n > y) \, dy$$

which claim can be shown by some care in the estimates (recall the earlier suggestions for these estimates), but, since both $\varepsilon > 0$ and $\eta > 0$ are arbitrary, it will follow that these values are of smaller order of magnitude than the left-hand side of (6.52) (the details are left to the reader). Thus, all significant contributions to $E(\exp(zS_n))$ come from

$$a_4 = z \int_{z(1-\eta)nV}^{z(1+\eta)nV} e^{zy} P(S_n > y) \, dy$$

We can certainly say that $a_4 > (1/2) E[\exp(zS_n)]$. Note that, $z(1-\eta)nV = y(n)$. Hence, by using the monotonicity of $P(S_n > y)$ and the inequality (6.52), we get

$$\tfrac{1}{2} \exp[\tfrac{1}{2} nz^2 V(1-\varepsilon)] \le a_4 \le 2\eta z^2 nVP(S_n > y(n)) \exp[z^2 nV(1+\eta)]$$

for large n, which yields that, for large n and arbitrary $\delta > 0$

$$P(S_n \ge y(n)) \ge \exp\left[-\frac{y^2(n)}{2nV} (1+\delta) \right]$$

Finally, we choose $n = N_k' \sim s^k$ and $y(n) = (1 - \varepsilon_1/2)b(k)$ and obtain (6.49) with $\gamma = (1+\delta)(1 - \varepsilon_1/2)^2$. Hence, with the proper choice of $\delta > 0$, $\gamma < 1$. Note that γ does not depend on s, so it can be chosen freely in (6.51). The proof is completed.   ∎

Exercises

1. Let A be an event in a random experiment and $p = P(A)$. Let f_n be the relative frequency of A in n independent repetitions of the experiment. Deduce from Lemma 1 that, for $0 < \varepsilon < p(1-p)$, $0 < p < 1$,

$$P(|f_n - p| \ge \varepsilon) \le 2 \exp\left[-\frac{n\varepsilon^2}{2p(1-p)(1 + \varepsilon/2p(1-p))^2} \right]$$

2. Compare the inequality of Exercise 1 with the asymptotic normality of f_n. Use numerical computations to demonstrate your conclusions. Give a case where the quoted inequality is superior to the asymptotic normality of f_n.

3. Restate the law of the iterated logarithm for f_n.

6.7 MULTIVARIATE EXTENSION OF STABLE DISTRIBUTIONS FOR SUMS AND EXTREMES

We start with multivariate stable distributions for sums. Let $\mathbf{X} = (X^{(1)}, X^{(2)}, \ldots, X^{(d)})$ be a random vector with distribution function $F(x_1, x_2, \ldots, x_d)$. Let $\mathbf{X}_1, \mathbf{X}_2, \ldots, \mathbf{X}_n$ be independent copies of $\mathbf{X}$ and set $\mathbf{S}_n = \mathbf{X}_1 + \mathbf{X}_2 + \cdots + \mathbf{X}_n$, where summation as well as all arithmetic operations with vectors are component by component. Let $\mathbf{a}_n$ and $\mathbf{b}_n > 0$ be two sequences of constant vectors. We say that $G(x_1, x_2, \ldots, x_d)$ is a d-dimensional stable distribution function if it is nondegenerate in the sense that all of its univariate marginals are nondegenerate distribution functions and G is the limiting distribution function of $(\mathbf{S}_n - \mathbf{a}_n)/\mathbf{b}_n$ with some $\mathbf{a}_n$ and $\mathbf{b}_n > 0$, and some F. Since operations are component by component, it immediately follows that each univariate marginal of G is a univariate stable distribution. However, not all multivariate G whose univariate marginals are stable are multivariate stable. The dependence structure of G, when stable, is restricted for which in fact there is no good characterization that would be applicable to conclude that, given a distribution function G, it is indeed stable. An abstract characterization does exist which is similar to the univariate case: $G(x_1, x_2, \ldots, x_d)$ is stable if it is nondegenerate in the previously stated sense and if for every $n \geq 1$ there are constants $\mathbf{a}_n$ and $\mathbf{b}_n > 0$ such that G is the exact distribution function of $(\mathbf{S}_n - \mathbf{a}_n)/\mathbf{b}_n$, $n \geq 1$ (note that $n = 1$ is included, so $F = G$ when the basic notation of this section is used). This latter characterization of stability can be utilized in two ways: (i) it leads to a functional equation for the characteristic function $\Psi(t_1, t_2, \ldots, t_d)$ of G, from which Ψ can fully be characterized just as in the univariate case. In particular it follows from such investigation that if each marginal of G has finite variance then not only each marginal is normal but G is a multivariate normal distribution (see Section 2.9). (ii) One can construct multivariate stable distributions as follows. Let $Y_1, Y_2, \ldots, Y_d$ be independent and identically distributed univariate stable random variables, and let c_{ij}, $1 \leq i \leq d$, $1 \leq j \leq d$, be constants. Define $X^{(j)} = c_{1j} Y_1 + c_{2j} Y_2 + \cdots + c_{dj} Y_d$. Then the vector $\mathbf{X}$ with components $X^{(j)}$, $1 \leq j \leq d$, is a stable vector (given that the mentioned abstract characterization of stability has been established, the stability of $\mathbf{X}$ just constructed is trivial). Even though these stable vectors $\mathbf{X}$ are very special, these are practically the only ones which are explicitly known (and do not forget that the choice of the distribution of Y_j is very limited when we want explicit forms).

Note that we can have d-dimensional stable distributions such that some of their marginals are normal and others are not. In such cases, some components have variance and others do not. It is rare that such a stable distribution would appear in applications. One may question that nonnormal stable distributions would appear at all in practice, since it automatically implies that the variance is infinite. But if one argues differently and says that a quantity that is subject to random fluctuations (such as stock prices) and its value at any given time is the initial value plus the aggregate of "gains" and "losses," which gains and losses are assumed independent and identically distributed random values, one gets that the actual distribution is stable. For example, stock prices are believed to have stable distributions. Now, if such a distribution is normal then it must be symmetric in all respects such as the magnitude of loss or gain as well as the frequency of such changes. However, this is what contradicts experience for stock prices, and thus the infinite variance is just a secondary consequence of this argument. If individual stocks have stable distributions, then a portfolio of d stocks must have a d-dimensional stable distribution. See P. L. Brockett (1977) and Chapters 6 and 12 in a book of Press (1982).

We now turn to multivariate extreme value distributions. The basic notation remains the same as in the previous case of stability. For the i.i.d. vectors $\mathbf{X}_j$, $1 \leq j \leq n$, we define the maximum vector $\mathbf{Z}_n$ by taking the maximum component by component: $Z_n^{(j)} = \max\{X_k^{(j)}: 1 \leq k \leq n\}$. This way, $\mathbf{Z}_n$ usually is none of the $\mathbf{X}_j$ but a new vector constructed out of the $\mathbf{X}_j$. As an application, the reader is advised to think of the highest water levels along a river at various points (cities) over the same period of time. We seek conditions on $F(x_1, x_2, \ldots, x_d)$ which guarantee the existence of constant vectors $\mathbf{a}_n$ and $\mathbf{b}_n > 0$ such that $(\mathbf{Z}_n - \mathbf{a}_n)/\mathbf{b}_n$ has a nondegenerate limiting distribution $H(z_1, z_2, \ldots, z_d)$. Here, nondegenerate means the same as in the case of stability for sums: each univariate marginal of H is required to be nondegenerate. It thus follows that the univariate marginals of H are extreme value distributions and the theory of Section 6.5 applies to them. Just as in the case of sums, the marginals, however, do not make an H a d-dimensional extreme value distribution (weak limit of $\mathbf{Z}_n$, when normalized). Here we are able to completely characterize a function H whether it is a limiting distribution of a normalized $\mathbf{Z}_n$ or not by studying the dependence functions of F and H (recall (3.49) in Section 3.8). This is in contrast to stable distributions for sums which are characterized only in an indirect way by their characteristic functions.

We denote by $F_j(x_j)$ and $H_j(z_j)$, respectively, the jth marginal distribution of $F(x_1, x_2, \ldots, x_d)$ and $H(z_1, z_2, \ldots, z_d)$. From the definitions, for $\mathbf{x} = (x_1, x_2, \ldots, x_d)$,

$$P(Z_n^{(j)} \leq x_j) = F_j^n(x_j) \quad \text{and} \quad P(\mathbf{Z}_n \leq \mathbf{x}) = F^n(x_1, x_2, \ldots, x_d) \quad (6.54)$$

and thus, with appropriate choice of $\mathbf{a}_n$ and $\mathbf{b}_n > 0$ (their components are to be chosen by the rules and results of Section 6.5; see Exercise 2),

$$P(Z_n^{(j)} \leq a_n^{(j)} + b_n^{(j)} z_j) = F_j^n(a_n^{(j)} + b_n^{(j)} z_j) \to H_j(z_j) \quad (6.55)$$

where $H_j(z_j)$ is one of the three possible types of distribution given in Theorem 16. So, the choice of the normalizing constant vectors $\mathbf{a}_n$ and $\mathbf{b}_n > 0$ is fully determined by the univariate marginals, and nothing new arises in the multivariate case. However, in order to determine H, and not only its marginals, we turn to the second equation of (6.54). By (3.49), there is a d-dimensional uniform distribution function $D_F(u_1, u_2, \ldots, u_d)$ on the unit cube $0 \le u_j \le 1$, $1 \le j \le d$, such that

$$F(x_1, x_2, \ldots, x_d) = D_F(F_1(x_1), F_2(x_2), \ldots, F_d(x_d)) \tag{6.56}$$

which, by the second equation of (6.54), yields

$$P(\mathbf{Z}_n \le \mathbf{x}) = D_F^n(F_1(x_1), F_2(x_2), \ldots, F_d(x_d))$$
$$= D_F^n(M_1^{1/n}(x_1), M_2^{1/n}(x_2), \ldots, M_d^{1/n}(x_d))$$

where the second equation utilizes the first equation of (6.54) with $M_j(x_j) = P(Z_n^{(j)} \le x_j)$, $1 \le j \le d$. In other words, if we set

$$M(x_1, x_2, \ldots, x_n) = P(\mathbf{Z}_n \le \mathbf{x}), \tag{6.57}$$

and $D_M(u_1, u_2, \ldots, u_d)$ for its dependence function as at (6.56) for F, we have

$$D_M(M_1(x_1), M_2(x_2), \ldots, M_d(x_d)) = D_F^n(M_1^{1/n}(x_1), M_2^{1/n}(x_2), \ldots, M_d^{1/n}(x_d)) \tag{6.58}$$

that is,

$$D_M(u_1, u_2, \ldots, u_d) = D_F^n(u_1^{1/n}, u_2^{1/n}, \ldots, u_d^{1/n}) \tag{6.59}$$

where we suppressed the dependence of M on n. The reader can now easily fill in the details in the following steps:

(a) by using the second equation of (6.54) with some appropriate constants $\mathbf{a}_n$ and $\mathbf{b}_n > 0$ and with $\mathbf{x} = \mathbf{a}_n + \mathbf{b}_n \mathbf{z}$, and pass to the limit as $n \to +\infty$, getting $H(\mathbf{z})$. Then replacing n by nk with a fixed integer k, and repeating the previous step, a routine extension of Lemma 2 of Section 4.6 to multivariate distributions will lead to the equation

$$H^k(\mathbf{A}_k + \mathbf{B}_k \mathbf{z}) = H(\mathbf{z}) \tag{6.60}$$

for every $k \ge 1$, with some constants $\mathbf{A}_k$ and $\mathbf{B}_k > 0$.
(b) in Exercise 2, we deduce from (6.60), that whenever $(\mathbf{Z}_n - \mathbf{a}_n)/\mathbf{b}_n$ converges weakly to $H(\mathbf{z})$, then each marginal converges to the marginals of H.
(c) we conclude from (6.60) that a limiting distribution H of $(\mathbf{Z}_n - \mathbf{a}_n)/\mathbf{b}_n$ satisfies for all $k \ge 1$

$$D_H^k(y_1^{1/k}, y_2^{1/k}, \ldots, y_d^{1/k}) = D_H(y_1, y_2, \ldots, y_d)$$

(d) from (6.56)–(6.59), one gets: to an F there are constants $\mathbf{a}_n$ and $\mathbf{b}_n > 0$ such that $\mathbf{Z}_n^* = (\mathbf{Z}_n - \mathbf{a}_n)/\mathbf{b}_n$ has a nondegenerate limiting distribution $H(\mathbf{z})$ if, and only if, the marginals of $H(\mathbf{z})$ are of the types of the three limiting distributions of univariate maxima, and the components of $\mathbf{Z}_n^*$ converge weakly to the marginals of H. Furthermore,

$$D_F^n(y_1^{1/n}, y_2^{1/n}, \ldots, y_d^{1/n}) \to D_H(y_1, y_2, \ldots, y_d).$$

The latter satisfies equation (c).

Examples for the theory and for the distributions discussed in the present section are provided by the exercises below.

Exercises

1. Give a special bivariate normal distribution and a bivariate stable Cauchy distribution by evaluating the density of the vector $(aX + bY, cX + dY)$, where X and Y are independent and identically distributed (i) normal (ii) Cauchy variables, and a, b, c, and d are constants.

2. Show that if the d-dimensional distribution functions $G_n(x)$ converge weakly to a continuous d-dimensional distribution function $G(x)$, then the tth marginals of G_n converge to the tth marginal of G (weakly). Show that $H(\mathbf{z})$ satisfying the equations in (a) of the text (following (6.59)) is continuous.

3. Show that both

   (i) $\exp(-\exp(-x) - \exp(-y))$ and   (ii) $\exp(-\exp(-\min(x, y)))$

   can occur as limiting distribution of properly normalized bivariate maxima.

4. Denote by $H^*(x, y)$ and $H^{**}(x, y)$, respectively, the distribution functions of Exercise 3(i) and 3(ii). Let $H(x, y)$ be an arbitrary bivariate limiting distribution function, with marginals $\exp(-e^{-x})$ and $\exp(-e^{-y})$, of Z_n, when properly normalized. Show that

   $$H^*(x, y) \le H(x, y) \le H^{**}(x, y)$$

5. Compute the marginals and the dependence function of

   $$H(x, y) = \begin{cases} \exp(-e^{-x} - (1/2)e^{x-2y}) & \text{if } x \le y \\ \exp(-e^{-y} - (1/2)e^{-x}) & \text{if } x > y \end{cases}$$

   Is this a limiting distribution function of normalized bivariate maxima?

REFERENCES

Aczél, J. (1989). Personal communication: letter to the author.

Brockett, P. L. (1977). Supports of infinitely divisible measures on Hilbert space. *Ann. Probability 5*, 1012–1017.

Castillo, E. (1988). *Extreme Value Theory in Engineering*. Academic Press, New York.

Elliott, P. D. T. A. (1979–80). *Probabilistic Number Theory*, Vols. I and II. Springer, New York.

Galambos, J. (1982). The role of functional equations in stochastic model building. *Aequationes Math. 25*, 21–41.

Galambos, J. (1987). *The Asymptotic Theory of Extreme Order Statistics*, 2nd ed. Krieger, Melbourne, Florida.

Galambos, J. and S. Kotz (1978). *Characterizations of Probability Distributions. Lectures Notes in Math. 675*. Springer, Heidelberg.

Galambos, J. and A. Obretenov (1987). Restricted domains of attraction of $\exp(-e^{-x})$. *Stochastic Proc. and Appl. 25*, 265–271.

Galambos, J. and Y. Xu (1990). Regularly varying expected residual life and domains of attraction of extreme value distributions. *Annales Univ. Sci. Budapest, Sectio Math. 33*, 105–108.

Galambos, J., J. Lechner and E. Simiu (eds.). (1994). *Extreme Value Theory and Applications*. Kluwer, Dordrecht.

Gnedenko, B. V. (1943). Sur la distribution limite du terme maximum d'une série aléatoire. *Ann. Math. 44*, 423–453.

Haan, L. de (1970). On regular variation and its application to the weak convergence of sample extremes. *Math. Centre Tracts 32*. Amsterdam.

Haan, L. de (1976). Sample extremes: an elementary introduction. *Statistica Neerlandica 30*, 161–172.

Halmos, P. R. (1974), *Measure Theory*. Springer, Berlin.

Hartman, P. and A. Wintner (1941). On the law of the iterated logarithm. *Amer. J. Math. 63*, 169–176.

Kac, M. (1937). Une remarque sur les équations fonctionnelles. *Comm. Math. Helvetici 9*, 170–171.

Mori, T. (1981). The relation of sums and extremes of random variables. *Bull. ISI, 43rd Session*: 879–894. Buenos Aires.

Press, S. J. (1982). *Applied Multivariate Analysis, 2nd ed.*, Krieger, Melbourne, Florida.

Seneta, E. (1976). *Regularly Varying Functions. Lecture Notes in Math. 508.* Springer, Heidelberg.

Varadhan, S. R. S. (1984). *Large Deviations and Applications.* SIAM, Philadelphia.

7
Conditional Expectation; Martingales

While most of the results of this book have so far been developed under the assumption of independence, our first fundamental result on the convergence of the relative frequency to the probability of a given event is established in Theorem 1.16 with no reference to independence. Since this theorem is the philosophical basis of the more general strong law of large numbers, one can expect far-reaching extensions of the previous results developed for the independent case. One possible way to utilize dependence is to use conditional probabilities rather than unconditional ones. However, the conditional probabilities introduced in Section 1.6 are unsuitable for the investigation of random variables in general, since no meaning is assigned to conditioning with respect to a random variable Y whose distribution function is continuous at a given point x, because $P(Y = x) = 0$ in such a case. The aim, therefore, is to extend the concepts of conditional probabilities and expectations so that conditioning will be allowed on events of zero probability. Here, for the most part, we reverse the process of development and so start with conditional expectations rather than probabilities. Conditional probabilities will then be defined via the conditional expectations of indicator variables. The rigorous development of this concept is contained in Kolmogorov's fundamental work (1933).

7.1 CONDITIONAL EXPECTATION, GIVEN A DISCRETE RANDOM VARIABLE

Recall the classical definition of conditional probability from Section 1.6. Starting with the probability space $(\Omega, \mathscr{A}, P)$, the conditional probability $P(A|B)$ of $A \in \mathscr{A}$, given $B \in \mathscr{A}$ such that $P(B) > 0$, is defined by

$$P(A|B) = \frac{P(A \cap B)}{P(B)} \tag{7.2}$$

Assume that, in addition, $P(B) < 1$, i.e., $P(B^c) > 0$. Then $P(A|B^c)$ is defined also. We can view the combination of $P(A|B)$ and $P(A|B^c)$ as a new random variable that is $P(A|B)$ on B and $P(A|B^c)$ on B^c. In this new interpretation, we write $P(A|I(B))$, where $I(B)$ is the indicator variable of B and

$$P(A|I(B)) = P(A|B)I(B) + P(A|B^c)I(B^c) \tag{7.2}$$

The definition (7.2) easily extends from the indicator variable $I(B)$ to an arbitrary discrete random variable. Let Z be a discrete random variable taking the values $z_1, z_2, \ldots$, and let $C_j = \{Z = z_j\}$. Then $P(C_j) = p_j > 0$ and $P(\cup C_j) = 1$, where the union is taken over all j. We define the conditional probability $P(A|Z)$ of $A \in \mathscr{A}$, given the random variable Z, as a random variable that takes the value $P(A|C_j)$ on C_j. That is,

$$P(A|Z) = \sum P(A|C_j)I(C_j) \tag{7.3}$$

where the summation is extended over all values of j (but only one term can differ from zero at any $\omega \in \Omega$). Next, we turn to conditional expectations. Since we want to preserve the relation between the probability of an event and the expectation of its indicator, we define

$$E(I(A)|Z) = P(A|Z) \tag{7.4}$$

Thus,

$$E(I(A)|Z) = \sum E(I(A)|C_j)I(C_j)$$

$$= \sum \frac{1}{P(C_j)} E[I(A)I(C_j)]I(C_j) \tag{7.5}$$

where $E(I(A)|C_j)$ is an ordinary expectation with respect to the probability $P(\cdot|C_j)$ (see (7.1) and Theorem 1.18). Equation (7.5) is convenient for defining the conditional expectations of more general random variables. However, before formulating the exact definition, let us recall that random variables are defined on Ω, except perhaps on a subset of it whose probability is zero. We add a new convention here: if $P(B) = 0$, we shall say that $E(I(A)|B)$ is undefined on B. This permits us to include those C_j in the definition of Z for which $P(C_j) = 0$, but does not affect any previous formula, as we want random variables defined only almost surely. This, however, permits us to condition with respect to $Z = I(\Omega)$, in which case all conditional expectations become unconditional ones a.s.

Definition Let X be a random variable with finite expectation. Then, for a discrete random variable Z, the conditional expectation of X, given Z,

$$E(X|Z) = \sum_j \frac{1}{P(C_j)} E[XI(C_j)]I(C_j) \text{ a.s.} \tag{7.6}$$

where the C_j are as defined in the previous paragraph.

The definition immediately implies, from the respective properties of the expectation (see Section 2.2),

(i) $E(c|Z) = c$ a.s. for a constant c
(ii) $E(X|Z) \geq 0$ a.s. if $X \geq 0$ (a.s.)
(iii) $E(c_1 X_1 + c_2 X_2 | Z) = c_1 E(X_1 | Z) + c_2 E(X_2 | Z)$ a.s.

where c_1 and c_2 are constants. One new property, one which one would anticipate from a meaningful definition, is that, if $g(x)$ is a function defined for all $x = x_j$, then

$$E[g(Z)|Z] = g(Z) \text{ a.s.} \tag{7.7}$$

or, more generally,

$$E[Xg(Z)|Z] = g(Z)E(X|Z) \text{ a.s.} \tag{7.8}$$

where it is assumed that the unconditional expectation of $Xg(Z)$ ($X = 1$ for (7.7)) is finite. Both (7.7) and (7.8) are direct consequences of (7.6) because $Z = z_j$, a constant, on C_j; thus

$$E(Xg(Z)I(C_j)) = g(z_j)E(XI(C_j))$$

which, together with $g(z_j)I(C_j) = g(Z)I(C_j)$, immediately yields (7.7) and (7.8). Note that properties (i)–(iii) are, to some extent, a set of characteristic properties of the definition (7.6). That is, if we assume that (i)–(iii) and (7.5) hold, then (7.6) follows. Its verification is left to the reader as Exercise 1.

Example 1 Let X be an exponentially distributed random variable with $E(X) = 1$. Let Z be the integer part of X. Let us determine the conditional expectation $E(X|Z)$.

Since Z is discrete, so is $E(X|Z)$, and the latter is constant on the same sets for which Z is. Let $C_j = \{Z = j\} = \{j \le X < j + 1\}$. Then (7.6) yields

$$E(X|Z) = \frac{1}{P(j \le X < j + 1)} \int_j^{j+1} x e^{-x}\, dx = j + \frac{e - 2}{e - 1}$$

on C_j; thus, $E(X|Z) = Z + (e - 2)/(e - 1)$ a.s.   ∎

Let us return to (7.6). We shall freely switch between expectations and integrals with the rules

$$E(X) = \int_\Omega X\, dP \qquad \text{and} \qquad E[XI(A)] = \int_A X\, dP$$

Now, let $B = \cup^* C_j$, where the asterisk signifies that the respective operation (union, in this case) is taken over a subset of the set of all possible j. Then, since the C_j are mutually exclusive, $I(B) = \sum^* I(C_j)$ and, thus, by (7.6),

$$E(X|Z)I(B) = \sum{}^* \frac{1}{P(C_j)} E[XI(C_j)]I(C_j) \text{ a.s.}$$

Because the sum of a finite number of the terms $E[XI(C_j)]$ is dominated by $E(|X|)$, the dominated convergence theorem (Theorem 2.16) ensures that the integral of the right-hand side will be finite and we can interchange summation and integration. We thus have

$$\int_B E(X|Z)\, dP = \int_B X\, dP \tag{7.9}$$

Since the smallest σ-field with respect to which Z is measurable, $\sigma(Z)$, is exactly the collection of $B = \cup^* C_j$, (7.9) is, in fact, valid for all $B \in \sigma(Z)$. This is the form that we shall use as a definition in Section 7.3, when we shall extend the concept of conditional expectation to nondiscrete random variables as condition.

Note the important difference between the two sides of (7.9). The random variable $E(X|Z)$ is measurable with respect to $\sigma(Z)$, and thus always discrete in this section, while X can be any kind of random variable on the original probability space (recall Example 1). In other words, if P_Z denotes the probability P restricted to $\sigma(Z)$, then the integral on the left-hand side can be computed with respect to P_Z, but the original P must be used on the right-hand side. In order to demonstrate this point, let us go back to Example 1.

Example 1a Continuing Example 1, we choose $B = \{Z \text{ is even}\}$. Then, since $E(X|Z) = Z + (e - 2)/(e - 1)$ a.s.,

$$\int_B E(X|Z)\,dP = \int_B \left(Z + \frac{e-2}{e-1}\right) dP_Z$$

$$= \sum_{j=0}^{+\infty} 2jP(Z = 2j) + \frac{e-2}{e-1} P(Z \text{ is even}) = \left(\frac{e}{e+1}\right)^2$$

On the other hand,

$$\int_B X\,dP = \sum_{j=0}^{+\infty} \int_{2j}^{2j+1} xe^{-x}\,dx$$

which, after integrating by parts, results (obviously) in the same value $e^2/(1 + e)^2$ as the first integral did. ∎

We shall record a few more consequences of the definition (7.6) and of (7.9). First, if we apply (7.9) with $B = \Omega$, we obtain

$$E[E(X|Z)] = E(X) \tag{7.10}$$

Next, let X and Z be independent. Then X and $I(C_j)$ are independent for all j and, thus, $E(XI(C_j)) = E(X)P(C_j)$ for all j (see Theorem 2.4). Hence, (7.6) yields $E(X|Z) = E(X)$. Finally, we note that if we extend the definition of $P(A|Z)$ in (7.3) by first adding $(\cup C_j)^c = D$ to the set of the C_j for which, necessarily, $P(D) = 0$ and if we define $P(A|D) = P(A)$ on D, then $P(A|Z)$ is defined for all $\omega \in \Omega$ and, for each $\omega \in \Omega$, $P(A|Z)$ is a probability on $A \in \mathscr{A}$ (due to Theorem 1.18). Therefore, for each $\omega \in \Omega$, we can speak of the expectation of a random variable X with respect to $P(\cdot|Z)$. Let us define the random variable Y (when finite) by

$$Y = \int_\Omega X\,dP(\cdot|Z)$$

By (7.3), if $\omega \in C_j$ such that $P(C_j) > 0$, then $P(A|Z) = P(A|C_j) = P(A \cap C_j)/P(C_j)$; thus,

$$Y = \frac{1}{P(C_j)} \int_\Omega XI(C_j)\,dP \qquad \omega \in C_j$$

(The reader may want to go through the steps of the definition of an integral as we did in Section 2.1, starting with indicators in place of X, then simple random variables, etc.) That is, Y is of the form (7.6), yielding

$$Y = \int_\Omega X\,dP(\cdot|Z) = E(X|Z) \text{ a.s.} \tag{7.11}$$

In other words, the conditional expectation a.s. equals the expectation with respect to the conditional probability.

Exercises

1. Assume that the conditional expectation $E(X|Z)$ is a random variable satisfying properties (i)–(iii) of this section and that (7.5) applies for indicator variables $X = I(A)$. Prove that (7.6) must hold.
2. Let $X_1, X_2, \ldots$ be i.i.d. discrete random variables with $E(X_j) = 0$. Set $S_n = X_1 + X_2 + \cdots + X_n$. Show that $E(S_{n+1}|S_n) = S_n$ a.s.
3. Let Y and Z be independent and identically distributed discrete random variables with finite expectation. Show that $E(Y|Y + Z) = (Y + Z)/2$ a.s.
4. Let Z_1 and Z_2 be two discrete random variables such that $\sigma(Z_1) \subseteq \sigma(Z_2)$. Let X be a random variable with finite expectation. Find the conditional expectations $U_{12} = E(Y_1|Z_2)$ and $U_{21} = E(Y_2|Z_1)$, where $Y_1 = E(X|Z_1)$ and $Y_2 = E(X|Z_2)$.

7.2   RADON–NIKODYM THEOREM

In order to further extend the concept of conditional expectation, we have to establish an important result in connection with integrals.

Let X be a random variable on a given probability space $(\Omega, \mathscr{A}, P)$ whose expectation is finite. Then, for any $A \in \mathscr{A}$, $XI(A)$ also has a finite expectation; thus, the set function

$$Q(A) = Q_X(A) = \int_\Omega XI(A)\, dP = \int_A X\, dP$$

is well defined for $A \in \mathscr{A}$. It follows immediately from the basic properties of the integral that if the A_j are disjoint members of $\mathscr{A}$, then,, for every n,

$$Q\left(\bigcup_{j=1}^n A_j\right) = \sum_{j=1}^n Q(A_j)$$

We now show that this additive property continues to hold if $n = +\infty$, i.e., $Q(A)$ is σ-additive. This property was, in fact, proved in (2.25a) if $X \geq 0$. Now, if X is an arbitrary integrable random variable, then we can decompose $X = X^+ - X^-$, where $X^+ = \max(0, X)$ and $X^- = -\min(0, X)$, and both $Q^+(A) = Q_{X^+}(A)$ and $Q^-(A) = Q_{X^-}(A)$ are finite for all $A \in \mathscr{A}$. Hence, applying (2.25a) to both $Q^+(A)$ and $Q^-(A)$, the relation $Q(A) = Q^+(A) - Q^-(A)$ entails the σ-additivity of $Q(A)$. In addition, in the above decomposition of $Q(A)$, both $Q^+(A)$ and $Q^-(A)$ are nonnegative. Among the basic properties, we finally note that if $P(A) = 0$, then $Q(A) = 0$ as well. We shall refer to this property by saying that $Q(A)$ is *absolutely continuous with respect to* $P(A)$.

The set function $Q(A) = Q_X(A)$, $A \in \mathcal{A}$, is called the indefinite integral of X on the space $(\Omega, \mathcal{A}, P)$. On the other hand, a set function $u(A)$ that is defined and finite for $A \in \mathcal{A}$ is called a signed measure (on $\mathcal{A}$) if it can be decomposed as $u(A) = u_1(A) - u_2(A)$, where each $u_i(A)$ $(i = 1, 2)$ is nonnegative and σ-additive. Therefore, an indefinite integral is an absolutely continuous signed measure. Theorem shows that such a property characterizes the indefinite integrals. But first we establish a fundamental property of signed measures.

Closure Theorem. Let $u(A)$ be a finite signed measure on a sigma field $\mathcal{A}$ of subsets of a set Ω. Then $u(A)$ is bounded and there are set D and D^* in $\mathcal{A}$ such that

$$u(D) = \inf\{u(A): A \in \mathcal{A}\} \quad \text{and} \quad u(D^*) = \sup\{u(A): A \in \mathcal{A}\}$$

Proof. Clearly, it is sufficient to show the claim for the infimum because, by arguing with $-u(A)$, one gets the claim for the supremum in the same way.

Let $\alpha = \inf\{u(A): A \in \mathcal{A}\}$. One can select a sequence $A_n \in \mathcal{A}$ such that $u(A_n) \to \alpha$. If $u(A) \geq 0$ for all $A \in \mathcal{A}$, i.e., if it is a measure, then, in view of $u(\varnothing) = 0$, $\alpha = 0$ and $D = \varnothing$ is a solution. Therefore, we assume $\alpha < 0$. Define $B = \cup A_n$ (the union is over all n), and decompose it by the identity

$$B = B \cap \left[\bigcap_{j=1}^{n} (A_j \cup A_j^c) \right] = \bigcup_{k=1}^{2^n} C_{k,n}, \quad \text{say}, \quad n \geq 1$$

where, in the expansion on the right hand side, the 2^n sets $C_{k,n}$ are disjoint sets of intersections of n sets A_j^*, where A_j^* is either A_j or A_j^c. Let us select those $C_{k,n}$ for which $u(C_{k,n}) \leq 0$, and let their union be C_n. Note the following structure of the sequence $C_{k,n}$ as n changes. Every $C_{k,n+1}$ intersects a $C_{t,n}$ either by A_{n+1} or A_{n+1}^c; consequently, $C_{k,n+1}$ is either a subset of $C_{t,n}$ or $C_{k,n+1}$ and $C_{t,n}$ are disjoint. We thus have that $C_n \cup C_{n+1} = C_n \cup E_n$, where C_n and E_n are disjoint and $u(E_n) \leq 0$, entailing

$$u(C_n) \geq u(C_n \cup C_{n+1}) \geq \cdots \geq u\left(\bigcup_{j=n}^{m} C_j \right) \quad m > n$$

Now, A_n is the union of all those $C_{k,n}$ which contain A_n. If we drop those $C_{k,n}$ from A_n for which the u-value is positive, $u(A_n)$ decreases, which is further decreased if we add all those negative $u(C_{k,n})$ which are not yet in $u(A_n)$, yielding $u(A_n) \geq u(C_n)$. That is, for every $m > n$,

$$u(A_n) \geq u\left(\bigcup_{j=n}^{m} C_j\right) \geq u\left(\bigcup_{j=n}^{+\infty} C_j\right)$$

where we first utilized our previous inequalities, and then appealed to Theorem 1.8 (note that Theorem 1.8 relies on Axiom 3 alone, that is, only the additivity of P is needed for it). by letting $n \to +\infty$, we get from the choice of A_n and from Theorem 1.8 again

$$\alpha \geq u\left(\bigcap_{n=1}^{n} \bigcup_{j=n}^{+} C_j\right)$$

which, first of all, implies that $\alpha > -\infty$. Second, we further decrease if we write α on the right hand side, entailing that

$$\alpha = u\left(\bigcap_{n=1}^{+} \bigcup_{j=n}^{+} C_j\right) \quad \text{and thus} \quad D = \bigcap_{n=1}^{+} \bigcup_{j=n}^{+} C_j$$

The closure theorem is established.   ∎

Corollary. With the notation and assumptions of the Closure Theorem, $u(A \cap D) \leq 0$ and $u(A \cap D^c) \geq 0$ for all $A \in \mathscr{A}$.

Proof. Assume $u(A \cap D) > 0$. Then $u(A^c \cap D) = u(D) - u(A \cap D) < u(D)$, contradicting the definition of D in the Closure Theorem. On the other hand, if $u(A \cap D^c) < 0$, then a contradiction with the meaning of D would come from

$$u(D \cup (A \cap D^c)) = u(D) + u(A \cap D^c) < u(D)$$

which completes the proof.   ∎

Theorem 1 (Radon–Nikodym) Let $(\Omega, \mathscr{A}, P)$ be a probability space and let $u(A)$ be an absolutely continuous signed measure on $\mathscr{A}$. Then there is a random variable X such that

$$u(A) = \int_A X \, dP \qquad A \in \mathscr{A} \tag{7.12}$$

and X is unique a.s. (with respect to P).

Proof. First, we prove that X is a.s. unique in (7.12). Let us assume the opposite. That is, that there are two random variables X_1 and X_2 for which (7.12) holds and $P(X_1 > X_2) > 0$. Then there is a number $a > 0$ such that $P(A(a)) > 0$, where $A(a) = \{X_1 - X_2 \geq a\}$, implying

$$\int_{A(a)} X_1 \, dP - \int_{A(a)} X_2 \, dP = \int_{A(a)} (X_1 - X_2) \, dP \geq aP(A(a)) > 0$$

which contradicts the assumption that both terms on the left-hand side equal $u(A(a))$. Therefore, it remains to prove the existence of a random variable X such that (7.12) holds.

In view of the assumption that $u(A)$ is the difference of two nonnegative set functions, it suffices to prove the existence of X under the additional assumption that $u(A) \geq 0$. Let R be the collection of all random variables Y such that

$$\int_A Y \, dP \leq u(A) \qquad \text{for all } A \in \mathcal{A} \tag{7.13}$$

Since $Y = 0$ a.s. is a member of R, R is not empty. Let

$$s = \sup_{Y \in R} \int_\Omega Y \, dP \leq u(\Omega) < +\infty$$

Thus, there is a sequence Y_n, $Y_n \in R$, such that

$$\lim \int_\Omega Y_n \, dP = s$$

as $n \to +\infty$. Set $Y_n^* = \max_{1 \leq k \leq n} Y_k$. Then $Y_k^* \in R$, because, upon introducing the events $B_1 = \{Y_1 = Y_n^*\}$ and $B_k = \{Y_k = Y_n^*, Y_j \neq Y_n^*, 1 \leq j \leq k - 1\}$, $2 \leq k \leq n$, $I(B_1) + I(B_2) + \cdots + I(B_n) = 1$ and we have, for every $A \in \mathcal{A}$,

$$\int_A Y_n^* \, dP = \sum_{k=1}^n \int_A Y_n^* I(B_k) \, dP = \sum_{k=1}^n \int_A Y_k I(B_k) \, dP$$

$$\leq \sum_{k=1}^n u(A \cap B_k) = u(A)$$

We thus have, for every n, $Y_n^* \in R$, $Y_n^* \leq Y_{n+1}^*$ and, for every $A \in \mathcal{A}$

$$\lim \int_A Y_n^* \, dP \leq u(A) \qquad \text{and} \qquad \lim \int_\Omega Y_n^* \, dP = u(\Omega)$$

as $n \to +\infty$, the latter equation to be proved below. Let $\lim Y_n^* = X$ $(n \to +\infty)$. The monotone convergence theorem (Theorem 2.14) then entails

$$\int_A X \, dP \le u(A) \qquad \text{all } A \in \mathscr{A} \qquad \text{and} \qquad \int_\Omega X \, dP = u(\Omega)$$

Now, assume that there is an $A \in \mathscr{A}$ such that

$$d(A) = u(A) - \int_A X \, dP > 0$$

That is, $0 = d(\Omega) = d(A) + f(A^c) \ge d(A) > 0$, which is a contradiction. Consequently, $d(A) = 0$ for all $A \in \mathscr{A}$, and it remains only to show that, as $n \to +\infty$,

$$\lim \int_\Omega Y_n^* \, dP = u(\Omega)$$

Let

$$v(A) = u(A) - \int_A X \, dP \qquad A \in \mathscr{A}$$

As was established above, $v(A) \ge 0$, and thus $v(A)$ is an absolutely continuous measure on $\mathscr{A}$ (with respect to P). Assume $v(\Omega) > 0$. Then, for some $k > 0$, $P(\Omega) - kv(\Omega) < 0$. Hence, $P^*(A) = P(A) - kv(A)$ is a signed measure. Therefore, we have from the Closure Theorem that there is a $D \in \mathscr{A}$ such that

$$P^*(D) = \inf\{P^*(A) : A \in \mathscr{A}\}$$

First, note that, for this D, $P(D) > 0$ because if $P(D) = 0$ then $v(D) = 0$ by absolute continuity, contradicting the second inequality of the Corollary to the Closure Theorem and our choice of k: with $A = \Omega$, we have $0 \le P^*(\Omega \cap D^c) = P^*(D^c) = P(D^c) - kv(D^c) = P(\Omega) - kv(\Omega) < 0$. Next, define $Z = /k$ on D and $Z = 0$ on D^c. Then, for all $A \in \mathscr{A}$,

$$\int_A Z \, dP = \frac{1}{k} P(A \cap D) \le v(A \cap D) \le v(A) = u(A) - \int_A X \, dP$$

where the first inequality is due to the Corollary to the Closure Theorem. The two extreme sides yield $\int_A (X + Z) \, dP \le u(A)$, which contradicts the construction of X by its supremum property via Y_n^*, since, as it may be recalled, $X + Z > X$ on D and $X + Z = X$ on D^c and $P(D) > 0$. Our initial assumption is therefore wrong, that is, $v(\Omega) = 0$, which, by the monotone convergence theorem, is equivalent to the equation we set out to prove. This concludes the proof.   ■

We shall make a slight generalization. First, note that the theorem obviously does not change if we multiply (7.12) by a constant $c > 0$, so we let P absorb c, resulting in a reformulation of the basic assumptions in such a way that P can be a nonnegative and σ-additive set function on $\mathscr{A}$ with $P(\Omega) < +\infty$ (which we call a *measure* on $\mathscr{A}$). Further generalization is possible. Assume $P(\Omega) = +\infty$, where P is a nonnegative and σ-additive set function on the σ-field $\mathscr{A}$ of subsets of the set Ω, and that Ω can be decomposed as the union of a denumerable number of disjoint sets Ω_j such that $P(\Omega_j) < +\infty$ for each j (in which case P is called a *σ-finite measure*). Then the just-extended Theorem 1 is applicable on $(\Omega_j, \mathscr{A}_j, P_j)$, where $\mathscr{A}_j$ is the σ-field of those members of $\mathscr{A}$ which are subsets of Ω_j and P_j is P restricted to $\mathscr{A}_j$. Since P and both sides of (7.12) are σ-additive, the conclusions of Theorem 1 remain valid. Hence, the following extension of Theorem 1 holds.

Theorem 1a   **(Radon–Nikodym)**   If P is a σ-finite measure on the σ-field $\mathscr{A}$ of subsets of a set Ω and if $u(A)$ is an absolutely continuous signed measure on $\mathscr{A}$, then there is a finite function X defined a.s. on Ω that is measurable with respect to $\mathscr{A}$ and (7.12) holds. The function X is unique a.s. (all qualifying properties are with respect to P).

Definition   A function X occurring in Theorem 1a (or in Theorem 1) is called a derivative of u with respect to P, for which we use the notation

$$X = X(\omega) = \frac{du}{dP}(\omega) = \frac{du}{dP} \quad \text{(a.s.)}$$

Any two derivatives of u are equal a.s. with respect to P.

Let P be Lebesgue measure on the whole real line Ω. Then P is finite on $\Omega_j = [j, j+1)$, so P is a σ-finite measure on the set $\mathscr{A}$ of Borel sets of the real line. Let $(\Omega^*, \mathscr{A}^*, P^*)$ be an arbitrary probability space and let Z be a random variable on this space. Let $F(x)$ be the distribution function of Z. Then $F(x)$ can be extended into a probability on the Borel sets $\mathscr{A}$ of the real line, which we also denote by $F(x)$. Now, if $F(x)$ is absolutely continuous with respect to Lebesgue measure (P), then $F(x)$ is a.s. differentiable with respect to P, which we denote by

$$f(x) = \frac{dF(x)}{dx} \quad \text{a.s.}$$

instead of using dP. This particular derivative is called a density function of $F(x)$ (compare it to the classical definition used in the earlier chapters). Theorem 1a states that

$$P(Z \in A) = \int_A f(x)\, dx$$

for all Borel sets A, where the integral is a Lebesgue integral. We also have that the derivatives of $F(x)$ can differ only on a set of Lebesgue measure zero.

Another particular case of Theorem 1 will form the basis of the general definition of conditional expectation, which will be discussed in the next section.

Theorem 1a makes it possible to transform an integral with respect to a (σ-finite) measure u into an integral with respect to another (σ-finite) measure P, which is the content of the next theorem.

Theorem 2 Let u be a measure and P be a σ-finite measure on the σ-field $\mathscr{A}$ of subsets of a set Ω. Assume that u is absolutely continuous with respect to P. Let X be a function defined a.s. (with respect to P) on Ω and measurable with respect to $\mathscr{A}$. Then, if X is integrable with respect to u,

$$\int_A X\, du = \int_A X \frac{du}{dP}\, dP \qquad \text{for all } A \in \mathscr{A}$$

Proof. Let B be a member of $\mathscr{A}$ and let $X = I(B)$. Then, by Theorem 1a,

$$\int_A I(B)\, du = u(A \cap B) = \int_{A \cap B} \frac{du}{dP}\, dP = \int_A \frac{du}{dP} I(B)\, dP$$

That is, the theorem is valid when X is an indicator variable. But then, by the additivity of integrals, the theorem is valid if X is a simple random variable. Next, the monotone convergence theorem implies that we can take X as a nonnegative measurable function. Finally, if X is an arbitrary measurable function whose integral is finite with respect to u, then the result obtained in the preceding sentence ensures that the theorem holds with both X^+ and X^- and thus with X, as well. The theorem is established. ∎

The reader can easily verify that Theorem 2 continues to hold if u is assumed to be σ-finite only.

Exercises

1. Let n be a fixed positive integer and let $\mathscr{A}$ be the smallest (σ-) field containing the intervals $[k/n, (k + 1)/n]$, $0 \le k \le n - 1$. Define

$$u(A) = \int_A x \, dx \qquad A \in \mathscr{A}$$

where the integral is either a Riemann or a Lebesgue integral. With P denoting Lebesgue measure restricted to $\mathscr{A}$, show that $u(A)$ is absolutely continuous with respect to $P(A)$ and find the derivatives du/dP.

2. Let u_1, u_2, and u_3 be σ-finite measures on the same σ-field and assume that u_2 is absolutely continuous with respect to u_1 and that u_3 is absolutely continuous with respect to u_2. Show that u_3 is absolutely continuous with respect to u_1 and that the following chain rule applies:

$$\frac{du_3}{du_1} = \frac{du_3}{du_2} \frac{du_2}{du_1}$$

7.3 CONDITIONAL EXPECTATION: THE GENERAL CASE

In Section 7.1, when conditioning was with respect to a discrete random variable Z, we arrived at the important equation (7.9), which was deduced as a consequence of the definition of $E(X|Z)$. With the Radon–Nikodym theorem at hand, we can now show that (7.9) is, in fact, equivalent to the definition (7.6) by noting that

$$u(B) = \int_B X \, dP \qquad B \in \sigma(Z)$$

is a signed measure on $\sigma(Z)$ that is absolutely continuous with respect to P_Z. Therefore, by Theorem 1, there is an a.s. (with respect to P_Z) unique $\sigma(Z)$-measurable function Y such that

$$u(B) = \int_B Y \, dP_Z \qquad B \in \sigma(Z)$$

Now, since $E(X|Z)$ of (7.6) is one such Y (on account of (7.9)), all possible Ys are a.s. equal to $E(X|Z)$. This is the reason that we can turn to (7.9) as the definition of conditional expectation. We make one more deviation from the approach of Section 7.1. Since we did not fully use the fact that Z is a random variable—instead, we used $\sigma(Z)$—we shall define conditional expectations, given a σ-field that is a sub-sigma-field of $\mathscr{A}$, i.e., a subset of $\mathscr{A}$.

Definition   Let $(\Omega, \mathscr{A}, P)$ be a probability space and let $\mathscr{B}$ be a sub-sigma-field of $\mathscr{A}$. That is, $\mathscr{B} \subset \mathscr{A}$ and $\mathscr{B}$ is a σ-field. Let X be a random variable on $(\Omega, \mathscr{A}, P)$ with finite expectation. Then the conditional expectation $E(X|\mathscr{B})$ of X, given $\mathscr{B}$, is any $\mathscr{B}$-measurable function such that

$$\int_B E(X|\mathscr{B}) \, dP_B = \int_B X \, dP \qquad B \in \mathscr{B} \tag{7.14}$$

where $P_{\mathscr{B}}$ is P restricted to $\mathscr{B}$. If X is the indicator variable $I(A)$ of the event $A \in \mathscr{A}$, then $E(I(A)|\mathscr{B}) = P(A|\mathscr{B})$ is the conditional probability of A, given $\mathscr{B}$. If $\mathscr{B}$ is the smallest σ-field $\sigma(Z)$ with respect to which the random variable Z is measurable, then $E(X|\mathscr{B}) = E(X|Z)$ is an alternative notation and we then speak of conditional expectation, given the random variable Z.

Just as in the first paragraph of this section, if we set

$$u(B) = \int_B X \, dP \qquad B \in \mathscr{B}$$

then $u(B)$ is a signed measure on $\mathscr{B}$ and is absolutely continuous with respect to P. Hence, by Theorem 1, there is a $P_{\mathscr{B}}$-a.s. unique $\mathscr{B}$-measurable function $E(X|\mathscr{B})$ such that (7.14) holds. We thus have that any two variants of the conditional expectation $E(X|\mathscr{B})$ can differ only on a set of $P_{\mathscr{B}}$-measure zero. It should be noted that, in view of (7.9), the new definition of conditional expectation is an extension of the definition of Section 7.1, where conditioning was restricted.

Next, we establish a number of the elementary properties of the newly introduced conditional expectation. These are listed under the label (CEj), $j \geq 1$.

(CE1) $E[E(X|\mathscr{B})] = E(X)$

This follows by choosing $B = \Omega$ in (7.14).

(CE2)   If $\mathscr{A} = \mathscr{B}$, or if X is $\mathscr{B}$-measurable, then $E(X|\mathscr{B}) = X$ a.s.

This is due to the a.s. uniqueness of $E(X|\mathscr{B})$ in (7.14).

(CE3)   If $X = c$, a constant, a.s., then $E(X|\mathscr{B}) = c$ a.s.

Since a constant is measurable with respect to any σ-field, this property is entailed by (CE2).

(CE4) $E(c_1 X_1 + c_2 X_2|\mathscr{B}) = c_1 E(X_1|\mathscr{B}) + c_2 E(X_2|\mathscr{B})$ a.s., where c_1 and c_2 are constants and both X_1 and X_2 have finite expectation.

Upon using the corresponding property of the integral on the right-hand side of (7.14), the a.s. uniqueness of the conditional expectation yields (CE4). A similar argument leads to the following property:

(CE5) $E(X|\mathscr{B}) = E(X^+|\mathscr{B}) - E(X^-|\mathscr{B})$ a.s.

(CE6) If $X \le Y$ a.s., then $E(X|\mathscr{B}) \le E(Y|\mathscr{B})$ a.s., where both X and Y are assumed to have finite expectation.

Once again, from the corresponding property of the integral, it follows via (7.14) that

$$\int_B E(X|\mathscr{B})\, dP_{\mathscr{B}} \le \int_B E(Y|\mathscr{B})\, dP_{\mathscr{B}} \qquad \text{all } B \in \mathscr{B}$$

Now, let $B = \{E(X|\mathscr{B}) > E(Y|\mathscr{B})\}\vdash$ Then

$$\int_B E(Y|\mathscr{B})\, dP_{\mathscr{B}} \le \int_B E(X|\mathscr{B})\, dP_{\mathscr{B}}$$

Thus, equality must hold for this B. Hence, by (CE4),

$$\int_B [E(X|\mathscr{B}) - E(Y|\mathscr{B})]\, dP_{\mathscr{B}} = 0 \tag{7.15}$$

and, on this particular B, the integrand is positive. We thus have $P_{\mathscr{B}}(B) = 0$ (otherwise, there would be a number $a > 0$ such that $P_{\mathscr{B}}[E(X|\mathscr{B}) - E(Y|\mathscr{B}) \ge a] > 0$ which contradicts (7.15)), as desired.

(CE7) $|E(X|\mathscr{B})| \le E(|X|\,\|\,\mathscr{B})$ a.s.

Because $-|X| \le X \le |X|$, (CE6) ensures (CE7).

(CE8) Let $\mathscr{B}_1 \subset \mathscr{B}_2 \subset \mathscr{A}$ be two σ-fields. Then

$$E[E(X|\mathscr{B}_1)|\mathscr{B}_2] = E(X|\mathscr{B}_1) \text{ a.s.}$$

Because $E(X|\mathscr{B}_1)$ is $\mathscr{B}_1$-measurable and $\mathscr{B}_1 \subset \mathscr{B}_2$, $E(X|\mathscr{B}_1)$ is $\mathscr{B}_2$-measurable as well. Hence, (CE2) entails (CE8).

(CE9) Once again, let $\mathscr{B}_1 \subset \mathscr{B}_2 \subset \mathscr{A}$ be two σ-fields. Then

$$E[E(X|\mathscr{B}_2)|\mathscr{B}_1] = E(X|\mathscr{B}_1) \text{ a.s.}$$

This is much less obvious than the previous property. Let $B \in \mathscr{B}_1$. Then, by (7.14),

$$\int_B E(X|\mathscr{B}_1)\, dP_{\mathscr{B}_1} = \int_B X\, dP$$

But $\mathscr{B}_1 \subset \mathscr{B}_2$, i.e., $B \in \mathscr{B}_2$; thus,

$$\int_B X\, dP = \int_B E(X|\mathscr{B}_2)\, dP_{\mathscr{B}_2} = \int_B E[E(X|\mathscr{B}_2)|\mathscr{B}_1]\, dP_{\mathscr{B}_1}$$

We thus have, for all $B \in \mathscr{B}_1$,

$$\int_B E(X|\mathscr{B}_1)\, dP_{\mathscr{B}_1} = \int_B E[E(X|\mathscr{B}_2)|\mathscr{B}_1]\, dP_{\mathscr{B}_1}$$

Since both integrands are $\mathscr{B}_1$-measurable, the uniqueness part of Theorem 1 gives the a.s. identity of the integrands.

(CE10)   Let X and $I(B)$ be independent for all $B \in \mathscr{B}$. Then

$$E(X|\mathscr{B}) = E(X) \text{ a.s.}$$

Since, for all $B \in \mathscr{B}$,

$$\int_B E(X|\mathscr{B})\, dP_{\mathscr{B}} = \int_B X\, dP = \int_\Omega XI(B)\, dP = E(X)E[I(B)]$$

$$= E(X)\int_B dP$$

where the penultimate equation is justified by Theorem 2.4. The uniqueness property associated with (7.14) completes the proof of (CE10).

(CE11)   Let $0 \le X_n \le X_{n+1}$ and $\lim X_n = X\ (n \to +\infty)$ be random variables with finite expectation. Then, as $n \to +\infty$,

$$\lim E(X_n|\mathscr{B}) = E(X|\mathscr{B}) \text{ a.s.}$$

Clearly, it suffices to assume that $E(X)$ is finite. Now, $0 \le X_n \le X_{n+1}$ implies $0 \le E(X_n|\mathscr{B}) \le E(X_{n+1}|\mathscr{B})$ a.s. (see (CE3) and (CE6)); thus, as $n \to +\infty$, $\lim E(X_n|\mathscr{B}) = Y \ge 0$ exists a.s., which is a $\mathscr{B}$-measurable function. Therefore, by the monotone convergence theorem, for every $B \in \mathscr{B}$

$$\int_B E(X_n|\mathscr{B})\, dP_{\mathscr{B}} \to \int_B Y\, dP_{\mathscr{B}} \qquad \int_B X_n\, dP \to \int_B X\, dP$$

as $n \to +\infty$. Since the left-hand sides are equal, so are the limits on the right-hand sides, and thus (7.14) and its uniqueness yields $Y = E(X|\mathscr{B})$ a.s., as desired.

(CE12)   Let Y be a $\mathscr{B}$-measurable random variable and assume that both X and XY have finite expectation. Then

$$E(XY|\mathscr{B}) = YE(X|\mathscr{B}) \text{ a.s.}$$

First, let $Y = I(A)$, $A \in \mathcal{B}$. Then, for every $B \in \mathcal{B}$,

$$\int_B E(XY|\mathcal{B}) \, dP_{\mathcal{B}} = \int_B XY \, dP = \int_{A \cap B} X \, dP = \int_{A \cap B} E(X|\mathcal{B}) \, dP_{\mathcal{B}}$$

$$= \int_B YE(X|\mathcal{B}) \, dP_{\mathcal{B}}$$

Thus, the desired equation follows for indicator variables. This, by the additivity property (CE4), ensures that the statement is valid for simple random variables Y as well. Next, let $Y \geq 0$ and $\mathcal{B}$-measurable. Then there are simple random variables $0 \leq Y_n \leq Y_{n+1}$ that are themselves $\mathcal{B}$-measurable and $\lim Y_n = Y$ as $n \to +\infty$. Since the property to be established has just been proved for the simple variables Y_n, (CE11) ensures that the property continuoues to hold after passage to the limit when applied separately to YX^+ and YX^- (see (CE5)). A simple application of the additivity property (CE4) now completes the proof. ∎

Example 1 Let $\Omega = [0, 1)$ and let $\mathcal{A}$ be the set of its Borel sets. Let $X = X(x) = x$, $x \in \Omega$. With Lebesgue measure as P, let us find $E(X|\mathcal{B}_n)$, where $\mathcal{B}_n$ is the smallest σ-field containing the intervals $B_{k,n} = [(k-1)/2^n$, $k/2^n)$, $1 \leq k \leq 2^n$, where $n \geq 1$ is a fixed integer.

Since $E(X|\mathcal{B}_n)$ is $\mathcal{B}_n$-measurable, it is a constant $c_{k,n}$ on $B_{k,n}$. Thus, by (7.14),

$$2^{-n}c_{k,n} = \int_{(k-1)/2^n}^{k/2^n} x \, dx = 2^{-2n-1}(2k-1) \quad \text{(a.s.)}$$

i.e., $E(X|\mathcal{B}_n) = (k - 1/2)/2^n$ if $x \in B_{k,n}$ (a.s.). Now, since $B_{k,n+1} \subset B_{j,n}$ and $B_{k-1,n+1} \cup B_{k,n+1} = B_{j,n}$ for $k = 2j$, $j \geq 1$, $\mathcal{B}_n \subset \mathcal{B}_{n+1}$. Therefore, $E(X|\mathcal{B}_n)$ is $\mathcal{B}_{n+1}$-measurable as well; thus, $E[E(X|\mathcal{B}_n)|\mathcal{B}_{n+1}] = E(X|\mathcal{B}_n)$ a.s. without further computation. However, additional computations are required to find $E[E(X|\mathcal{B}_{n+1})|\mathcal{B}_n]$ if the definition (7.14) is to be used. On the other hand, by (CE9), we once again obtain $E(X|\mathcal{B}_n)$ a.s. ∎

Note that, in the preceding example, the continuous variable X was "discretized" by $E(X|\mathcal{B}_n)$, whose values are simply the averages of X on the "smallest" elements $B_{k,n}$ of $\mathcal{B}_n$, which we call the *atoms* of $\mathcal{B}_n$. In general, a set B is an *atom* of a σ-field $\mathcal{B}$ if $P_{\mathcal{B}}(B) > 0$ and if we have either $P_{\mathcal{B}}(B_1) = 0$ or $P_{\mathcal{B}}(B_1) = P_{\mathcal{B}}(B)$ for every $B_1 \subset B$ such that $B_1 \in \mathcal{B}$. The averaging on atoms by the conditional expectation $E(X|\mathcal{B})$ just observed is generally true; the verification of this is left to the reader. A more important observation that we can make in the preceding example is that $|X - E(X|\mathcal{B}_n)| \leq 1/2^n$, and, thus, $\lim E(X|\mathscr{B}_n) = X$ a.s. as $n \to +\infty$. This, again, is a special case of a general limit theorem that we shall prove in the next section.

The next example is a demonstration of the use of some of the basic properties (CEj).

Example 2 Let X and Z be independent random variables (on the same probability space) and assume that X^2 and Z^2 have finite expectations. Then

$$E(3X^2 - 5XZ + 2Z^2 - 5|Z) = 3E(X^2|Z) - 5E(XZ|Z) + 2E(Z^2|Z) - 5$$
$$= 3E(X^2) - 5ZE(X) + 2Z^2 - 5 \text{ a.s.}$$

where we first used (CE4) and (CE3) and then (CE12) and (CE10). ∎

In the preceding example, as well as in all the examples—including the problems—of Section 7.1, $E(X|Z)$ was always found to be a function of Z. This is generally true. To be precise, let $(\Omega, \mathscr{A}, P)$ be the underlying probability space and let Z be an n-dimensional random vector. Let R^n be the n-dimensional Euclidean space and let $\mathscr{B}^n$ be the set of its Borel sets. Let $\sigma(Z) \subset \mathscr{A}$ and $\mathscr{B}_Z^n$ be the smallest σ-fields generated by Z in $\mathscr{A}$ and $\mathscr{B}^n$, respectively; i.e., B_Z^n consists of all sets $D \in \mathscr{B}^n$ such that $\{\omega : \omega \in \Omega, Z \in D\} = \{Z \in D\} \in \mathscr{A}$—these latter events form $\sigma(Z)$. As before, we denote the probability P restricted to $\sigma(Z)$ by P_Z and define

$$P_Z^{(n)}(D) = P_Z(Z \in D) \qquad D \in \mathscr{B}_Z^n$$

Theorem 3 If X is a random variable with finite expectation, then $E(X|\sigma(Z)) = E(X|Z)$ can be chosen as $h(Z)$, where $h(z)$ is a Borel-measurable function on R^n.

Proof. Let us define

$$u(A) = \int_A X \, dP \qquad A \in \mathscr{A}$$

and $u_n(D) = u(Z \in D)$ for $D \in \mathscr{B}_Z^n$. Then $u_n(D)$ is a σ-additive signed measure on $\mathscr{B}_Z^n$ and absolutely continuous with respect to $P_Z^{(n)}$. Hence, by the Radon–Nikodym theorem (Theorem 1), there is a Borel-measurable function $h(z)$ on R^n such that

$$\int_D h \, dP_Z^{(n)} = u_n(D) = u(Z \in D) = \int_{\{Z \in D\}} X \, dP$$

Thus, by (7.14),

$$\int_D h \, dP_Z^{(n)} = \int_{\{Z \in D\}} E(X|Z) \, dP_Z \tag{7.16}$$

Next, we show that

$$\int_D h \, dP_Z^{(n)} = \int_{\{Z \in D\}} h(Z) \, dP_Z \tag{7.17}$$

which, when combined with (7.16), yields the desired form for $E(X|Z)$ in view of the a.s. uniqueness of the Radon–Nikodym derivative (i.e., the a.s. uniqueness of $E(X|Z)$).

First, we prove (7.17) for $h(z) = 1$ if $z \in D_1$ and $h(z) = 0$ outside of D_1, where D_1 is an arbitrary member of $\mathscr{B}_Z^n$. Indeed,

$$\int_D h \, dP_Z^{(n)} = P_Z^{(n)}(D \cap D_1) = P_Z(Z \in D \cap D_1) = \int_{D \cap D_1} dP_Z = \int_D h(Z) \, dP_Z$$

The additivity of integrals thus yields that (7.17) is valid if $h(z)$ is a simple function on R^n that is measurable with respect to $\mathscr{B}_Z^n$. Next, we appeal to the monotone convergence theorem and conclude that (7.17) is valid for nonnegative functions $h(z)$ on R^n that are measurable with respect to $\mathscr{B}_Z^n$ and, thus, it is valid in general, since the decomposition of $h(z)$ into its positive and negative parts only requires the use of the additivity of integrals. The theorem is established. ∎

Example 3 We show that, for an arbitrary event A,

$$P(A) = \int_{R^n} P(A|Z = z) \, dF_Z(z)$$

where $F_Z(z)$ is the distribution function of the vector Z. Furthermore, if $F_Z(z)$ is absolutely continuous with respect to Lebesgue measure, then

$$P(A) = \int_{R^n} P(A|Z = z) f_Z(z) \, dz$$

where $f_Z(z)$ is a density of $F_Z(z)$.

We first appeal to (CE1) with $X = I(A)$. We get

$$P(A) = E[E(I(A)|Z)] = \int_\Omega P(A|Z) \, dP = \int_\Omega h(Z) \, dP$$

where the last equation uses Theorem 3. The desired formula now follows from Theorem 2.22, while the special case when a density exists is an application of Theorem 2. ∎

We shall conclude this section with an analysis of conditional probabilities. While $P(A|\mathscr{B})$ is a special case of $E(X|\mathscr{B})$, and thus all properties established in this section apply to $P(A|\mathcal{B})$, one additional problem arises in this connection. Let A_j, $j \geq 1$, be an infinite sequence of mutuallly exclusive events. Then, for a given σ-field $\mathcal{B}$, (CE4) and (CE11) ensure that

$$P\left(\bigcup_{j=1}^{+\infty} A_j \Big| \mathcal{B}\right) = \sum_{j=1}^{+\infty} P(A_j|\mathcal{B}) \text{ a.s.}$$

but the exceptional set depends on the sequence A_j, and thus the σ-additivity of $P(A|\mathcal{B})$ is not necessarily valid a.s. In other words, $P(A|\mathcal{B})$ is not guaranteed to be a probability, not even a.s. When it is a probability a.s., we speak of a regular conditional probability. That is, we say that a function $P(A, \omega)$, $\omega \in \Omega$, $A \in \mathcal{A}$, is a regular conditional probability, given $\mathcal{B}$, if (i) for every $\omega \in \Omega$, $P(A, \omega)$ is a probability on $\mathcal{A}$ and (ii) for every $A \in \mathcal{A}$, $P(A, \omega)$, as a random variable, is a variant of the conditional probability $P(A|\mathcal{B})$, i.e., $P(A, \omega) = P(A|\mathcal{B})$ a.s.

Theorem 4 Let $P(A, \omega)$ be a regular conditional probability, given $\mathcal{B}$, and let X be a random variable with finite expectation. Then

$$E(X|\mathcal{B}) = \int_\Omega X(\omega^*) \, dP(\omega^*, \omega) \text{ a.s.}$$

Proof. The proof, by now, must be routine. We first note that the theorem is valid if $X = I(A)$, $A \in \mathcal{A}$. That is, in such a case, the two sides above become

$$P(A|\mathcal{B}) = P(A, \omega) \text{ a.s.}$$

which is part (ii) of the definition of a regular probability. Now, by the additivity of integrals and by (CE4), the conclusion of the theorem extends from indicators to simple random variables. This ensures, via the monotone convergence theorem, which is valid for both integrals and conditional expectations (see (CE11)), that the theorem is true if $X \geq 0$. Thus, by additivity (again), the proof is concluded. ∎

   In principle, the same kind of problem arises in connection with conditional distribution functions. That is, (i) the random variable $P(Z \leq z|\mathcal{B})$ is not necessarily a distribution function in z a.s., or (ii) $P(Z \in D|\mathcal{B})$, $D \in \mathcal{B}_Z^n$ (the notation introduced prior to Theorem 3 applies), is not necessarily a measure a.s., because the respective properties are valid a.s. only; thus, the exceptional sets again depend on z or on D, respectively. Therefore, we again speak of regular conditional distributions and distribution functions. We say that a function $F(z, \omega)$, $\omega \in \Omega$, $z \in R^n$, is a *regular conditional distribution function* of Z, given $\mathcal{B}$, if (a) for every $\omega \in \Omega$, $F(z, \omega)$ is a (multivariate) distribution function in z and (b) for each $z \in R^n$, $F(z, \omega) = P(Z \le z | \mathscr{B})$ a.s. Similarly, a function $P_Z(D, \omega)$, $\omega \in \Omega$, $D \in \mathscr{B}_Z^n$, is a *regular conditional distribution* of Z, given $\mathscr{B}$, if (c) for each $\omega \in \Omega$, $P_Z(D, \omega)$ is a probability measure on $\mathscr{B}_Z^n$ and (d) for each $D \in \mathscr{B}_Z^n$, $P_Z(D, \omega) = P(Z \in D | \mathscr{B})$ a.s. Now, the next theorem shows that the regularity assumption for conditional distributions and distribution functions is not a real restriction, but rather a direction towards a proper choice among the variants of these conditional probabilities. (In this paragraph and below, multidimensional inequalities are meant component by component.)

Theorem 5 Given a σ-field $\mathscr{B}$, every random vector Z has a regular conditional distribution function $F(z, \omega)$ and a regular conditional distribution $P_Z(D, \omega)$. Furthermore, the relation

$$P_Z(D, \omega) = \int_D dF(z, \omega) \qquad D \in \mathscr{B}_Z^n$$

applies.

Proof. We say that an n-dimensional vector r is rational if each of its components is a rational number. Let us consider $F_1(r, \omega) = P(Z \le r | \mathscr{B})$, where r is rational. Each of the properties required for $F_1(r, \omega)$ to be an n-dimensional distribution function (see the discussion around (1.37)) holds a.s., which follows immediately from the properties (CEj), $j = 4, 6$, and 11. Since the rational vectors r form a denumerable set, the aggregate of the exceptional sets still has probability zero. That is, there is a set A such that $P(A) = 0$ and $F_1(r, \omega)$ satisfies all the properties required for a (multivariate) distribution function on A^c, where r is restricted to rational vectors (including the case when passing to limits). Now, let z be an arbitrary vector in R^n and define $F(z, \omega) = \lim F_1(r, \omega)$ as $r \to z$ decreasingly (thus, $r \ge z$) if $\omega \in A^c$. On the other hand, if $\omega \in A$, $F(z, \omega) = G(z)$, where $G(z)$ is an arbitrary n-variate distribution function (say, that of n independent exponential variables). We shall show that this $F(z, \omega)$ is a regular conditional distribution function of Z, given $\mathscr{B}$.

First we show that $F(z, \omega)$ is a distribution function for each $\omega \in \Omega$. Indeed, if $\omega \in A$, then we have chosen it so, If $\omega \in A^c$, then $F(z, \omega) \le F_1(r, \omega)$, $z \le r$; thus, since $F_1(r, \omega) \to 0$ as at least one component of r goes to $-\infty$. a similar property also applies to $F(z, \omega)$. Moreover, if $z_1 < z_2$ then $F(z_1, \omega) \le F(z_2, \omega)$, because we could choose rationals $z_1 \le r_1 < z_2 \le r_2$, where r_2 is so close to z_2 that the inequalities $F(z_2, \omega) \le F_1(r_2, \omega) < F(z_1, \omega) \le F_1(r_1, \omega)$ hold, which is a contradiction for rationals $r_1 < r_2$ on A^c. The right continuity of $F(z, \omega)$ follows in a similar manner. The monotonicity of $F(z, \omega)$ obviously implies that $F(z, \omega) \to 1$ as $z \to +\infty$ because $F_1(r, \omega)$ does so as

$r \to +\infty$. Finally, the expression in (1.37) is nonnegative because it is nonnegative when all variables involved are rationals; this condition is not disturbed by passage to limits.

Next, we have to show that $F(z, \omega)$ is a variant of $P(Z \leq z|\mathscr{B})$ a.s. for every z. But this is true by definition for $z = r$ rational. If z is arbitrary, we find that this relation continues to hold by taking a sequence r of rationals such that $r \geq z$ and $r \to z$ decreasingly. Then $F(r, \omega) = F_1(r, \omega) \to F(z, \omega)$ on A^c by definition again and $P(Z \leq r|\mathscr{B}) \to P(Z \leq z|\mathscr{B})$ by (CE11). Hence, $F(z, \omega)$ is a regular conditional distribution function of Z, given $\mathscr{B}$.

Let $P_Z(D, \omega)$ now be defined as in the theorem. Then $P_Z(D, \omega)$ is a measure on $\mathscr{B}_Z^n$ for each $\omega \in \Omega$. It is also a variant of $P(Z \in D|\mathscr{B})$ for each $D \in \mathscr{B}_Z^n$, which can be seen as follows. Let $\mathscr{D}$ be the collection of all D for which $P_Z(D, \omega)$ is a variant of $P(Z \in D|\mathscr{B})$ a.s. Then, by definition, the sets $(-\infty, z]$ (again, meant component by component) are in $\mathscr{D}$, which entails, by additivity (CE4), that all intervals, as well as unions of disjoint intervals, are in $\mathscr{D}$. That is, $\mathscr{D}$ contains the smallest field containing all intervals. Moreover, if $D_1 \subset D_2 \subset \cdots$ are in $\mathscr{D}$, then (CE11) implies that $\cup D_j \in \mathscr{D}$, yielding that $\mathscr{D}$ contains $\mathscr{B}^n$. The theorem is established.   ∎

Exercises

1. Let Z be an exponentially distributed random variable and let X be the integer part of Z. Determine $E(X|Z)$. (Recall Example 1 of Section 7.1).

2. Let $X_1, X_2, \ldots$ be independent random variables, each with zero expectation. Set $S_n = X_1 + X_2 + \cdots + X_n$, $n \geq 1$. Determine

$$E(S_{n+1}|S_n, S_{n-1}, \ldots, S_1) \qquad \text{and} \qquad E(S_{n+1}|S_n).$$

3. Let (X_1, X_2) be a bivariate normal vector with $E(X_j) = 0$ for $j = 1$ and 2. That is, the bivariate density function of (X_1, X_2) exists (not just a.s., but for all $(x_1 x_2) \in R^2$) and equals

$$f(x_1, x_2) = \frac{1}{2\pi\sigma_1\sigma_2(1 - \rho^2)^{1/2}} \exp[-Q(x_1, x_2)]$$

where

$$Q(x_1, x_2) = \frac{1}{2(1 - \rho^2)} \left(\frac{x_1^2}{\sigma_1^2} - 2\rho \frac{x_1 x_2}{\sigma_1 \sigma_2} + \frac{x_2^2}{\sigma_2^2} \right)$$

Show that $E(X_2|X_1) = \rho(\sigma_2/\sigma_1)X_1$ a.s.

4. Let $X_1, X_2, \ldots$ be i.i.d. exponential variables with $E(X_j) = 1/a$. Set $S_n = X_1 + X_2 + \cdots + X_n$. Find the conditional density (a.s.) of the vector $(S_1, S_2, \ldots, S_n)$, given $S_{n+1} = y$.

7.4 MARTINGALES

Let X denote the gain in a random game. If the game is fair in any reasonable sense, we must have $E(X) = 0$. Now, if the game is repeated in independent matches and X_j is X in the jth match, then $S_n = X_1 + X_2 + \cdots + X_n$ is the aggregate gain in n matches and $E(S_n) = 0$. However, at any time n, one can be in a winning position $S_n > 0$ (or a losing position $S_n < 0$), and one more reference to fairness allows the player to expect to keep such gains. That is, as the game progresses, and thus $S_1, S_2, \ldots, S_n$ become given values, one wants $E(S_{n+1}|S_n, S_{n-1}, \ldots, S_1) = S_n$ a.s. We demonstrated this property for independent repetitions of matches in Exercise 2 of the preceding section. Now, by this argument, the independence is not essential. The essential part of the preceding discussion is that $E(X_1) = 0$ and, for $j \geq 2$,

$$E(X_j|X_1, X_2, \ldots, X_{j-1}) = 0 \text{ a.s.} \tag{7.18}$$

which ensures both $E(X_j) = 0$ and

$$
\begin{aligned}
E(S_{n+1}|S_n, S_{n-1}, \ldots, S_1) &= E(X_{n+1} + S_n|S_n, S_{n-1}, \ldots, S_1) \\
&= E(X_{n+1}|S_n, S_{n-1}, \ldots, S_1) + S_n \\
&= E(X_{n+1}|X_n, X_{n-1}, \ldots, X_1) + S_n = S_n
\end{aligned}
$$

where each equation is valid a.s. Here, we used the basic properties of the conditional expectation established in the previous section and the fact that $\sigma(S_j, 1 \leq j \leq n) = \sigma(X_j, 1 \leq j \leq n)$. This property of S_n is called a martingale property; it can replace independence in several theorems we had established for sums in the previous chapters. One such extension of the results is as follows. First, note that the above-mentioned martingale property and (7.18) are equivalent. Since the implication of the martingale property by (7.18) has been shown, we have to establish that $E(S_{n+1}|S_n, S_{n-1}, \ldots, S_1) = S_n$ a.s. implies (7.18). Now,

$$
\begin{aligned}
E(X_j|X_1, X_2, \ldots, X_{j-1}) &= E(S_j - S_{j-1}|S_1, S_2, \ldots, S_{j-1}) \\
&= E(S_j|S_1, S_2, \ldots, S_{j-1}) - S_{j-1} \\
&= S_{j-1} - S_{j-1} = 0 \text{ a.s.}
\end{aligned}
$$

So, let us assume that (7.18) is valid. Then, for $k < j$,

$$E(X_k X_j|X_1, X_2, \ldots, X_{j-1}) = X_k E(X_j|X_1, X_2, \ldots, X_{j-1}) = 0 \text{ a.s.}$$

entailing $E(X_k X_j) = 0$ for all $j \neq k$. Therefore,

$$V(S_n) = V(X_1) + V(X_2) + \cdots + V(X_n)$$

from which, by Chebyshev's inequality, the weak law of large numbers follows immediately. Furthermore, if $A_{n-1} \in \sigma(X_1, X_2, \ldots, X_{n-1})$,

$$E(X_n I(A_{n-1})S_{n-1} | X_1, X_2, \ldots, X_{n-1})$$
$$= I(A_{n-1})S_{n-1} E(X_n | X_1, \ldots, X_{n-1}) = 0 \text{ a.s.}$$

and thus $E(X_n I(A_{n-1})S_{n-1}) = 0$. This is essentially (2.15), which is the only form in which independence is used in the proof of Kolmogorov's inequality. In other words, Kolmogorov's inequality, together with all of its direct consequences, remains valid if independence is replaced by (7.18). For example, we have the following result (see Section 2.3).

Theorem 6 If (7.18) holds then, for arbitrary $a > 0$,

$$P\left(\max_{1 \le k \le n} |S_k| \ge a \right) \le [V(X_1) + V(X_2) + \cdots + V(X_n)]/a^2$$

Furthermore, if (7.18) holds and $\sum V(X_j) < +\infty$, then $\sum X_j$ converges a.s.

The asymptotic normality of S_n, when properly normalized, can also be reobtained in a more general form along the lines we have seen in the case of independence. These possibilities of directly extending results on sums with independent terms to those whose summands satisfy (7.18) induced the initial study of sums with a martingale property, although the name martingale was introduced only at an advanced stage of the theory by Ville (1939). It was the work of J. L. Doob, culminating with his book of 1953, which made the theory of martingales a prominent branch of probability theory. Several of the theorems which we shall present in this section are due to Doob.

We now introduce the basic concepts of the theory of martingales. As always, we denote the underlying probability space by $(\Omega, \mathscr{A}, P)$. Let $\mathscr{A}_n \subset \mathscr{A}$, $n \in I$, where I is either a finite or an infinite (open, closed, or semiclosed) interval of the set $\{-\infty, \ldots, -1, 0, 1, 2, \ldots, +\infty\}$, be σ-fields. We say that $\{\mathscr{A}_n\}$ is an increasing sequence if $\mathscr{A}_n \subset \mathscr{A}_{n+1}$ and a decreasing one if $\mathscr{A}_n \supset \mathscr{A}_{n+1}$ for all permissible n. Let $\mathscr{A}_n$, $n \in I$, be an increasing sequence. Let S_n, $n \in I$, be a sequence of random variables satisfying (i) S_n is measurable with respect to $\mathscr{A}_n$, (ii) the expectation of S_n is finite, and (iii) $E(S_n | \mathscr{A}_m) = S_m$ a.s. for all $m < n$, where both m and n belong to I. Then the sequence S_n, $n \in I$, is called a *martingale* with respect to the sequence $\mathscr{A}_n$, $n \in I$, which, for brevity, we also write as "$\{S_n, \mathscr{A}_n, n \in I\}$ is a martingale." If conditions (i), (ii), and (iv) $E(S_n | \mathscr{A}_m) \ge S_m$ a.s. for all $m < n$ with $m, n \in I$ hold, then $\{S_n, \mathscr{A}_n, n \in I\}$ is called a *submartingale*. If conditions (i), (ii), and (v) $E(S_n | \mathscr{A}_m) \le S_m$ a.s. for all $m < n$ with $n, m \in I$ are satisfied, then the sequence $\{S_n, \mathscr{A}_n, n \in I\}$ is said to be a *supermartingale*. Finally, if the sequence $\mathscr{A}_n$ is decreasing and conditions (i), (ii), and (vi) $E(S_n|\mathscr{A}_m) = S_m$ a.s. for all $m > n$ with $m, n \in I$, are satisfied, we call $\{S_n, \mathscr{A}_n, n \in I\}$ a *reverse martingale* or *backward martingale*. Note that, while there is an obvious transformation of a martingale into a reverse martingale when I is finite, there is not always an obvious transformation of a limit theorem for martingales into one for reverse martingales.

As an introduction to this section, we have seen that if $X_j, j \geq 1$, are independent random variables with $E(X_j) = 0$, then the sequence $S_n = X_1 + X_2 + \cdots + X_n, n \geq 1$, forms a martingale with respect to the sequence $\mathscr{A}_n = \sigma(X_1, X_2, \ldots, X_n), n \geq 1$. The reader can easily verify that $\{S_n, \mathscr{A}_n, n \geq 1\}$ is a submartingale if $E(X_j) \geq 0$ (finite) and the X_j are independent.

To some extent, the concept of a martingale can unify the study of sums and products of independent random variables. Indeed, let $X_j \geq 0$ be independent random variables with $E(X_j) = 1, j \geq 1$. Set $S_n = X_1 X_2 \cdots X_n$. Then

$$E(S_n|S_{n-1}, S_{n-2}, \ldots, S_1) = \left(\prod_{j=1}^{n-1} X_j\right) E(X_n|S_{n-1}, \ldots, S_1)$$

$$= S_{n-1} E(X_n) = S_{n-1}$$

the last but one equation being due to the assumed independence. So, $\{S_n, \sigma(X_1, X_2, \ldots, X_n), n \geq 1\}$ is a martingale. We note that, just as in the case of sums, independence is not essential here. What we utilized above is the property

$$E(X_n|X_1, X_2, \ldots, X_{n-1}) = E(X_n|S_1, S_2, \ldots, S_{n-1}) = 1$$

for every $n \geq 2$. Note also that S_n can be written as the sum of the differences $S_j - S_{j-1}, 1 \leq j \leq n$ with $S_0 = 0$, yielding

$$S_n = \sum_{j=2}^n X_1 X_2 \cdots X_{j-1}(X_j - 1) + X_1$$

where the summands are no longer independent even if the X_j are (but, of course, (7.18) holds if we replace X_j there with $Y_j = S_j - S_{j-1}, j \geq 1, S_0 = 0$). Further examples are listed below.

Example 1 Let X be an arbitrary random variable with finite expectation. Let $\mathscr{A}_n$ be an increasing sequence of sub-σ-fields of $\mathscr{A}$. Then $\{E(X|\mathscr{A}_n), \mathscr{A}_n, n \in I\}$ is a martingale, where $I = \{1, 2, \ldots, +\infty\}$, $\mathscr{A}_{+\infty} = \mathscr{A}$; thus,

$$E(X|\mathscr{A}_{+\infty}) = E(X|\mathscr{A}) = X \text{ a.s.}$$

In particular, if $Y_1, Y_2, \ldots$ is an arbitrary sequence of random variables, then $\{E(X \mid Y_1, Y_2, \ldots, Y_n), \sigma(Y_1, Y_2, \ldots, Y_n), n \geq 1\}$ is a martingale. One can again include X as well as the last number of this martingale, which we label by $n = +\infty$.

The martingale property of the above sequences is easily obtained. Set $S_n = E(X \mid \mathscr{A}_n)$. Then S_n is $\mathscr{A}_n$-measurable, $E(S_n) = E(X)$ is finite by assumption, and, for $m < n$,

$$E(S_n \mid \mathscr{A}_m) = E[E(X \mid \mathscr{A}_n) \mid \mathscr{A}_m] = E(X \mid \mathscr{A}_m) = S_m \text{ a.s.}$$

where we used (CE9) of the previous section. The computation is also valid for $n = +\infty$. ∎

The example is a typical one for a martingale closed on the right. We say that a martingale is *closed on the right* if I is an interval closed on the right, i.e., if it has a last member.

Example 2 Let X be a random variable with a finite expectation. Let $Y_1, Y_2, \ldots$ be an infinite sequence of random variables and let $\mathscr{A}_n = \sigma(Y_n, Y_{n+1}, \ldots)$. Then, with $I = \{\ldots, -2, -1\}$, $\{E(X \mid \mathscr{A}_n), \mathscr{A}_n, n \in I\}$ is again a martingale.

The verification is the same as in Example 1. ∎

Example 3 Let $\mathscr{A}_n$ be an increasing sequence of sub-σ-fields of $\mathscr{A}$ and let $X_1, X_2, \ldots$ be a sequence of random variables with finite expectations such that, for $j \geq n + 1$, $E(X_j \mid \mathscr{A}_n) \geq 0$ and $S_n = X_1 + X_2 + \cdots + X_n$ is $\mathscr{A}_n$-measurable. Then $\{S_n, A_n, n \geq 1\}$ is a submartingale. Moreover, if T is a positive integer-valued random variable such that, for all n, the event $\{T > n\} \in \mathscr{A}_n$ and if $T_n = \min(T, n)$, then $\{S_{T_n}, \mathscr{A}_n, n \geq 1\}$ is a submartingale.

The first part of this example can be verified as in the case of independence. Concerning S_{T_n}, note that, if $n > 1$,

$$E(S_{T_n} \mid \mathscr{A}_{n-1}) = S_{T_{n-1}} + E(X_n I(T \geq n) \mid \mathscr{A}_{n-1})$$
$$S_{T_{n-1}} + I(T \geq n) E(X_n \mid \mathscr{A}_{n-1}) \geq S_{T_{n-1}} \text{ a.s.}$$

where (CE12) of the previous section is used in the last equation. The result in Exercise 1 now ensures that the submartingale property holds. ∎

Example 4 Let X_j, $j \geq 1$, be a sequence of random variables with $E(X_j \mid \mathscr{A}_n) = 0$ a.s. for $j > n$, where $\mathscr{A}_n$, $n \geq 1$, is an increasing sequence of sub-σ-fields of $\mathscr{A}$. Assume that the partial sum S_n of the X_j is $\mathscr{A}_n$-measurable. Let $c > 0$ be a fixed number and define $T_c = \min\{j : S_j > c\}$ if such a j exists and $T_c = +\infty$ otherwise. Then, with $T(n) = \min(T_c, n)$, $\{S_{T(n)}, \mathscr{A}_n, n \geq 1\}$ is a martingale.

The verification is similar to that in the preceding example. Let $n > 1$; then,

$$
\begin{aligned}
E(S_{T(n)} | \mathscr{A}_{n-1}) &= S_{T(n-1)} + E(X_n I[T(n) \geq n] | \mathscr{A}_{n-1}) \\
&= S_{T(n-1)} + I[T(n) \geq n] E(X_n | \mathscr{A}_{n-1}) = S_{T(n-1)} \text{ a.s.}
\end{aligned}
$$

and Exercise 1 completes the argument. ∎

The variables T and $T(n)$ of the last two examples are particular cases of *stopping variables* which frequently appear in renewal theory and sequential analysis (of statistics). A measurable function T (on Ω) that takes the values $1, 2, \ldots, +\infty$, is called a *stopping time* relative to an increasing sequence of sub-σ-fields $\mathscr{A}_n \subset \mathscr{A}$ if the event $\{T = n\} \in \mathscr{A}_n$. If $P(T = +\infty) = 0$, then T is called a *finite* stopping time or, alternatively, a *stopping variable* or *stopping rule*. The variables T, T_n, and $T(n)$ of Examples 3 and 4 are finite stopping times. Clearly, for finite stopping times T,

$$
\{T \geq n\} = \bigcap_{j=1}^{n-1} \{T = j\}^c \in \mathscr{A}_{n-1}
$$

Conversely, if $\{T \geq n\} \in \mathscr{A}_{n-1}$, then $\{T = n\} = \{T \geq n\} \cap \{T \geq n + 1\}^c \in \mathscr{A}_n$. In Exercise 3, Example 4 is extended to arbitrary finite stopping times.

Next, we establish relations between martingales and submartingales. In order to demonstrate that if $\{S_n, \mathscr{A}_n, n \in I\}$ is a martingale and $E(|S_n|^p) < +\infty$ for some $p \geq 1$, then $\{|S_n|^p, \mathscr{A}_n, n \in I\}$ is a submartingale, we prove the following general inequality.

Theorem 7 (Jensen's Inequality) Let $g(x)$ be a convex function on the real line R. Let X be a random variable for which both $E(X)$ and $E(g(X)]$ are finite. Let $\mathscr{B}$ be a σ-field of members of $\mathscr{A}$. Then

$$
g[E(X | \mathscr{B})] \leq E[g(X) | \mathscr{B}] \text{ a.s.}
$$

Proof. By definition, the slope $[g(x) - g(a)]/(x - a)$ of the secant of the graph of $g(y)$ between the points $(x, g(x))$ and $(a, g(a))$, $x < a$, is an increasing function of x(for fixed a); thus, its limit $g'_1(a)$ exists as $x \to a-$. Furthermore, $g'_1(a)$ is finite and nondecreasing on R. Moreover, the secants through $(a, g(a))$ are above the (one-sided) tangent at a, and thus

$$
g(x) \geq g(a) + (x - a)g'_1(a) \qquad \text{for all } x \text{ and } a
$$

This, upon choosing $x = X$ and $a = E(X|\mathscr{B})$, yields

$$g(X) \geq g[E(X|\mathscr{B})] + [X - E(X|\mathscr{B})]g_1'[E(X|\mathscr{B})] \text{ a.s.} \tag{7.19}$$

In order to make sure that we can integrate every term in this inequality, we truncate the variables by multiplying every term by the indicator variable $I(K) = I[|E(X|\mathscr{B})| \leq K]$, with $K > 0$ fixed. Since $I(K)$ is measurable with respect to $\mathscr{B}$, so is $Y = I(K)g_1'(E(X|\mathscr{B}))$. In addition, Y is bounded, and thus integrable. Hence, by (CE12),

$$E\{[X - E(X|\mathscr{B})]Y|\mathscr{B}\} = YE[X - E(X|\mathscr{B})|\mathscr{B}] = 0 \text{ a.s.}$$

Thus, (7.19) yields

$$E[I(K)g(X)|\mathscr{B}] \geq E\{I(K)g[E(X|\mathscr{B})]|\mathscr{B}\} \text{ a.s.}$$

By once more appealing to (CE12), $I(K)$ can be removed from behind the (conditional) expectation sign on both sides and, since $I(K) \to 1$ a.s. as $K \to +\infty$, the desired inequality follows. ∎

Corollary If $\{S_n, \mathscr{A}_n, n \in I\}$ is a martingale and $g(x)$ is any convex function on the real line such that $g(S_n)$ has finite expectation, then $\{g(S_n), \mathscr{A}_n, n \in I\}$ is a submartingale. In particular, if $E(|S_n|^p) < +\infty$ for some $p \geq 1$, then, for every martingale $\{S_n, \mathscr{A}_n, n \in I\}$, $\{|S_n|^p, \mathscr{A}_n, n \in I\}$ is a submartingale.

Proof. By Jensen's inequality, for $m < n$ such that $m, n \in I$,

$$E(g(S_n)|\mathscr{A}_m) \geq g(E(S_n|\mathscr{A}_m)) = g(S_m) \text{ a.s.}$$

The particular case follows by observing that the function $g(x) = |x|^p$, $p \geq 1$, is convex on the whole real line. ∎

It is clear from the definitions that every martingale is a submartingale as well. A close relation in the opposite direction is the content of the next theorem.

Theorem 8 (Doob's Decomposition) Let $\{S_n, \mathscr{A}_n, n \in I\}$ be a submartingale. Then there is a martingale $\{M_n, \mathscr{A}_n, n \in I\}$ and a nondecreasing sequence Y_n of random variables such that

$$S_n = M_n + Y_n \text{ a.s.}$$

provided that either $N = \inf\{n: n \in I\}$ is finite or, if $N = -\infty$, $\lim E(S_M) > -\infty$ as $M \to -\infty$.

Proof. First, let N be finite, i.e., $N > -\infty$, and assume that $N \in I$ (otherwise replace N by $N + 1$ below). Define $Y_N = 0$ and, for $n > N$, $n \in I$, let

$$Y_n = \sum_{j=N+1}^{n} [E(S_j|\mathscr{A}_{j-1}) - S_{j-1}] \text{ a.s.}$$

Then Y_n is $\mathscr{A}_{n-1}$ measurable and, by the submartingale property, each term in the sum defining Y_n is nonnegative a.s. Hence, $Y_n \leq Y_{n+1}$ a.s. Next, let $M_N = S_N$ and $M_n = S_n - Y_n$ for $n > N$, $n \in I$. Then, for $N \leq m < n$, $n \in I$,

$$\begin{aligned}
E(M_n|\mathscr{A}_m) &= E(S_n|\mathscr{A}_m) - E(Y_n|\mathscr{A}_m) \\
&= E(S_n|\mathscr{A}_m) - E(Y_m|\mathscr{A}_m) \\
&\quad - \sum_{j=m+1}^{n} E[E(S_j|\mathscr{A}_{j-1}) - S_{j-1}|\mathscr{A}_m]
\end{aligned}$$

Now, by (CE9), for $m < j \leq n$,

$$E[E(S_j|\mathscr{A}_{j-1}) - S_{j-1}|\mathscr{A}_m] = E(S_j|\mathscr{A}_m) - E(S_{j-1}|\mathscr{A}_m)$$

Thus, the last sum on the right-hand side above equals $E(S_n|\mathscr{A}_m) - E(S_m|\mathscr{A}_m)$. That is, since Y_m is $\mathscr{A}_{m-1} \subset \mathscr{A}_m$ measurable,

$$E(M_n|\mathscr{A}_m) = E(S_m|\mathscr{A}_m) - E(Y_m|\mathscr{A}_m) = S_m - Y_m = M_m \text{ a.s.}$$

which establishes the martingale property for $\{M_n, \mathscr{A}_n, n \in I\}$ under the restriction $N > -\infty$. When $N = -\infty$, we proceed as above except that now we have to show that the infinite sum defining Y_n is convergent a.s. For this we write, for $n \in I$ finite,

$$Y_n = \lim_{M=-\infty} \sum_{j=M}^{n} [E(S_j|\mathscr{A}_{j-1}) - S_{j-1}] = \lim_{M=-\infty} U_M(n)$$

say. Since the summands are nonnegative, $U_M(n) \leq U_{M-1}(n)$ and $E(U_M(n)) = E(S_n) - E(S_{M-1})$. Furthermore, by the submartingale property, $E(S_{M-1}) \geq E(S_{M-2})$; thus, both $\lim U_M(n)$ (a.s.) and $\lim E(U_M(n)) = E(S_n) - \lim E(S_{M-1})$ exist as $M \to -\infty$. Since the latter is assumed to be finite, it follows, by the nonotone convergence theorem (Theorem 2.14), that $Y_n = \lim U_M(n)$, $M \to -\infty$, is integrable, implying that it is a.s. finite. The theorem is established. ∎

The following upper bound on the expected number of times that a submartingale moves from a value $\leq a$ to one $\geq b$, where $a < b$ are finite, is instrumental in proving the basic theorem on the convergence of submartingales. Both the inequality and the convergence theorem are due to Doob.

Theorem 9   Let $\{S_n, \mathscr{A}_n, 1 \leq n \leq r\}$ be a submartingale. Let $[a, b]$ be a finite closed interval on the real line and let T be the number of times that the sequence S_n crosses from a value $\leq a$ to one $\geq b$. Then

$$(b - a)E(T) \leq E[(S_r - a)^+]$$

Proof.   Let us define the integer-valued random variables $0 = t_0 < t_1 < t_2 < \cdots < t_d \leq r$ sequentially as follows. Assume that t_{2k} has been defined. Then let t_{2k+1} be the smallest n with $t_{2k} < n \leq r$ such that $S_n \leq a$ and t_{2k+2} be the smallest n with $t_{2k+1} < n \leq r$ such that $S_n \geq b$, $k \geq 0$. Clearly, the last index d with $t_d \leq r$ is a random variable and d equals either $2T$ or $2T + 1$. Next, we introduce the indicator variables I_j, $1 \leq j \leq r$, by letting $I_j = 1$ if the largest t_u among the integers $1, 2, \ldots, j-1$ has an even subscript u (and $I_j = 0$ otherwise).

First, consider those sample points for which $T > 0$. If $d = 2T + 1$, then

$$\sum_{j=2}^{r} I_j(S_j - S_{j-1}) = \sum_{k=1}^{T} (S_{t_{2k+1}} - S_{t_{2k}}) \leq T(a - b)$$

while, for $d = 2T$,

$$\sum_{j=2}^{r} I_j(S_j - S_{j-1}) = \sum_{k=1}^{T-1} (S_{t_{2k+1}} - S_{t_{2k}}) + (S_r - S_{t_{2T}})$$

$$\leq (T - 1)(a - b) + (S_r - b)$$

$$= T(a - b) + (S_r - a) \leq T(a - b) + (S_r - a)^+$$

and both inequalities trivially hold if $T = 0$. By taking expectations, the theorem will be established if we show that

$$\sum_{j=2}^{r} E[I_j(S_j - S_{j-1})] \geq 0 \qquad (7.20)$$

Now, by definition, the event $\{I_j = 1\} \in \sigma(S_1, S_2, \ldots, S_{j-1})$, i.e., I_j is $\mathscr{A}_{j-1}$ measurable. Hence, by (CE1), (CE12), and the submartingale property,

$$E[I_j(S_j - S_{j-1})] = E\{E[I_j(S_j - S_{j-1})|\mathscr{A}_{j-1}]\}$$

$$= E\{I_j[E(S_j|\mathscr{A}_{j-1}) - S_{j-1}]\}$$

$$\geq E[I_j(S_{j-1} - S_{j-1})] = 0$$

This obviously implies (7.20), which completes the proof of the theorem. ∎

Theorem 10   Let $\{S_n, \mathscr{A}_n, n \geq 1\}$ be a submartingale. If $\sup E(|S_n|) < +\infty$, then S_n converges a.s. to a random variable X with $E(|X|) \leq \sup E(|S_n|)$.

Proof. Let T_r be the random variable T of the previous theorem for the submartingale $\{S_n, \mathscr{A}_n, 1 \le n \le r\}$. Then $T_r \le T_{r+1}$, and thus, as $r \to +\infty$, $T^* = \lim T_r$ exists (possibly infinite). Now, by Theorem 9, for all finite $a < b$,

$$(b - a)E(T_r) \le E[(S_r - a)^+] \le E(|S_r|) + |a| \le |a| + \sup_{r \ge 1} E(|S_r|)$$

and thus, by letting $r \to +\infty$,

$$(b - a)E(T^*) \le |a| + \sup E(|S_r|) < +\infty$$

Therefore, T^* is finite a.s. for all finite $a < b$, implying

$$P(\liminf S_n < a < b < \limsup S_n) = 0 \qquad (n \to +\infty)$$

If we sum these probabilities for all rational $a < b$, we obtain

$$P(\liminf S_n < \limsup S_n) = 0 \qquad (n \to +\infty)$$

i.e., $\lim S_n = X$ exists a.s., which may be infinite. However, by Fatou's lemma (Theorem 2.15), $E(|X|) \le \sup E(|S_n|) < +\infty$; thus, X is finite a.s. The proof is completed. ∎

From the details of the preceding proof one can see that the following modified version of Theorem 10 is valid.

Theorem 10a Let $\{S_n, \mathscr{A}_n, n \ge 1\}$ be a submartingale. Assume $\sup E(S_n^+) < +\infty$. Then S_n converges to a measurable function $Y < +\infty$ a.s.

Proof. If in the proof of Theorem 10 we use the estimate

$$E((S_r - a)^+) \le E(S_r^+) + |a|$$

with the rest of the proof unchanged, we obtain that $S_n \to Y$ a.s., and $Y < +\infty$ a.s. follows from Fatou's lemma when applied with S_n^+, $n \ge 1$. ∎

We now deduce from Theorem 10a the following statement.

Theorem 10b Let $\{S_n, \mathscr{A}_n, n \ge 1\}$ be a martingale and denote $S_j - S_{j-1} = X_j$, $j \ge 2$, with $S_1 = X_1$. Assume $E(\sup X_j) < +\infty$ and $\sup S_n < +\infty$ a.s. Then S_n converges a.s. to a Y with $Y < +\infty$ a.s.

Proof. Let $M > 0$ be a fixed number. Define $T(M)$ as the first j such that $S_j > M$ if such a j exists and $T(M) = +\infty$ otherwise. Then $T(M)$ is a stopping variable and, by Example 4, $\{S_{\min(T(M),n)}, \mathscr{A}_n, n \geq 1\}$ is a martingale. Now, since

$$S_{\min(T(M),n)} = S_n \quad \text{if} \quad T(M) > n \quad \text{(and then } S_n \leq M)$$
$$= S_{T(M)-1} + X_{T(M)} \leq M + \sup X_j \quad \text{if} \quad T(M) \leq n$$

we have

$$\sup E(S^+_{\min(T(M),n)}) \leq E(\sup S^+_{\min(T(M),n)}) \leq M + E(\sup X_j) < +\infty$$

Consequently, Theorem 10a is applicable to $S_{\min(T(M),n)}$, yielding that, a.s., $S_{\min(T(M),n)} \to Y < +\infty$. However,

$$\{\sup S_n < +\infty\} = \bigcup_{M=1}^{+\infty} \{T(M) = +\infty\}$$

and, on $T(M) = +\infty$, $S_{\min(T(M),n)} = S_n$, implying that, by the assumption of the theorem, the above convergence is about S_n. The theorem is established. ∎

Note that if the differences $X_j \leq K < +\infty$ for all j, then the sole condition in Theorem 10b is $\sup S_n < +\infty$ a.s., that is, if martingale differences are uniformly bounded above then the underlying martingale $S_n \to Y < +\infty$ a.s. See also the forthcoming Example 6.

Remark At the beginning of the present section we have seen that $E(X_j) = 0$, $j \geq 2$, in the case of a martingale, entailing $E(S_n) = E(X_1)$ for all $n \geq 1$. Hence, if $E(X_1) = 0$ by assumption, then $E(S_n^+) = E(S_n^-)$ and $E(|S_n|) = 2E(S_n^+)$. Therefore, Y is a random variable in Theorem 10a under the slight additional assumption $E(X_1) = 0$. This remark applies to Theorem 10b as well since, although Theorem 10a is utilized in the proof of Theorem 10b with the martingale $S_{\min(T(M),n)}$, $S_{\min(T(M),1)} = S_1$ whatever $M > 0$.

Theorem 10 provides a tool for strengthing the second part of Theorem 6.

Theorem 10c Let $\{S_n, \mathscr{A}_n, n \geq 1\}$ be a martingale and set $X_j = S_j - S_{j-1}$, $j \geq 2$, $X_1 = S_1$. Assume that X_j^2 is integrable, $j \geq 1$, and

$$\sum_{j=2}^{+\infty} E(X_j^2|\mathscr{A}_{j-1}) < +\infty \quad \text{a.s.}$$

Then S_n converges a.s. to a random variable.

Proof. Let $c > 0$ be a fixed number and define $T(c)$ as the first n such that

$$E(X_1^2) + \sum_{j=2}^{n+1} E(X_j^2 | \mathscr{A}_{j-1}) > c$$

and $T(c) = +\infty$ otherwise. Then, just as in the previous proof, $T(c)$ is a stopping variable and $S_{\min(T(c), n)}$ is a martingale. Let $I(T(c) \geq j)$ be the indicator variable of the event $T(c) \geq j$. Then

$$S_{\min(T(c), n)} = \sum_{j=1}^{n} I(T(c) \geq j)X_j$$

and since, for $k < j$,

$$E(I(T(c) \geq k)I(T(c) \geq j)X_k X_j | \mathscr{A}_{j-1})$$
$$= I(T(c) \geq k)I(T(c) \geq j)X_k E(X_j | \mathscr{A}_{j-1})$$

which equals zero, we have

$$E(S_{\min(T(c), n)}^2) = E\left(\sum_{j=1}^{n} I(T(c) \geq j)X_j^2 \right)$$

If, in the jth term, we condition $\mathscr{A}_{j-1}$ and take expectations, the jth term becomes $E[I(T(c) \geq j)E(X_j^2 | \mathscr{A}_{j-1})]$, which, when summed over j, yields

$$E(S_{\min(T(c), n)}^2) = E\left[\sum_{j=1}^{\min(T(c), n)} E(X_j^2 | \mathscr{A}_{j-1}) \right] \leq c$$

where the last inequality is due to the definition of $T(c)$. Hence, by the inequality $|a| \leq a^2 + 1$,

$$E(|S_{\min(T(c), n)}|) \leq E(S_{\min(T(c), n)}^2) + 1 \leq c + 1$$

that is,

$$\sup E(|S_{\min(T(c), n)}|) \leq c + 1$$

We thus have from Theorem 10 that $S_{\min(T(c), n)}$ converges a.s. to a random variable Y with $E(|Y|) \leq \sup E(|S_{\min(T(c), n)}|) \leq c + 1$. But, by the a.s. convergence of the sum involved in the definition of $T(c)$, $T(c) = +\infty$ for some c and $S_n = S_{\min(T(c), n)}$ whenever $T(c) = +\infty$, we established that S_n converges a.s. to a random variable. ∎

In the following examples we apply Theorem 10, the results of which go back to Lévy and Kolmogorov.

Example 5 Let X be a random variable with finite expectation and Z_1, $Z_2, \ldots$ be a sequence of random variables. Then $E(X|Z_1, Z_2, \ldots, Z_n)$ converges a.s. to a random variable Y as $n \to +\infty$.

Recall Example 1. The sequence $S_n = E(X|Z_1, Z_2, \ldots, Z_n)$ is a martingale (with respect to $\sigma(Z_1, Z_2, \ldots, Z_n), n \geq 1$). Moreover, $E(|S_n|) \leq E(|X|) < +\infty$ (combine the Corollary to Theorem 7 with the full extent of Example 1), i.e., $\sup E(|S_n|) < +\infty$. Hence, Theorem 10 is applicable. ∎

In the special case of Example 1 of Section 7.3, we made the observation that S_n not only converges a.s., but also converges to X itself. This is always true if X is measurable with respect to $\sigma(Z_1, Z_2, \ldots)$ (which is satisfied in the quoted example). In order to see this, let $B_n \in \sigma(Z_1, Z_2, \ldots, Z_n)$ be an arbitrary set. Then, by the definition of conditional expectations and the martingale property,

$$\int_{B_n} X \, dP = \int_{B_n} S_n \, dP = \int_{B_n} S_{n+m} \, dP \qquad \text{all } m \geq 1$$

By letting $m \to +\infty$ and interchanging limit and integration (this is justified by the results of the next section), we get

$$\int_{B_n} X \, dP = \int_{B_n} Y \, dP \qquad \text{for all} \qquad B_n \in \sigma(Z_1, Z_2, \ldots, Z_n), \qquad n \geq 1$$

The Radon–Nikodym theorem (Theorem 1) entails $X = Y$ a.s. if X is measurable with respect to $\sigma(Z_1, Z_2, \ldots)$ (because Y is).

Example 6 Let $\{S_n, \mathscr{A}_n, n \geq 1\}$ be a martingale with a.s. uniformly bounded differences $X_1 = S_1$, $X_j = S_j - S_{j-1}, j \geq 2$. Then $\limsup S_n = +\infty$ a.s. and $\liminf S_n = -\infty$ a.s. as $n \to +\infty$ on the set on which $\sum X_j$ diverges.

We go back to Example 4, where we demonstrated that, with $T(n) = \min(T_c, n)$, where $c > 0$ is a fixed number and $T_c = \inf\{n: S_n > c\}$ if such an n exists and $T_c = +\infty$ otherwise, $\{S_{T(n)}, \mathscr{A}_n, n \geq 1\}$ is a martingale also. Now, since

$$S^+_{T(n)} \leq S^+_{T(n)-1} + X^+_{T(n)} \leq c + K$$

where K is the common bound on the X_j and $S_{T(n)-1} = 0$ if $T(n) = 1$, we have that $\sup E(|S_{T(n)}|) < +\infty$. (Note that, by the martingale property, $E(S_{T(n)}) = E(S_{T(1)})$; thus,

$$E(|S_{T(n)}|) \leq E(|X_1|) + 2E(S^+_{T(n)}) \leq 2c + 3K$$

for all $n \geq 1$). Hence, by Theorem 10, $S_{T(n)}$ converges a.s. to a finite limit. Since, on $\{T_c = +\infty\} = \{\sup S_n \leq c\}$, $S_n = S_{T(n)}$, $\lim S_n$ exists and is finite a.s. on this same set. But $c > 0$ is arbitrary, so we have that $\lim S_n$ exists and is finite a.s. on the set $\{\sup S_n < +\infty\} = \{\lim \sup S_n < +\infty\}$. On the other hand, if we argue with the martingale $\{-S_n, \mathscr{A}_n, n \geq 1\}$, we get that $\lim S_n$ exists and is finite a.s. on the set $\{\lim \inf S_n > -\infty\}$. Therefore, on the set $\{S_n \text{ diverges}\}$, $\lim \sup S_n = +\infty$ a.s. and $\lim \inf S_n = -\infty$ a.s. ∎

Even if the differences X_j are independent, Example 6 adds new information to the peculiar behavior of sums of independent random variables. We know from the three series theorem (Theorem 4.2) that the partial sums S_n of independent and uniformly bounded random variables X_j with $E(X_j) = 0$ either converge a.s. or diverge a.s. and that the latter occurs if and only if $\sum V(X_j) = +\infty$. This last condition thus entails, by Example 6, that $\lim \sup S_n = +\infty$ a.s. and $\lim \inf S_n = -\infty$ a.s.

Example 7 (Extended Borel–Cantelli Lemma) Let $A_1, A_2, \ldots$ be a sequence of events and $\mathscr{A}_n$ be the smallest σ-field containing A_j, $1 \leq j \leq n$. Then, a.s., the number v of the A_n, $n \geq 1$, that occur is finite or infinite depending on whether $\sum P(A_{n+1}|\mathscr{A}_n)$ converges or diverges.

Let $X_j = I(A_j) - P(A_j|\mathscr{A}_{j-1})$, $j \geq 1$, where $\mathscr{A}_0$ is the trivial σ-field (consisting of Ω and $\varnothing$ only) and let $S_n = X_1 + X_2 + \cdots + X_n$. Then $\{S_n, \mathscr{A}_n, n \geq 1\}$ is a martingale (since (7.18) holds) with uniformly bounded differences X_j, $j \geq 1$. Thus, the conclusion of Example 6 applies, i.e., if S_n diverges, then, concurrently, $\lim \sup S_n = +\infty$ a.s. and $\lim \inf S_n = -\infty$ a.s. But this is equivalent to

$$\sum_{n=1}^{+\infty} I(A_n) = +\infty \quad \text{and} \quad \sum_{n=1}^{+\infty} P(A_n|\mathscr{A}_{n-1}) = +\infty$$

In other words, it cannot happen with positive probability that one of the above sums is finite while the other is infinite. This entails the desired result, because the summands in both sums are nonnegative and, thus, both are either finite or infinite a.s. ∎

In the following example we return to the Radon–Nikodym derivative, and bring it into a more familiar form for monotonic functions.

Example 8 Let $F(x)$ be continuous and nondecreasing for $a \leq x \leq b$, a, b finite. Divide the interval $[a, b]$ by the points $a = x_{0,n} < x_{1,n} < \cdots < x_{N,n} = b$, which we shall refer to as division D_n. Let D_{n+1} be a subdivision of D_n, that is, every point $x_{j,n}$ is a point $x_{t,n+1}$ in D_{n+1}. Let $\mathscr{A}_n$ be the smallest (sigma) field containing the intervals $(x_{j,n}, x_{j+1,n}]$, $0 \le j \le N - 1$, $n \ge 1$. Evidently, $\mathscr{A}_n \subset \mathscr{A}_{n+1}$ by the relation of D_n and D_{n+1}. Define

$$f_n(x) = \frac{F(x_{j,n}) - F(x_{j-1,n})}{x_{j,n} - x_{j-1,n}} \quad \text{if} \quad x_{j-1,n} < x \le x_{j,n}, \qquad 1 \le j \le N$$

Then $f_n(x)$ is $\mathscr{A}_n$—measurable and, with probability proportional to Lebesgue measure (we drop the proportionality constant and write every integral as Lebesgue integral with dx for measure),

$$E(f_n | \mathscr{A}_m) = f_m \qquad \text{a.s.} \qquad \text{for } m < n$$

i.e.

$$\int_B f_m(x)\, dx = \int_B f_n(x)\, dx \qquad \text{for } B \in \mathscr{A}_m \subset \mathscr{A}_n, m < n$$

Since $f_n \ge 0$ and $E(f_n) = F(b) - F(a)$ for all n, Theorem 10 is applicable, yielding that there is a function $f(x)$, $a \le x \le b$, Lebesgue measurable and finite a.s., and, as $\max(x_{j,n} - x_{j-1,n}: 1 \le j \le N) \to 0$ with $n \to +\infty$, $f_n(x) \to f(x)$ a.s. Note that if $a \le x \le b$ is such that there are n and j with $x_{j,n} = x$, then for every $s \ge n$, $x = x_{t,s}$ as well with some t, and

$$F(x) - F(a) = \int_a^x f_s(u)\, du \qquad \text{all } s \ge n$$

One easily obtains from this that, by passing to the limit $f(x)$, the above relation remains to hold for all x specified before. Since such x-points are evidently dense in (a, b) and since $F(x)$ is continuous by assumption, we proved that, for all x,

$$F(x) - F(a) = \int_a^x f(u)\, du \qquad a \le x \le b$$

We call $f(x)$ the derivative of $F(x)$, which evidently extends to the whole real line if $F(x)$ is nondecreasing and continuous there (place disjoint finite intervals $(a_r, b_r]$ on the real line), and $f(x)$ a.s. equals the Radon–Nikodym derivative of $F(x)$ with respect to Lebesgue measure. (Let us remark that the possibility of interchanging integration and passage to the limit with s above is proved in a general framework in the forthcoming Theorem 13 of the next section.) ∎

Exercises

1. Let $\{S_n, -\infty < n < +\infty\}$, be a sequence of random variables and $\mathscr{A}_n$ an increasing sequence of sub-σ-fields of $\mathscr{A}$. Assume that $E(S_n)$ is finite and that S_n is $\mathscr{A}_n$-measurable. Show that (i) if $E(S_{n+1}|\mathscr{A}_n) = S_n$ a.s. for all n, then $\{S_n, \mathscr{A}_n, \text{all } n\}$ is a martingale and (ii) if $E(S_{n+1}|\mathscr{A}_n) \geq S_n$ a.s., then $\{S_n, \mathscr{A}_n, \text{all } n\}$ is a submartingale.

2. Show that if $\{S_n, \mathscr{A}, n \in I\}$ is a submartingale and if $\mathscr{B}_1 \subset \mathscr{B}_2 \subset \cdots$ is another sequence of σ-fields such that S_n is $\mathscr{B}_n$ measurable and $\mathscr{B}_n \subset \mathscr{A}_n$, then $\{S_n, \mathscr{B}_n, n \in I\}$ is a submartingale.

3. Let $\{S_n, \mathscr{A}_n, n \geq 1\}$ be a martingale and T be a finite stopping time. Show that $\{S_{T(n)}, \mathscr{A}_n, n \geq 1\}$ is a martingale, where $T(n) = \min(T, n)$.

4. In the proof of Theorem 8 it is remarked that the monotonic component Y_n of a submartingale $\{S_n, \mathscr{A}_n, n \in I\}$ can always be chosen to be $\mathscr{A}_{n-1}$ measurable. Now, let $\{R_n, \mathscr{A}_n, n \geq 1\}$ be a martingale and let $S_n = M_n + Y_n$ be an a.s. Doob decomposition of the submartingale $\{S_n = R_n^2, \mathscr{A}_n, n \geq 1\}$, where Y_n is $\mathscr{A}_{n-1}$ measurable. Show that

$$Y_n = \sum_{i=1}^{n} E(X_i^2|\mathscr{A}_{i-1}) \qquad \text{(conditional variance)}$$

a.s., where $X_1 = R_1$, $X_j = R_j - R_{j-1}, j \geq 2$, and $\mathscr{A}_0$ is the trivial σ-field.

5. Let $\{S_n, \mathscr{A}_n, n \in I\}$ be a martingale. Show that $\{R_n, \mathscr{A}_n, n \in I\}$ is a submartingale, where a is an arbitrary real number and $R_n = \exp[-(S_n - a)^+]$.

6. Show that if $\{S_n, \mathscr{A}_n, n \in I\}$ is a submartingale and $g(x)$, $x \geq 0$, is a nonincreasing convex function and if $S_n \geq 0$ a.s., then $\{g(S_n), \mathscr{A}_n, n \in I\}$ is a submartingale, provided that $E[g(S_n)]$ is finite.

7. Prove that if $\{S_n, \mathscr{A}_n, n \geq 1\}$ is a reverse martingale, then S_n converges a.s. to a random variable X.

7.5 L_p SPACES

In several instances, more is revealed about a sequence of random variables by knowing that certain integrals converge than by a.s. convergence. In this regard, the most systematic studies are available for random variables that belong to the L_p space defined below.

Definition Let $p > 0$. The set of all random variables X on a given probability space $(\Omega, \mathscr{A}, P)$ with $\int_\Omega |X|^p \, dP < +\infty$ is called the L_p space (of the given probability space). The norm of X belonging to L_p is defined as

$$\|X\|_p = \left(\int_\Omega |X|^p \, dP \right)^{1/p}$$

Finally, we say that X_n converges to X in L_p mean (in our notation: $X_n \to X$ in L_p) if

$$\lim \int_\Omega |X_n - X|^p \, dP = 0$$

as $n \to +\infty$.

Before we proceed, let us recall several simple inequalities from Section 2.2. These, and additional simple properties of L_p, will be given the labels (Lj). First, by Theorem 2.3,

(L1) $X \in L_p$ if and only if $\displaystyle\sum_{j=1}^{+\infty} P(|X| > j^{1/p}) < +\infty$

Next, by Exercise 5 of Section 2.2,

(L2)   If $X \in L_p$, $p > 0$, then $X \in L_r$ with $0 < r \le p$

Furthermore, by Exercise 6 of Section 2.2 (see also the Hints for Solutions of the Exercises), we get from Hölder's inequality the following two results:

(L3) $\|X\|_p$ is nondecreasing in p

(L4) $\log E(|X|^p)$ is a convex function of p over p with $X \in L_p$

Let us strengthen (L3) as follows.

(L5)   If $X \in L_p$ and $0 < r_j \le p$ with $r_j \le r_{j+1} \to p$, then $\|X\|_{r_j} \to \|X\|_p$

Since $|X|^{r_j} \le \max(1, |X|^p)$ and $|X|^p$ is integrable by assumption, the dominated convergence theorem ensures that $E(|X|^{r_j}) \to E(|X|^p)$, from which the desired convergence of the norms follows.

(L6)   (Minkowski's inequality) If $p \ge 1$, $\|X + Y\|_p \le \|X\|_p + \|Y\|_p$

We start with the obvious inequality

$$E(|X + Y|^p) \le E(|X| |X + Y|^{p-1}) + E(|Y| |X + Y|^{p-1})$$

Next, we apply Hölder's inequality to both terms of the right-hand side. We get

$$E(|X + Y|^p) \le E^{1/q}(|X + Y|^{(p-1)q})[E^{1/p}(|X|^p) + E^{1/p}(|Y|^p)]$$

where $1/p + 1/q = 1$, i.e., $(p - 1)q = p$. Now, if $\|X + Y\|_p = 0$, then (L6) is trivially true, while, for $\|X + Y\|_p > 0$ finite, the last inequality, after simplication by $\|X + Y\|_p$, yields the desired inequality. Therefore, it remains only to show that if both $\|X\|_p$ and $\|Y\|_p$ are finite, then so is $\|X + Y\|_p$. This is the content of (L7):

(L7)   If both X and $Y \in L_p$, then $X + Y \in L_p$

We actually prove a more precise statement by establishing the inequality

$$\int_\Omega |X + Y|^p \, dP \le c_p\left(\int_\Omega |X|^p \, dP + \int_\Omega |Y|^p \, dP\right) \tag{7.21}$$

where $c_p = 1$ if $p \le 1$ and $c_p = 2^{p-1}$ if $p \ge 1$ and the inequality is meant to be valid if the right-hand side is finite.

To prove (7.21), note that, for $p > 0$ and for arbitrary real numbers u and v.

$$|u + v|^p \le (|u| + |v|)^p \le c_p(|u|^p + |v|^p)$$

with $c_p = 1$ if $p \le 1$. If $p > 1$, then, dividing the last inequality by $(|u| + |v|)^p$, we have to determine a value of c_p such that

$$1 \le c_p(a^p + (1 - a)^p) \qquad 0 \le a \le 1$$

Since the function $g(a) = a^p + (1 - a)^p, 0 \le a \le 1$, has a single minimum at $a = 1/2$, $c_p g(1/2) \ge 1$ will ensure $c_p g(a) \ge 1$ for all $0 \le a \le 1$. Thus, we can choose $c_p = 2^{p-1}$ for $p > 1$. Now, by the choice $u = X$ and $v = Y$ and by integration, (7.21) obtains, from which (L7) follows.

We close this list of elementary properties by observing that

(L8) $\|X\|_p = 0$ if and only if $X = 0$ a.s.

which is an immediate consequence of the concept of the integral.

It is occasionally convenient to extend the concept of an L_p space to $p = 0$ and $p = +\infty$. Since, for every random variable $X^0 = 1$ a.s., L_0 is, in fact, the set of all random variables. On the other hand, L_∞ can be introduced via (L3). Since, as $p \to +\infty$, $\|X\|_p \to s$, possibly $+\infty$, we say that $X \in L_\infty$ if s is finite, which we then denote by $\|X\|_\infty$. Note that s is the essential supremum of $|X|$ (to be abbreviated as ess.sup$|X|$) in the following sense: $P(|X| \le s) = 1$ but, for every $0 < a < s$, $P(|X| \ge a) > 0$. Indeed, let $s =$ ess.sup$|X|$. Then, for every $p > 0$ and every $0 < a < s$ (possibly infinite),

$$s \ge E^{1/p}(|X|^p) \ge E^{1/p}[|X|^p I(|X| \ge a)] \ge aP^{1/p}(|X| \ge a)$$

Now, since $P(|X| \ge a) > 0$, $P^{1/p}(|X| \ge a) \to 1$ as $p \to +\infty$, so $s \ge \|X\|_\infty \ge a$ for every $0 < a < s$. By letting $a \to s$, the equation ess.sup$|X| = \|X\|_\infty$ obtains. In other words, L_∞ is the set of all a.s. bounded random variables. Note that the previously established properties (Lj) do not necessarily apply to L_0 or L_∞.

We now turn to analyzing the concept of convergence in L_p. First, note that if $g(x)$, $x \ge 0$, is a nondecreasing and nonnegative function, then Markov's inequality (Theorem 2.7) yields

$$P(|X| \ge a) \le \frac{E[g(X)]}{g(a)} \qquad \text{all } a > 0 \tag{7.22}$$

By the choice $g(x) = |x|^p$, $0 < p < +\infty$, we thus have that

$$\text{if } X_n \to X \text{ in } L_p, \qquad \text{then } X_n \to X \text{ in probability} \tag{7.23}$$

The converse is not true. Even if $X_n \to X$ a.s., it does not follow that $X_n \to X$ in L_p, as was demonstrated in Remark 2 of Section 2.4. Theorem 13 below shows the precise relation between convergence in L_p and in probability. However, first let us prove the following important result for $0 < p < +\infty$.

Theorem 11   Let $X_n \in L_p$. If $X_n \to X$ in L_p, then $X \in L_p$ and $X_m - X_n \to 0$ in L_p as n and $m \to +\infty$. Conversely, if $X_m - X_n \to 0$ in L_p as n and $m \to +\infty$, there exists a random variable $X \in L_p$ such that $X_n \to X$ in L_p.

Proof.   If $E(|X_n - X|^p) \to 0$ as $n \to +\infty$, then it is certainly finite for all n sufficiently large. Thus, for $X_n \in L_p$, (7.21) ensures $X \in L_p$ (replace X by X_n and Y by $X - X_n$ in (7.21)). By one more appeal to (7.21), we find that if both $E(|X_n - X|^p) \to 0$ and $E(|X - X_m|^p) \to 0$ as n and $m \to +\infty$, then $E(|X_n - X_m|^p) \to 0$ as well.

Conversely, if $E(|X_n - X_m|^p) \to 0$ as n and $m \to +\infty$, then, by (7.23), $X_n - X_m \to 0$ in probability. Therefore, there is a subsequence $X_{n(k)}$ and a random variable X such that $X_{n(k)} \to X$ a.s. as $n(k) \to +\infty$ (see Theorem 2.11 and the Remark following its proof). Since $E(|X_m - X_{n(k)}|^p) \to 0$ as m and $n(k) \to +\infty$ and, clearly, $X_m - X_{n(k)} \to X_m - X$ a.s. as $n(k) \to +\infty$ for every m, then, by Fatou's lemma (Theorem 2.15),

$$E(|X_m - X|^p) \le \liminf_{n(k) = +\infty} E(|X_m - X_{n(k)}|^p)$$

where the right-hand side is arbitrarily small if m is large. That is, $X_m \to X$ in L_p as $m \to +\infty$, which completes the proof.   ∎

To formulate the next theorem, we redefine the concept of uniform integrability introduced in connection with Theorem 2.27 (Section 2.8). We say that the random variables $|X_n|$ are uniformly integrable if

$$\int_{|X_n| \ge a} |X_n|\, dP \to 0 \qquad \text{uniformly in } n, \text{ as } a \to +\infty \tag{7.24}$$

Next, let A_t be a sequence of events such that $P(A_t) \to 0$ as $t \to +\infty$. Set $B_n = B_n(a) = \{|X_n| \ge a\}$. Then

$$\int_{A_t} |X_n|\, dP = \int_{A_t \cap B_n} |X_n|\, dP + \int_{A_t \cap B_n^c} |X_n|\, dP$$

$$\le \int_{B_n} |X_n|\, dP + aP(A_t) \tag{7.25}$$

Now, first let $t \to +\infty$ and then $a \to +\infty$. We proved that if the $|X_n|$ are uniformly integrable, then, for any sequence A_t such that $P(A_t) \to 0$ as $t \to +\infty$,

$$\int_{A_t} |X_n| \, dP \to 0 \qquad \text{uniformly in } n \tag{7.26}$$

Since $B_n(a)$ is a permissible sequence for A_t with $t = a$, (7.26) is both necessary and sufficient for uniform integrability.

Theorem 12 Let $X_n \in L_p$, $0 < p < +\infty$. If $X_n \to X$ in L_p, then $\|X_n\|_p \to \|X\|_p$, $n \to +\infty$. Furthermore, $\sup_n \|X_n\|_p < +\infty$ and the $|X_n|^p$ are uniformly integrable.

Proof. By Theorem 11, $X \in L_p$. Let $0 < p \le 1$. If $X_n \to X$ in L_p, then the double application of (7.21) in the forms

$$E(|X|^p) \le E(|X_n - X|^p) + E(|X_n|^p) \tag{7.27a}$$

and

$$E(|X_n|^p) \le E(|X_n - X|^p) + E(|X|^p) \tag{7.27b}$$

yields

$$E(|X|^p) \le \liminf E(|X_n|^p) \le \limsup E(|X_n|^p) \le E(|X|^p)$$

as $n \to +\infty$, i.e., $\|X_n\|_p \to \|X\|_p$. For $p > 1$, Minkowski's inequality will replace (7.21), but the argument remains the same; this establishes the first part of the theorem. Furthermore, it follows from (7.27b) and from the corresponding application of Minkowski's inequality that, for $n \ge n_0$,

$$E(|X_n|^p) \le \varepsilon + E(|X|^p) \qquad 0 < p \le 1$$

and

$$\|X_n\|_p \le \varepsilon + \|X\|_p \qquad p > 1$$

where $\varepsilon > 0$ is any number. That is, $\sup_n \|X_n\|_p < +\infty$. Finally, since $X_n \to X$ in L_p implies $X_n I(A) \to X I(A)$ in L_p for an arbitrary indicator variable $I(A)$, we have, by the first part of the theorem and by an estimate similar to the one in (7.25), where we replace B_n by $B = \{|X| \ge a\}$, that

$$\int_{A_t} |X_n|^p \, dP \le \varepsilon + \int_{A_t} |X|^p \, dP \le \varepsilon + \int_B |X|^p \, dP + aP(A_t)$$

for all $n \ge n_0$, where the right-hand side does not depend on n. Hence, (7.26) applies, which is equivalent to uniform integrability. The theorem is established. ∎

We are now in a position to prove the main convergence theorem for L_p spaces.

Theorem 13   Let every $X_n \in L_p$. Then $X_n \to X$ in L_p if and only if (i) $X \in L_p$, (ii) $X_n \to X$ in probability, and either of the two conditions (iii) $\|X_n\|_p \to \|X\|_p$ or (iv) the $|X_n|^p$ are uniformly integrable hold.

Proof.   It has been established in (7.23) and in Theorems 11 and 12 that $X_n \to X$ in L_p implies (i)–(iv). Hence, only the converse statement needs proof. That is, let $X_n \to X$ in probability, where $X \in L_p$. First, we assume that (iii) holds as well. Then, setting $R_{n,p} = c_p|X_n|^p + c_p|X|^p$, $R_{n,p} \to 2c_p|X|^p$ in probability,

$$0 \le \int_\Omega R_{n,p} \, dP \to 2c_p \int_\Omega |X|^p \, dP < +\infty$$

and $|X_n - X|^p \le R_{n,p}$ because of (7.21). Now, the proofs of Fatou's lemma and the dominated convergence theorem (Theorems 2.15 and 2.16) reveal that the domination of $|X_n - X|^p$ by $R_{n,p}$ with the stated properties leaves these theorems valid (note that, when an a.s. convergent subsequence of X_n was chosen in those proofs, it was automatically an a.s. convergent subsequence of $R_{n,p}$ as well). Thus, by this modified dominated convergence theorem,

$$\int_\Omega |X_n - X|^p \, dP \to 0 \qquad \text{i.e. } X_n \to X \text{ in } L_p$$

Next, we assume (iv) in place of (iii). Then, using the fact that $X_n \to X$ in probability is equivalent to $X_n - X_m \to 0$ in probability as n and $m \to +\infty$, i.e., for arbitrary $\varepsilon > 0$, $P(|X_n - X_m| \ge \varepsilon) \to 0$, we have, with $A_{nm} = \{|X_n - X_m| \ge \varepsilon\}$,

$$\int_\Omega |X_n - X_m|^p \, dP \le \int_{A_{nm}} |X_n - X_m|^p \, dP + \varepsilon^p$$

$$\le \int_\Omega I(A_{nm})(|X_n| + |X_m|)^p \, dP + \varepsilon^p$$

$$\le c_p \left\{ \int_\Omega I(A_{nm})|X_n|^p \, dP + \int_\Omega I(A_{nm})|X_m|^p \, dP \right\} + \varepsilon^p$$

where A_{nm} is a permissible sequence for A_t in (7.25). Thus, the uniform integrability of $|X_n|^p$ implies that the right-hand side converges to zero as first n and $m \to +\infty$ and then $\varepsilon \to 0$. Therefore, by Theorem 11, X_n converges to some X^* in L_p. However, $X_n \to X$ in probability, so $X_n \to X^*$ in L_p entails $X_n \to X^*$ in probability as well ((7.23)), which ensures $X = X^*$ a.s. The proof is completed. ∎

Exercises

1. Let X_n be a binomial variable with parameters n and p, $0 < p < 1$, $n \geq 1$. Show that $X_n/n \in L_r$ for all r and that $X_n/n \to p$ in L_r, $r \geq 1$.
2. Let X_n, $n \geq 1$, be identically distributed random variables. Assume that $X_n \in L_1$ and that $S_n/n \to Y$ in probability, with $Y \in L_1$, where $S_n = X_1 + X_2 + \cdots + X_n$. Show that $S_n/n \to Y$ in L_1.

7.6 FURTHER LIMIT THEOREMS FOR MARTINGALES

We start with some results on convergence in L_p.

Theorem 14   Let $\{S_n, \mathscr{A}_n, 1 \leq n \leq +\infty\}$ be a martingale. If $S_n \in L_p$, $1 \leq p < +\infty$, then there is a random variable X such that $S_n \to X$ in L_p and a.s. as $n \to +\infty$. The random variable X closes the martingale $\{S_n, \mathscr{A}_n, 1 \leq n < +\infty\}$ on the right.

Proof. Set $I = \{n \colon 1 \leq n \leq +\infty\}$. Let $B_n = \{|S_n| \geq a\}$, where $a > 0$ is a fixed number and let

$$B = \bigcup_{n \in I} B_n = \left\{ \sup_{n \in I} |S_n| \geq a \right\} = \bigcup_{j \in I} C_j$$

where $C_j = \{|S_j| \geq a, |S_n| < a, 1 \leq n < j\}$, $j \in I$. Clearly, the C_j are mutually exclusive. Hence, since $S_n \in L_p$, $p \geq 1$, implies that $\{|S_n|^p, \mathscr{A}_n, n \in I\}$ is a submartingale (Corollary to Theorem 7),

$$\int_B |S_\infty|^p \, dP = \sum_{j \in I} \int_{C_j} |S_\infty|^p \, dP = \sum_{j \in I} \int_{C_j} E(|S_\infty|^p | \mathscr{A}_j) \, dP$$

$$\geq \sum_{j \in I} \int_{C_j} |S_j|^p \, dP \geq a^p \sum_{j \in I} P(C_j) = a^p P(B) \tag{7.28}$$

Furthermore, by the submartingale property,

$$\int_{B_n} |S_n|^p \, dP \leq \int_{B_n} |S_\infty|^p \, dP \leq \int_B |S_\infty|^p \, dP \tag{7.29}$$

Now, since $S_\infty \in L_p$, the passage by a to $+\infty$ in (7.28) yields $P(B) \to 0$. Therefore, the right-hand side of (7.29) converges to zero as $a \to +\infty$ (see Remark 1 following the proof), which convergence does not depend on n.

Consequently, the $|S_n|^p$ are uniformly integrable ((7.24)). This is sufficient to allow the conclusion that $S_n \to X$ in L_p (Theorem 13), because, by (7.29)

$$E(|S_n|^p) = \int_{B_n} |S_n|^p \, dP + \int_{B_n^c} |S_n|^p \, dP \le E(|S_\infty|^p) + a^p$$

i.e., in view of (L3) of the previous section,

$$E(|S_n|) \le E^{1/p}(|S_n|^p) \le [E(|S_\infty|^p) + a^p]^{1/p}$$

Thus, Theorem 10 is applicable, which ensures that S_n converges a.s. to a random variable X.

It remains to prove that X closes the martingale $\{S_n, \mathscr{A}_n, 1 \le n < +\infty\}$ on the right. That is, we have to show that, for every $n \ge 1$ and every event $A_n \in \mathscr{A}_n$,

$$\int_{A_n} X \, dP = \int_{A_n} S_n \, dP \qquad \text{(i.e., } E(X|\mathscr{A}_n) = S_n \text{ a.s.)} \tag{7.30}$$

Let $N \ge 1$ be arbitrary. Then

$$\int_{A_n} S_{n+N} \, dP = \int_{A_n} S_n \, dP \tag{7.31}$$

Now, we have established that $S_{n+N} \to X$ in L_p as $N \to +\infty$. This, in view of the monotonicity of the L_p norm, implies that $S_{n+N} \to X$ in L_1 as well; thus, by Exercise 1, we can interchange passage to the limit and integration in (7.31) as $N \to +\infty$, yielding (7.30). The theorem is established. ∎

Remark 1 Let $Y \in L_1$ and let $B(a)$ be events such that $P(B(a)) \to 0$ as $a \to +\infty$. For arbitrary events A, define

$$P^*(A) = \frac{1}{E(|Y|)} \int_A |Y| \, dP$$

Then $P^*(A)$ is a probability on $\mathscr{A}$. If $B(a)$ is a decreasing sequence in a and if $B(+\infty) = \lim B(a)$ as $a \to +\infty$, then, by the continuity of probability (Theorem 1.6), $\lim P(B(a)) = P(B(+\infty))$ and $\lim P^*(B(a)) = P^*(B(+\infty))$. Since $P(B(+\infty)) = 0$ by assumption, the definition of integral ensures that $P^*(B(+\infty)) = 0$ as well. This is the form we used in the preceding proof. The monotonicity of $B(a)$ in a is apparently not necessary for the conclusion, since one can appeal to Theorem 1.8 instead of Theorem 1.6.

Remark 2 Note that (7.28) is valid with $p = 1$ for any submartingale closed on the right. In particular, if $\{S_n, \mathscr{A}_n, 1 \leq n \leq N\}$ is a submartingale, then (7.28) becomes

$$\int_B S_N \, dP \geq aP(B) \tag{7.32}$$

where $B = \{\max_{1 \leq n \leq N} S_n \geq a\}$ and a is arbitrary.

By integrating (7.32) with respect to different measures, we can obtain a large variety of inequalities from which further limit theorems can be deduced without assuming that a martingale is closed on the right. We shall exploit this possibility after the introduction of a class of functions to serve as the above-mentioned measures. The idea came from Doob's works; he used $g(x) = |x|^p$, $p \leq 1$, and the measure generated by it (for $x \geq 0$). It was discovered later that the convexity of this function and some simple growth properties of it were the crucial elements in those inequalities. This discovery led to wide applications of the following type of function in martingale theory.

Definition   Let $g(x)$ be a nondecreasing and right-continuous function on $[0, +\infty)$ such that $g(0) = 0$ and $g(+\infty) = +\infty$. Then the function

$$G(x) = \int_0^x g(t) \, dt \qquad x \geq 0$$

is called a Young function. Furthermore, for $y \geq 0$, let the inverse function $u(y)$ of $g(x)$ be defined as $u(y) = \inf\{x: g(x) \geq y\}$. Then

$$U(x) = \int_0^x u(y) \, dy \qquad x \geq 0$$

is called the conjugate of $G(x)$.

Clearly, $U(x)$ is also a Young function and a Young function is nondecreasing, continuous, and convex. The following properties are left to the reader for verification:

(Y1) $xy = G(x) + U(y)$ for $g(x - 0) \leq y \leq g(x)$

(Y2) $xy \leq G(x) + U(y)$ for all $x \geq 0$ and $y \geq 0$

which is known as Young's inequality. Furthermore, for continuity points of $g(x)$ and $u(y)$,

(Y3) $xg(x) = G(x) + U(g(x))$

(Y4) $\displaystyle\int_0^x t \, du(t) = G(u(x))$

and, finally,

(Y5) $G(ax) \leq aG(x)$ for all $x \geq 0$ and all $0 < a \leq 1$

This last inequality is evidently not true in general if $a > 1$. Following Seneta (1976), we say that a positive, nondecreasing function $G(x)$, $x \geq A$, is of *dominated variation* if there exists a number $a > 1$ such that

$$\lim \sup G(ax)/G(x) < +\infty$$

as $x \to +\infty$. Proceeding in the same way as for regularly varying functions (Section 6.3), we find that dominated variation is equivalent to a representation of the form

$$G(x) = \exp\left[\eta(x) + \int_A^x \frac{\varepsilon(t)}{t} dt \right] \tag{7.33}$$

where both $\eta(t)$ and $\varepsilon(t)$ are bounded functions and the latter can always be taken to be positive. We always assume that a Young function is of dominated variation. Since we are dealing with conjugate pairs $G(x)$ and $U(x)$, we want both to be of dominated variation, which imposes further conditions on $g(x)$. First note that if $g(x)$ is continuous, then $\log G(x)$ is defined and differentiable for all sufficiently large x; thus, we have

$$\frac{G(ax)}{G(x)} = \exp\left[\int_x^{ax} \frac{tg(t)}{G(t)} \frac{1}{t} dt \right] \qquad \text{all large } x$$

from which we find that

$$\lim \sup \frac{xg(x)}{G(x)} = p \tag{7.34}$$

must be finite for $G(x)$ under the assumption of dominated variation as $x \to +\infty$. The number p in (7.34) is called the *power* of $G(x)$. As a side result, we obtain that we can choose

$$\varepsilon(t) = \frac{tg(t)}{G(t)}$$

in the representation (7.33). It also follows from the above argument that

$$\frac{G(ax)}{G(x)} \leq a^p \qquad \text{as} \quad x \to +\infty \tag{7.35}$$

(note that the continuity assumption on $g(x)$ is not required, but was assumed for simplicity only). Now, returning to the discussion of the requirement that both $U(x)$ and $G(x)$ be of dominated variation, we assume that the power

p of $G(x)$ is finite and estimate the power q of $U(x)$. By changing from x to $g(x)$, we get

$$q = \limsup \frac{xu(x)}{U(x)} \geq \limsup \frac{g(x)x}{U(g(x))} \qquad x \to +\infty$$

Thus, by (Y3),

$$q \geq \limsup \frac{xg(x)}{xg(x) - G(x)} = \limsup \frac{xg(x)/G(x)}{xg(x)/G(x) - 1}$$

Now, the denominator is smaller than $p - 1 + \varepsilon$ for all large x, where $\varepsilon > 0$ is arbitrary. Hence, by letting $x \to +\infty$ and then $\varepsilon \to 0$, we get

$$q \geq \frac{p}{p - 1} \quad \text{i.e.,} \quad \frac{1}{p} + \frac{1}{q} \leq 1 \tag{7.36}$$

From the displayed inequality preceding (7.36) we get, by

$$q \geq \limsup \frac{1}{1 - G(x)/xg(x)}$$

the estimate

$$\limsup \frac{G(x)}{xg(x)} \leq \frac{q - 1}{q} \tag{7.37}$$

We are now in a position to establish a necessary and sufficient condition for the powers of conjugate Young functions to be finite. The following result is due to Mogyoródi and Móri (1983).

Theorem 15 Let $G(x)$ be a Young function with finite power p. Then its conjugate $U(x)$ has finite power if and only if

$$\limsup \frac{1}{g(x)} \int_1^x \frac{g(t)}{t} \, dt = \beta < +\infty \tag{7.38}$$

as $x \to +\infty$.

Proof. Upon integrating by parts, we get

$$\frac{1}{g(x)} \int_1^x \frac{g(t)}{t} \, dt = \frac{G(x)}{xg(x)} - \frac{G(1)}{g(x)} + \frac{1}{g(x)} \int_1^x \frac{G(t)}{tg(t)} \frac{g(t)}{t} \, dt$$

Note that if we change the path of integration, $[1, x]$, on the right-hand side to $[K, x]$ with a large but fixed K, the error is of the magnitude $O[1/g(x)]$ and we find that (7.37) can be used under the integral. After (7.37) has been used, we change back the integration from 1 to x, by which we further increase. That is, we have, for all large x and for arbitrary $\varepsilon > 0$,

$$\frac{1}{g(x)} \int_1^x \frac{g(t)}{t} \, dt \le \left(\frac{q-1}{q} + \varepsilon\right)\left[1 + \frac{1}{g(x)} \int_1^x \frac{g(t)}{t} \, dt\right] + O\left[\frac{1}{g(x)}\right]$$

By passing to limits we get $\beta \le q - 1$ whenever q is finite, yielding that if q is finite, then β must be finite as well.

Conversely, let us assume that (7.37) holds. Then, with $y = u(2x)$,

$$\frac{1}{g(y)} \int_1^y \frac{g(t)}{t} \, dt \ge \frac{1}{2x} \int_{u(x)}^{u(2x)} \frac{g(t)}{t} \, dt \ge \frac{g(u(x))}{2x} \int_{u(x)}^{u(2x)} \frac{dt}{t}$$

$$\ge \frac{1}{2} \log \frac{u(2x)}{u(x)}$$

i.e.,

$$\limsup \frac{u(2x)}{u(x)} \le e^{2\beta}$$

as $x \to +\infty$, which means that $u(x)$ is of dominated variation. But then so is $U(x)$, since, for large x,

$$U(2x) = \int_0^{2x} u(t) \, dt = U(x) + \int_x^{2x} u(t) \, dt = U(x) + \frac{1}{2} \int_{x/2}^x u(2y) \, dy$$

$$= U(x) + \frac{1}{2} \int_{x/2}^x \frac{u(2y)}{u(y)} \, dy \le U(x) + \frac{1}{2}(e^{2\beta} + \varepsilon)U(x)$$

ensuring that $U(2x)/U(x)$ remains bounded. This completes the proof. ∎

Let us return to martingales. Let $\{S_n, \mathscr{A}_n, n \ge 1\}$ be a nonnegative submartingale and set

$$S_N^* = \max_{1 \le n \le N} S_n$$

The following inequality is essentially due to Garcia (1973), except that it incorporates an improvement of a constant in it by Mogyoródi (1979). We follow Mogyoródi's proof.

Theorem 16 Let $G(x)$ be a Young function and assume that both $G(x)$ and its conjugate $U(x)$ have finite powers. Then

$$E[G(S_N^*)] \le E[G(qS_N)]$$

where $q = \sup\{xu(x)/U(x): x > 0\}$, a modification of the power of $U(x)$.

Proof. Let $b > 0$ be a large number and define $T_N^* = \min(S_N^*, b)$. It follows immediately that Doob's inequality (7.32) remains valid in the form

$$E[S_N I(C)] \geq a E[I(C)]$$

where $C = \{T_N^* \geq a\}$. Now, denoting the (σ-finite) measure generated by $g(x)$ by $g(x)$, we get from the above inequality that

$$\int_0^{+\infty} a E[I(T_N^* \geq a)]\, dg(a) \leq \int_0^{+\infty} E[S_N I(T_N^* \geq a)]\, dg(a)$$

which becomes

$$E\left[\int_0^{T_N^*} a\, dg(a)\right] \leq E[S_N g(T_N^*)]$$

by interchanging the orders of integration. The integral on the left-hand side equals $U[g(T_N^*)]$ (obtained by partial integration; see (Y3) or (Y4)); thus,

$$E\{U[g(T_N^*)]\} \leq E[q S_N g(T_N^*)]/q$$

With an appeal to Young's inequality (Y2), we now have

$$E\{U[g(T_N^*)]\} \leq \frac{1}{q} E\{U[g(T_N^*)]\} + \frac{1}{q} E[G(q S_N)]$$

Because T_N^* is bounded, the left-hand side is finite, so the inequality can be rearranged to

$$\left(1 - \frac{1}{q}\right) E\{U[g(T_N^*)]\} \leq \frac{1}{q} E[G(q S_N)] \qquad (7.39)$$

Now, by (Y3) and the definition of q,

$$q = \sup_{x>0} \frac{x u(x)}{U(x)} \geq \sup_{x>0} \frac{x g(x)}{U(g(x))} \geq \frac{x g(x)}{U(g(x))} = \frac{G(x)}{U(g(x))} + 1$$

i.e., $G(x) \leq (q - 1) U[g(x)]$, which, by (7.39), entails

$$E[G(T_N^*))] \leq E[G(q S_N)]$$

Finally, since, by letting $b \to +\infty$ monotonically, $T_N^* \to S_N^*$ a.s. (and T_N^* is monotonic in b), we get the desired inequality because of the monotone convergence theorem. This concludes the proof. ∎

The special case $G(x) = x^p$, $x \geq 0$, $p > 1$, where $q = p/(p - 1)$, is due to Doob. The inequality of Theorem 16 can serve to establish the a.s. convergence of martingales, as well as their convergence in some metric (which can be finer than convergence in L_p if $G(x)$ deviates from the above special choice). First, we establish a result with the special choice $G(x) = x^p$, $x \geq 0$, $p > 1$. When comparing the theorem below with Theorem 14, note that the martingale is not required to be closed on the right.

Theorem 17 Let $1 < p < +\infty$. If $\{S_n, \mathscr{A}_n, n \geq 1\}$ is a martingale and $\sup\{E(|S_n|^p): n \geq 1\} < +\infty$, then there is a random variable X such that $S_n \to X$ a.s. and in L_p.

Proof. Because $\|Y\|_p$ is nondecreasing in p (see (L3)), the assumption of the theorem ensures that Theorem 10 is applicable. Thus, there is a random variable X such that $S_n \to X$ a.s. In order to extend this convergence to one in L_p, we have to show that $|S_n|^p$, $n \geq 1$ are uniformly integrable (Theorem 13), i.e., that

$$E[|S_n|^p I(|S_n| \geq a)] \to 0 \qquad \text{uniformly in } n, \qquad \text{as } a \to +\infty$$

But

$$E[|S_n|^p I(|S_n| \geq a)] \leq E\left[\left(\sup_{k \geq 1} |S_k|^p\right) I(|S_n| \geq a)\right] \tag{7.40}$$

on the right-hand side of which we can use the fact that, by Theorem 16,

$$E\left(\max_{1 \leq k \leq N} |S_k|^p\right) \leq q^p E(|S_N|^p)$$

i.e., in view of the assumption of the theorem,

$$E\left(\sup_{n \geq 1} |S_n|^p\right) < +\infty$$

Furthermore, by Markov's inequality,

$$P(|S_n| \geq a) \leq a^{-p} E(|S_n|^p) \leq a^{-p} \sup_{k \geq 1} E(|S_k|^p)$$

which converges to zero uniformly in n as $a \to +\infty$. Hence, (7.40) takes the place of (7.29) in the proof of Theorem 14. We see that the $|S_n|^p$ are indeed uniformly integrable, which completes the proof. ∎

Let us go back to general Young functions $G(x)$. In order to properly define G-norms (in place of L_p-norm generated by $G(x) = x^p$), let us redefine the L_p-norm as follows. Since

$$a(p) = \|X\|_p = \left(\int_\Omega |X|^p \, dP\right)^{1/p}$$

is equivalent to saying that, with $G(x) = x^p$, $x > 0$, $a(p)$ is the unique number such that

$$\int_\Omega G\left(\frac{|X|}{c}\right) dP \leq 1 \qquad \text{if } c > a(p)$$

$$> 1 \qquad \text{if } c < a(p)$$

i.e., $a(p) = \inf\{c: E[G(|X|/c)] \leq 1\}$, we use this expression as the definition of the G-norm: If $G(x)$, $x > 0$, is a Young function, the G-norm of the random variable X is defined as

$$|X|_G = \inf\{c: E[G(|X|/c)] \leq 1\}$$

It is left to the reader to see that this indeed is a norm. It is known (see J. Neveu (1975)) that $|X|_G$ is a complete metric (i.e. that Cauchy convergence in G-metric or G-norm implies the existence of a limit random variable with finite G-norm). We do not want to discuss this aspect of G-norms. Rather, in order to get familiar with the concept we go through some simple computations.

Example 1 Let the underlying probability space be Lebesgue measure on the measurable subsets of $(0, 1)$. Let $G(t) = e^t - 1 - t$, $t > 0$, which is a Young function. Let $X = -\log(1 - x)$, $0 < x < 1$. Then X is a unit exponential variable and

$$G(X/a) = e^{X/a} - 1 - X/a$$

and

$$E\left[G\left(\frac{X}{a}\right)\right] = E(e^{X/a}) - 1 - \frac{1}{a} = \int_0^1 \frac{1}{(1-x)^{1/a}} dx - 1 - \frac{1}{a}$$

which is divergent if $0 < a \leq 1$, and, for $a > 1$, it equals

$$\frac{a}{a-1} - 1 - \frac{1}{a} = \frac{1}{a(a-1)}$$

which is ≤ 1 if $a \geq (1/2)(1 + \sqrt{5})$, i.e., $|X|_G = (1/2)(1 + \sqrt{5})$.

Example 2 With the notation of the preceding example, let $\mathscr{A}_n$, $n \geq 1$, be an increasing sequence of subsigma-fields of the Lebesgue measurable subsets of $(0, 1)$. Define $S_n = E(X|\mathscr{A}_n)$. Then $\{S_n, \mathscr{A}_n, n \geq 1\}$ is a martingale, and we show that

$$\sup[|S_n|_G: n \geq 1] \leq (1 + \sqrt{5})/2$$

Indeed, since $G(x)$ is convex (concave up), Jensen's inequality entails

$$E(G(X_n/a)) = E[G(E(X/a|\mathscr{A}_n))] \leq E[E(G(X/a)|\mathscr{A}_n)] = E(G(X/a))$$

which was found in Example 1 to be ≤ 1 for all $a \geq (1 + \sqrt{5})/2$. This establishes our claim.   ▣

Example 3   Let now the underlying probability space be arbitrary. Let $\mathscr{A}_n$, $n \geq 1$, be an increasing sequence of subsigma-fields of the set of events, and let X be a bounded random variable. Define $S_n = E(X|\mathscr{A}_n)$, $n \geq 1$. Assume that $S_n \to X$ a.s. (we know from Theorem 10 that S_n has an a.s. limit which is a random variable). Then $S_n \to X$ in any G-norm.

First, note that, since $|X| \leq K$, we have $|S_n| \leq K$ for all $n \geq 1$. Then by monotonicity of a Young function G,

$$G\left(\frac{|S_n - X|}{c}\right) \leq G\left(\frac{2K}{c}\right)$$

and, by the continuity of G, the left hand side above converges to $G(0) = 0$ a.s. Hence, for n sufficiently large, the dominated convergence theorem entails that, whatever the value $c > 0$,

$$E\left[G\left(\frac{|S_n - X|}{c}\right)\right] \leq 1,$$

that is, $\lim|S_n - X|_G = 0$ as $n \to +\infty$.   ▣

We are now in the position to prove the following general theorem, which we do without the inequality of Theorem 16, even though it is closely related to Theorem 17 with a general Young function in the place of x^p.

Theorem 17a   With the notation and assumptions of Example 3 except that X is assumed integrable but not bounded, if G is an arbitrary Young function (with dominated variation) and if $\sup\{|S_n|_G : n \geq 1\} < +\infty$, then $|S_n - X|_G \to 0$.

Proof.   Truncate X by the number K (arbitrary):

$$X^* = \begin{cases} X & \text{if } |X| \leq K \\ 0 & \text{if } |X| > K \end{cases}$$

and set $X^{**} = X - X^*$. Then $X = X^* + X^{**}$ and

$$S_n - X = E(X^*|\mathscr{A}_n) - X^* + E(X^{**}|\mathscr{A}_n) - X^{**}$$

By the triangular inequality,

$$|S_n - X|_G \le |E(X^*|\mathscr{A}_n) - X^*|_G + |E(X^{**}|\mathscr{A}_n) - X^{**}|_G$$

The first term on the right-hand side converges to zero by Example 3. In the second term on the right-hand side we once again utilize Jensen's inequality, yielding

$$|E(X^{**}|\mathscr{A}_n)|_G \le |X^{**}|_G$$

that is, by one more application of the triangular inequality,

$$|E(X^{**}|\mathscr{A}_n) - X^{**}|_G \le 2|X^{**}|_G$$

Now,

$$E\left[G\left(\frac{|X^{**}|}{c}\right)\right] = \int_{|X|>K} G\left(\frac{|X|}{c}\right) dP$$

Set $s = \sup|S_n|_G$. For $c \ge s$, write above $1/c = (s/c)/s$, and utilize the monotonocity of G. We have that we increase if we replace c by s on the right-hand side, and then further increase by integrating on Ω. By $|X|_G \le \sup|S_n|_G = s$, the just modified right-hand side is bounded by one.

Next, let $s > c$. By the dominated variation of G, with some $a > 1$, $G(ax) \le dG(x)$; thus, if k is defined as $a^{k-1} < s/c \le a^k$, we have $G(x(s/c)) \le G(a^k x) \le dG(a^{k-1}x) \le d^2 G(a^{k-2}x) \le \cdots \le d^k G(x)$. Hence, upon putting $D = d^k$,

$$E\left[G\left(\frac{|X^{**}|}{c}\right)\right] = \int_{|X|>K} G\left(\frac{s}{c}\frac{|X|}{s}\right) dP \le D \int_{|X|>K} G\left(\frac{|X|}{s}\right) dP$$

$$= D \int_\Omega I(|X| > K)G\left(\frac{|X|}{s}\right) dP$$

which converges to zero, and thus ultimately becomes smaller than one, as $K \to +\infty$, whatever $c > 0$. So, the G-norm of X^{**} equals zero as $K \to +\infty$, which completes the proof. ∎

We conclude this section with a few comments on weak convergence of martingales. Emphasizing the partial sum character of martingales, let us write $S_n = X_1 + X_2 + \cdots + X_n$ and assume that $E(X_1) = 0$ and $\{S_n, \mathscr{A}_n, n \ge 1\}$ is a martingale. Then one might expect that, in the general case, one could find normalizing constants $b_n > 0$ such that S_n/b_n would be asymptotically normal. One indeed can find simple sufficient conditions for this, such as the Lindeberg condition for the differences X_j when normalized by the square root of the variance of S_n, if, in addition, the conditional variances $E(X_j^2|\mathscr{A}_{j-1})$

are "very close" to the unconditional ones $E(X_j^2)$. For example, if $\mathscr{A}_0$ is the trivial σ-field, then

$$\sum_{j=1}^{n} E[|E(X_j^2|\mathscr{A}_{j-1}) - E(X_j^2)|] = o(s_n^2)$$

is sufficient, besides the Lindeberg condition, for the asymptotic normality of S_n/s_n, where $s_n^2 = E(X_1^2) + E(X_2^2) + \cdots + E(X_n^2)$. The reader can easily reproduce the proof of such a statement along the lines of Section 3.3 (first using conditional expectations rather than unconditional ones). But such results are neither general nor natural. We learned from the a.s. results that the a.s. limit of S_n is usually a random variable and not a constant; thus, normalization is more natural by random variables than by sequences of constants. Then, by proper conditioning, one can get general weak convergence results, whose unconditional forms are then mixtures of a normal distribution with some other distribution. Since such results can now be proved under the more general set up of mixing random variables, we shall not discuss this further—see Hall and Heyde (1980).

Exercises

1. Show that if $X_n \to X$ in L_1, then $E(X_n I(A)) \to E(XI(A))$ uniformly in $A \in \mathscr{A}$.
2. Show that if, in the definition of a Young function $G(x)$, $g(x)$ is convex, then the conjugate $U(x)$ of $G(x)$ has finite power.
3. Make the inequality of Theorem 16 specific by choosing several $g(x)$. Is this theorem applicable when $g(x) = \log(x + 1)$?

7.7 EXCHANGEABILITY, DE FINETTI'S THEOREM

The random variables $\{X_n, n \geq 1\}$ are called *exchangeable* if the joint distribution of every finite subset of k, $k \geq 1$, of these random variables depends only on k, not on the particular subset. In particular, exchangeable random variables are identically distributed. When the X_n are indicators, we speak of *exchangeable events*.

Clearly, every subset of exchangeable random variables is exchangeable. The converse question, Does a finite set of exchangeable variables come from a larger set of exchangeable ones?, is a difficult one to decide, and very little is known about the subject. That there are finite sequences of exchangeability is demonstrated by the following example.

Example 1 Let $A_1, A_2, \ldots, A_n$ be exchangeable events with $P(A_j) = 0.5$ and $P(A_i \cap A_j) = 0.2$, $i \neq j$. Then $n \leq 6$.

Let $I_j = I(A_j)$ and set $N_n = I_1 + I_2 + \cdots + I_n$. Then, by exchangeability, $E(N_n) = nE(I_1) = 0.5n$ and, since $I_j = I_j^2$,

$$E(N_n^2) = nE(I_1) + n(n-1)E(I_1 \cap I_2) = 0.5n + 0.2n(n-1)$$

Hence,

$$0 \le V(N_n) = 0.5n + 0.2n(n-1) - (0.5n)^2 = 0.3n - 0.05n^2$$

i.e., $n \le 6$ as stated.   ∎

If we repeat the calculations of the example with $P(A_1) = p_1$ and $P(A_1 \cap A_2) = p_2$, we immediately find that, in order for a sequence A_1, $A_2, \ldots, A_n$ to be a segment of an infinite sequence of exchangeable events, it is necessary that

$$p_2 \ge p_1^2 \tag{7.41}$$

Since independent events are obviously exchangeable if $P(A_j) = p_1$ for all j, equality can indeed occur in (7.41). The condition (7.41), however, is not sufficient, since a sufficient condition must involve all p_k, $k \ge 1$, where

$$p_k = P(A_1 \cap A_2 \cap \cdots \cap A_k) = P(A_{i_1} \cap A_{i_2} \cap \cdots \cap A_{i_k}) \tag{7.42}$$

with distinct integers i_j. We now develop one condition that ensures that a set p_k, $1 \le k \le n$, of numbers can be associated with n exchangeable events, as in (7.42).

Note that, by the elementary identity $P(C \cap D^c) = P(C) - P(C \cap D)$, exchangeability yields

$$P(A_{i_1} \cap A_{i_2} \cap \cdots \cap A_{i_k} \cap A_{i_{k+1}}^c) = p_k - p_{k+1} = \delta p_k$$

say, where (and everywhere below) the subscripts i_j are assumed to be distinct. Upon repeating this argument, we have

$$P(A_{i_1} \cap A_{i_2} \cap \cdots \cap A_{i_k} \cap A_{i_{k+1}}^c \cap \cdots \cap A_{i_{k+t}}^c) = \delta^t p_k \tag{7.43}$$

by induction, where $\delta^t p_k = \delta(\delta^{t-1} p_k)$ for $t \ge 2$ and $\delta p_k = \delta^1 p_k$. Let $N = N_n$ be the number of the A_j, $1 \le j \le n$, which occur. Since

$$P(N = k) = \sum P(A_{i_1} \cap A_{i_2} \cap \cdots \cap A_{i_k} \cap A_{i_{k+1}}^c \cap \cdots \cap A_{i_n}^c) \tag{7.44}$$

where summation is over all integers $1 \le i_1 < i_2 < \cdots < i_k \le n$ and

$$\sum_{k=0}^{n} P(N = k) = 1 \tag{7.45}$$

we have from (7.43) that

$$\delta^t p_k \ge 0 \qquad \text{for all} \quad 0 \le k \le n, 0 \le t \le n - k \tag{7.46}$$

and

$$\sum_{k=0}^{n} \binom{n}{k} \delta^{n-k} p_k = 1 \tag{7.47}$$

where $p_0 = 1$ and $\delta^0 p_k = p_k$. The next result is due to D. G. Kendall (1967).

Theorem 18   A set $1 = p_0 \geq p_1 \geq p_2 \geq \cdots \geq p_n \geq 0$ of numbers can be associated with n exchangeable events $A_1, A_2, \ldots, A_n$ as in (7.42) if and only if (7.46) and (7.47) hold. Moreover, if $N = N_n$ is the number of the A_j that occur, then the conditional probabilities

$$P(A_{i_1} \cap A_{i_2} \cap \cdots \cap A_{i_k} | N) = \frac{N(N-1)\cdots(N-k+1)}{n(n-1)\cdots(n-k+1)}$$

Proof.   We have seen that exchangeability leads to (7.46) and (7.47). Conversely, if (7.46) and (7.47) hold for the numbers p_j, $0 \leq j \leq n$, of the theorem, then take the set of dyadic rationals

$$\sum_{j=1}^{n} \frac{e_j}{2^j} \qquad e_j = 0 \text{ or } 1$$

as Ω. Assign $P(\{\omega\}) = \delta^r p_{n-r}$ to $\omega \in \Omega$, where r is the number of those e_j in ω which are equal to zero. Then (7.46) and (7.47) ensure that P is a proper probability on the set of all subsets of Ω. Let $A_j = \{e_j = 1\}$. Then the A_j are clearly exchangeable. Furthermore,

$$P(A_{i_1} \cap A_{i_2} \cap \cdots \cap A_{i_k}) = \sum_{r=0}^{n-k} \delta^r p_{n-r} \binom{n-k}{r} = p_k$$

the last equation being valid by an "inversion formula" of the difference operation δ^r (since the $\delta^r p_{n-r}$ are probabilities assigned to the members of Ω, the reader can work backwards from (7.43) to establish this inversion). This completes the proof of the first part of the theorem.

Next, we return to a general probability space, where the A_j are exchangeable with p_k, as in (7.42). Then, by (7.43) and (7.44),

$$P(N = j) = \binom{n}{j} \delta^{n-j} p_j = r_j$$

say. Now, N is a simple random variable, so conditioning on N takes us back to computing traditional conditional probabilities, given $\{N = j\}$. But the event $A_{i_1} \cap A_{i_2} \cap \cdots \cap A_{i_k} \cap \{N = j\}$ equals

$$\cup \{A_{i_1} \cap \cdots \cap A_{i_k} \cap A_{i_{k+1}} \cap \cdots \cap A_{i_j} \cap A_{i_{j+1}}^c \cap \cdots \cap A_{i_n}^c\}$$

where the subscripts beyond k are arbitrary distinct integers that differ from the first k subscripts. Hence, by (7.43),

$$P(A_{i_1} \cap A_{i_2} \cap \cdots \cap A_{i_k} | N) = \binom{n-k}{N-k} \delta^{n-N} p_N / r_N = \binom{n-k}{N-k} \Big/ \binom{n}{N}$$

which, when written in detail, is the desired formula. The theorem is established. ∎

Recall formula (1.21) in connection with the hypergeometric distribution. By the similarities of the results of Theorem 18 and (1.21), we have a new interpretation of finite exchangeability: every system of n exchangeable events is equivalent to a random selection from a lot of two types of items, in which the number of type I items is a random variable N. In fact, this interpretation of Theorem 18 immediately yields the following formula (compare (1.20) and (1.21)).

Corollary Let $A_1, A_2, \ldots, A_n$ be exchangeable events and let $m \le n$ be an integer. Let N_m be the number of the A_j, $1 \le j \le m$, that occur. Then

$$P(N_m = s) = E\left[\binom{N_n}{s} \binom{n - N_n}{m - s} \Big/ \binom{n}{m} \right]$$

The following Poisson approximation to the distribution of N_m, due to Ridler–Rowe (1967), is remarkable in that its assumptions are limited to $P(A_1)$ and $P(A_1 \cap A_2)$ only. The case $N_n = +\infty$ is due to D. G. Kendall (1967).

Theorem 19 Consider a sequence of probability spaces and, on the nth space, let $A_1^{(n)}, A_2^{(n)}, \ldots, A_n^{(n)}$ be exchangeable events. Assume that $m = m(n) \le n$ is a sequence of integers such that $m(n) \to +\infty$ with n and that, for some $0 < a < +\infty$

$$m(n) P_n(A_1^{(n)}) \to a \qquad m^2(n) P_n(A_1^{(n)} \cap A_2^{(n)}) \to a^2 \tag{7.48}$$

as $n \to +\infty$. If $N_m = N_m^{(n)}$ denotes the number of the $A_j^{(n)}$, $1 \le j \le m$, that occur and if $m(n)/n \to 0$ as $n \to +\infty$, then

$$\lim P_n(N_m = s) = \frac{a^s e^{-a}}{s!}, \qquad s = 0, 1, 2, \ldots$$

as $n \to +\infty$.

Proof. By Theorem 18,

$$P_n(A_1^{(n)}) = E\left(\frac{N_n}{n}\right) \qquad P_n(A_1^{(n)} \cap A_2^{(n)}) = E\left[\frac{N_n(N_n-1)}{n(n-1)}\right]$$

Hence, (7.48) implies that there are numbers $\varepsilon_1 \to 0$ and $\varepsilon_2 \to 0$ as $n \to +\infty$ such that

$$E\left[\frac{m(n)}{n} N_n\right] = (1 + \varepsilon_1)a \qquad V\left[\frac{m(n)}{n} N_n\right] = \varepsilon_2 a^2$$

Then, by Chebyshev's inequality, the probability of the event

$$C_n = \left\{ \left| \frac{m(n)}{n} N_n - (1 + \varepsilon_1)a \right| \geq \varepsilon_2^{1/2} a \right\}$$

converges to zero as $n \to +\infty$. Therefore, since the (random) numbers

$$h_{n,s}(N_n, m) = \binom{N_n}{s}\binom{n-N_n}{m-s} \Big/ \binom{n}{m}$$

are probabilities, and thus bounded, the dominated convergence theorem ensures that

$$\lim E[I(C_n)h_{n,s}(N_n, m)] = 0 \qquad (n \to +\infty)$$

On the other hand, $N_n m(n)/n \to a$ and $m(n)/n \to 0$ on C_n^c as $n \to +\infty$, in which case it is known (and quite simple to prove in the light of the closeness of the hypergeometric distribution to the binomial; see Exercise 2 of Section 1.5) that $h_{n,s}(N_n, m) \to a^s e^{-a}/s!$ By one more appeal to the dominated convergence theorem, the theorem now follows from the Corollary. The proof is completed. ∎

Next, we turn to infinite sequences X_j, $j \geq 1$, of exchangeable random variables. The fundamental result of de Finetti (1930) and (1937) is given below.

Theorem 20   (de Finetti)   An infinite sequence $X_j, j \geq 1$, of random variables is exchangeable if and only if there is a σ-field $\mathcal{B} \subset \mathcal{A}$ such that, given $\mathcal{B}$, the X_j are conditionally independent and identically distributed. That is, for every $1 \leq i_1 < i_2 < \cdots < i_k$, almost surely,

$$P(X_{i_1} \leq x_1, X_{i_2} \leq x_2, \ldots, X_{i_k} \leq x_k | \mathcal{B}) = F(x_1)F(x_2)\cdots F(x_k) \qquad (7.49)$$

where $F(x)$ is a $\mathcal{B}$-measurable random variable that is a.s. a distribution function in x.

Proof. Clearly, if (7.49) holds, then, by taking expectations, exchangeability obtains. Hence, only the converse statement needs proof. That is, we assume that the X_j are exchangeable. For every real x, let

$$Y_n(x) = \frac{1}{n} \sum_{j=1}^{n} I(X_j \le x) \qquad n \ge 1$$

Then, by exchangeability, for $m < n$, $m \to +\infty$,

$$E[(Y_n - Y_m)^2] = \frac{n-m}{nm} [P(X_1 \le x) - P(X_1 \le x, X_2 \le x)] \to 0$$

Hence, there is a random variable $Y(x)$ such that $Y_n(x) \to Y(x)$ in L_2 (Theorem 11), and thus in probability as well (recall (7.23)). Therefore, since the $Y_n(x)$ are bounded by 0 and 1, the dominated convergence theorem entails

$$E[Y_n(x_1)Y_n(x_2)\cdots Y_n(x_k)I(B)] \to E[Y(x_1)Y(x_2)\cdots Y(x_k)I(B)]$$

for any fixed real numbers $x_1, x_2, \ldots, x_k$, and for an arbitrary event B. Now, if B is in the tail σ-algebra

$$\mathscr{B} = \bigcap_{n=1}^{+\infty} \sigma(X_n, X_{n+1}, \ldots)$$

then, by exchangeability,

$$E[Y_n(x_1)Y_n(x_2)\cdots Y_n(x_k)I(B)] \to E[I(X_1 \le x_1, \ldots, X_k \le x_k)I(B)]$$

because the product of $Y_n(x_t)$, $1 \le t \le k$, consists of n^k terms of the form

$$I(X_{i_1} \le x_1)I(X_{i_2} \le x_2)\cdots I(X_{i_k} \le x_k) = I(X_{i_1} \le x_1, \ldots, X_{i_k} \le x_k)$$

in which, when all subscripts i_j are distinct, we can switch to $I(X_1 \le x_1, X_2 \le x_2, \ldots, X_k \le x_k)$ behind the expectation sign (by exchangeability), while there are terms with repeated indices $i_u = i_v$ for some $u \ne v$ in fewer than n^{k-1} terms. Hence, since we divide by n^k, their contribution tends to 0 as $n \to +\infty$. The two forms of limit yield

$$P(X_1 \le x_1, X_2 \le x_2, \ldots, X_k \le x_k | \mathscr{B}) = E[Y(x_1)\cdots Y(x_k)|\mathscr{B}]$$

a.s., but each $Y(x_j)$ is $\mathscr{B}$ measurable and thus

$$P(X_1 \le x_1, X_2 \le x_2, \ldots, X_k \le x_k | \mathscr{B}) = Y(x_1)Y((x_2)\cdots Y(x_k) \text{ a.s.}$$

for every finite $k \ge 1$. It is clear from the definition of $Y_n(x)$ that its limit $Y(x)$ has a.s. the properties of a distribution function. In particular, upon letting any $k - 1$ of the variables x_j, $1 \le j \le k$, tend to $+\infty$, we get

$$P(X_j \le x_j | \mathscr{B}) = Y(x_j) \text{ a.s.}$$

and thus the desired conditional independence is established. This completes the proof. ∎

It should be noted that we can repeat the argument of Theorem 5 and obtain that $Y(x)$ can always be replaced by a regular conditional distribution function. Another remark worth noting is that there are other σ-fields on which one can condition to get conditional independence of (and identical distribution for) the X_j. We shall not go into details, but the reader should recognize that the preceding proof requires no changes when conditioning is on the following σ-field. Let $\mathscr{B}_n^*$ be the smallest σ-field of events from $\sigma(X_1, X_2, \ldots)$ that are not affected by the permutation of the first n subscripts $1, 2, \ldots, n$; it is a decreasing sequence of σ-fields. Let $\mathscr{B}^*$ be the intersection of all $\mathscr{B}_n^*$. Members of $\mathscr{B}^*$ are called *permutable events*. Now, we could have conditioned on $\mathscr{B}^*$ in the preceding theorem. Obviously, $\sigma(X_{n+1}, X_{n+2}, \ldots) \subset \mathscr{B}_n^*$; thus, the tail σ-field $\mathscr{B} \subset \mathscr{B}^*$. Note also that $S_n = X_1 + X_2 + \cdots + X_n$ is $\mathscr{B}_n^*$-measurable (but not with respect to $\sigma(X_{n+1}, X_{n+2}, \ldots)$). In the following theorem we use this notation and these facts.

Theorem 21 Let $X_n, n \geq 1$, be an infinite sequence of exchangeable random variables with $E(X_1)$ finite. Then $n^{-1}S_n \to E(X_1 | \mathscr{B}^*)$ a.s. and in L_1.

Proof. Because, by exchangeability, for every $n \geq 1$, $E(X_j | \mathscr{B}_n^*) = E(X_1 | \mathscr{B}_n^*)$ for all $1 \leq j \leq n$, we have

$$\frac{1}{n} S_n = \frac{1}{n} E(S_n | \mathscr{B}_n^*) = \frac{1}{n} \sum_{j=1}^n E(X_j | \mathscr{B}_n^*) = E(X_1 | \mathscr{B}_n^*) \text{ a.s.}$$

Thus, $n^{-1}S_n$ is a reverse martingale. By Exercise 7 of Section 7.4, there is a random variable U such that $n^{-1}S_n \to U$ a.s. The convergence in L_1 follows from the fact that the absolute values of moments decrease in n (by the reverse submartingale property). That $U = E(X_1 | \mathscr{B}^*)$ a.s. follows in the same way in which it was shown in a similar case for a martingale following Example 5 in Section 7.4. The theorem is established. ∎

Exercises

1. Extend (7.41) to arbitrary exchangeable random variables $X_j \in L_2$ by showing that if $X_j, j \geq 1$, is an infinite sequence of exchangeable variables, then $E(X_i X_j) \geq E^2(X_1)$, $i \neq j$.
2. With the notation of Theorem 19, show that if $m(n)/n$ fails to converge to zero as $n \to +\infty$, then $N_m^{(n)}$ is not, in general, asymptotically Poisson.
3. Use the Corollary to Theorem 18 to obtain a variety of limit theorems for (normalized) sums of indicators of exchangeable events.

REFERENCES

Adler, R. J. and D. J. Scott (1975). Martingale central limit theorems without uniform asymptotic negligibility. *Bull. Austral. Math. Soc. 13*, 45–56 (Corrigendum: *ibid. 18* (1978), 311–319).

Burkholder, D. L. (1973). Distribution function inequalities for martingales. *Ann. Probability 1*, 19–42.

Chow, Y. S. and H. Teicher (1978) *Probability Theory. Independence, Interchangeability, Martingales.* Springer, New York.

Diaconis, P. and D. Freedman (1980). Finite exchangeable sequences. *Ann. Probability 8*, 745–764.

Doob, J. L. (1940). Regular properties of certain families of chance variables. *Trans. Amer. Math. Soc. 47*, 455–486.

Doob, J. L. (1953). *Stochastic Processes.* Wiley, New York.

Eagleson, G. K. (1975). Martingale convergence to mixtures of infinitely divisible laws. *Ann. Probability 3*, 557–562.

Eagleson, G. K. and N. C. Weber (1978). Limit theorems for weakly exchangeable arrays. *Math. Proc. Cambridge Philos. Soc. 84*, 123–130.

Finetti, B. de (1930). Funzione caratteristica di uno fenomeno aleatorio. *Memorie R. Accad. Lincei 4*, 86–133.

Finetti, B. de (1937). Le prévision: ses lois logiques, ses sources subjectives. *Ann. Inst. H. Poincaré 7*, 1–68.

Galambos, J. (1987). *The Asymptotic Theory of Extreme Order Statistics*, 2nd ed. Krieger, Melbourne, Florida.

Garcia, A. M. (1973). *Martingale Inequalities.* Benjamin, Reading, Mass.

Hall, P. and C. C. Heyde (1980). *Martingale Limit Theory and its Applications.* Academic Press, New York.

Halmos, P. R. (1974). *Measure Theory.* Springer, Berlin.

Hewitt, E. and L. J. Savage (1955). Symmetric measures on Cartesian products. *Trans. Amer. Math. Soc. 80*, 470–501.

Kendall, D. G. (1967). On finite and infinite sequences of exchangeable events. *Studia Sci. Math. Hungar. 2*, 319–327.

Kingman, J. F. C. (1978). Uses of exchangeability. *Ann. Probability 6*, 183–197.

Koch, G. and F. Spizzichino (1982). *Exchangeability in Probability and Statistics*. North-Holland, Amsterdam.

Kolmogorov, A. N. (1933). *Grundbegriffe der Wahrscheinlichkeitsrechnung*. Springer, Berlin. (English translation: *Foundations of the Theory of Probability*. Chelsea, New York, 1956.)

Lévy, P. (1937). *Théorie de l'addition des variables aléatoires*. Gauthier-Villars, Paris.

Loeve, M. (1963). *Probability Theory*, 3rd ed. Van Nostrand, New York.

Mogyoródi, J. (1979). On an inequality of Marcinkiewicz and Zygmund. *Publ. Math. Debrecen 26*, 267–274.

Mogyoródi, J. and T. F. Móri (1983). Necessary and sufficient condition for the maximal inequality of convex Young functions. *Acta Sci. Math. 45*, 325–332.

Neveu, J. (1965). *Mathematical Foundations of the Calculus of Probability*. Holden-Day, San Francisco.

Neveu, J. (1975). *Discrete Parameter Martingales*. North-Holland, Amsterdam.

Philipp, W. and W. F. Stout (1975). Almost sure invariance principles for partial sums of weakly dependent random variables. *Memoirs Amer. Math. Soc.*, No. 161.

Rényi, A. (1963). On stable sequences of events. *Sankhyā A 25*, 293–302.

Rényi, A. (1970). *Foundations of Probability*. Holden-Day, San Francisco.

Rényi, A. and P. Révész (1963). A study of sequences of equivalent events as special stable sequences. *Publ. Math. Debrecen 10*, 319–325.

Ridler-Rowe, C. J. (1967). On two problems on exchangeable events. *Studia Sci. Math. Hungar. 2*, 415–418.

Seneta, E. (1976). *Regularly Varying Functions. Lecture Notes in Math. 508*. Springer, Heidelberg.

Stout, W. F. (1974). *Almost Sure Convergence*. Academic Press. New York.

Takács, L. (1965). On the distribution of the supremum of stochastic processes with interchangeable increments. *Trans. Amer. Math. Soc. 199*, 367–379.

Ville, J. (1939). *Etude critique de la notion de collectif*. Gauthier-Villars, Paris.

8

Topics in the Theory of Stochastic Processes

The mathematical description of a random phenomenon as it changes in time is a stochastic process. That is, a stochastic process is a function of both a random element ω and a time parameter t. More accurately, given a probability space $(\Omega, \mathscr{A}, P)$ and a set T on the real line, a function $X_t(\omega)$, $t \in T$, $\omega \in \Omega$, is a *stochastic process* if, for each $t \in T$, $X_t(\omega)$ is a random variable. If $T = \{1, 2, \ldots,\}$, then the stochastic process X_t, $t \in T$, is just a sequence of random variables and, in such a case, we speak of a stochastic process of *discrete time*. When T is a continuous set (an interval), we speak of stochastic processes of *continuous time*. The aim of this chapter is to give the flavor and the richness of the theory of stochastic processes (without repeating the material for sequences covered in the earlier chapters). Section 8.1 is devoted to some basic concepts; subsequent sections give details for some special processes.

8.1 FOUNDATIONS AND BASIC CONCEPTS

In this section, we shall introduce a number of concepts, but we shall not develop any theory. In particular, we do not plan to state or prove theorems. Rather, the motivations behind the concepts will be stressed.

So, we have a probability space and a set T on the real line which we call the *parameter space* or *time space*. Then X_t, $t \in T$, is called a *stochastic process*

353

if, for every $t \in T$, $X_t = X_t(\omega)$ is a random variable. On the other hand, for a given ω, X_t is a function on T which we call a *sample function*. If T is a discrete set, then a sample function is a sequence of numbers (finite or infinite). Examples for continuous stochastic processes are given below.

Example 1 Let X be an exponential random variable and let $X_t = \exp(-tX)$, $0 \le t < +\infty$. Then, for each $t \in T = [0, +\infty)$, X_t is a random variable and, for every ω, X_t is the exponential function $\exp(-tX)$, $t \ge 0$ (which are the sample functions). We encountered this "stochastic process" in connection with Laplace transforms.   ∎

Example 2 Let $T = (0, 60)$ and define X_t, $t \in T$, as the number of customers a specific shop has serviced up to time t (measured in minutes). Then, with an appropriate probability space, X_t is a random variable and, for every ω, X_t as a function of t on T is a step function whose jumps are at certain points (when a customer leaves the shop after service) and an actual jump equals one.   ∎

Example 3 Let $T = [0, 1]$ and let X_t, $t \in T$, be an abstract stochastic process about which we know only that the differences $X_{t+s} - X_t$, $s > 0$, are normal random variables with zero expectation and with variance s. Furthermore, if the intervals $(t_j, t_j + s_j)$, $s_j > 0$, $1 \le j \le k$, are nonoverlapping, then the differences $X_{t_j+s_j} - X_{t_j}$ are independent.

If such a process can be constructed (we shall see that it can), then, for each t, X_t is normal, but the sample functions are difficult to describe.   ∎

The three examples exhibit the different approaches to constructing stochastic processes and possess several characteristics that can be used to develop a theory. These are described below.

Example 1 is a well-defined process and its construction is purely mathematical. Example 2 is a well-formulated applied problem; a reasonable mathematical model should be set up in order to make it accurate. Example 3 is an abstract model for which some properties are postulated; then the question arises, Can such properties be ensured by any mathematical model? When a stochastic process is sought by the approach of Example 3, the properties we want to guarantee might come from either theoretical or practical considerations. In fact, Example 3 will be arrived at in both ways; we call it *Brownian motion*. Example 2 will lead us to the *Poisson process* and will be extended to several other processes whose common characteristics are that they represent a number of occurrences of a particular event as time progresses. A general name for such processes is *point processes*; the particular one that we shall study is the family of *renewal processes*. These will be defined accurately in subsequent sections.

The independence property postulated in Example 3—that is, that the differences $X_{t_j+s_j} - X_{t_j}$, $s_j > 0$, $1 \le j \le k$, $k \ge 2$ are independent when the intervals $(t_j, t_j + s_j)$ are nonoverlapping (and, of course, every t_j and $t_j + s_j \in T$)—is referred to by saying that the process X_t, $t \in T$, has *independent increments*. If the distribution of the vector X_{j_i+s}, $1 \le i \le k$, with all $j_i + s \in T$, is independent of s for all $k \ge 1$, then the process is called *stationary*. Clearly, the Brownian motion defined in Example 3 is a stationary process with independent increments.

A large variety of processes can be obtained by assigning a particular family of distributions to the vectors X_{j_i}, $1 \le i \le k$, $k \ge 1$, $j_i \in T$. For example, if all these vectors are (multivariate) normal, then the process X_t, $t \in T$, is called a Gaussian process. Similarly, stable, extremal, and other processes were introduced on the basis of distributional properties (among which is the property that the univariate, or *marginal, distribution* of X_t for each $t \in T$ must be of the same type of stable distribution, or extreme value distribution, etc.). Usually, a distribution property is combined with some other properties in order to obtain a meaningful result for the process. However, for Gaussian processes, the distribution alone contains a large amount of information about the process, due to the special characteristics of the normal distributions. Recall from Section 2.9 that a multivariate normal distribution, and thus a Gaussian process is completely determined by the expectations $E(X_j)$ and by $E(X_i X_j)$ for all $i, j \in T$. In particular, if $E(X_j) = 0$ for all j and $E(X_i X_j)$ depends only on $r = |i - j|$, then a Gaussian process becomes stationary. Due to this property of the Gaussian process, an arbitrary process is sometimes called *stationary in the wide sense* if the second moment $E(X_t^2)$ is finite for each $t \in T$ and the "covariances" $E(X_{t+s} X_t)$, t, $t + s \in T$, are functions of s only. As a distinction, the stationarity introduced earlier is then called "*strictly stationary*." We shall not need these distinctions below.

Let us return to the property of independent increments. It implies that X_t can be decomposed into sums of independent random variables; thus, several results of Chapters 4 and 5 are readily applicable to X_t. We can express this close relation of X_t to sums of independent random variables by means of conditional probabilities or expectations. We can, in fact, proceed as in the case of the introduction of martingales, but we wish to stress another point this time. Let $Y_1, Y_2, \ldots$ be independent random variables and set $X_n = Y_1 + Y_2 + \cdots + Y_n$. Then $P(X_{n+1} \le x | X_1, X_2, \ldots, X_n) = P(X_{n+1} \le x | X_n)$ a.s. In other words, in regard to the future X_{n+1}, the present X_n alone gives just as much information as all the past $X_1, X_2, \ldots, X_{n-1}$ and the present X_n combined. Such a property is called the Markovian property of a process. In a general and precise form, the process X_t, $t \in T$, is called *Markovian*, or a *Markov process*, if

$$P(X_{t_{n+1}} \leq x | X_{t_1}, X_{t_2}, \ldots, X_{t_n}) = P(X_{t_{n+1}} \leq x | X_{t_n}) \text{ a.s.} \tag{8.1}$$

for all $n \geq 1$ and all parameter values $t_1 < t_2 < \cdots < t_{n+1}$. An immediate consequence of the Markovian property (8.1) is the conditional independence of the "future" and the "past," given the "present." That is, for any positive integers m and n and for all parameter values $u_1 < u_2 < \cdots < u_m < t < t_1 < t_2 < \cdots < t_n$,

$$P(X_{u_j} \leq x_j, 1 \leq j \leq m; X_{t_k} \leq y_k, 1 \leq k \leq n | X_t)$$
$$= P(X_{u_j} \leq x_j, 1 \leq j \leq m | X_t) P(X_{t_k} \leq y_k, 1 \leq k \leq n | X_t) \text{ a.s.} \tag{8.2}$$

In order to see that (8.1) indeed implies (8.2), let us introduce the indicator variables I_1 and I_2 of the events $\{X_{u_j} \leq x_j, 1 \leq j \leq m\}$ and $\{X_{t_k} \leq y_k, 1 \leq k \leq n\}$, respectively. Then, by (CE12) of Section 7.3,

$$E[I_1 E(I_2 | X_t) | X_t] = E(I_2 | X_t) E(I_1 | X_t) \text{ a.s.} \tag{8.3}$$

On the other hand, by the Markov property (8.1),

$$E(I_2 | X_t) = E(I_2 | X_t, X_{u_j}, 1 \leq j \leq m) \text{ a.s.}$$

Thus, by another appeal to Section 7.3 ((CE12) and (CE9)),

$$E[I_1 E(I_2 | X_t) | X_t] = E[E(I_1 I_2 | X_t, X_{u_j}, 1 \leq j \leq m) | X_t]$$
$$= E(I_1 I_2 | X_t) \text{ a.s.} \tag{8.4}$$

Clearly, (8.3) and (8.4) yield

$$E(I_1 I_2 | X_t) = E(I_1 | X_t) E(I_2 | X_t) \text{ a.s.} \tag{8.5}$$

which is just another form of (8.2).

Recall Theorem 7.5, in which we established that the conditional distributions (distribution functions) occurring in (8.2) (and which are defined only a.s.) can always be assumed to be regular; thus, we do so whenever we speak of conditional distributions. In particular, all regular versions of the conditional distributions $P(X_t \in A | X_u)$, $u < t$, where X_t, $t \in T$, is a Markov process, $u, t \in T$, and A goes through the Borel sets of the real line are called the *transition probabilities* of the process. Since, for a Markov process, for any partameter values $s < u < t$,

$$P(X_t \in A | X_s, X_u) = P(X_t \in A | X_u) \text{ a.s.}$$

one more appeal to (CE9) yields

$$P(X_t \in A | X_s) = E[P(X_t \in A | X_u) | X_s] \text{ a.s.} \tag{8.6}$$

This equation is known as the *Chapman–Kolmogorov equation* of the transition probabilities. If T has a smallest value t_0, then the distribution $P(X_{t_0} \in A)$ is called the *initial distribution* of the process. A Markov process is well described by its initial distribution (when it exists) and by its transition probabilities via the Chapman–Kolmogorov equation (8.6). This fact will be demonstrated below for the special case when both T and the X_t are discrete. If the set S of values of X_t for each $t \in T$ is the set $\{1, 2, \ldots\}$ or a subset of it, then the Markov process X_t, $t \in T$, is called a *Markov chain*; when T is discrete also, we speak of a discrete time Markov chain. Note that a subset X_t, $t \in T_1$, of a Markov chain (or process) X_t, $t \in T$, $T_1 \subset T$, is a Markov chain (process) as well.

In introducing all these concepts, we have avoided any discussion of their implications, so we have not faced the following major problem of continuous-time stochastic processes. Let $S = \sup\{X_t: a < t \le b\}$. Then $\{S \le z\}$, z real, is the union of $\{X_t \le z\}$ over all $a < t < b$, which is not denumerable. Hence, S is not guaranteed to be measurable, and thus a random variable when finite. The same is true for the operations of lim inf, lim sup, and lim as t approaches a parameter in an interval. Since we want all these operations to lead to measurable functions, in most cases we impose the condition that a continuous-time stochastic process should have the following property. Let X_t, $t \in T$, be a stochastic process. We call it *separable* if there is an event B with $P(B) = 0$ and a sequence t_j of parameters such that, for every closed set A (including the whole real line) and open interval I, the sets

$$C = \{\omega: X_t(\omega) \in A, t \in I \cap T\} \qquad D = \{\omega: X_{t_j}(\omega) \in A, t_j \in I \cap T\}$$

satisfy

$$C^c \cap D \subset B$$

In other words, the sets C and D differ only on a set that is a subset of an event of zero probability. Thus, if the underlying probability space is complete, then this difference is of zero probability. We thus have that, for separable processes, sets of type C always reduce to the same type (D) but involve a denumerable set of random variables only. In particular, $\sup X_t$, $\inf X_t$, $\limsup X_t$, and $\liminf X_t$ are always measurable for separable stochastic processes.

Note that another way to express separability is that we require the existence of an event B with $P(B) = 0$ and a sequence t_j of parameter values such that, for all open intervals I and for all $\omega \notin B$,

$$\inf\{X_t: t \in I \cap T\} = \inf\{X_{t_j}: t_j \in I \cap T\} \tag{8.7}$$

and

$$\sup\{X_j: t \in I \cap T\} = \sup\{X_{t_j}: t_j \in I \cap T\} \tag{8.8}$$

The reader is invited to verify the equivalence of the joint validity of (8.7) and (8.8) to the previous definition of separability. The equations (8.7) and (8.8) are quite convenient methods for checking whether or not a process is separable. There is a very general case (see Exercise 5) when the sequence t_j can be limited to an arbitrary dense set in T (for example, to rational points).

Exercises

1. Assume the stochastic process X_t, $0 \le t < +\infty$, with $X_0 = 0$ a.s., satisfies the conditions that (i) X_t is a nonnegative integer-valued random variable, (ii) the process has independent increments, and (iii) the distribution of $X_{t+s} - X_t$, $s > 0$, depends only on s. Show that if X_t is not identically zero, then $G(s) = P(X_{t+s} - X_t = 0) = \exp(-as)$, $s \ge 0$, with some $a > 0$.
2. Let X_t, $t \in T$, be a Gaussian process with $E(X_t) = 0$ for all t. Show that if $E(X_{t+s}X_t) = 0$, then X_t and X_{t+s} are independent. This, however, is not a characteristic property of Gaussian variables. Why do some exchangeable sequences of random variables exhibit similar properties?
3. Let Y_1, Y_2,... be independent random variables and set $X_n = Y_1 + Y_2 + \cdots + Y_n$. Let $F_{nm}(x)$ be the distribution function of $X_n - X_m$, $n > m \ge 1$. Show that one version of the conditional distribution function $P(X_n \le x | X_1, X_2, \ldots, X_u)$, $1 \le u < n$, is $F_{nu}(x - X_u)$.
4. Write out the Chapman–Kolmogorov equation in detail if X_n, $n \ge 1$, is a discrete time Markov chain, each X_n taking the integers 0, 1, 2, ..., M only. Calling the transition probabilities $P(X_t = j | X_{t-1} = i) = p_{ij,t}$ one-step transitions, show that if $p_{ij,t}$ does not depend on t, then the initial distribution $P(X_1 = j)$ and the one-step transitions p_{ij} uniquely determine all distributions $P(X_n = j)$, $n \ge 1$, $0 \le j \le M$.
5. Let X_t, $t \in T$, be a separable stochastic process and assume that, for every $t \in T$, $X_s \to X_t$ in probability as $s \to t$, $s \in T$. Show that any sequence t_j, dense in T, can be chosen in the definition of separability (i.e., in (8.7) and (8.8), say).

8.2 POISSON PROCESS

Guided by Example 2 of the previous section, we want to develop a model to describe the (random) arrivals and departures of customers at a service station. However, in order to make the model more widely applicable, we shall speak of the distribution of random points in a collection of sets. This can then be translated to the service problem mentioned, as well as to the number of incoming telephone calls at a large switchboard over an interval of time, earthquakes at a specific location, the distribution of insects in a field, or the stars visible by simple instruments. Purely mathematical applications are also possible. Common to all these applications is the fact that a large number of "individuals" act independently of each other in time or in space and, as the time interval or volume of space becomes very small, the most likely case becomes that in which no individual is observable, and it is very unlikely that more than one individual would be present in such small sets (of time or space). This is made precise in the following model. Both the model and the theorem that follow are due to A. Prékopa (1958).

Let T be a set where a random number of points is distributed. Assume that these points are observable in each of a collection $\mathscr{B}$ of subsets of T and, thus, that there is a probability space $(\Omega, \mathscr{A}, P)$ such that, for $B \in \mathscr{B}$, the number $X(B)$ of points in $\mathscr{B}$ is a random variable. The following assumptions are made: (i) $\mathscr{B}$ is a field of subsets of T, (ii) the set $\{t\} \in \mathscr{B}$ for each $t \in T$ and $P[X(\{t\}) = 0] = 1$, (iii) the random variables $X(B_j)$, $j > 1$, are independent for disjoint sets $B_j \in \mathscr{B}$, and (iv) for every $n \geq n_0$, each $B \in \mathscr{B}$ can be decomposed as a finite union of disjoint sets $B_j^{(n)} \in \mathscr{B}$ such that

$$P[X(B_j^{(n)}) = 0] \geq 1 - 2^{-n} P(X(B) = 0) \tag{8.9}$$

and, with some constant $c > 0$,

$$P[X(B_j^{(n)}) \geq 2] \leq c\{1 - P[X(B_j^{(n)}) = 0]\}^2 \tag{8.10}$$

Remark   It can be seen from the proof of the main result below (Theorem 1) that 2^{-n} in (8.9) can be replaced by any function of n that goes to zero as $n \to +\infty$. In addition, the quadratic function on the right-hand side of (8.10) can be replaced by any function that goes faster to zero than cx, as $x \to 0$.

Theorem 1   Let T, $\mathscr{B}$, and the distributions of $X(B)$, $B \in \mathscr{B}$, satisfy (i)–(iv). Then

$$P_k(B) = P(X(B) = k) = \frac{\lambda^k(B) e^{-\lambda(B)}}{k!} \qquad k = 0, 1, \ldots \tag{8.11}$$

where $\lambda(B)$ is a finite and nonnegative additive set function on $\mathscr{B}$.

Definition   The random variables $X(B)$, $B \in \mathscr{B}$, are called a Poisson process on $(T, \mathscr{B})$ with intensity measure $\lambda(B)$ whenever condition (iii) and formula (8.11) apply. If T is an arbitrary bounded Borel set of the k-dimensional Euclidean space and $\mathscr{B}$ is the set of Borel subsets of T and if, furthermore, $\lambda(B)$ is proportional to Lebesgue measure, then the corresponding Poisson process is called a (k-dimensional) Poisson process.

Proof. Note that (iii) and (8.9) ensure that $P_0(B) > 0$ for all $B \in \mathscr{B}$. Set

$$\lambda(B) = -\log P_0(B) \qquad B \in \mathscr{B}$$

Then $\lambda(B) \geq 0$, finite, and, on account of (iii) again, additive. Furthermore, by (ii), $\lambda(\{t\}) = 0$ for all $t \in T$.

Now, for $B \in \mathscr{B}$, and for the decomposition in (iv), define

$$Y(B_j^{(n)}) = \begin{cases} 1 & \text{if } X(B_j^{(n)}) = 1 \\ 0 & \text{otherwise} \end{cases}$$

Then $\sum_j Y(B_j^{(n)}) \neq X(B)$ means that, for at least one j, $X(B_j^{(n)}) \geq 2$. Hence, in view of (8.9) and (8.10),

$$P\left(\sum_j Y(B_j^{(n)}) \neq X(B)\right) \leq \sum_j P(X(B_j^{(n)}) \geq 2)$$

$$\leq c2^{-n} \sum_j [1 - P_0(B_j^{(n)})]$$

However, $1 - z \leq -\log z, \ 0 < z \leq 1$; thus,

$$\sum_j [1 - P_0(B_j^{(n)})] \leq \sum_j \lambda(B_j^{(n)}) = \lambda(B)$$

where the additivity of $\lambda(\cdot)$ is used. Hence,

$$\lim_{n = +\infty} \sum_j Y(B_j^{(n)}) = X(B) \text{ in probability} \tag{8.12}$$

Consequently, we can determine the distribution of $X(B)$ as the (weak) limit of the distribution of $Y_n = \sum_j Y(B_j^{(n)})$, where, by (iii), the summands $Y(B_j^{(n)})$ are independent. We turn to characteristic functions. Since

$$\varphi_{j,n}(t) = E\{\exp[it\,Y(B_j^{(n)})]\} = 1 - P_1(B_j^{(n)}) + e^{it} P_1(B_j^{(n)})$$

by independence,

$$\varphi_n(t) = E[\exp(it\,Y_n)] = \prod_j [1 + P_1(B_j^{(n)})(e^{it} - 1)] \tag{8.13}$$

Next, we write

$$P_1(B_j^{(n)}) = 1 - P_0(B_j^{(n)}) - P(X(B_j^{(n)}) \geq 2) = \lambda(B_j^{(n)}) + \eta\lambda^2(B_j^{(n)})$$

where the extreme right-hand side is justified by (8.10) and by the Taylor formula $(0 < z \leq 1)$

$$\log z = \log(1 - (1 - z)) = -(1 - z) + \vartheta(1 - z)^2$$

where $|\vartheta| \leq 1$, if $z \geq 1/2$ (and thus $|\eta| \leq 1 + c$ in the previous equation).

Hence, because of (8.9), we can take the logarithm of (8.13). We have, by Taylor's expansion,

$$
\begin{aligned}
\log \varphi_n(t) &= \sum_j \lambda(B_j^{(n)})(e^{it} - 1) + \delta \sum_j \lambda^2(B_j^{(n)}) \\
&= \lambda(B)(e^{it} - 1) + \delta * 2^{-n} \sum_j \lambda(B_j^{(n)}) \\
&= \lambda(B)(e^{it} - 1 - \delta * 2^{-n})
\end{aligned}
$$

where δ and $\delta*$ are bounded values. With an appeal to the continuity theorem of characteristic functions, we obtain (8.11) by (8.12) and (8.13). The theorem is established. ∎

The question arises: Could we have obtained the same Poisson distribution (8.11) under essentially weaker assumptions than those in our model? We have remarked that we could have replaced 2^{-n} in (8.9) and the square function in (8.10) by other functions, which, however, would not change the emphasis of those conditions. Note also that the independence postulated in (iii) was not fully used in the proof. Instead, we used the independence of the events $\{X(B_j) = 0\}$ and $\{X(B_j) = 1\}$ for some sets B_j for distinct j. The next theorem shows that the weakening of the independence assumption of (iii) to the independence of types of events just mentioned would not result in a generalization. The following result is due to A. Rényi (1967).

Theorem 2 Let T be an arbitrary finite interval $(C, D]$ on the real line and let $\mathscr{B}$ be the family of finite unions of finite subintervals $(a, b]$ of T. If (8.11) and condition (ii) of our model hold and if $\lambda((a, b]) \to 0$ as $b - a \to 0$, then condition (iii) holds as well, i.e., the process $X(B)$, $B \in \mathscr{B}$, is Poisson (for arbitrary T).

Proof. Let $I_i = (a_i, b_i]$, $1 \le i \le n$, be disjoint subintervals of T and let B be their union. Then $B \in \mathscr{B}$ and

$$
\lambda(B) = \sum_{i=1}^n \lambda(I_i)
$$

Hence, by (8.11),

$$
P_0(B) = e^{-\lambda(B)} = \prod_{i=1}^n e^{-\lambda(I_i)} = \prod_{i=1}^n P_0(I_i) \tag{8.14}
$$

i.e., the events $\{X(I_i) = 0\}$ are independent.

Next, let B and $B*$ be two disjoint sets from $\mathscr{B}$. By assumption, we can decompose B and B^* as the union of a finite number of disjoint intervals I_i and I_i^*, respectively, with the property that I_i, $I_i^* \in \mathcal{B}$, and

$$\max_i \lambda(I_i) < \varepsilon \qquad \text{and} \qquad \max_i \lambda(I_i^*) < \varepsilon \qquad (8.15)$$

For every I_i and I_i^*, we introduce the indicator variables

$$Y_i = \begin{cases} 1 & \text{if } X(I_i) \geq 1 \\ 0 & \text{otherwise} \end{cases} \qquad Y_i^* = \begin{cases} 1 & \text{if } X(I_i^*) \geq 1 \\ 0 & \text{otherwise} \end{cases}$$

Now, arguing as in the previous proof,

$$X(B) \neq \sum_i Y_i \qquad \text{implies} \qquad \max_i X(I_i) \geq 2 \qquad (8.16)$$

and

$$X(B^*) \neq \sum_i Y_i^* \qquad \text{implies} \qquad \max_i X(I_i^*) \geq 2$$

But, by (8.11),

$$P(X(I_i) \geq 2) = \sum_{k=2}^{+\infty} \frac{\lambda^k(I_i) e^{-\lambda(I_i)}}{k!} \leq \lambda^2(I_i)$$

Thus, by (8.15) and (8.16),

$$P\left(X(B) \neq \sum_i Y_i \right) \leq \sum_i \lambda^2(I_i) < \varepsilon \sum_i \lambda(I_i) = \varepsilon \lambda(B) \qquad (8.17)$$

Similarly,

$$P\left(X(B^*) \neq \sum_i Y_i^* \right) \leq \varepsilon \lambda(B^*) \qquad (8.18)$$

But, as demonstrated at (8.14), the indicator variables Y_i and Y_j^* are independent for all i and j; thus, $\sum Y_i$ and $\sum Y_i^*$ are independent of each other. Consequently, (8.17) and (8.18) entail

$$|P(X(B) = u, X(B^*) = v) - P(X(B) = u)P(X(B^*) = v)| \leq 2\varepsilon\lambda(B \cup B^*)$$

Since $\varepsilon > 0$ is arbitrary, we have that $X(B)$ and $X(B^*)$ are independent. No change is required when repeating this argument to show that if B_1, $B_2, \ldots, B_s$ are disjoint sets from $\mathcal{B}$, then the random variables $X(B_j)$, $1 \leq j \leq s$, are independent. This completes the proof. ∎

We now specialize. We choose $T = (0, D]$, with $D > 0$ arbitrary and $\mathcal{B}$ as in Theorem 2, i.e., $\mathcal{B}$ contains all finite unions of disjoint subintervals $(a_i, b_i]$ of T. Furthermore, we assume that $\lambda(B)$ is $\lambda m(B)$, where $m(\cdot)$ is Lebesgue measure and $\lambda > 0$ is a constant not depending on T (i.e., on D). Hence, the role of T is completely eliminated, so long as complements of the sets B are not considered. This leads to the introduction of the random variables $X(B)$ for all finite unions of disjoint finite intervals $(a_i, b_i]$. We assume that condition (iii) and formula (8.11) apply, that is, $X(B)$ is a Poisson process on the (just specified subsets of the) nonnegative real line. Note that the independence assumption in (iii) and (8.11) completely determine $X(B)$ if the process

$$X_t = X((0, t]) \qquad t > 0 \tag{8.19}$$

is given. Formula (8.11) reduces to

$$P(X_t = k) = \frac{(\lambda t)^k e^{-\lambda t}}{k!} \qquad k = 0, 1, \ldots \tag{8.20}$$

We set $X_0 = 0$, which is in agreement with condition (ii). Thus, the process X_t, $t \geq 0$, defined by (8.19), (8.20), and condition (iii) is a process with stationary independent increments. We shall refer to this family of processes as *homogeneous Poisson processes on the positive real line*. Note that every such process satisfies condition (iv) of our general model, where, in (8.10), the constant c is meant to be constant on any finite interval of t. To see this, first observe that it suffices to prove (iv) for finite intervals I only, for which, by (8.20),

$$P_0(I) = P(X(I) = 0) = e^{-\lambda m(I)}$$

Now, if I is divided into 2^{2n} equal parts, the jth of which being denoted by $B_j^{(n)}$, then (8.9) follows by observing that $m(B_j^{(n)}) = 2^{-2n} m(I)$ and, thus,

$$P_0(B_j^{(n)}) = \exp[-\lambda 2^{-2n} m(I)] \geq 1 - \lambda 2^{-2n} m(I) > 1 - 2^{-n} P_0(I)$$

where the last inequality is valid for all large n and in the last but one inequality we utilized $1 - x \leq e^{-x}$. On the other hand, (8.10) has actually been demonstrated in the course of the last proof (just before (8.17)) because, for $I \subset (0, T]$ with $T > 0$,

$$\lambda^2(B_j^{(n)}) = (\lambda 2^{-2n} m(I))^2 \leq 4\{1 - \exp[-\lambda 2^{-2n} m(I)]\}^2$$

for all large n, uniformly in $m(I) \leq T$. The last inequality comes from the elementary inequality $1 - x/2 \geq e^{-x}$, $0 \leq x \leq 1$.

Let us turn to the investigation of the structure of random points in a homogeneous Poisson process on the positive real line. Let S_1 be the first point of this process. Then, since $\{S_1 > t\} = \{X_t = 0\}$, (8.11) yields

$$P(S_1 \leq t) = 1 - P(X_t = 0) = 1 - e^{-\lambda t} \qquad t > 0 \tag{8.21}$$

that is, the first "arrival" in this Poisson process is an exponential variable with expectation $1/\lambda$. Let us denote the successive points of the real line which are being "counted" by the process X_t by $0 < S_1 < S_2 < S_3 < \cdots$; for easier reference, we call the differences $S_j - S_{j-1} = D_j$, $j \geq 1$, where $S_0 = X_0 = 0$, *interarrival times.*

Theorem 3 The interarrival times D_j, $j \geq 1$, in a homogeneous Poisson process on the positive real line are independent and identically distributed exponential variables with $E(D_j) = 1/\lambda$.

Proof. We prove the theorem by induction. For D_1, the theorem has been proved at (8.21). We now calculate the joint distribution of D_1 and D_2. After conditioning on $D_1 = x$ and taking expectation, we get

$$P(D_1 \leq u, D_2 > v) = \int_0^u P(S_2 > v + x | D_1 = x)\lambda e^{-\lambda x}\, dx$$

However, given $D_1 = S_1 = x$, $\{S_2 > v + x\} = \{X((x, v + x]) = 0\}$. Hence, by (8.11), the conditional probability behind the integral equals $e^{-\lambda v}$, that is,

$$P(D_1 \leq u, D_2 > v) = e^{-\lambda v} \int_0^u \lambda e^{-\lambda x}\, dx = e^{-\lambda v}(1 - e^{-\lambda u})$$

which is the statement of the theorem for D_1 and D_2. Next, assume that the theorem has been proved for $D_1, D_2, \ldots, D_k$. Then the joint distribution of D_j, $1 \leq j \leq k + 1$, calculated by first conditioning on $D_j = x_j$, $1 \leq j \leq k$ and then taking the expectation,

$$P(D_j \leq u_j, 1 \leq j \leq k, D_{k+1} > x_{k+1})$$

$$= \lambda^k \int_0^{u_1} \cdots \int_0^{u_k} P\left(S_{k+1} > \sum_{j=1}^{k+1} x_j \middle| D_i = x_i, 1 \leq i \leq k\right)$$

$$\times \exp\left(-\lambda \sum_{i=1}^k x_i\right) dx_1 \cdots dx_k$$

But the condition is the same as $S_1 = x_1$, $S_2 = x_1 + x_2, \ldots$, $S_k = x_1 + x_2 + \cdots + x_k$; thus, the conditional probability behind the integral once more becomes $P(X((a_k, a_k + x_{k+1}]) = 0)$, where a_k is the sum of the x_i, $1 \leq i \leq k$. Hence, (8.11) is applicable, by which the above multiple integral immediately yields the desired formula for the joint distribution of the D_j, $1 \leq j \leq k + 1$. The proof is completed. ■

The theorem gives a new interpretation of the Poisson processes (on the positive real line). Assume that a machine is put into service at time $t = 0$. It functions for a random length of time S_1, breaks down, is repaired (or replaced) instantaneously, then functions, independently of S_1, for another length of time D_2, and so on. If $S_1 = D_1, D_2, D_3, \ldots$ are identically distributed exponential variables, then the random epochs of repairs form a Poisson process (although a converse to Theorem 3 is used in this discussion, we shall establish in Exercise 2 that the distribution property of the interarrivals $D_1, D_2, \ldots$, stated in Theorem 3, is both necessary and sufficient for $0 < S_1 < S_2 < \cdots$ to form a homogeneous Poisson process on the positive real line). In the next section, we shall generalize the concept of a Poisson process by considering random points $0 < S_1 < S_2 < \cdots$ on the positive real line whose differences are independent and identically distributed but not necessarily exponential. Such processes are called *renewal processes*, a term suggested by the repair (or renewal) model of this paragraph.

Exercises

1. Let X_t, $t > 0$, be a nonnegative integer-valued stationary stochastic process with independent increments. Assume that X_t is not identically zero for any $t > 0$. Show that if X_{t_0} is a Poisson variable for some $t_0 > 0$, then the distribution of X_t is Poisson for each $t > 0$. (Recall Exercise 2 of Section 4.2.)

2. Let D_j, $j \geq 1$, be i.i.d. exponential variables and let $S_k = D_1 + D_2 + \cdots + D_k$, $k \geq 1$. Let $X_t, t > 0$, be the number of S_k that do not exceed t. Show that $X_t, t > 0$, is a homogeneous Poisson process on the positive real line.

3. Let $0 < S_1 < S_2 < \cdots$ be the random points of a homogeneous Poisson process X_t, $t > 0$, on the positive real line. Show that the conditional distribution of $S_1, S_2, \ldots, S_n$, given $X_t = n$, coincides with the joint distribution of the n order statistics of n independent random variables, each of which is uniformly distributed on $(0, t)$.

4. Let $g(x)$ be a positive, continuous, and bounded function for $x > 0$. For an interval $I = (a, b]$, $0 \leq a < b < +\infty$, define

$$\lambda(I) = \int_a^b g(x)\, dx$$

and extend $\lambda(B)$ by additivity to all B that are finite unions of intervals of the type of I. Let $X(B)$ be a random variable for each such set B and assume that (a) $X(B)$ is nonnegative and integer valued, (b) $P(X(B) = 0) = \exp[-\lambda(B)]$, and (c) $P(X(I) \geq 2) \leq \alpha \lambda^2(I)$, $\alpha > 0$ for all intervals. Show that the conditions of Theorem 1 are satisfied with an appropriate choice of T ($\mathscr{B}$ is the collection of the sets B above) and, thus, that (8.11) applies. Find the distribution of S_1 and the joint distribution of S_1 and S_2, where $S_1 < S_2 < \cdots$ are the random points of the process $X_t = X((0, t])$, $t > 0$.

5. Let X_t, $t > 0$, be a homogeneous Poisson process on the positive real line. For a fixed value $t_0 > 0$, let $N = N(t_0)$ be the smallest integer n such that $t_0 \leq S_n$, where S_n is defined as in Exercise 3. Show that the distribution of (the waiting time) $S_N - t_0$ is the same exponential distribution as the common distribution of the $D_j = S_j - S_{j-1}$, $j \geq 1$.

8.3 RENEWAL PROCESSES

We introduced the concept of a renewal process in the last paragraph of the previous section. That is, $0 < S_1 < S_2 < \cdots$ form the random points of a renewal process if $S_0 = 0$ and

$$S_n = Y_1 + Y_2 + \cdots + Y_n \qquad n \geq 1$$

where the Y_j are independent random variables with a common distribution function $F(x)$ such that $F(0) = 0$. The associated stochastic process X_t, $t \geq 0$, where $X_t = k$ if $S_k \leq t < S_{k+1}$, is called a renewal process. For easier reference, we call the points S_k, $k \geq 0$, the *renewal points*; $Y_j = S_j - S_{j-1}$, $j \geq 1$, the *interarrival times*; and their distribution $F(x)$ the *interval distribution* of the process. Clearly, if $F(x)$ is exponential, then a renewal process becomes a Poisson process.

Because the distribution function $G_n(x)$ of S_n is the n-fold convolution of $F(x)$, we also use the alternative notation $G_n(x) = F^{(n*)}(x)$. Furthermore, we set $P_k(t) = P(X_t = k)$.

Since $P(X_t \geq n) = P(S_n \leq t)$, we have the formal equation

$$E(X_t) = \sum_{k=1}^{+\infty} kP_k(t) = \sum_{k=1}^{+\infty} P(X_t \geq k) = \sum_{n=1}^{+\infty} F^{(n*)}(t)$$

Hence, with the notation

$$R(x) = \sum_{n=0}^{+\infty} F^{(n*)}(x) \tag{8.22}$$

where $F^{(0*)}(x) = 1$ for all $x \geq 0$ (and 0 for $x < 0$), we proved that

$$E(X_t) = R(t) - 1 \tag{8.23}$$

provided that we establish that $R(t)$ is finite for all t. We call the function $R(x)$ the *renewal function* of the process.

To prove that $R(t)$ is finite for all t, first note that if $F(x)$ is degenerate at a point $c > 0$, then both sides of (8.23) are clearly finite for every fixed t. On the other hand, if $F(x)$ is not degenerate, then, since $F(0) = 0$ and

$F(+\infty) = 1$, there is a $y > 0$ such that $0 < F(y) < 1$. Hence, for every $m \geq 1$, $F^{(m*)}(my) < 1$ because

$$1 - F^{(m*)}(my) = P(S_m > my) \geq P(Y_j > y, 1 \leq j \leq m) = [1 - F(y)]^m > 0$$

Furthermore, for any $z > 0$,

$$F^{(n*)}(z) = \int_0^z F^{(k*)}(z - x)\, dF^{(n-k,*)}(x) \leq F^{(k*)}(z) F^{(n-k,*)}(z)$$

i.e., $F^{(n*)}(z) \leq (F^{(m*)}(z))^d$, where $n \geq md$. Now, if $x > 0$ is arbitrary, we choose (and fix) $m \geq 1$ so that $my \geq x$. Then, with d the integer part of n/m,

$$F^{(n*)}(x) \leq F^{(n*)}(my) \leq [F^{(m*)}(my)]^d$$

The definition of $R(x)$ at (8.22) now entails that $R(x)$ is indeed finite.

Since $R(x)$ is always finite, (8.23) extends to finite intervals $(a, b]$, $0 \leq a < b$, as

$$R(b) - R(a) = \sum_{n=1}^{+\infty} [F^{(n*)}(b) - F^{(n*)}(a)] = E(X_b - X_a)$$

which is the expected number of renewal points in the interval $(a, b]$. In particular, if the Y_j are integer valued, then, with $a = k - 1/2$, say, and $b = k$, where k is an integer,

$$R(b) - R(a) = \sum_{n=1}^{+\infty} P(S_n = k) = E(X(k)) \tag{8.24}$$

where $X(k)$ is the number of renewals equal to k. Since $F(0) = 0$, i.e., $P(Y_j = 0) = 0$, $X(k)$ is either 1 or 0 depending on whether $S_n = k$ for one n or S_n is never k; thus, $R(b) - R(a)$ in (8.24) is the probability that the points $S_1 < S_2 < \cdots$ ever hit the value k. There are obvious cases for which this probability is zero; for example, if each Y_j is even, say, then S_n can never be odd. But when it is not assured a priori that certain kinds of integer cannot be hit by the sequence S_n, then, as we shall establish in Theorem 4, it will hit every large value k with the same asymptotic probability $1/\mu$, where $\mu = E(Y_j)$, assumed positive and finite.

Note that if we take the convolution of $F(x)$ and $R(x)$ in (8.22), we get the equation

$$R(x) = 1 + R(x) * F(x) \qquad x > 0 \tag{8.25}$$

This equation can be used as the definition of $R(x)$ since its solution is unique among the functions which are identically zero for negative x and which are bounded over finite intervals of the positive real line (Exercise 1). Hence, (8.25) is called the renewal equation.

Recall from Section 8.1 that (strict) stationarity of X_t entails that (X_0, X_s) (that is, X_s since $X_0 = 0$) and (X_t, X_{t+s}), $s > 0$, have the same distribution, yielding that $X_{t+s} - X_t$ is distributed as X_s. That is, $E(X_{t+s} - X_t) = E(X_s)$. Hence, for $t, s > 0$,

$$E(X_{t+s}) = E(X_{t+s} - X_t) + E(X_t) = E(X_s) + E(X_t)$$

which is Cauchy's functional equation (6.2) for $E(X_t)$ as a function of t. By the finiteness and measurability of $R(t) - 1 = E(X_t)$, Theorem 6.2 implies that for a strictly stationary renewal process X_t, $R(t) = 1 + R(1)t$, $t > 0$. We thus have that, in view of (8.20) and (8.23), a Poisson process on the positive real line X_t is strictly stationary (the statements above can be reversed) but X_t is not stationary for a general renewal process. However, a very little modification of X_t guarantees its stationarity. Let us modify X_t as follows. Let $F(x)$ be as before but we assume that

$$a = \int_0^{+\infty} (1 - F(x)) \, dx < +\infty$$

Set

$$F_1(x) = \frac{1}{a} \int_0^x (1 - F(y)) \, dy$$

Let us replace Y_1 by another random variable Y_1^* whose distribution function is $F_1(x)$ and which is independent of Y_j, $j \geq 2$. Let

$$S_n^* = Y_1^* + Y_2 + \cdots + Y_n \qquad n \geq 1$$

Finally, let X_t^* be the process counting the number of points S_n^* in the interval $(0, t)$; i.e., $X_t^* = k$ if $S_k^* \leq t < S_{k+1}^*$. Put $m(t) = E(X_t^*)$. A formula similar to (8.22–8.23) still applies to $m(t)$ except that in the n-fold convolution at (8.22) the first distribution is F_1 while the rest are equal to F. Hence, by separating the term $n = 1$, we have

$$m(t) = F_1(t) + m(t) * F(t)$$

Let us turn to Laplace transforms. Upon denoting the Laplace transforms of $m(t)$, $F_1(t)$ and $F(t)$ by $\text{Lap}_m(z)$, $\text{Lap}_1(z)$ and $\text{Lap}(z)$, respectively, we have from (3.43) and the equation above

$$\text{Lap}_m(z) = \text{Lap}_1(z) + \text{Lap}_m(z) \, \text{Lap}(z)$$

that is, $\text{Lap}_m(z) = \text{Lap}_1(z)/[1 - \text{Lap}(z)]$, $z > 0$. But, from the definition of $F_1(z)$, one gets by easy calculation that

$$\text{Lap}_1(z) = [1 - \text{Lap}(z)]/az$$

and thus $\text{Lap}_m(z) = 1/az$. By the uniqueness of Laplace transforms (Theorem 3.24) we thus proved $m(t) = t/a$ which is equivalent to the stationarity of X_t^* (see also the discussion following the proof of Theorem 6).

In order to formulate the next theorem, we introduce the following concept. Assume that the random variable U takes nonnegative integers only and $P(U = k) = 0$, $k \geq 0$ integer, unless $k = md$, $m = 0, 1, 2, \ldots$, where $d > 1$ is a fixed integer. Then we say that U has a periodic distribution. If d is the largest integer with the above-mentioned property, then d is called the period of the distribution of U. Otherwise, the distribution of U is called aperiodic.

Theorem 4 Let p_k, $k \geq 0$, be an aperiodic probability distribution whose expected value is μ, possibly $+\infty$. Let r_k and z_k be nonnegative numbers such that r_k is bounded and $B = \sum z_k < +\infty$. Assume that the extended renewal equation

$$r_n = z_n + \sum_{k=0}^{n} p_{n-k} r_k \qquad n \geq 0 \tag{8.26}$$

holds. Then, as $n \to +\infty$, $\lim r_n$ exists and is zero if $\mu = +\infty$ and B/μ if $\mu < +\infty$.

Remark Note that if Y_j has an aperiodic discrete distribution p_k and $r_k = R(k) - R(k - 1/2)$, say, then, with $z_n = 1$ or 0 depending on whether $n = 0$ or $n > 0$, (8.26) becomes (8.25); thus, the conclusion of the theorem contains what was said about (8.24). Because of this relation, the theorem is called a Renewal Theorem; in its present form, it is due to Erdös, Feller, and Pollard (1949). A number of extensions of this theorem are known; the first major generalization of it is due to Blackwell (1953), who showed that the conclusion of Theorem 4 continues to hold if one works directly with (8.25), in which the distribution of Y_j does not have to be discrete (only its expectation is assumed to be positive). Recent generalizations of Theorem 4 in the form of (8.24), with a higher dimensional index replacing n, show interesting relations of this problem to both the central limit theorem with good error terms and to deep problems of number theory. Due to this latter relation, we shall not go into these generalizations; rather, the reader is referred to the papers of Maejima and Mori (1984) and Galambos, Indlekofer, and Kátai (1987).

Proof. This proof also has a number-theoretic flavor, but it is elementary. Let $\lim \sup r_n = \gamma$ as $n \to +\infty$. By assumption, $\gamma < +\infty$. Let n_j be a subsequence of the integers such that $r_{n_j} \to \gamma$. Let k^* be such that $p_{k^*} > 0$. Then $r_{n_j - k^*} \to \gamma$ as well. Namely, if $r_{n_j - k^*} \leq \delta < \gamma$ for infinitely many j, then, by (8.26) and by the fact that $\{p_t\}$ is a probability distribution, furthermore, by $r_n \leq K$, finite, we would have, for all large j,

$$r_{n_j} = z_{n_j} + \sum_{k=0}^{n_j} p_k r_{n_j-k} \leq \varepsilon + \sum_{k=0}^{N} p_k r_{n_j-k} + K \sum_{k=N+1}^{+\infty} p_k$$

$$< 2\varepsilon + (\gamma + \varepsilon) \sum_{\substack{k \geq 0 \\ k \neq k^*}} p_k + p_{k^*}\delta < 3\varepsilon + \gamma(1 - p_{k^*}) + \delta p_{k^*}$$

which contradicts the assumption that $r_{n_j} \to \gamma$. We can now choose k^* as any of the subscripts for which $p_k > 0$ and any of their integer multiples, finally, their sums. That is, if

$$s = a_1 k_1 + a_2 k_2 + \cdots + a_u k_u \qquad p_{k_j} > 0, \, a_j \geq 0 \text{ integer} \qquad (8.27)$$

then $r_{n_j-s} \to \gamma$. But, since the aperiodicity of the distribution $\{p_k\}$ ensures that the largest common divisor of those integers k for which $p_k > 0$ is one, a result of number theory states (the proof of which is given at the end of this proof) that all large s can be written in the form of (8.27). In turn, this implies, via (8.26), that $r_{n_j-m} \to \gamma$ for all m. That is, if $r_{n_j-s} \to \gamma$ for all $s > m$, then

$$r_{n_j-m}(1 - p_0) = z_{n_j-m} + \sum_{k=1}^{n_j-m} p_k r_{n_j-m-k} \to \gamma \sum_{k=1}^{+\infty} p_k = \gamma(1 - p_0)$$

i.e., $r_{n_j-m} \to \gamma$ because $1 - p_0 = 0$ is not possible due to the aperiodicity of the distribution $\{p_k\}$.

Next, we set $q_k = p_k + p_{k+1} + \cdots$. Then

$$\sum_{k=1}^{+\infty} q_k = \sum_{k=1}^{+\infty} k p_k = \mu \qquad (8.28)$$

Furthermore, (8.26) can be rewritten as

$$r_n - p_0 r_n - \sum_{k=1}^{n} (q_k - q_{k+1}) r_{n-k} = z_n$$

i.e., since $1 - p_0 = q_1$,

$$\sum_{k=0}^{n} q_{k+1} r_{n-k} = \sum_{k=1}^{n} q_k r_{n-k} + z_n$$

Upon setting $Q_n = \sum_{k=0}^{n} q_{k+1} r_{n-k}$, we thus have

$$Q_n = Q_{n-1} + z_n \qquad n \geq 1, \text{ with } Q_0 = q_1 r_0 = (1 - p_0) r_0 = z_0$$

By summing these equations, we have $Q_n = z_0 + z_1 + \cdots + z_n$. Now, we return to the subsequence n_j for which we have established $r_{n_j - m} \to \gamma$ for all $m \geq 0$. We have

$$B \geq \sum_{k=0}^{n_j} z_k = Q_{n_j} = \sum_{k=0}^{n_j} q_{k+1} r_{n_j - k} \geq \sum_{k=0}^{M} q_{k+1} r_{n_j - k} \qquad (8.29)$$

where M is an arbitrary fixed number and n_j is assumed to be large. By letting $n_j \to +\infty$, we obtain $B \geq \gamma(q_1 + q_2 + \cdots + q_M)$. Hence, if $\mu = +\infty$ (recall (8.28)), $\gamma = 0$, and the corresponding part of the theorem is proved. On the other hand, if $\mu < +\infty$, then we obtain from this same last estimate

$$\gamma \leq B/\mu \qquad (8.30)$$

In order to show that the right-hand side is, in fact, the limit of r_n, we show that a reversed inequality is valid with $\liminf r_n = \rho$ whenever $\mu < +\infty$. Let m_j be another subsequence of the integers such that $r_{m_j} \to \rho$. Just as in the first part, we get $r_{m_j - m} \to \rho$ for all $m \geq 0$. Recall the notation $r_n \leq K$. Then, estimating in the opposite direction in (8.29), we get

$$\sum_{k=0}^{m_j} z_k \leq \sum_{k=0}^{M} q_{k+1} r_{m_j - k} + K \sum_{k=M+1}^{+\infty} q_{k+1} \qquad (8.31)$$

Let $m_j \to +\infty$ and then $M \to +\infty$ in this last estimate (8.31). It indeed results in $\rho \geq B/\mu$, which, together with (8.30), entails that the limit of r_n exists and equals B/μ, as desired.

It remains to prove the statement at (8.27). Write all the linear combinations

$$c_1 k_1 + c_2 k_2 + \cdots + c_u k_u \qquad p_{k_j} > 0, \; c_j \text{ integer} \qquad (8.32)$$

We show that all integers are represented in (8.32), where we permit c_j to be negative. First, assume that there are two consecutive integers m and $m + 1$ in (8.32). Then $(m + 1) - m$ also has a representation in (8.32), and so does every positive integer v, since they all can be written as v times the representation of one. Next, we assume that there are no consecutive integers in (8.32). Let g be the smallest gap among all integers in (8.32). Then, we have $g, 2g, 3g, \ldots$ in (8.32) and no others; otherwise, we would have a smaller gap if we had others. Let k_j be such an integer that $p_{k_j} > 0$ and not all divisors of g are divisors of k_j. Such a k_j must exist by the assumption of aperiodicity. But then $g + k_j$ is in (8.32), which is not of the form mg, which contradiction yields that $g = 1$. Now, to see that c_j can be taken to be nonnegative for all large s, as in (8.27), let the representation of the number one at (8.32) be $1 = c_1^* k_1 + c_2^* k_2 + \cdots + c_t^* k_t$ and set $d = k_1 + k_2 + \cdots + k_t$.

Furthermore, let $m = \max(d|c_j^*|: 1 \le j \le t)$. Then, for $s \ge (m+1)d$,

$$s = Ad + C = A \sum_{j=1}^{t} k_j + C \sum_{j=1}^{t} c_j^* k_j = \sum_{j=1}^{t} (A + Cc_j^*)k_j$$

where the division by d was done so that $0 \le C \le d - 1$; thus, each coefficient $A + Cc_j^* \ge (m+1) - m = 1$, which completes the proof of (8.27) and thus the theorem as well.   ■

Let us return to the Remark following the statement of Theorem 4. There we observed that, if Y_j is positive integer valued with an aperiodic distribution and finite expectation a, then Theorem 4 ($a = \mu$ in this connection), via (8.25), entails that, as $k \to +\infty$, $R(k) - R(k-1) \to 1/a$. From this the (weaker) asymptotic formula $R(t) - R(t-b) = b/a + o(b)$, for $0 < b < t$, follows, where b is either bounded or tends to infinity with t. The actual error term for $b = t - 1$, say, can be calculated by the following coupling method, due to T. Lindvall (1992) (the coupling method is a way of simplifying calculations by introducing an additional variable which, at the end, can be eliminated from the conclusions).

We start with a general renewal process X_t, and we count $S_0 = 0$ in $X_t' = 1 + X_t$. With our previous notation, we add Y_1^* to the sequence Y_j, $j \ge 1$, by possibly enlarging the probability space. Define $S_n^* = Y_1^* + S_n$, $n \ge 0$. Then the process X_t^* counting the points S_n^* in the interval $(0, t)$ is stationary, for which $m(t) = E(X_t^*) = t/a$. Hence,

$$R(t) - t/a = R(t) - m(t) = E(X_t') - E(X_t^*) = E(X_t' - X_{t-Y_1^*}')$$

Now, by the independence of Y_1^* and X_t', the conditional expectation

$$E(X_t' - X_{t-Y_1^*}' | Y_1^*) = R(t) - R(t - Y_1^*) \text{ a.s.}$$

where $R(s) = 0$ for $s < 0$. By taking expectation, we thus have

$$R(t) - t/a = E[R(t) - R(t - Y_1^*)] \le E(R(Y_1^*))$$

which is generally true, and meaningful only if the extreme right hand side is finite. Such is the case if $R(t)$ is bounded by a linear function in t and if $E(Y_j^2) < +\infty$, implying $E(Y_1^*) < +\infty$. In particular, when a renewal theorem of the form discussed earlier in the discrete case is valid then $R(t) - t/a$ remains bounded as $t \to +\infty$ whenever $E(Y_j^2) < +\infty$. (With very little effort the reader can find the exact error term under the validity of a renewal theorem and the finiteness of the second moment of Y_j by appealing to the dominated convergence theorem and the renewal theorem with $b = Y_1^*$.)

Note that, from another point of view, the question of stationarity is about the relation of the renewal measure (based on $R(t)$ or $m(t)$) to Lebesgue measure, and the renewal theorem guarantees an approximation by the renewal measure to a measure proportional to Lebesgue measure.

In the next theorems we express certain distributions by means of the renewal function. As before, let $S_1 < S_2 < \cdots$ be the positive renewal points of the process. For a fixed $t > 0$, let $N = N(t)$ be the smallest integer n such that $t < S_n$. Hence, the random variables $w(t) = S_N - t$ and $u(t) = t - S_{N-1}$ are nonnegative. While $w(t) + u(t) = S_N - S_{N-1}$, its distribution is not $F(x)$ due to the random index N (recall Exercise 5 of the previous section).

Theorem 5 For a renewal process with interval distribution $F(x)$ and renewal function $R(x)$,

$$W_t(x) = P(w(t) \le x) = 1 - \int_0^t [1 - F(t + x - y)] \, dR(y)$$

(where the R-measure of zero is one).

Proof. By the total probability rule

$$P(w(t) > x) = \sum_{n=0}^{+\infty} P(w(t) > x, X_t = n) = \sum_{n=0}^{+\infty} P(w(t) > x, S_n \le t < S_{n+1})$$

$$= \sum_{n=0}^{+\infty} P(S_n \le t, S_{n+1} > t + x)$$

where $S_0 = 0$. By conditioning on $S_n = y$ and taking expectation, we get (note that the formula below is valid even for $n = 0$)

$$P(S_n \le t, S_{n+1} > t + x) = \int_0^t P(S_{n+1} - S_n > t + x - y) \, dF^{(n*)}(y)$$

whose summation over n yields the desired formula. ∎

Theorem 6 With the notation and assumptions of the previous theorem, for $x \ge 0$ and $0 < z < t$,

$$P(w(t) > x, u(t) \le z) = \int_{t-z}^t [1 - F(t + x - y)] \, dR(y)$$

Proof. Starting with the total probability rule again, we first note that $P(u(t) \le z, X_t = 0) = 0$ for $z < t$. Hence,

$$P(w(t) > x, u(t) \le z) = \sum_{n=1}^{+\infty} P(w(t) > x, u(t) \le z, X_t = n)$$

We once more evaluate the terms under the summation sign by conditioning on $S_n = y$. Note that such conditional probabilities are equal to zero if $y \notin (t - z, t)$, while, for $y \in (t - z, t)$, the meaning of such conditional probabilities is that the process has no points in the interval $(y, t + x)$, i.e., $S_{n+1} - S_n > t + x - y$. Therefore, by the previous formula,

$$P(w(t) > x, u(t) \leq z) = \sum_{n=1}^{+\infty} \int_{t-z}^{t} [1 - F(t + x - y)] \, dF^{(n*)}(y)$$

which yields the desired formula by interchanging summation and integration.

∎

The substitution $x = 0$ in Theorem 6 clearly gives the distribution function $U_t(z) = P(u(t) \leq z)$ for $0 < z < t$, while, for $z \geq t$, $U_t(z) = 1$.

As was remarked earlier, for a homogeneous Poisson process on the positive real line, $R(t) = E(X_t) + 1 = \lambda t + 1$ by formulas (8.23) and (8.20). Thus, by Theorems 5 and 6, $w(t)$ is exponential with $E(w(t)) = E(S_{n+1} - S_n)$ (we saw this in Exercise 5 of the previous section) and $w(t)$ and $u(t)$ are independent. The distribution of $u(t)$ is a truncated exponential distribution. We thus have what was alluded to in the paragraph preceding Theorem 5: the expectation of $w(t) + u(t)$ is larger than that of $S_{n+1} - S_n$. This is known as the waiting time paradox, since it means that if service times are exponential, then the service time of the particular person who is just "before me" always takes longer than an "ordinary service."

When deriving the alternative formula of Exercise 3 for the distribution function of $w(t)$, the same formula obtains if we work with the stationary points S_n^*, $n \geq 1$, where $R(y)$ becomes $m(y)$. Since we proved $m(y) = y/a$, the substitution $u = t + x - y$ immediately yields $P(w(t) \leq x) = F_1(x)$, as it should be for stationary renewal processes.

Exercises

1. Assume that $R(x)$ is a variable function in the equation (8.25), which is bounded over finite intervals and identically zero for negative x. Show that $R(x)$ is necessarily the renewal function (8.22).

2. Let X_t be a renewal process with interval distribution $F(x)$ satisfying $F(0) = 0$. Show that $E(X_t^k)$ is finite for all $k \geq 1$ and that

$$E(X_t^k) = \sum_{n=0}^{+\infty} [(n + 1)^k - n^k] F^{(n+1,*)}(t)$$

3. Derive the alternative formula

$$P(w(t) \le x) = \int_t^{t+x} [1 - F(t + x - y)] \, dR(y) \qquad t > 0, x \ge 0$$

by arguing that $w(t) \le x$ means that there is at least one renewal point in the interval $(t, t + x]$. If the largest such point is S_n, then $P(w(t) \le x) = \sum P(t < S_n \le t + x < S_{n+1})$, where summation is for all $n \ge 1$.

4. Let the Laplace transforms of $F(x)$ and $W_t(x) = P(w(t) \le x)$ be $L_F(s)$ and $L_W(s)$, respectively. Show that

$$L_W(s) e^{-st} = [1 - L_F(s)] \int_t^{+\infty} e^{-sy} \, dR(y)$$

5. Use the relation $P(X_t < n) = P(S_n > t)$ to show that if the interval distribution $F(x)$ has finite expectation $\mu \ne 0$ and variance $V > 0$, then $(X_t - t/\mu)(Vt/\mu^3)^{-1/2}$ is asymptotically normally distributed as $t \to +\infty$.

6. Show that if the interval distribution $F(x)$ of a renewal process is continuous and if $E(w(t))$ is a finite constant for $t > 0$, then the process is Poisson (Cinlar and Jagers (1973) and Holmes (1974)).

7. Assume that customers arrive at a store according to a renewal process and that, independently of the process and of each other, every customer makes a purchase with the same probability p. Show that those customers who make a purchase also form the "renewal points" of a renewal process. Let the time scale be expanded by a factor of $1/p$. Find the interval distribution of the new process in terms of the original interval distribution.

8.4 THE GALTON–WATSON PROCESS; BUSY PERIODS IN QUEUES

In the last century, Galton became interested in the probability of a family name's extinction. Since a family name is carried to the next generation by the sons, we formulate the problem in the conservative terminology of considering the generations of males in a family tree. Later, we shall relate the problem to finding the distribution of the length of time a server is busy by continuously serving customers who joined a queue.

Let X_n be the number of males in the nth generation of a family tree. We assume that each member of the nth generation X_n has j sons with probability p_j, independent of the number of individuals in his, or previous, generations. We introduce the generating function

$$G(z) = \sum_{j=0}^{+\infty} p_j z^j \tag{8.33}$$

If $X_n = k$, then X_{n+1} is the sum of k independent random variables, each with the generating function $G(z)$. Hence

$$g_n(z) = E(z^{X_n}) \tag{8.34}$$

satisfies (see (3.48))

$$g_{n+1}(z) = g_n(G(z)) \tag{8.35}$$

By differentiating and setting $z = 1$, we get

$$E(X_{n+1}) = E(X_n)G'(1) = E(X_n) \sum_{j=1}^{+\infty} jp_j$$

where we assume that $\alpha = G'(1) < +\infty$, and which, by an earlier assumption, does not depend on generations, i.e., it is a characteristic of the family. Thus, by iteration,

$$E(X_n) = \alpha^n E(X_0) \tag{8.36}$$

Therefore, for $\alpha > 1$, the family can expect to expand and for $\alpha < 1$, the family can anticipate extinction, while $\alpha = 1$ is not a clear-cut case at this stage. A more accurate picture develops by investigating $\rho_n = P(X_n = 0) = g_n(0)$. This is the probability of extinction by the nth generation; i.e., at, or before the nth generation. Clearly, $0 \le \rho_n \le \rho_{n+1} \le 1$, and thus

$$\rho = \lim \rho_n \qquad (n \to +\infty) \tag{8.37}$$

exists. The value ρ is interpreted as the probability of extinction in finite time. We determine ρ under the assumptions (i) $X_0 = 1$ and (ii) $G(z) \neq z$ for all z, i.e., it is not an a.s. event that every member of any generation has exactly one son.

Theorem 7   Under the preceding assumptions on the process X_n, $n \ge 0$, $\rho = 1$ if $\alpha \le 1$. On the other hand, if $\alpha > 1$, $\rho < 1$, which is the unique solution of the equation $G(z) = z$ in the interval $[0, 1)$.

Proof.   By the assumption $X_0 = 1$, $g_1(z) = G(z)$, and thus, by (8.35), $g_n(z) = G_n(z)$, defined by the iteration $G_2(z) = G(G(z))$ and $G_n(z) = G_{n-1}(G(z))$. Since this is a symmetric operation, we also have $G_n(z) = G(G_{n-1}(z))$, $n \ge 2$. Hence,

$$\rho_{n+1} = g_{n+1}(0) = G_{n+1}(0) = G(G_n(0)) = G(\rho_n)$$

Now, $G(z)$ is a continuous function (a power series); thus, by letting $n \to +\infty$ we get (see (8.37))

$$\rho = G(\rho)$$

i.e., ρ is a solution of the equation $G(z) = z$. Since $G(z)$ is the generating function of a probability distribution, $G(1) = 1$, so $z = 1$ is always a solution of $G(z) = z$. Next, we analyze all possible solutions of $G(z) = z$ with $0 \le z < 1$. Referring once again to the fact that $G(z)$ is a probability generating function, we see that $G(z)$ is finite for all $0 \le z \le 1$ and is differentiable, with positive and increasing derivative $G'(z)$, for $0 \le z \le 1$. Hence, $G(z)$, $0 \le z \le 1$, is a convex function that can have at most one intersection with the line $z = x$ (since it has one at $z = 1$) that satisfies $0 \le z < 1$. Assume that there is one z_0 such that $0 \le z_0 < 1$ and $z_0 = G(z_0)$. Then, by the convexity of $G(z)$, the slope $\alpha = G'(1)$ of the tangent line at $z = 1$ is a bound on the slope of the secant connecting the points $(z_0, G(z_0)) = (z_0, z_0)$ and $(1, G(1)) = (1, 1)$, i.e., $\alpha > 1$. We thus have that, if $\alpha \le 1$, no such z_0 exists, and thus $\rho = 1$. On the other hand, if $\alpha > 1$, such a z_0 might exist, for which case we show that $\rho = z_0$. We already know that ρ is either z_0 or 1.

First, note that if $\rho_1 = G(0) = p_0 = 0$, then, by induction, $\rho_{n+1} = G(\rho_n) = 0$ for all n, i.e., $\rho = 0$. Clearly, this is the case when $z_0 = 0$. Hence, from now on, we shall assume that $\rho_1 = G(0) = p_0 > 0$, implying $z_0 > 0$. In fact, by the monotonicity of $G(z)$, $\rho_1 = p_0 = G(0) < G(z_0) = z_0$. Next, we apply the mean value theorem of calculus in the form $[G(z) - z_0]/(z - z_0) = G'(u)$, where u is a number between z and z_0, and choose $z = G_n(0) = \rho_n$. Then $G(z) = G(\rho_n) = \rho_{n+1}$, so

$$\frac{\rho_{n+1} - z_0}{\rho_n - z_0} = G'(u) \tag{8.38}$$

with some u between z_0 and ρ_n. Since $G'(u) > 0$ for any $u > 0$, this means that both ρ_n and ρ_{n+1} are either to the left or to the right of z_0. But we have seen that ρ_1 is to the left of z_0, so ρ_2 is also and, by induction, ρ_n for all n as well. Hence, between $z_0 < 1$ and 1, only z_0 can be the limit ρ of ρ_n, which completes the proof. ∎

Note that we automatically proved a speed of convergence result as well. Indeed, since $G'(z)$ is strictly increasing and $u \le z_0$ in (8.38), $G'(u) \le G'(z_0) < 1$, which is due to the convexity of $G(z)$ ($G'(z_0)$ is less than the slope of the secant connecting the points $(z_0, G(z_0))$ and $(1, G(1))$, i.e., (z_0, z_0) and $(1, 1)$). Thus, we have, by the iteration of (8.38),

$$0 \le z_0 - \rho_{n+1} \le G'(z_0)(z_0 - \rho_n) \le [G'(z_0)]^n(z_0 - p_0)$$

Example 1 The modified geometric distribution $p_j = cp^j, j \ge 1, c > 0$ such that $p_0 > 0$, is believed to be a good representation of the offspring distribution p_j, $j \geq 0$. For such a case,

$$G(z) = cpz \frac{1}{1-pz} + 1 - cp \frac{1}{1-p}$$

Now, $p_0 > 0$ is equivalent to $cp \neq 1 - p$. Since $G(z) = d + c/(1 - pz)$ with a constant d, $G'(z) = cp/(1 - pz)^2$, yielding $\alpha = cp/(1 - p)^2$. If this value does not exceed one, then we conclude that $\rho = 1$, i.e., that such a family will become extinct with probability one, while the extinction probability $\rho < 1$ if $\alpha > 1$. In this case, ρ is the solution of the simple equation $G(z) = z$.

Let us take two numerical cases. Let $c = 0.4$ and $p = 0.6$. Then $p_0 = 0.4$ and $p_j = 0.4(0.6)^j$, $j \geq 1$. Next, we compute $\alpha = G'(1) = cp/(1 - p)^2 = 0.24/0.16 > 1$. So, we proceed to $G(z) = z$, i.e.,

$$\frac{0.24z}{1-0.6z} + 0.4 = z \qquad \text{or} \qquad 0.6z^2 - z + 0.4 = 0$$

whose two solutions are 1 and $z_0 = 2/3$. Hence, by Theorem 7, $\rho = 2/3$.

In the next numerical example we keep $p = 0.6$, but choose $c = 0.2$. Now, $p_0 = 0.7$ and $p_j = 0.2(0.6)^j$, $j \geq 1$. By the general formula $\alpha = cp/(1 - p)^2 = 0.12/0.16 < 1$, so no further computation is required and we conclude from Theorem 7 that $\rho = 1$. ∎

Example 2 (Planned population growth) Assume each couple is limited to one child: $G(z) = p_0 + p_1 z$ with $p_0 > 0$. Then $G'(1) = p_1 < 1$, and thus extinction is guaranteed if the rule is not reversed. This rule can be reversed with much restriction still in place. Indeed, by limiting each married couple to 2 children, $G(z) = p_0 + p_1 z + p_2 z^2$. Then, $G'(1) = p_1 + 2p_2 = 1 + p_2 - p_0$, and growth is guaranteed if $p_2 > p_0$ (but still extinction follows if $p_2 \leq p_0$).

Example 3 (An application to algebra) Let $G(z) = a_0 + a_1 z + a_2 z^2 + a_3 z^3 + a_4 z^4$ with $G(1) = 0$, all coefficients $a_j \geq 0$ with the exception of a_1, $G'(1) > 0$, and $G(z)$ be at least quadratic. Then $G(z) = 0$ has one and only one root z_0 with $0 \leq z_0 < 1$.

Indeed, if we first multiply $G(z)$ by an appropriate number $c > 0$ such that $|a_j|c < 1$, then $cG(z) + z = G^*(z)$ is the generating function of a probability distribution. Clearly, $G(z) = 0$ is equivalent to $G^*(z) = z$. The assumptions imply that $G^*(z)' > 1$ at $z = 1$, and the claim follows from Theorem 7. ∎

The process X_n, $n \geq 0$, of this section is a typical example of a branching process; with its current restrictive assumptions, it is called a Galton–Watson process. A branching process is a sequence of positive integer-valued variables X_n, $n \geq 0$, and at any stage (generation) n, a random number Y_{nj}, $1 \leq j \leq X_n$, of members of the next generation $(n + 1)$ is associated with every one of the X_n members. Hence,

$$X_{n+1} = Y_{n1} + Y_{n2} + \cdots + Y_{nX_n}$$

If the distribution of Y_{nj} does not depend on j, then, given the generating function $G_n(z)$ of the common distribution of Y_{nj}, formula (8.35) still applies. Instead of the iteration of a single function, the iteration of a sequence of functions enters the investigation, and the theory would be hopeless in such a general case. However, with restrictions on the functions $G_n(z)$, the theory became both sufficiently general to treat a variety of problems and reasonably restrictive, so as to obtain nice and applicable results.

Branching processes can obviously be applied to the study of population growth, both human and animal, as well as the spread of infectious disease. The theory is considerably enriched by the realization that several problems of the theory of queues overlap with it. We shall discuss only one problem, which is related to Theorem 7.

Assume that customers arrive at a counter according to a Poisson process and that service is on a first-come-first-served basis. We call the first customer X_0, so service starts with the arrival of X_0. Those who come during the service time of X_0 form the first generation X_1. The next generation, X_2 comprises all the arrivals during the aggregate service times of the first generation X_1, etc. The (first) busy period is the total time from the start of service (the arrival of X_0) until the time the server becomes idle for the first time. This is equivalent to the extinction time of the population in the branching process formulation. Our assumptions for the Galton–Watson process are equivalent to the assumption that service times are independent and identically distributed random variables. Let this common distribution function of service time be $H(z)$. After determining $G(z)$, we can now apply Theorem 7. If X_0 arrives at time S_1 and if the service time of X_0 is Y, then

$$p_j = P(j \text{ arrivals in the interval } (S_1, S_1 + Y))$$

If we condition on $S_1 = x$ and $Y = y$ and recall that, in a Poisson process, S_1 is exponential and that the number of arrivals in a fixed interval is Poisson, we get

$$p_j = \int_0^{+\infty} \int_0^{+\infty} P(j \text{ arrivals in } (x, x + y)) \lambda e^{-\lambda x} \, dx \, dH(y)$$

$$= \int_0^{+\infty} \frac{(\lambda y)^j e^{-\lambda y}}{j!} \, dH(y)$$

because the integration with respect to x contributes a factor of one. Hence,

$$G(z) = \sum_{j=0}^{+\infty} p_j z^j = \int_0^{+\infty} e^{-\lambda y} \sum_{j=0}^{+\infty} \frac{(\lambda y)^j}{j!} z^j \, dH(y)$$

$$= \int_0^{+\infty} e^{-\lambda y(1-z)} \, dH(y)$$

In particular, if service is exponential, i.e., if $H(y) = 1 - e^{-by}$, $y \geq 0$, then we get

$$G(z) = \frac{b}{b + \lambda(1 - z)}$$

Thus, the equation $G(z) = z$ becomes $b = (b + \lambda)z - \lambda z^2$, whose solutions are 1 and $z_0 = b/\lambda$. Theorem 7 now says that the first busy period will eventually end with probability one if and only if $b/\lambda \geq 1$. If $b < \lambda$, then, with probability $1 - b/\lambda$, the busy period never ends. Note that we argued with the roots of the equation $G(z) = z$ only, which is, of course, equivalent to what one gets from first computing $\alpha = G'(1) = \lambda/b$.

Exercises

1. In the first numerical case of Example 1, we found that $\rho_n \leq 2/3$ for all n. Find n_0 such that $0.65 \leq \rho_n$ for all $n \geq n_0$.
2. Find $G(x)$ in each of the following cases: p_j is (i) Poisson, (ii) binomial, and (iii) negative binomial (the sum of N independent geometric variables). In each case verify that $G(z)$ satisfies a differential equation of the form

$$(c + dz)G'(z) = a + bG(z)$$

where a, b, c, and d are constants. Conversely, if such a differential equation holds for $G(z)$, show that the following recursive formula applies: if $P(X_n = k + 1) > 0$, then, for $k \geq 0$ and $n = 1, 2, \dots$.

$$(c + d\rho_{n-1})P(X_n = k + 1)$$

$$= aP(X_{n-1} = k + 1) + \frac{1}{k + 1} \sum_{j=0}^{k} [b(k + 1 - j) - dj]$$

$$\times P(X_n = j)P(X_{n-1} = k + 1 - j)$$

(Adès, Dion, Labelle and Nanthi (1982))

3. Let $\alpha > 1$. Assume that $X_0 = 1$ and $0 < E(X_1^2) < +\infty$. Show that the sequence $X_n/E(X_n)$ is an a.s. convergent martingale.

(Harris (1963))

8.5 MARKOV CHAINS

The Markov property of a stochastic process was introduced in (8.1). We established there that, given the "present," the "past" and the "future" are conditionally independent. Furthermore, the Chapman–Kolmogorov equation was established in (8.6), which, for a special case, was further discussed in Exercise 4 of Section 8.1. In subsequent sections, we introduced some special stochastic processes, from which the reader can easily verify that the Poisson process is a continuous-time Markov process; that the branching processes (the Galton–Walton process in particular) are discrete-time Markov processes, which we also called Markov chains; and that the renewal processes are, in general, non-Markovian. Since a queue length process can be interpreted as a branching process, the random queue length also forms a Markov chain.

In this section, we shall deal with discrete time Markov chains (the branching processes and the queue length are examples of these), for which both the definition of the Markovian property and the Chapman–Kolmogorov equation become simpler than in the general case. Therefore, we shall begin by reintroducing the concept of a discrete time Markov chain.

Definition 1 Let X_n, $n \geq 1$, be a sequence of discrete random variables with values in the set S (finite or denumerable). We say that X_n forms a Markov chain if, for every $k \geq 1$ and for all values of the subscripts $1 \leq n_1 < n_2 < \cdots < n_k < n_{k+1}$,

$$P(X_{n_{k+1}} = s_{k+1} | X_{n_j} = s_j, \; 1 \leq j \leq k) = P(X_{n_{k+1}} = s_{k+1} | X_{n_k} = s_k)$$

where the s_j are arbitrary members of S and the conditional probabilities above are assumed to be meaningful. The set S is called the *state space* of the chain and the conditional probabilities

$$p_{ij,m}^{(k)} = P(X_{m+k} = s_j | X_m = s_i)$$

are called the *transition probabilities in k steps*. The one-step transition probabilities are simply called *transition probabilities* and denoted $p_{ij,m}$.

Example 1 Consider the following special Galton–Watson process. Let X_n be the size of the population at time n and assume that each member of the population, independently of the others (including both the present and past members of the population), has either 0 or 2 offspring with probabilities $1 - p$ and p, respectively. Then the X_n, $n \geq 1$, form a Markov chain, whose transition probabilities

$$p_{ij,m} = p_{ij} = \begin{cases} \binom{i}{j/2} p^{j/2}(1-p)^{i-j/2} & \text{if } 0 \le j \le 2i \text{ and even} \\ 0 & \text{otherwise} \end{cases}$$

Even though X_n is finite for every n, it is unbounded in n; thus, S is the set of all nonnegative integers. ∎

Example 2 **(Random walk with an absorbing barrier at the origin)** Let X_n, $n \ge 1$, be nonnegative integer points on the real line. Assume that $X_n = U_1 + U_2 + \cdots + U_n$ as long as this sum is positive, where the U_j are independent random variables taking the values 1 and -1 only, with probability p and $1 - p$, respectively. On the other hand, $X_{n+1} = 0$ if $X_n = 0$. Clearly, we again have a Markov chain, where $p_{00,m} = 1$ and

$$p_{ij,m} = \begin{cases} p & \text{if } j = i+1 \\ 1-p & \text{if } j = i-1 \\ 0 & \text{otherwise} \end{cases}$$

for $i \ge 1$. ∎

Example 3 **(Random walk with two reflecting barriers)** Let the notation be the same as in the preceding example. Let $M > 0$ be an integer. Now, $X_n = U_1 + U_2 + \cdots + U_n$ as long as this sum is strictly between 0 and M, while $X_{n+1} = 1$ if $X_n = 0$ and $X_{n+1} = M - 1$ if $X_n = M$.

The Markov property is again obvious and $p_{ij} = p_{ij,m}$ is the same as in the preceding example if $0 < i < M$. For $i = 0$ and M, we have $p_{01,m} = 1$ and $p_{0k,m} = 0$ for all $k \ne 1$ and $p_{M,M-1,m} = 1$ and $p_{Mk,m} = 0$ for $k \ne M - 1$. ∎

Example 4 Assume that a machine can be in one of two states with the following properties. If it is in state 1 at time n, then it will remain in that state until time $n + 1$, regardless of past performance, with probability a_n and will change to state 2 with probability $1 - a_n$. Similarly, regardless of the past, a machine in state 2 at time n will remain in that state until time $n + 1$ with probability b_n and will change to state 1 with probability $1 - b_n$. (Think of the states 1 and 2 as "functioning" and "broken down," respectively.) Denoting the state of the machine at time n by X_n, X_n, $n \ge 1$, is a Markov chain by assumption, whose transition probabilities are

$$p_{11,n} = 1 - p_{12,n} = a_n \qquad p_{22,n} = 1 - p_{21,n} = b_n$$ ∎

Notice that the transition probabilities $p_{ij,m}$ did not depend on m in Examples 1–3. A Markov chain whose transition probabilities $p_{ij,m} = p_{ij}$ do not depend on m, is called *homogeneous*. Hence, Examples 1–3 are homogeneous Markov chains, while Example 4 is not.

From now on, we shall identify s_j of S with j, $j \geq 0$; thus, S will always mean either the set $\{0, 1, 2, \ldots\}$ or the set $\{0, 1, \ldots, M\}$ with some finite M. Furthermore, we shall use the matrix notation $\mathbf{P}_m = (p_{ij,m})$ and $\mathbf{P} = (p_{ij})$, the latter in the case of a homogeneous chain. We refer to these matrices as *transition matrices*. The k-step transition matrices are $\mathbf{P}_m^{(k)} = (p_{ij,m}^{(k)})$, which reduce to $\mathbf{P}^{(k)} = (p_{ij}^{(k)})$ in the homogeneous case. For most proofs, including an alternate proof for the Chapman–Kolmogorov equation, the following elementary lemma will suffice instead of a reference to properties of conditional expectations and probabilities.

Lemma   Let A and B be two events such that $P(B) > 0$. Then

$$P(A|B) = \sum P(A, X_n = j|B) \tag{8.39}$$

and

$$P(A|B) = \sum P(A|X_n = j, B)P(X_n = j|B) \tag{8.40}$$

where X_n is a discrete random variable and the summations are over its possible values j. In (8.40), $P(A|X_n = j, B) = 0$ if $P(X_n = j|B) = 0$.

Proof.   Both formulas are almost trivial. The first one is a simple application of the third axiom of probability to the conditional probability $P(\cdot|B)$. The second formula, on the other hand, follows immediately after substituting the elementary form of the conditional probability into each term (for both cases, recall Section 1.6). The proof is complete.   ∎

Note that, with $B = \Omega$, (8.40) reduces to the total probability rule (Section 1.6 again).

If X_n, $n \geq 1$, is a Markov chain, then the choices $A = \{X_{m+k+t} = s\}$. $B = \{X_m = i\}$, and $n = m + k$ in (8.40) yield

$$p_{is,m}^{(k+t)} = \sum_{j \in S} p_{ij,m}^{(k)} p_{js,m+k}^{(t)} \tag{8.41}$$

which are the Chapman–Kolmogorov equations. Another important special case of (8.40) is obtained by choosing $B = \Omega$ and $A = \{X_{n+t} = s\}$. We get

$$P(X_{n+t} = s) = \sum_{j \in S} p_{js,n}^{(t)} P(X_n = j) \tag{8.42}$$

We call the distribution of X_1 the *initial distribution* of the chain. Formula (8.42) with $n = 1$ thus states that the initial distribution and the transition matrices uniquely determine the (unconditional) distributions $P(X_{t+1} = s)$, $s \in S$, $t \ge 1$, of the chain.

Note that (8.41) can be written in the matrix form

$$\mathbf{P}_m^{(k+t)} = \mathbf{P}_m^{(k)} \mathbf{P}_{m+k}^{(t)} \tag{8.43}$$

which, in the homogeneous case, becomes

$$\mathbf{P}^{(k)} = \mathbf{P}^k \tag{8.44}$$

We should mention that, in each of these transition matrices, every entry is nonnegative (being probabilities) and the row sums are equal to one. This latter property follows immediately from (8.39) upon choosing $A = \Omega$ and $B = \{X_{n-k} = i\}$, $k \ge 1$. In principle, the distributional problems of a Markov chain are solved by (8.42) and (8.43). However, since the actual computations involved in these formulas can be quite tedious, one can benefit by developing limit theorems. The analysis leading to such limit theorems will reveal some very interesting structural properties of Markov chains.

From now on, unless otherwise stated, it will be assumed that the Markov chain X_n, $n \ge 1$, is homogeneous. We shall discuss the structure of a Markov chain by classifying its states according to different properties.

We say that the state j is *reachable* from the state i if, for some $n \ge 1$, $p_{ij}^{(n)} > 0$. If j is reachable from i and i is reachable from j, we say that i and j *intercommunicate*, for which the notation $i \leftrightarrow j$ will be used. Note that if $i \leftrightarrow j$ and $j \leftrightarrow s$, then $i \leftrightarrow s$. Indeed, if there exist integers k and t such that $p_{ij}^{(k)} > 0$ and $p_{js}^{(t)} > 0$ then, by (8.41), $p_{is}^{(k+t)} > 0$, that is, s is reachable from i whenever j is reachable from i and s is reachable from j. This, when applied in the opposite direction as well, yields the claimed conclusion $i \leftrightarrow s$. Consequently, intercommunicating states form closed subsets of S from which it is impossible to escape and every pair intercommunicates within such a closed subset. In other words, S can be decomposed into a finite or denumerable number of disjoint subsets C_m and one additional set T such that passage out of any C_m is impossible; if i and j belong to the same C_m, then $i \leftrightarrow j$; and it is possible to leave T but impossible to return to it.

Example 5 Let the transition matrix of a homogeneous chain be

$$\mathbf{P} = \begin{bmatrix} \frac{1}{2} & \frac{1}{2} & 0 & 0 & 0 \\ \frac{1}{3} & \frac{2}{3} & 0 & 0 & 0 \\ 0 & 0 & \frac{1}{8} & \frac{7}{8} & 0 \\ 0 & 0 & 1 & 0 & 0 \\ \frac{1}{5} & \frac{2}{5} & 0 & \frac{1}{5} & \frac{1}{5} \end{bmatrix}$$

Its state space $S = \{0, 1, 2, 3, 4\}$, $C_1 = \{0, 1\}$ and $C_2 = \{2, 3\}$ are closed subsets of S, and $T = \{4\}$ is a state to which return is impossible once it is left. Members of both C_1 and C_2 intercommunicate. ∎

Movement from a C-set is impossible, while movement from a T-set is possible, although that is not a necessary condition for it.

Example 6 Let S be the infinite set $\{0, 1, 2, \ldots\}$, let $\mathbf{P} = (p_{ij})$ be defined as $p_{ii} = p_{i,i+1} = 1/2$ for all $i \geq 0$, and let all other entries of $\mathbf{P}$ be zero. Then S has no intercommunicating closed subset, because movement is possible only towards higher values. That is, $S = T$, even though T "cannot be left." If we change only the second row to $p_{10} = p_{11} = 1/2$ and $p_{1j} = 0$ for $j \geq 2$, then $C_1 = \{0, 1\}$ is a closed subset of intercommunicating states and all the rest $\{2, 3, \ldots\}$ belong to T. Once again, T cannot be left, but its members do not intercommunicate. ∎

It should be noted that the states are not always as neatly arranged as in Example 5. However, by relabeling the states, we can always ensure that the closed intercommunicating states form disjoint square matrices (and possibly one additional infinite block) along the main diagonal of $\mathbf{P}$, as in Example 5. Since, in that example, every power of the matrix $\mathbf{P}$ is of this same form, there will be no loss of generality when a limit theorem is discussed under the assumption that it has a single closed set of intercommunicating states.

To further classify the states, we introduce the following probabilities. Let $f_{ij}^{(0)} = 0$ for all $i, j \in S$ and, for $n \geq 1$, let

$$f_{ij}^{(n)} = P(X_{m+n} = j, X_t \neq j \text{ for } m < t < m + n | X_m = i)$$

That is, $f_{ij}^{(n)}$ is the conditional probability that it requires exactly n steps to reach the state j from the state i. In particular, $f_{ii}^{(n)}$ is the probability that the *first return* to the state i occurs in n steps. We also set

$$f_{ij}^{*} = \sum_{n=1}^{+\infty} f_{ij}^{(n)}$$

which is the (conditional) probability of ever reaching j from i. A state j is said to be *persistent* if $f_{jj}^{*} = 1$ and *transient* if $f_{jj}^{*} < 1$.

There is a strong relation between the transition probabilities $p_{ij}^{(n)}$ and the newly introduced probabilities $f_{uv}^{(k)}$. In fact, if we write $p_{ij}^{(0)} = 0$ if $i \neq j$ and 1 if $i = j$, then

$$p_{ij}^{(n)} = \sum_{k=0}^{n} f_{ij}^{(k)} p_{jj}^{(n-k)} \qquad n \geq 1 \tag{8.45}$$

To prove (8.45), note that

$$p_{ij}^{(n)} = P(X_{n+1} = j \mid X_1 = i)$$

$$= \sum_{k=1}^{n+1} P(X_{n+1} = j, X_{k+1} = j, X_t \neq j, 1 < t \le k \mid X_1 = i)$$

where the probability under the summation sign can also be written as

$$\sum{}^* P(X_{n+1} = j, X_{k+1} = j, X_t = u_t, 1 < t \le k \mid X_1 = i)$$

where $\sum^*$ signifies summation over all vectors $(u_2, u_3, \ldots, u_k)$ with $u_t \in S$ and $u_t \neq j$. Now, the general term in $\sum^*$ equals the product of the conditional probabilities

$$P(X_{n+1} = j \mid X_{k+1} = j, X_t = u_t, 1 < t \le k, X_1 = i) = p_{jj}^{(n-k)}$$

and

$$P(X_{k+1} = j, X_t = u_t, 1 < t \le k \mid X_1 = i)$$

We note that the first factor does not depend on the u_t and that the sum $\sum^*$ of the second factors is exactly $f_{ij}^{(k)}$. This now proves (8.45), since $f_{ij}^{(0)} = 0$. It is convenient to transform the equation (8.45) into an equation of generating functions. Setting

$$P_{ij}(z) = \sum_{n=0}^{+\infty} p_{ij}^{(n)} z^n \quad \text{and} \quad F_{ij}(z) = \sum_{n=0}^{+\infty} f_{ij}^{(n)} z^n$$

where z is real or complex with $|z| < 1$, we get from (8.45) that

$$P_{ii}(z) = \frac{1}{1 - F_{ii}(z)} \qquad |z| < 1 \tag{8.46}$$

and

$$P_{ij}(z) = F_{ij}(z) P_{jj}(z) \qquad |z| < 1, i \neq j \tag{8.47}$$

where the difference between (8.46) and (8.47) stems from the fact that $p_{ii}^{(0)} = 1$, but $p_{ij}^{(0)} = 0$ for $i \neq j$. (Recall that similar relations were obtained in Section 3.7 in connection with a random walk problem, which we now know not to be an accident.)

By letting $z \to 1$ in (8.46), we get

$$\text{state } i \text{ is persistent if and only if } \sum_{n=0}^{+\infty} p_{ii}^{(n)} = +\infty \tag{8.48}$$

This, in turn, easily leads to

$$\text{if } i \leftrightarrow j \text{ and if } i \text{ is persistent, then so is } j \tag{8.49}$$

Indeed, since $i \leftrightarrow j$ implies the existence of integers n and m such that $p_{ji}^{(n)} > 0$ and $p_{ij}^{(m)} > 0$, the evident inequality

$$p_{jj}^{(n+k+m)} \geq p_{ji}^{(n)} p_{ii}^{(k)} p_{ij}^{(m)} \tag{8.50}$$

yields, when summed over k, (8.49) via (8.48).

Before stating our major limit theorem, one more classification of states is needed. Note that we cannot have a limiting form of the transition matrix P^n if $p_{jj}^{(n)} = 0$ unless n is a multiple of an integer $d > 1$. Such a state is called periodic; its period is d if d is the largest integer such that $p_{jj}^{(n)} = 0$ for all $n \neq md$ with some integer m. A state that is not periodic is called aperiodic. Intercommunicating states share periodicity as well. That is,

if $i \leftrightarrow j$ and if j is periodic, then so is i (8.51)

To establish (8.51), we start with (8.50) again. Let the period of j be $d > 1$. With $k = 0$ in (8.50), we find that $n + m$ is a multiple of d. Now let k be arbitrary, for which $p_{ii}^{(k)} > 0$. Then the left-hand side of (8.50) is positive, i.e., $n + k + m$ is a multiple of d, which is possible only if k is a multiple of d. Hence, we have proved (8.51) with the side result that the period of i is at least as large as that of j. Since $i \leftrightarrow j$ is symmetric, the converse also follows, yielding that i and j have the same period in (8.51).

By (8.49) and (8.51), if one state is persistent in a closed set of intercommunicating states, then so are all states, and if one state is aperiodic, then so are all states. We can thus speak of persistent and aperiodic closed sets.

Theorem 8 Let X_n, $n \geq 1$, be a homogeneous Markov chain and assume that every pair of states intercommunicates. Let the chain be persistent and aperiodic. Let $E_j = \sum k f_{jj}^{(k)}$, where summation is for all $k \geq 1$. Then

$$\lim p_{ij}^{(n)} = 1/E_j \qquad \text{for all } i \in S \tag{8.52}$$

as $n \to +\infty$, where $1/E_j = 0$ if $E_j = +\infty$. If E_j is finite for one j, then it is finite for all j, and then $1/E_j$, $j \in S$, is the unique set of solutions of the equations

$$x_j = \sum_{i \in S} x_i p_{ij} \qquad \sum_{i \in S} x_i = 1 \qquad x_j > 0 \tag{8.53}$$

Proof. The main part of the theorem follows from Theorem 4, which we proved in connection with renewal processes. Indeed, by setting $p_n = f_{jj}^{(n)}$, $r_n = p_{jj}^{(n)}$, and $b_n = 1$ or 0 depending on whether $n = 0$ or $n \geq 1$, (8.45) with $i = j$ becomes (8.26) when we see that the equation corresponding to (8.45) for $n = 0$ is $p_{jj}^{(0)} = 1$. The assumptions of Theorem 4 on p_n, r_n, and b_n for our choice are satisfied, since $r_n = p_{jj}^{(n)}$ is clearly bounded, $B = \sum b_n = 1$, and

$p_n = f_{jj}^{(n)}$, $n \geq 1$, is an aperiodic probability distribution because aperiodicity in the two cases coincides (the reader can easily verify that the aperiodicity of $f_{jj}^{(n)}$ is the same as that of $p_{jj}^{(n)}$). Finally, that $f_{jj}^{(n)}$, $n \geq 1$, is a probability distribution is ensured by every state's being persistent. Hence, the limits in (8.52) follow from Theorem 4 for $i = j$. Next, let $i \neq j$. Then, by (8.45), we can choose an integer M such that

$$\left| p_{ij}^{(n)} - \sum_{k=1}^{M} f_{ij}^{(k)} p_{jj}^{(n-k)} \right| < \varepsilon$$

for all $n \geq n_0$, where $\varepsilon > 0$ is arbitrary. Thus, if $E_j = +\infty$, $p_{jj}^{(n-k)} \to 0$ implies that $p_{ij}^{(n)} \to 0$ for all $i \in S$, while, for $E_j < +\infty$, we have,

$$\lim p_{ij}^{(n)} = \frac{1}{E_j} \sum_{k=1}^{+\infty} f_{ij}^{(k)} = \frac{f_{ij}^*}{E_j}$$

as $n \to +\infty$. Therefore, in order to complete the proof of (8.52), we have to show that $f_{ij}^* = 1$, knowing that $i \leftrightarrow j$ and that both are persistent. This, however, easily follows from the observation that the probability of not returning to j, given that the chain starts at j—which is zero for a persistent state—can be estimated as follows. Since $i \leftrightarrow j$, $a = P(i$ is reached $\mid$ chain starts at $j) > 0$ and the probability of not returning to j once the chain is at i equals $1 - f_{ij}^*$. Thus, $0 \geq a(1 - f_{ij}^*)$, yielding $f_{ij}^* = 1$. This completes the proof of (8.52).

Now, if $E_i < +\infty$ for one i, then the inequality (8.50) and the limits (8.52) ensure that $1/E_j > 0$, i.e., $E_j < +\infty$ for any other $j \in S$.

The remaining part of the theorem is almost obvious if S is finite, but it requires detailed proof if S is infinite. First, observe that, since row sums in $\mathbf{P}^n$ are equal to one, for any fixed finite M, $\sum_{j=1}^{M} p_{ij}^{(n)} \leq 1$, from which (8.52) yields

$$\sum_{j=0}^{M} (1/E_j) \leq 1 \qquad \text{i.e.,} \qquad \sum_{j \in S} (1/E_j) \leq 1 \tag{8.54}$$

Similarly, from the Chapman–Kolmogorov equation (8.41) with $k = n$ and $t = 1$, we get, via (8.52),

$$1/E_s \geq \sum_{j=0}^{M} (1/E_j) p_{js} \qquad \text{i.e.} \quad 1/E_s \geq \sum_{j \in S} (1/E_j) p_{js} \tag{8.55}$$

By summing over $s \in S$ and interchanging the order of summation on the right-hand side, (8.55) entails

$$\sum_{s \in S} (1/E_s) \geq \sum_{j \in S} (1/E_j) \sum_{s \in S} p_{js} = \sum_{j \in S} (1/E_j)$$

since the second sum in the middle is a row sum in $\mathbf{P}$, which equals one. That is, we obtained an equation that is possible only if the inequalities of (8.55) are, in fact, equations as well. Hence, the first equation of (8.53) holds with $x_i = 1/E_i$. In order to see that the second equation of (8.53) holds also, we start with the first one, bearing in mind that $x_i = 1/E_i$. We have

$$x_j = \sum_{i \in S} x_i p_{ij} = \sum_{i \in S} \left(\sum_{t \in S} x_t p_{ti} \right) p_{ij}$$

$$= \sum_{t \in S} x_t \left(\sum_{i \in S} p_{ti} p_{ij} \right) = \sum_{t \in S} x_t p_{tj}^{(2)}$$

By continuing in this manner, we get by induction that

$$x_j = \sum_{t \in S} x_t p_{tj}^{(n)} \qquad \text{all } n \geq 1 \tag{8.56}$$

Hence, with an appeal to (8.54), for every $\varepsilon > 0$ there is a finite M such that

$$\left| x_j - \sum_{t=0}^{M} x_t p_{tj}^{(n)} \right| < \varepsilon$$

By letting $n \to +\infty$ and then $\varepsilon \to 0$, we get, from (8.52),

$$x_j = \sum_{t \in S} x_t x_j$$

which yields the second equation of (8.53), since $x_j = 1/E_j > 0$ by assumption. Finally, to see that (8.53) has no other solution when $E_j < +\infty$, take $\{x_j\}$ as any set of numbers satisfying (8.53). Then (8.56) still applies, from which, by letting $n \to +\infty$, the argument above leads to $x_j = \sum_{t \in S} x_t (1/E_j)$. The second equation of (8.53), which is now an assumption, gives $x_j = 1/E_j$. The proof is completed. ∎

Next, we prove a simple statement whose value lies in its relation to Theorem 8.

Theorem 9 If, for a Markov chain, $\lim p_{js}^{(t)} = u_s$ for all $j \in S$ as $t \to +\infty$, then $P(X_t = s) \to u_s$. In particular, under the conditions of Theorem 8, X_t has a limiting distribution $u_s = 1/E_s$, $s \in S$, which is a proper probability distribution if $E_s < +\infty$ for one s.

Proof. Simply apply (8.42) with $n = 1$ and arbitrary t. We get, by the usual technique,

$$\left| P(X_{t+1} = s) - \sum_{j=0}^{M} p_{js}^{(t)} P(X_1 = j) \right| < \varepsilon$$

from which the double passages to the limit as $t \to +\infty$ and then $\varepsilon \to 0$ yield the desired limit for $P(X_{t+1} = s)$. Since the conditions of the first part of the theorem are satisfied under the conditions of Theorem 8, the theorem is established.   ■

For an example where Theorem 8 does not apply but Theorem 9 does, see Example 7.

Earlier we decomposed the states S into disjoint closed sets of inter-communicating states and one set called T. By the definition of T, either the chain could escape from T and could not return or there is an $i \in T$ for any $j \in T$ that does not intercommunicate with j (because T has no closed subset of intercommunicating states). In either case, if the chain starts at $j \in T$, then it has a positive probability of not returning to j, i.e., $f_{jj}^* < 1$. Hence, members of T are transient. Now, for transient states j, (8.48) and (8.47) with $z \to 1$ yield $\sum_{i \in S} p_{ij}^{(n)} < +\infty$, implying

if j is transient, then $p_{ij}^{(n)} \to 0$ for all $i \in S$ as $n \to +\infty$ (8.57)

For finite Markov chains, we can now draw the following conclusions. First, it must have a persistent state—otherwise, by (8.57), $\mathbf{P}^n$ would converge to an identically zero matrix, which is impossible since row sums must remain one. For easier reference, we shall call a matrix *stochastic* if its entries are nonnegative and its row sums equal one. For this same reason, it cannot consist entirely of aperiodic persistent states j with $E_j = +\infty$. In fact, the reader is asked in Exercise 5 to verify that a finite Markov chain cannot have any such state. So it has at least one closed set of persistent states j with $E_j < +\infty$. We can assume that these correspond to the states 0, 1, 2, ..., M. Let the corresponding transition matrix be $\mathbf{P}_1$. This occupies the upper left-hand corner of $\mathbf{P}$ and $p_{ij} = 0$ for all $j \in S$ such that $j > M$ and $0 \le i \le M$. In $\mathbf{P}^n$, these latter entries remain 0 and the upper left-hand corner becomes $\mathbf{P}_1^n$. Since $\mathbf{P}_1^n$ does not have a limit if a corresponding state i, i.e., $0 \le i \le M$, is periodic (note that if their period is d, then $\mathbf{Q} = \mathbf{P}_1^d$ represents an aperiodic Markov chain to which Theorem 8 applies), so $\mathbf{P}^n$ can have a limit only if its persistent states are aperiodic. An important special case, whose direct proof goes back to Markov, which proof had much influence on the later development of the whole theory, is formulated in Exercise 7.

While some of our previous examples are discussed among the exercises, let us look at a new example for an illustration of the main results of this section.

Example 7 Let X_n, $n \geq 1$, be a homogeneous Markov chain with the transition matrix

$$
P = \begin{bmatrix}
0.2 & 0.3 & 0 & 0.5 \\
0.4 & 0 & 0 & 0.6 \\
0 & 0.2 & 0.5 & 0.3 \\
0.2 & 0.8 & 0 & 0
\end{bmatrix}
$$

Hence, $S = \{0, 1, 2, 3\}$. Since state 2 is not reachable from any other state but can be left for other states, $T = \{2\}$ is transient. All other states intercommunicate; thus, $C = \{0, 1, 3\}$ is a closed set of persistent states. From $p_{00} = 0.2 > 0$, we conclude that state 0 is aperiodic and, thus, so are 1 and 3 (see (8.51)). Consequently, we can apply Theorem 8 to the submatrix

$$
P_1 = \begin{bmatrix}
0.2 & 0.3 & 0.5 \\
0.4 & 0 & 0.6 \\
0.2 & 0.8 & 0
\end{bmatrix}
$$

which corresponds to C and, in P^n, the corresponding entries are those of P_1^n. By Theorem 8, $P_1^n \to R$ with $r_{ij} = x_j$, satisfying (8.53). That is, to find R, we have to solve the system of equations

$$x_0 + x_1 + x_3 = 1$$
$$x_0 = 0.2x_0 + 0.4x_1 + 0.2x_3$$
$$x_1 = 0.3x_0 + 0.8x_3$$
$$x_3 = 0.5x_0 + 0.6x_1$$

where $x_j > 0$. We get $x_0 = 26/95$, $x_1 = 35/95$, and $x_3 = 34/95$. Thus,

$$p_{i0}^{(n)} \to 26/95 \qquad p_{i1}^{(n)} \to 35/95 \qquad p_{i3}^{(n)} \to 34/95 \qquad i \in C \tag{8.58}$$

as $n \to +\infty$. Furthermore, by (8.57), $p_{i2}^{(n)} \to 0$. Next, note that, just as in the proof of (8.52), $p_{2j}^{(n)} \to (1/E_j)f_{2j}^*$, $j \in C$. But, since, for a finite chain, $f_{2j}^* = 1$ for $j \in C$, we have that (8.58) continues to hold with $i = 2$ as well. This makes Theorem 9 applicable and thus

$$\{P(X_t = s): s = 0, 1, 2, 3\} \to (26/95, 35/95, 0, 34/95) \qquad \blacksquare$$

Exercises

1. Classify the states of Examples 2 and 3.
2. Show that if X_n, $n \geq 1$, is a Markov chain then, for integers $n_1 < n_2 < \cdots < n_k < n_{k+1} < n_{k+2}$

$$P(X_{n_{k+2}} = u_{k+2}, X_{n_{k+1}} = u_{k+1} | X_{n_j} = u_j, 1 \leq j \leq k)$$
$$= P(X_{n_{k+2}} = u_{k+2}, X_{n_{k+1}} = u_{k+1} | X_{n_k} = u_k)$$

   whenever these conditional probabilities are defined.
3. Let the transition matrix $\mathbf{P} = (p_{ij})$ of a finite homogeneous Markov chain X_n, $n \geq 1$, be such that $p_{ij} > 0$ for all i and j and both its row sums and its column sums equal one (double stochastic). Find the limit of $\mathbf{P}^n$ and conclude that X_n is asymptotically uniformly distributed as $n \to +\infty$.
4. Prove that if, in our standard notation, $p_{jj}^{(n)} \to w_j$ as $n \to +\infty$, then $p_{ij}^{(n)} \to w_j f_{ij}^*$ for all $i \in S$.
5. Show that there is no persistent state j with $E_j = +\infty$ in a finite homogeneous Markov chain.
6. Assume that the conditions of Theorem 8 hold and that $E_j < +\infty$ for one j. Let the initial distribution of the chain be the limit distribution $1/E_j$, $j \in S$, obtained in the theorem. Find the distribution of X_n for an arbitrary n.
7. Let $\mathbf{P} = (p_{ij})$ be the transition matrix of a finite homogeneous Markov chain. Assume that there is a j such that $0 < p_{ij} < 1$ for all $i \in S$. Show that the limit of $\mathbf{P}^n$ exists as $n \to +\infty$.
8. Find the limit of $\mathbf{P}^n$ in Example 5.

8.6 CONTINUOUS-TIME MARKOV CHAINS

In this section we develop properties of Markov chains X_t with values in a set $S \subset \{0, 1, 2, \ldots\}$ whose parameter set $t \in T = [0, +\infty)$. Recall (8.1) and (8.6): we have, for all $0 \leq t_1 < t_2 < \cdots < t_n < t_{n+1}$,

$$P(X_{t_{n+1}} = j | X_{t_1} = i_1, X_{t_2} = i_2, \ldots, X_{t_n} = i_n) = P(X_{t_{n+1}} = j | X_{t_n} = i_n)$$

and, for $0 \leq s < u < t$,

$$P(X_t = j | X_s = i) = \sum_{k \in S} P(X_u = k | X_s = i) P(X_t = j | X_u = k)$$

which we call the Chapman–Kolmogorov equations. We shall deal with homogeneous processes only which are defined by the stationarity property

(P:1) $P(X_t = j | X_s = i) = P(X_{t-s} = j | X_0 = i)$ $0 \leq s < t$

The functions $P_t(i, j) = P(X_t = j | X_0 = i)$, $t \geq 0$, $i, j \in S$, are called the transition functions of the process X_t, $t \geq 0$, which we also write in the square matrix form $\mathbf{P}_t$ whose (i, j)th entry is $P_t(i, j)$. The matrix $\mathbf{P}_t$, $t \geq 0$, can of course be infinitely large.

In addition to (P:1) we record the following basic properties and conventions.

(P:2) $P_t(i, j) \geq 0$ for all $i, j \in S$ and all $t \geq 0$, and

$$\sum_{j \in S} P_t(i, j) = P(X_t \in S | X_0 = i) = 1$$

(P:3)   We adopt the convention that $P_0(i, i) = 1$ and $P_0(i, j) = 0$ if $i \neq j$, for all $i, j \in S$.

(P:4)   We make the continuity assumption: for every $j \in S$, $\lim P_t(j, j) = 1$ as $t \searrow 0$,

and the Chapman–Kolmogorov equations take the simpler form

(P:5)   For all $u, v \geq 0$, $P_{u+v}(i, j) = \sum_{k \in S} P_u(i, k) P_v(k, j)$, or, in matrix form, $\mathbf{P}_{u+v} = \mathbf{P}_u \mathbf{P}_v$.

Some simple consequences of (P:1–P:5) are contained in the following statement.

Theorem 10   (i)   For every integer $n \geq 1$, $\mathbf{P}_{nt} = \mathbf{P}_t^n$, $t \geq 0$;

(ii) $\lim P_t(i, j) = 0$ as $t \searrow 0$ for every $i, j \in S$, $i \neq j$;

(iii) $P_t(j, j) > 0$ for all $t \geq 0$ and $j \in S$. Furthermore, if $P_t(j, j) = 1$ for some $t > 0$ then $P_t(j, j) = 1$ for all $t \geq 0$;

(iv)   if $P_t(i, j) > 0$ for some $t > 0$ $(i \neq j)$ then $P_s(i, j) > 0$ for all $s \geq t$. Furthermore, $P_t(i, j)$ is a uniformly continuous function of t; in fact, for every c, $|P_{t+c}(i, j) - P_t(i, j)| \leq 1 - P_{|c|}(i, i)$.

Proof.   (i) is immediate from (P:5). First choose $t = u = v$, and the claim obtains for $n = 2$. Next, by choosing $u = (n - 1)t$ and $v = t$, the claim follows by induction. For (ii), combine (P:2) and (P:4) and conclude that, for every $i \in S$, uniformly in j, $P_t(i, j) \to 0$ as $t \searrow 0$. Upon turning to (iii), we observe that, by (P:4), $P_u(j, j) > 0$ for sufficiently small $u > 0$ and for $j \in S$. But then, with $u = t/n$, where n is a large integer, part (i) of the theorem yields the claim for arbitrary $t > 0$. Part (i) also implies the second part of the statement at (iii) for $s \geq t$. Indeed, assume that $P_u(j, j) = 1$ has been established for all $u \leq t$. Then for any $s > t$ and large integer n such that $u = s/n < t$, part (i) in the form $P_s = P_{s/n}^n$, that is, $P_s(j, j) \geq P_{s/n}^n(j, j) = P_u^n(j, j) = 1$ yields $P_s(j, j) = 1$. We thus have to show $P_u(j, j) = 1$ for all $u < t$. Let $t = u + v$. Then, by

(P:5), $P_{u+v}(j, i) \geq P_u(j, i)P_v(i, i)$, entailing via (P:2)

$$0 = 1 - P_{u+v}(j, j) = \sum_{i \neq j} P_{u+v}(j, i) \geq \sum_{i \neq j} P_u(j, i)P_v(i, i)$$

which, in view of the first part of (iii), is possible only if $P_u(j, i) = 0$ for all $j \neq i$. Consequently, (P:2) with the parameter value u yields $P_u(j, j) = 1$. This completes the proof of part (iii). The first part of part (iv) is immediate from (P:5) and the first part of (iii). Let $s = t + u$. Then $P_s(i, j) \geq P_t(i, j)P_u(j, j)$, where the right-hand side is guaranteed to be positive. For the uniform continuity we assume $c > 0$. This will suffice for the complete proof, since, for $c < 0$, we can interchange the roles of t and $t + c$. Now, by (P:5),

$$P_{t+c}(i, j) - P_t(i, j) = \sum_{k \in S} P_c(i, k)P_t(k, j) - P_t(i, j)$$
$$= \sum_{\substack{k \in S \\ k \neq i}} P_c(i, k)P_t(k, j) - P_t(i, j)[1 - P_c(i, i)]$$

from which we get

$$-[1 - P_c(i, i)] \leq -P_t(i, j)[1 - P_c(i, i)] \leq P_{t+c}(i, j) - P_t(i, j)$$
$$\leq \sum_{k \neq i} P_c(i, k)P_t(k, j) \leq \sum_{k \neq i} P_c(i, k) = 1 - P_c(i, i).$$

which is the claimed inequality. The theorem is established. ∎

Theorem 11 For every $i \in S$, as $t \searrow 0$, $\lim(P_t(i, i) - 1)/t = -q_i$ exists, possibly infinite. When $q_i < +\infty$, $P_0'(i, i) = -q_i$.

Proof. We proved in Theorem 10 (iii) that $P_t(i, i) > 0$ for every $i \in S$ and all $t \geq 0$, so we can turn to $p_t(i) = \log P_t(i, i)$. Set $r_i = \inf\{p_t(i)/t : t > 0\}$. We shall show that, in fact, $r_i = -q_i$. Clearly, $\liminf p_t(i)/t \geq r_i$ as $t \searrow 0$, so we have to show the converse inequality for the limsup. Let r^* be arbitrary with $r^* > r_i$. Choose $u > 0$ such that $p_u(i)/u < r^*$. To $t > 0$, that goes down to zero, choose n (integer) and $0 \leq v < t$ such that $u = nt + v$. Then, since $P_{s+h}(i, i) \geq P_s(i, i)P_h(i, i)$ and $P_t(i, i) \to 1$ as $t \searrow 0$ (see (P:5) and (P:4)),

$$r^* > \frac{p_u(i)}{u} \geq \frac{np_t(i) + p_v(i)}{u} = \frac{nt}{u} \frac{p_t(i)}{t} + \frac{p_v(i)}{u}$$

As $t \searrow 0$, $(nt)/u \to 1$, $v \to 0$ and u is fixed. Hence, the last term on the extreme right hand side tends to zero, and we have

$$\limsup p_t(i)/t \leq r^* \quad (t \searrow 0), \text{ where } r^* > r_i \text{ arbitrary}$$

This completes the proof. ∎

From the proof we can see that, because $-q_i = r_i$, if $q_i = 0$ then $p_t(i)$ must be zero for all $t \geq 0$, that is, $q_i = 0$ if, and only if, $P_t(i, i) = 1$ for all $t \geq 0$. Once again, from the definition of r_i (which turned out to be $-q_i$) we have, for $i \in S$ and all $t \geq 0$,

$$P_t(i, i) \geq \exp(-q_i t) \geq 1 - q_i t$$

the last inequality being an elementary property of the exponential function.

The inequality just established above implies that, if $0 < q_i < +\infty$, and if the process X_t once reached the state i, then it will still be there after an exponential length of time. The following theorem makes this conclusion more accurate. We shall use the terms absorbing and stable states. State i is called absorbing if there is a $t > 0$ (and then for all $t \geq 0$ by Theorem 10 (iii)) such that $P_t(i, i) = 1$. This is equivalent to $q_i = 0$. An absorbing state is well understood by us: if the chain gets there it stays there forever. On the other hand, a state i is called stable if $q_i < +\infty$.

Theorem 12 Let $i \in S$ be a stable and nonabsorbing state. Assume $X_0 = i$. Define $T_i = \inf\{t > 0 : X_t \neq i\}$. Then

$$P(T_i > t \mid X_0 = i) = \exp(-q_i t) \qquad t \geq 0$$

Proof. We shall show that the conditional survival function

$$W_i(t) = W(t) = P(T_i > t \mid X_0 = i)$$

satisfies Cauchy's functional equation (6.3), from which it follows that $W(t)$ is exponential. We then determine its parameter. We have

$$
\begin{aligned}
W(t + s) &= P(X_u = i, 0 < u \leq t + s \mid X_0 = i) \\
&= P(X_u = i, t < u \leq t + s \mid X_u = i, 0 \leq u \leq t) \\
&\quad \times P(X_u = i, 0 < u \leq t \mid X_0 = i) \\
&= P(X_u = i, t < u \leq t + s \mid X_t = i) P(T_i > t \mid X_0 = i) \\
&= P(X_u = i, 0 < u \leq s \mid X_0 = i) W(t) = W(s) W(t)
\end{aligned}
$$

where the Markov property and the fact that X_t is homogeneous were repeatedly used. We thus have from Section 6.1 that $W(t) = e^{-at}$ with some $a > 0$. Evidently,

$$
\begin{aligned}
a &= -(1/t) \log P(T_i > t \mid X_0 = i) \\
&= -(1/t) \log P(X_u = i, 0 < u \leq t \mid X_0 = i)
\end{aligned}
$$

where the right-hand side is continuous in t. Hence, instead of $X_u = i$ for all $0 < u \leq t$, we may consider $X_u = i$ for u in a dense set of $(0, t]$. Such a set can be

$$D_t = \bigcup_{n=1}^{+\infty} \{kt2^{-n} : 1 \leq k \leq 2^n\} = \bigcup_{n=1}^{+\infty} A_{t,n}, \text{ say}$$

The sets $A_{t,n}$, $n \geq 1$, are increasing, and thus (see Theorem 1.6)

$$P(X_u = i, u \in D_t | X_0 = i) = \lim P(X_u = i, u \in A_{t,n} | X_0 = i)$$

as $n \to +\infty$. On $A_{t,n}$, we can apply the Markov property and stationarity, obtaining

$$P(X_u = i, u \in A_{t,n} | X_0 = i) = P_{t/N}^N(i, i) \qquad N = 2^n$$

We thus have

$$a = -(1/t)\log P(X_u = i, u \in D_t | X_0 = i)$$
$$= -(1/t)\lim \log P_{t/N}^N(i, i) = -(N/t)\lim \log P_{t/N}(i, i) = q_i$$

where in the limits $N \to +\infty$, and where we utilized

$$\log P_{t/N}(i, i) = \log\{1 - [1 - P_{t/N}(i, i)]\} = P_{t/N}(i, i) - 1 + o(t/N)$$

incorporating into $o(t/N)$ a multiple of $[1 - P_{t/N}(i, i)]^2$, as Taylor's expansion would imply; finally, Theorem 11 was applied. This concludes the proof.   ∎

The proofs of the following two theorems are very technical and we omit them.

Theorem 13   If $P_t(i, j)$ is a transition function, then $P_t'(i, j) = \dfrac{d}{dt} P_t(i, j)$ exists and is finite for all $i \neq j$ and all $t \geq 0$. Furthermore, $P_t'(i, j)$ is continuous for $t > 0$.

Theorem 14   (i)   Let $i \in S$ be a stable state. Then, for all $s > 0$, $t \geq 0$, and $j \in S$,

$$P_{s+t}'(i, j) = \sum_{k \in S} P_s'(i, k)P_t(k, j)$$

(ii)   Let $j \in S$ be a stable state. Then, for all $s \geq 0$, $t > 0$, and $i \in S$,

$$P_{s+t}'(i, j) = \sum_{k \in S} P_s(i, k)P_t'(k, j)$$

(iii)   If either i or j is stable then $P_t'(i, j)$ is continuous for $t \geq 0$.

Set $q_{ij} = P'_0(i, j)$ and $\mathbf{Q}$ for the matrix with entries q_{ij}. Note that the diagonal entries $q_{ii} = -q_i$ of $\mathbf{Q}$ may be infinite. When these diagonal entries are finite, they are nonpositive (Theorem 11), and since q_{ii} is finite for a stable state $i \in S$, the matrix $\mathbf{Q}$ is called stable if all of its diagonal entries are finite. The off-diagonal elements of $\mathbf{Q}$ are always finite (Theorem 13), and nonnegative since $P_0(i, j) = 0$ for $i \neq j$. We shall show below that all row sums of $\mathbf{Q}$ are nonpositive. If all row sums of $\mathbf{Q}$, in fact, equal zero, $\mathbf{Q}$ is called conservative.

Theorem 15   Let $P_t(i, j)$ be a transition function and let $i \in S$ be a stable state. Then $\sum q_{ij} \leq 0$, the summation being over $j \in S$. If S is finite, then the preceding sum is zero.

Proof. Note that if i is an absorbing state then each $q_{ij} = 0$ ($q_i = 0$ by definition, i.e., $P_t(i, i) = 1$ and thus $P_t(i, j) = 0$ for all $i \neq j \in S$ and all $t \geq 0$), and the (in)equality becomes evident. Hence, let us assume that $q_i > 0$. Define

$$g_t(i, j) = (e^{-q_i t}/q_i)[e^{q_i t} P_t(i, j)]' = (1/q_i)[P'_t(i, j) + q_i P_t(i, j)]$$

If we show that $g_t(i, j) \geq 0$, which we delay to the end of the proof, we get that $P_t(i, j)\exp(q_i t)$ is an increasing function of t. Therefore, their sum over j can be differentiated term by term for almost all t (with respect to Lebesgue measure—just apply the uniqueness of the Radon–Nikodym derivatives for almost all t and (2.25) for summations), entailing, for almost all t,

$$e^{q_i t} \sum_{j \in S} g_t(i, j) = \frac{1}{q_i}\left[e^{q_i t} \sum_{j \in S} P_t(i, j)\right]' = \frac{1}{q_i}(e^{q_i t})' = e^{q_i t}$$

that is, $\sum_{j \in S} g_t(i, j) = 1$ for almost all t. In order to see that this is, in fact, true for all $t > 0$, use the second form of $g_t(i, j)$ (the extreme right-hand side in its definition) and apply the Chapman–Kolmogorov equation as well as Theorem 14 (i) to $g_{t+s}(i, j)$, yielding for all $s > 0$ and $t \geq 0$

$$g_{s+t}(i, j) = \sum_{k \in S} g_s(i, k) P_t(k, j)$$

Upon summing these equations over j and interchanging the order of summations, we obtain

$$\sum_{j \in S} g_{s+t}(i, j) = \sum_{k \in S} g_s(i, k) \qquad \text{all } t \geq 0 \qquad \text{and} \qquad s > 0$$

Since the left-hand side equals one for almost all $t \geq 0$, the right-hand side is one by continuity for all $s > 0$. By appealing once more to the definition of $g_t(i, j)$ by the extreme right-hand side, summation over j yields

$$\sum_{j \in S} P'_t(i, j) = 0 \qquad \text{for all} \quad t \geq 0$$

and for absolute values we get

$$\sum_{j \in S} |P'_t(i, j)| \leq q_i \sum_{j \in S} g_t(i, j) + q_i \sum_{j \in S} P_t(i, j) = 2q_i$$

Let $t \to 0$. Recall that $q_{ii} \leq 0$ and $q_{ij} \geq 0$ for $i \neq j$ and that $q_{ii} = -q_i$. Hence, by dissolving the absolute value signs the claimed inequality follows. For finite S, one can pass to the limit in the above equation without violating the equation, so for finite S, the equation always holds.

In order to complete the proof we have to show that $g_t(i, j) \geq 0$. Let $0 \leq s < t$. Then, since

$$P_t(i, j) - P_s(i, j) \geq P_{t-s}(i, i)P_s(i, j) - P_s(i, j)$$
$$= P_s(i, j)[P_{t-s}(i, i) - 1]$$

the inequality $P_{t-s}(i, i) - 1 \geq -q_i(t - s)$, established at the end of the proof of Theorem 11, entails

$$P_t(i, j) - P_s(i, j) \geq -q_i(t - s)P_s(i, j)$$

which, when applied over successive (small) intervals (s_u, t_u), provides bounds on Riemann sums, ultimately leading to the inequality

$$P_t(i, j) - P_s(i, j) \geq -q_i \int_s^t P_u(i, j) \, du \qquad 0 \leq s < t$$

By letting $s \to 0$, we have that

$$G_t(i, j) = P_t(i, j) + q_i \int_0^t P_u(i, j) \, du - \delta(i, j) \geq 0,$$

where $\delta(i, j)$ is Kronecker's delta $(= 0$ if $i \neq j$ and $= 1$ if $i = j)$. It thus follows from the equation of Exercise 3 that $G_t(i, j)$ is a nondecreasing function of t. Therefore, $G'_t(i, j) = q_i g_t(i, j) \geq 0$, that is, $g_t(i, j) \geq 0$. This completes the proof.   ■

We close the analysis of the structure of a continuous time Markov process X_t by the following remark. We know from Theorem 12 that the random holding time T_i in the state i is an exponential variable; that is, given that $X_t = i$, the process satisfies $X_{t+u} = i$ for $u < T_i$ and then it moves to a state j. In other words, when a change occurs at $t + T_i$ we know that $X_s = i$ but $X_{s+h} \neq i$ for an arbitrarily small $h > 0$ $(s < t + T_i < s + h)$. Thus, the probability that the transition is from i to j can be approximated by $\lim P(X_{s+h} = j \mid X_s = i, X_{s+h} \neq i)$ as $h \to 0$ and $s < t + T_i < s + h$. But, by stationarity, for fixed s and $h > 0$,

$$P(X_{s+h} = j \mid X_s = i, X_{s+h} \neq i) = \frac{P(X_{s+h} = j \mid X_s = i)}{P(X_{s+h} \neq i \mid X_s = i)} = \frac{P(X_h = j \mid X_0 = i)}{P(X_h \neq i \mid X_0 = i)}$$

$$= \frac{P_h(i, j)/h}{[1 - P_h(i, i)]/h} \to \frac{q_{ij}}{q_i}$$

as $h \to 0$ and $i \neq j$.

Our information on a continuous time Markov chain X_t is adequate for having a new look at Poisson processes. We call a homogeneous Markov chain $X_t, t \geq 0$, a Poisson process if its q-matrix has identical diagonal entries $q_{ii} = -\lambda$, $\lambda > 0$, its state space is the set of positive integers and $q_{i,i+1} = \lambda$. Then, by Theorem 15, $q_{ij} = 0$ for all (i, j). Thus, a Poisson process stays in any state for an exponential time with parameter λ, and the only transition it can make is from state i to $i + 1$. The independence of distinct holding times follows from the general independence property at (8.2). Consequently, such an approach to Poisson processes on the positive real line leads to the same structure as in Section 8.2 (see Theorem 3 and Exercise 2 in that section). An even more striking similarity between the present definition of a Poisson process and the one in Section 8.2 can be observed by looking at the meaning of q_{ij}. We have $P(X_{t+h} = i \mid X_t = i) = 1 - \lambda h + o(h)$ and $P(X_{t+h} = i + 1 \mid X_t = i) = \lambda h + o(h)$, while, for any $k \geq 2$,

$$P(X_{t+h} = i + k \mid X_t = i) = o(h)$$

as $h \to 0$. That is, if we translate "no change," "change from i to $i + 1$," and "change from i to $i + k$" as having no point, one point, and k points, respectively, in the interval $(t, t + h)$, the above asymptotic properties are exactly those which characterize (8.19).

If we have a Markov chain with a q-matrix similar to the one for a Poisson process with the only change that in the ith row the parameter λ depends on i, λ_i, say, then we speak of a pure birth process. Transitions are again from i to $i + 1$, but, depending on λ_i, $i \geq 1$, it is possible in finite time t that infinitely many changes occur since an infinite sum $\sum T_j$ of independent exponential variables can be finite (see Section 4.1).

By making one step further in the above description in that a transition from i to any state j is allowed, whose probability is q_{ij}/q_i, we arrive at a general continuous time homogeneous (stationary) Markov chain: we have a sequence T_i, T_j, $T_k, \ldots$ of independent exponential variables, indexed by the elements of the space S in some order, and if the consecutive sums of these exponential variables are denoted $S_1 < S_2 < S_3 < , \ldots$, then changes of X_t occur at the (random) points S_1, S_2, S_3, etc., and X_t is in a constant state between these points (in our previous notation, $X_0 = i$ then at $S_1 = T_i$ it changed to state j, remained in j up to $S_2 = T_i + T_j$, when it changed to state k and $X_t = k$ for $S_2 < t < S_3 = T_i + T_j + T_k$, and then again a change occurred, etc.).

Another way to discretize a Markov chain X_t is to introduce the sequence $X_n^{(s)}$, $n \geq 1$, where $s > 0$ is a fixed number, $X_n^{(s)} = X_{ns}$ and the one-step transition matrix of $X_n^{(s)}$ is $\mathbf{P} = (P_s(i, j))$. Clearly, if X_t is homogeneous, then $X_n^{(s)}$, $n \geq 1$, is a discrete time homogeneous Markov chain whose n-step transition matrix is $(P_{ns}(i, j))$. Limit theorems of Section 8.5 therefore provide information on $P_t(i, j)$, $t \geq 0$, as well. Also, we can classify the states of X_t, $t \geq 0$, which are preserved by the chain $X_n^{(s)}$, $n \geq 1$, which chain we call the s-skeleton of X_t.

Based on the criterion (8.48) for a state i in a discrete time Markov chain to be persistent, we say that state i of a continuous time homogeneous Markov chain X_t, $t \geq 0$, is persistent if $\int_0^{+\infty} P_t(i, i)\, dt = +\infty$. Otherwise, state i is called transient. In addition, similarly to discrete time Markov chains, we say that j can be reached from i if, for some $t > 0$, $P_t(i, j) > 0$, and that i and j intercommunicate if j can be reached from i and i can be reached from j.

Theorem 16 Assume i and j intercommunicate. Then i is persistent if, and only if, j is persistent. Furthermore, if i is persistent then i is persistent for the s-skeleton.

Proof. By assumption, there are numbers u and t, both positive, such that $P_u(j, i) > 0$ and $P_t(i, j) > 0$. Then, for every $s \geq 0$,

$$P_{t+s+u}(i, i) \geq P_t(i, j)P_s(j, j)P_u(j, i)$$

which, when integrated with respect to s, yields that if j is persistent, so is i. Upon reversing the roles of i and j, the claimed equivalence follows.

In order to prove that persistency is preserved by any s-skeleton, first write

$$\int_0^{+\infty} P_t(i, i)\, dt = \sum_{n=0}^{+\infty} \int_{ns}^{(n+1)s} P_t(i, i)\, dt$$

and then estimate

$$s \min P_{ns+u}(i, i) \leq \int_{ns}^{(n+1)s} P(i, i)\, dt \leq s \max P_{ns+u}(i, i)$$

where both min and max are over $0 \leq u \leq s$. Upon appealing to

$$P_{ns+u}(i, i) \geq P_{ns}(i, i)P_u(i, i) \qquad \text{and} \qquad P_{(n+1)s}(i, i) \geq P_{ns+u}(i, i)P_{s-u}(i, i)$$

we have, with $r = \min\{P_u(i, i): 0 \le u \le s\}$,

$$\min P_{ns+u}(i, i) \ge rP_{ns}(i, i) \quad \text{and} \quad P_{ns+u}(i, i) \le P_{(n+1)s}(i, i)/r$$

By summing over n we get that persistency of i occurs concurrently for the process X_t and for its s-skeleton. The proof is completed. ∎

In view of Theorem 10 (iv), the limiting properties of $P_t(i, j)$ can be deduced from the behavior of the s-skeletons via Exercise 1. The reader is advised to work out the details of specific cases, even though Exercise 5 covers a very general case.

Exercises

1. Let the real valued function P_t, $t \ge 0$, be uniformly continuous. Let $s > 0$, and assume that, as $n \to +\infty$, $\lim P_{ns} = r_s$ exists and is finite. Show that r_s does not depend on s, and, as $t \to +\infty$, $\lim P_t = r$ ($= r_s$).
2. For the Markov chain X_t, $t \ge 0$, define the process $Y_t = X_{T-t}$, $0 \le t \le T$, where $T > 0$ is a fixed number. Show that Y_t, $0 \le t \le T$, is a Markov chain.
3. Show that the function $G_t(i, j)$ defined in the proof of Theorem 15 satisfies the equation

$$G_{s+t}(i, j) - G_t(i, j) = \sum_{k \in S} G_s(i, k)P_t(k, j)$$

4. Let $i \in S$ be a stable state. Show that $P_t'(i, j) = \sum_{k \in S} q_{ik} P_t(k, j)$ assuming that $\sum_{k \in S} q_{ik} = 0$.
5. Assume that every pair i, j of states intercommunicates (the chain is called irreducible just as in the discrete time case). Show that $p_j = \lim P_t(i, j)$ exists as $t \to +\infty$, where p_j does not depend on i. If $p_j > 0$ for one j, then $p_j > 0$ for all j and $\sum_{j \in S} p_j = 1$. Furthermore, the distribution p_j, $j \in S$, satisfies $p_j = \sum_{i \in S} p_i P_t(i, j)$ for all $j \in S$ and all $t \ge 0$. There is no other probability distribution that could satisfy the preceding equations for a single $t > 0$.

8.7 AN INVARIANCE PRINCIPLE

This section deals with a special case of the following general problem that is known as an invariance principle. Let $G_n(t; F) = G_n(t, \omega; F)$, $n \ge 1$, $t \in T$, $\omega \in \Omega$, be a sequence of stochastic processes defined by means of a sequence X_j, $j \ge 1$, of independent and identically distributed random variables with a common distribution function $F(x)$. Assume that $G_n(t; F)$ converges in some sense, as $n \to +\infty$, to a stochastic process $B(t) = B(t, \omega)$, $t \in T$, $\omega \in \Omega$, that is invariant under F within some set of distribution functions. Then we can find the characteristics of $B(t)$ by choosing the simplest possible $F(x)$, which, in turn, provides general limit theorems for all F in the class.

Our specific model assumes that the X_j satisfy $E(X_j) = 0$ and $0 < V(X_j) = \sigma^2 < +\infty$. Then, as usual, we set $S_n = X_1 + X_2 + \cdots + X_n$ and define

$$G_n(t; F) = \frac{1}{\sigma\sqrt{n}} S_{j-1} + (nt - j + 1) \frac{1}{\sigma\sqrt{n}} X_j \quad \text{if} \quad \frac{j-1}{n} \le t \le \frac{j}{n} \quad (8.59)$$

where $t \in T = [0, 1]$. Note that, with $t = 1$, $G_n(1; F) = S_n/\sigma\sqrt{n}$, which is known, via the classical central limit theorem, to be asymptotically normally distributed, independently of F in our class (having finite and positive variance). We would like to extend this result to establish limit theorems for $G_n(t; F)$, $t \in T$, that will, we hope, be independent of F. We thus face the task of extending some of the concepts and results we developed for random variables to $G_n(t; F)$. In particular, we want to speak of the distribution and the weak convergence of $G_n(t; F)$. Let us first have a new look at these concepts for (ordinary) random variables.

If Y is a random variable (on a probability space), then its distribution function $U(x)$ is defined by $U(x) = P(Y \in A(x))$, where $A(x) = (-\infty, x]$. This uniquely determines the distribution $P_Y(A) = P(Y \in A)$, where A is an arbitrary Borel set on the real line. Now, if Y_n is a sequence of random variables with distribution functions $U_n(x)$, then Y_n is said to converge weakly to Y if $U_n(x) \to U(x) = P(Y \le x)$ for all continuity points x of $U(x)$. We have seen that this restriction on x cannot be dropped and leave a meaningful limit theory of distribution functions. However, weak convergence does not imply the convergence of measures $P(Y_n \in A) \to P(Y \in A)$, not even for simple Borel sets. For example, with $Y_n = S_n/\sigma\sqrt{n}$, where S_n is as in (8.59) and with A consisting of the points $(k - np)/\sigma\sqrt{n}$, $k = 0, 1, \ldots, n$, $n \ge 1$, where $\sigma^2 = p(1 - p)$, $P(Y_n \in A) = 1$ if the summands X_j of S_n take the values $-p$ and $1 - p$ with probabilities $1 - p$ and p, respectively, while the limit $P(Y \in A) = 0$, because A is a denumerable set and Y is normal. What is common both to $A(x)$ in the definition of weak convergence, when x is a discontinuity point of $U(x)$, and to the special A we have just constructed in connection with the asymptotic normality of the binomial distribution is that the boundary points of $A(x)$ and A have positive probability with respect to the limiting distribution $U(x)$. In general, the set ∂J is the boundary of a set J if, for every $x \in \partial J$, there are sequences $x_n \in J$ and $y_n \in J^c$ such that $x_n \to x$ and $y_n \to x$ as $n \to +\infty$. Now $\partial A(x) = \{x\}$, whose U-probability is positive if it is a point of discontinuity of $U(x)$, and ∂A is the whole real line, since it is a dense set. Therefore, as $P(Y_n \in A) \to P(Y \in A)$ we shall consider only those Borel sets for which $P_Y(\partial A) = 0$. Such a set A will be called Y-continuous. (Here, P_Y is the probability generated by Y.)

Now, to introduce distributions and weak convergence for $G_n(t; F)$, $t \in T$, we have to introduce a distance in the set of possible values of $G_n(t; F)$ in order to enable us to speak of Borel sets of these values. Note that, for almost all ω, $G_n(t; F)$ is a continuous function (consisting of linear segments) on T; thus, we want to speak of Borel sets of continuous functions on T. For easier reference, we denote the set of all continuous functions on T by $C = C(T)$. We define the distance

$$\rho(u, v) = \sup_{0 \le t \le 1} |u(t) - v(t)| \qquad u = u(t), v = v(t) \in C \qquad (8.60)$$

Open and closed spheres, and thus open sets, and finally Borel sets can be defined in the same manner as in Euclidean spaces. This now makes it meaningful to define

$$P_n(A) = P(G_n(t; F) \in A) \qquad A \subset C \text{ and } A \text{ Borel set} \qquad (8.61)$$

Let $Q(A)$ be a probability measure on the Borel subsets of C. We say that $P_n(A) \to Q(A)$ weakly if $P_n(A) \to Q(A)$ for all A with $Q(\partial A) = 0$.

The aim of this section is to establish the following result, generally known as Donsker's theorem. In an equivalent form, it was first established by Donsker (1951). Its seeds are present in Kolmogorov (1931). Billingsley (1968) provides a systematic theory of weak convergence on metric spaces. This section deviates from the method of the rest of the book, in that only basic ideas, not all the details of the proofs, will be given. The interested reader can fill in the details by consulting Billingsley's book.

Theorem 17 Let the assumptions on X_m and $F(x)$ of (8.59) be satisfied. There is a probability measure $W(A)$ on the Borel subsets A of C such that $P_n(A)$ of (8.59) and (8.61) converges weakly to $W(A)$, which is independent of F.

Before outlining the proof of Theorem 17, let us deduce a few of its consequences. These deductions will represent the way in which an invariance principle works. We call $W(A)$ Wiener measure and a stochastic process $B(t) = B(t, \omega)$, $t \in T$, $\omega \in \Omega$, such that $P(B(t) \in A) = W(A)$ Brownian motion. Here, and in what follows, it is assumed that $P(B(t) \in C) = 1$ and $A \subset C$ goes through the Borel sets of C. Furthermore, $T = [0, 1]$.

Corollary 1 The distribution function of $B(t)$ is normal with zero expectation and variance t. The increments $B(t_r) - B(t_{r-1})$, $1 \le r \le k$, $0 \le t_{r-1} < t_r \le 1$, are independent normal variables.

Proof. By Theorem 17, we can choose F to be standard normal. Then $G_n(t; F)$ is normally distributed with zero expectation and variance $V_n = \{[nt] + (nt - [nt])^2\}/n$, where $[y]$ signifies the integer part of y. Hence, the asymptotic distribution of $G_n(t; F)$ is normal with the desired parameters. Next, consider the increments $G_n(t_r; F) - G_n(t_{r-1}; F)$, $1 \leq r \leq k$. Since, for large n, they involve mutually exclusive sets of the basic random variables X_m, which are assumed to be normal, they are independent normal variables. The corollary is established. ∎

Corollary 2 $W\left(\sup_{0 \leq t \leq 1} B(t) \leq x\right) = 2N(x) - 1$, $x \geq 0$ where $N(x)$ is the standard normal distribution function.

Proof. It is readily seen that $G_n(t; F) \to B(t)$ weakly implies $g(G_n(t; F)) \to g(B(t))$ weakly, as the function $g(u) = \sup_{0 \leq t \leq 1} u(t)$ is a continuous function on C. Now, since

$$\sup_{0 \leq t \leq 1} G_n(t; F) = \max_{0 \leq j \leq n} (S_j/\sigma\sqrt{n})$$

Theorem 17 implies that the corollary will be proved if we show that, with any special $F(x)$,

$$P\left(\max_{0 \leq j \leq n} S_j \leq x\sigma\sqrt{n}\right) \to 2N(x) - 1 \qquad \text{as } n \to +\infty \tag{8.62}$$

Let us choose a simple random walk, i.e., let X_m take 1 and -1 only, each with probability $1/2$, and let $X_0 = S_0 = 0$. By an elementary argument, one finds that, for $y \geq 0$ integer,

$$P\left(\max_{0 \leq j \leq n} S_j \geq y\right) = 2P(S_n > y) + P(S_n = y)$$

from which the classical central limit theorem (choose $y = y_n \sim x\sqrt{n}$) immediately yields the desired limit. ∎

Corollary 3 Let the assumptions on X_m and $F(x)$ of (8.59) be satisfied. Then

$$P\left(\max_{0 \leq j \leq n} S_j \leq x\sigma\sqrt{n}\right) \to 2N(x) - 1$$

as $n \to +\infty$.

Proof. It follows immediately from Theorem 17 and Corollary 2 by the observation made in the first sentence of the preceding proof. ∎

Note the method of proving Corollary 3. We first proved the same limit relation for a special case in (8.62) and then Theorem 17 permitted us to conclude that it remains valid for all F in our class (finite variance). Several other properties of the simple random walk S_n, where $P(X_m = 1) = 1 - P(X_m = -1) = 1/2$, extend to all of our S_n via Theorem 17. For example, those familiar with the arc sine law (see Chapter 6 in the introductory volume) will find that a similar law applies to all S_n where the variance is finite and positive.

Our definition of Brownian motion is inconvenient for proving that a particular stochastic process is a Brownian motion because the definition includes that for every Borel set in C, the distribution is given by $W(.)$. However, since Wiener measure W is uniquely determined by finite dimensional distributions, we can give the simpler definition: we say that a stochastic process $B(t)$, $t \in T$, where T is either $[0, 1]$ or the whole nonnegative real line, is a Brownian motion if (i) $B(t)$ is almost surely a continuous function over T, (ii) $B(t)$ has independent increments, that is, for nonoverlapping subintervals (u_j, t_j), $1 \le j \le n$, of T, the differences $B(t_j) - B(u_j)$ are independent, (iii) for every $t \in T$, $B(t)$ is a normal random variable with $E(B(t)) = 0$ and $V(B(t)) = t$, and (iv) $P(B(0) = 0) = 1$. Note that (iv) would be a part of (iii) if we define a general normal random variable as $sX + m$, where X is a standard normal variable and $s \ge 0$ and m are constants. Then (iii) would take the form $s = \sqrt{t}$, $m = 0$ and $X = X(t)$ standard normal for every t, and (iv) would follow. We must also record that, upon writing

$$[B(t) - B(u)] + B(u) = B(t) \qquad u < t \qquad u, t \in T$$

and observing that $B(u) = B(u) - B(0)$ by (iv), (ii) and (iii) entail that the increment $B(t) - B(u)$ is a summand in a decomposition of the normal variable $B(t)$ into independent terms, where

$$V(B(t) - B(u)) + u = t, \qquad \text{i.e.} \qquad V(B(t) - B(u)) = t - u > 0$$

Hence, by Cramer's theorem (Theorem 4.5), $B(t) - B(u)$, $u < t$, $u, t \in T$, is a normal variable with zero expectation and variance $t - u$.

Given a Brownian motion $B(t)$, $0 \le t \le 1$, let us define

$$B_b(t) = B(t) - tB(1), \qquad 0 \le t \le 1$$

Clearly, $P(B_b(0) = 0) = P(B_b(1) = 0) = 1$. Mainly for this reason, $B_b(t)$ is called a Brownian bridge or tied-down Brownian motion. Now, if we rewrite

$$B_b(t) = -t(B(1) - B(t)) - (t - 1)B(t), \qquad B(t) = B(t) - B(0)$$

we get from properties (ii)–(iv) of $B(t)$ that $B_b(t)$ is normal for $0 < t < 1$ with $E(B_b(t)) = 0$ and $V(B_b(t)) = t^2(1 - t) + (1 - t)^2 t = t(1 - t)$.

Starting again from a Brownian motion $B(t)$, $0 \le t \le 1$, let us define

$B^*(t) = (1 + t)B_b(t/(1 + t)), t \geq 0$. We show that $B^*(t)$ is a Brownian motion on the nonnegative real line. Indeed, by definition,

$$B^*(t) = (1+t)\left[B\left(\frac{t}{1+t}\right) - \frac{t}{1+t} B(1) \right] = B\left(\frac{t}{1+t}\right) - t\left[B(1) - B\left(\frac{t}{1+t}\right) \right]$$

where the two terms on the extreme right hand side are independent normal variables by (ii) and (iii) for $B(t)$. We also have

$$E(B^*(t)) = 0 \quad \text{and} \quad V(B^*(t)) = \frac{t}{1 + t} + t^2\left(1 - \frac{t}{1 + t}\right) = t$$

Thus, we established (iii) for $B^*(t)$. Since (i) and (iv) are evident from the definition, it remains to prove (ii). For $0 \leq t < s$ write $t' = t/(1 + t)$ and $s' = s/(1 + s)$. Then $t' < s' \leq 1$, and $E(B(s')B(t')) = t' = E(B(1)B(t'))$, while $E(B(1)B(s')) = s'$, all on account of (ii) and (iii) for $B(t)$. We thus have $E(B^*(t)B^*(s)) = t$ which is sufficient for (ii) because, as is easily verified, the finite dimensional distributions of the $B^*(t_j)$ are multivariate normal which are uniquely determined by the correlations given above. (Write $B^*(s) = B^*(s) - B^*(t) + B^*(t)$. If $B^*(t)$ and $B^*(s) - B^*(t)$ are independent, then, under (iii) for $B^*(.)$, the previously computed correlation must be t. By the uniqueness for normal distributions, this statement can also be reversed.) This completes the proof that $B^*(.)$ is a Brownian motion on the nonnegative real line.

By the same argument as above we also have that if $B(t)$, $t \geq 0$, is a Brownian motion then so is $tB(1/t)$. The details of computation are left to the reader as an exercise.

Recall the concept of bounded variation introduced in Exercise 6 of Section 2.7. Formulated for a Brownian motion $B(t)$, $0 \leq t \leq 1$, the total variation $v(B)$ of $B(t)$ is the supremum over all choices of $0 \leq t_1 < t_2 < \cdots < t_n \leq 1$ of the sum

$$\sum_{j=2}^{n} |B(t_j) - B(t_{j-1})|$$

Hence, $v(B)$ is larger than the above sum for any particular choice of the points t_j and of n. Let us therefore estimate the above sum for large $n = 2^m + 1$ and $t_j = (j - 1)/2^m$, $1 \leq j \leq n$. Denoting the resulting sum by $v_m(B)$, we note that $v_m(B)$ is the sum of independent random variables, each being the absolute value of a normal variable. We also have from (ii) and (iii) that each summand in $v_m(B)$ has expectation $c_1/2^{m/2}$ and variance $c_2/2^m$, where $c_1 > 0$ and $c_2 > 0$ are to be computed from the absolute value of a standard normal variable. Finally, we have from the elementary triangular inequality that $v_{m+1}(B) \geq v_m(B)$. Hence, we have that $\lim v_m(B)$ exists almost surely as $m \to +\infty$, and Chebyshev's inequality implies that this limit must be infinity almost surely. In other words, $B(t)$ is of unbounded variation almost surely, which translates to very wild oscillations of the continuous function $B(t)$ of t. This is particularly surprising if we recall that $B(t)$ was generated by the simple piecewise linear functions $G_n(t; F)$ of (8.59). For this same reason that $B(t)$ can be generated from the central limit theorem via the functions $G_n(t; F)$, Brownian motions are frequently used in physics and chemistry as the underlying process for describing the movement of particles in liquid or gas. It is rather pleasing to experimental scientists in the mentioned fields that mathematics produces the same conclusion of wild oscillation of particles as observed in laboratories.

We now turn to the proof of Theorem 17. The proof will be split up into a number of lemmas, some of which are simple; others, such as Prokhorov's result, constitute the most important results in this area.

Lemma 1 Let $0 \le t_0 < t_1 < \cdots < t_k \le 1$ be fixed numbers. The limiting distribution of the vector $G_n(t_r; F)$, $0 \le r \le k$, exists and, independently of F, it is the distribution of the vector $B(t_r)$, $0 \le r \le k$, as given in Corollary 1.

Proof. Just as in the proof of Corollary 1, we observe that the differences $G_n(t_r; F) - G_n(t_{r-1}; F)$, $r \ge 1$, are defined by means of mutually exclusive groups of X_m when n is sufficiently large. Indeed, apart from a term that tends to zero in probability, these differences are of the form

$$(S_{j(r)} - S_{j(r-1)})/\sigma\sqrt{n} \qquad 1 \le r \le k$$

where $j(r) \sim nt_r$. Hence, by the classical central limit theorem, these differences are asymptotically independent normal variables with zero expectation and variance $t_r - t_{r-1}$. The proof is complete. ∎

The collection of all distributions of Lemma 1 is called the finite-dimensional distribution of the process $G_n(t; F)$ and the limiting forms are called the finite-dimensional distributions of $B(t)$. In general, the convergence of finite-dimensional distributions to those of a limiting process does not imply weak convergence. However, when one analyzes the reasons behind the limit theorems for ordinary random variables further, it turns out that the crucial parts of the proofs are (i) every sequence of distributions has a subsequence that converges weakly, which is then guaranteed to be a proper distribution in particular limit theorems and (ii) the establishment of property (i) for ordinary random variables fully uses the fact that the real line has a denumerable dense set (the set of rationals). The first property is called the compactness of distributions and the second one, the separability of the real line. Fortunately, the space C is separable. Simply take continuous functions on T that are linear between the points i/n and $(i + 1)/n$ for all $0 \le i < n$ and whose coefficients are rational numbers. This is a denumerable collection of members of C; they are dense in C in view of the uniform continuity of members of C.

The compactness of distributions is not easy to guarantee. Remember that we want to guarantee the existence of a subsequence of any sequence of distributions that converges to a proper distribution. The following result of Prokhorov (1956) proved to be an easily applicable tool in this direction.

Lemma 2   A set $\mathscr{P}$ of probability distributions on the Borel sets of C is compact if and only if there is a compact set $D \subset C$ for every $\varepsilon > 0$ such that $P(D) \ge 1 - \varepsilon$ for all $P \in \mathscr{P}$.

Even though we shall not prove Lemma 2, it should be pointed out that the separability (and the completeness) of C are important for the result. The existence of a compact D with the property of Lemma 2 is the requirement for calling $\mathscr{P}$ tight. Now, for tightness on C, one can develop criteria step by step that reduce to the following simple criterion in the case of $G_n(t; F)$.

Lemma 3   The probability measures $P(G_n(t; F) \in A), n \ge 1$, are tight because, for each $\varepsilon > 0$, there exists a $\lambda > 1$ and an integer n_0 such that, for $n \ge n_0$,

$$P\left(\max_{j \le n} |S_{n+j} - S_k| \ge \lambda \sigma \sqrt{n} \right) \le \frac{\varepsilon}{\lambda^2} \qquad \text{for all } k$$

These proofs are not simple, but two purposes are served by having the sequence of statements needed for the proof of Theorem 17. One can see the difficulties one has to overcome when moving away from Euclidean spaces and, at the same time, the simplicity of the statements shows that an invariance principle, whose value is evident from the applications in the corollaries, is not too abstract a generalization of the simplest form of the central limit theorem.

Finally having the compactness of the distributions $P(G_n(t; F) \in A)$ guaranteed and knowing that the finite-dimensional distributions do converge to those of $B(t)$, we require only the observation that a measure on C is fully determined by the finite-dimensional distributions for the completion of the proof. This is due to the fact that an open set of C can be decomposed into a denumerable set of open or closed spheres (one can easily obtain such a decomposition by taking those members of C which figure in the separability of C as centers and taking radii to be rational). Hence, the above distributions have a subsequence that converges weakly, but each such subsequence has the same finite-dimensional limiting distribution. Since these finite-dimensional distributions uniquely determine the limiting measure, each convergent subsequence has the same limit; that is, the whole sequence converges.

REFERENCES

Adès, M., Dion, J.-P., Labelle, G., and K. Nanthi (1982). Recurrence formula and the maximum likelihood estimation of the age in a simple branching process. *J. Appl. Probability 19*, 776–784.

Anderson, W. J. (1991). *Continuous-Time Markov Chains*. Springer-Verlag, Heidelberg.

Billingsley, P. (1968). *Convergence of Probability Measures*. Wiley, New York.

Blackwell, D. (1953). Extension of a renewal theorem *Pacific J. Math. 3*, 315–320.

Cinlar, E., and P. Jagers (1973). Two mean values which characterize the Poisson process. *J. Appl. Probability 10*, 678–681.

Donsker, M. (1951). An invariance principle for certain probability limit theorems. *Memoirs Amer. Math. Soc.* No. 6.

Doob, J. L. (1953). *Stochastic Processes*. Wiley, New York.

Erdős, P., Feller, W., and H. Pollard (1949). A property of power series with positive coefficients. *Bull. Amer. Math. Soc. 55*, 201–204.

Feller, W. (1966). *An Introduction to Probability Theory and Its Applications*. Vol. II. Wiley, New York.

Galambos, J. (1984). *Introductory Probability Theory*. Marcel Dekker, New York.

Galambos, J. and S. Kotz (1978). *Characterizations of Probability Distributions. Lecture Notes in Math., 675*. Springer-Verlag, Heidelberg.

Galambos, J., Indlekofer, K.-H., and I. Kátai (1987). A renewal theorem for random walks in multidimensional time. *Trans. Amer. Math. Soc. 300*, 759–769.

Harris, T. E. (1963). *The Theory of Branching Processes*. Springer-Verlag, Berlin.

Holmes, P. T. (1974). Another characterization of the Poisson process. *Sankhyā, A 36*, 449–450.

Isaacson, D. L. and R. W. Madsen (1985). *Markov Chains. Theory and Applications.* Krieger, Melbourne, Florida.

Kallenberg, O. (1976). *Random Measures.* Academic Verlag, Berlin.

Kolmogorov, A. N. (1931). Eine Verallgemeinierung des Laplace–Liapunoff-schen Satzes. *Izv. Akad. Nauk SSSR Ser. Fiz-Mat.,* 959–962.

Lindvall, T. (1992). A simple coupling of renewal processes. *Adv. Appl. Probability 24,* 1010–1011.

Maejima, M. and T. Mori (1984). Some renewal theorems for random walks in multidimensional time. *Math. Proc. Cambridge Philos. Soc. 95,* 149–154.

Prékopa, A. (1958). On secondary processes generated by a random point distribution of Poisson type. *Ann. Univ. Sci. Budapest, Sectio Math. 1,* 153–170.

Prokhorov, Yu. V. (1956). Convergence of random processes and limit theorems in probability theory. *Theory Prob. Appl. 1,* 157–214.

Rényi, A. (1967). Remarks on the Poisson process. *Studia Sci. Math. Hungar. 2,* 119–123.

Seneta, E. (1981). *Non-negative Matrices and Markov Chains,* 2nd ed. Springer-Verlag, Berlin.

Takács, L. (1967). *Combinatorial Methods in the Theory of Stochastic Processes.* Wiley, New York.

Hints for Solutions of Exercises

CHAPTER 1
Section 1.1

In Exercises 1 and 2, show that all members of the left-hand side also belong to the right-hand side and vice versa. Alternatively, apply the rules of p. 2. With the latter method, 1(i) and 2(ii) are worked out below.

1. (i) $A \cup B = (A \cup B) \cap \Omega = (A \cup B) \cap (A \cup A^c) = A \cup (B \cap A) \cup (B \cap A^c) = A \cup (B \cap A^c)$

2. (ii) $B \triangle (A_1 \cup A_2) = [B \cap (A_1 \cup A_2)^c] \cup [B^c \cap (A_1 \cup A_2)] = (B \cap A_1^c \cap A_2^c) \cup \{B^c \cap [(A_1 \cap A_2) \cup (A_1 \triangle A_2)]\} = (B \triangle A_1) \cap (B \triangle A_2) \cup [B^c \cap (A_1 \triangle A_2)]$

3. If ω belongs to the left-hand side, then $\omega \in A_j$ for all $j \geq n$, for some n. But then, for all n, $\omega \in \bigcup_{j=n}^{+\infty} A_j$. Case (i): $\bigcap_{j=n}^{+\infty} A_j = A_n$ and $\bigcup_{j=n}^{+\infty} A_j = \bigcup_{j=1}^{+\infty} A_j$. Hence, both sides equal $\bigcup_{n=1}^{+\infty} A_n$. Case (ii): $\bigcup_{j=n}^{+\infty} A_j = A_n$ and $\bigcap_{j=n}^{+\infty} A_j = \bigcap_{j=1}^{+\infty} A_j$, yielding $\bigcap_{n=1}^{+\infty} A_n$ for both sides. Case (iii): $\bigcap_{j=n}^{+\infty} A_n = [1, 2 - 1/n)$ and $\bigcup_{j=n}^{+\infty} A_j = (1 - 1/n, 2)$. Hence, both sides equal $[1, 2)$.

4. Since $\mathscr{B} \subset \sigma(\mathscr{B})$, $\mathscr{B} \cap A \subset \sigma(\mathscr{B}) \cap A$, and one can easily verify that $\sigma(\mathscr{B}) \cap A$ is a σ-field of subsets of A. For the converse, we must show that $B \cap A \in \sigma(\mathscr{B} \cap A)$ for all $B \in \sigma(\mathscr{B})$. Show first that if $\mathscr{B}^*$ is the set of those

$B \in \sigma(\mathcal{B})$ for which $B \cap A \in \sigma(\mathcal{B} \cap A)$, then $\mathcal{B}^*$ is a σ-field. But $\mathcal{B} \subset \mathcal{B}^*$, hence $\sigma(\mathcal{B}) \subset \mathcal{B}^*$, implying $\sigma(\mathcal{B}) = \mathcal{B}^*$.

5. Set $A_{n+1} = (\bigcup_{j=1}^{n} A_j)^c$. Then $\mathcal{F}(\mathcal{B})$ is the set of the empty set and the unions $\bigcup_{t=1}^{k} A_{j_t}$, $1 \le j_1 < j_2 < \cdots < j_k \le n+1$, $1 \le k \le n+1$. $\sigma(\mathcal{B}) = \mathcal{F}(\mathcal{B})$.

6. Follow the instructions on p. 3 for constructing the smallest field containing a given collection of sets. Note that every intersection B of some C_j (or their complements) can also be written as a union of some A_t by applying $B = B \cap \Omega = B \cap [\bigcap_{j=1}^{n} (C_j \cup C_j^c)]$. Now apply Exercise 5. Finally, for (i), take all A_t for which $m = s$ in the definition of A_t. For (ii), take all A_t with associated $m \ge s$.

Section 1.2

1. Axioms 1 and 2 are ensured by (i) and (ii). Axiom 3 is to be checked by direct computation.
2. By assumption, $P(I) > 0$ for $I = (0, 1)$. Hence, $P(\Omega) > P(\bigcup_{k=1}^{+\infty} I_k) = \sum_{k=1}^{+\infty} P(I_k) = \sum_{k=1}^{+\infty} 1/k(k+1) = 1$, contradicting axiom 2.
3. Since $P(A \cap B) \le P(A)$ or $P(B)$ and $P(A) + P(B) - P(A \cap B) = P(A \cup B) \le 1$, (i) and (ii) are necessary conditions. For sufficiency, make the following construction: let $\Omega = (0, 1)$ and let A and B be subintervals of $(0, 1)$ with lengths p_1 and p_2 such that their intersection is of length p_{12}. Condition (ii) ensures that this construction is possible. Now, let $\mathcal{A} = \{\Omega, \varnothing, A, B, A^c, B^c, A \cup B, A^c \cup B, \ldots\}$ and let P be determined by "length." Extension to 3 events follows along this same line.
4. (i) Write $A \triangle C = (A \triangle C) \cap (B \cup B^c)$, $A \triangle B = (A \triangle B) \cap (C \cup C^c)$, and $B \triangle C = (B \triangle C) \cap (A \cup A^c)$. After expansion and taking probabilities (term by term), all terms of $P(A \triangle C)$ will appear in $P(A \triangle B) + P(B \triangle C)$.
   (ii) Write $P(A) = P(A \cap B) + P(A \cap B^c)$ and $P(B) = P(A \cap B) + P(A^c \cap B)$. Hence, $|P(A) - P(B)| \le P(A \cap B^c) + P(A^c \cap B) = P(A \triangle B)$.
5. Simple rules of calculus yield the first part. To show that $d(A)$ is not σ-additive, take finite disjoint sets A_k whose union is the set of all positive integers. Then $d(A_k) = 0$ for all k and $d(\bigcup_{k=1}^{+\infty} A_k) = 1$.
6. Since the A_j are disjoint, $\liminf A_n = \varnothing$. Now, $\bigcup_{j=k}^{+\infty} A_j = (0, 1/k]$; thus, $\limsup A_n = \varnothing$ as well.
7. (i) Apply de Morgan's law from p. 2.
   (ii) Apply the meaning of $\limsup A_n$, as explained on pp. 9–10.
8. From the decimal expansion algorithm, it easily follows that $P(A_n^{(T)}) = 0.9^n$. Apply the Borel–Cantelli lemma (Theorem 10). The result means that the digit 2 must occur infinitely often for almost all x.

Section 1.3

1. (i) Immediate from definitions.
    (ii) $A = \cup A_n$ and $B = \cup B_n$ with $A_n, B_n \in \mathscr{F}(\mathscr{B})$; thus, $A \cup B = \cup(A_n \cup B_n)$ and $A \cap B = \cup(A_n \cap B_n)$ with $A_n \cup B_n$ and $A_n \cap B_n \in \mathscr{F}(\mathscr{B})$. Furthermore, together with A_n and B_n, $A_n \cup B_n$ and $A_n \cap B_n$ are monotonic. The desired formula follows by letting $n \to +\infty$ in $P(A_n \cup B_n) = P(A_n) + P(B_n) - P(A_n \cap B_n)$.
    (iii) Immediate from the definition of $P(\cdot)$ on $\mathscr{U}(\mathscr{B})$.
    (iv) With $A_j = \cup_n A_{j,n}$, $A_{j,n} \subset A_{j,n+1}$, and $A_{j,n} \in \mathscr{F}(\mathscr{B})$, write $B_m = \cup_{j \le m} A_{j,m}$. Then $B_m \subset B_{m+1}$, $A_{j,m} \subset B_m \subset A_m$, $j \le m$. Hence, $P(A_{j,m}) \le P(B_m) \le P(A_m)$ for $j \le m$, from which we get $P(A_j) \le \lim P(B_m) \le \lim P(A_m)$ $(m \to +\infty)$. By letting $j \to +\infty$, we get both $\cup A_m = \cup B_m$, $B_m \in \mathscr{F}(\mathscr{B})$ and $\lim P(A_j) = P(\cup A_m)$.
2. (i) Unique, and $\sigma(\mathscr{B})$ is the set of Borel sets.
    (ii) Not unique. For $A = (0, 1/3)$ and $B = (1/4, 1/2)$, $P(A^c \cap B)$ and $P(A \cap B)$ can be chosen in several ways without changing $P(A)$ and $P(B)$.
3. Go through the instructions of Remark 3 on p. 18.
4. The argument is a part of the exercise.

Section 1.4

1. View the problem as a selection of n items with replacement from two kinds of item: "$+1$" and "-1." $C_{r,n}$ occurs if $(n+r)/2$ "$+1$"s and $(n-r)/2$ "-1"s are obtained. So, $P(C_{r,n}) = 0$ if $(n+r)/2$ is not an integer; otherwise, with $k = k_n(r)$, $P(C_{r,n}) = \binom{n}{k} 2^{-n}$. Now, $k_{2n}(0) = n$, and among all $\binom{2n}{k}$, $\binom{2n}{n}$ is the largest one. For the monotonicity in n, apply

$$\binom{2n+2}{n+1} = \frac{(2n+2)(2n+1)}{(n+1)^2}\binom{2n}{n} = 4\frac{2n+1}{2n+2}\binom{2n}{n}$$

2. In case (i), $p_2 = \binom{50}{2}\binom{50}{3}/\binom{100}{5} = 0.319$ and for (ii), $p_2 = \binom{5}{2} 2^{-5} = 0.313$.
3. Write (1.20) in detail in the form obtained when selection is one by one. Use in each factor $u - j = u(1 - j/u)$ (with u being either M or $R - M$ or R). Then, upon putting $p_{s,h}$ and $p_{s,b}$, respectively, for $P(B_s)$ at (1.20)

and (1.20a), we get

$$
p_{s,h} - p_{s,b} = p_{s,b} \left\{ \frac{\prod_{j=1}^{s-1} (1 - j/M) \prod_{j=1}^{n-s-1} [1 - j/(R-M)]}{\prod_{j=1}^{n-1} (1 - j/R)} - 1 \right\}
$$

which can be utilized for a variety of estimates depending on the aims with the estimates. We get one such estimate, which will help us with the second part, if we replace M and $R - M$ by R above. We get

$$
p_{s,h} - p_{s,b} \leq p_{s,b} \left\{ \left[\prod_{j=1}^{s-1} (1 - j/R) \right] \Big/ \prod_{j=n-s}^{n-1} (1 - j/R) - 1 \right\}
$$

For a lower estimate, replace M and $R - M$ by the smaller value of the two, and the denominator can be replaced by one, or kept unchanged. The stated limit is immediate.

4. There are 6! possible arrangements for the 6 numbers, of which 5! favor (i), and 4! favor (iii). Hence, the respective probabilities are $5!/6! = 1/6$ and $4!/6! = 1/30$. For (ii), introduce the events A_j = number j is in the jth position. We look for $P(\bigcup_{j=1}^{6} A_j)$. Extend Theorem 4 to 6 events, or in general, to n events (or refer to Exercise 1 of the next section), and observe that $P(A_{j_1} \cap A_{j_2} \cap \ldots \cap A_{j_k}) = (n-k)!/n!$ in the general case. With $n = 6$, one gets $P(\bigcup_{j=1}^{6} A_j) = 1 - 1/2! + 1/3! - 1/4! + 1/5! - 1/6!$, while, for general n, $P(\bigcup_{j=1}^{n} A_j) = \sum_{k=1}^{n} (-1)^{k-1} 1/k!$, which, by Taylor's formula, converges to $1 - e^{-1}$.

Section 1.5

1. Let $I(i_1, i_2, \ldots, i_k)$ be the indicator variable of $A_{i_1} \cap A_{i_2} \cap \cdots \cap A_{i_k}$. Then $I(i_1, i_2, \ldots, i_k) = 1$ or 0, depending on whether all or only some of the subscripts $i_1 < i_2 < \cdots < i_k$ belong to those f_n of the A_j which occur. Hence, their sum is $\binom{f_n}{k}$. Now, take the expectation and apply Theorem 14(i).

(a) If $f_n \leq k$, then $b_{k+1,n} = 0$. If $f_n > k$, then

$$
\binom{f_n}{k} \Big/ \binom{n}{k} = \frac{n-k}{f_n - k} \binom{f_n}{k+1} \Big/ \binom{n}{k+1} \geq \binom{f_n}{k+1} \Big/ \binom{n}{k+1}
$$

since $f_n \leq n$. By taking the expectation, the claim follows.

(b) Note that $\binom{k+r}{r}\binom{a}{k+r} = \binom{a}{r}\binom{a-r}{k}$. Next, prove by induction that $\sum_{k=0}^{t}(-1)^k\binom{a-r}{k} = (-1)^t\binom{a-r-1}{t}$, $a > r$. Set $a = f_n$ and take expectations.

2. Write

$$\frac{2}{k+1}f_n - \frac{2}{k(k+1)}\frac{f_n(f_n-1)}{2} = \frac{f_n(2k-f_n+1)}{k(k+1)}$$

which equals zero if $f_n = 0$. On the other hand, it is ≤ 1 because $x(2k+1-x)$ is a parabola in x whose maximum is at $x = k + 1/2$, but, since our $x = f_n$ is an integer, the largest value of $f_n(2k + 1 - f_n)$ is at $f_n = k$ or $k + 1$. We thus proved that the expression above $\leq I_n$, where I_n is the indicator of the union of A_j, $1 \leq j \leq n$. The desired inequality follows by taking the expectation. Finally, for given $b_{1,n}$ and $b_{2,n}$, show that $2b_{1,n}/(k+1) - 2b_{2,n}/k(k+1)$, as a function of k, increases up to $k_0 = 1 + [2b_{2,n}/b_{1,n}]$, where $[y]$ signifies the integer part, and decreases thereafter. Thus, the optimal k is k_0.

3. Simply prove

$$\frac{2}{k_0+1}b_{1,n} - \frac{2}{k_0(k_0+1)}b_{2,n} \geq \frac{b_{1,n}^2}{2b_{2,n}+b_{1,n}}$$

4. One way is to take the expectation. The other way follows by choosing all A_j as either Ω or $\varnothing$. To give new proofs for Exercises 1 and 2, write up those inequalities with the indicators and prove the resulting combinatorial inequalities. Exercises 3 cannot be proved directly by this method, but has to be deduced from Exercise 2.

5. Easy computations yield $P(A_j) = 1/10$ and $P(A_i \cap A_j) = 1/100$ for all $i \neq j$. Hence, the last part of Theorem 17 is applicable.

Section 1.6

1. On $(\Omega, \mathscr{A}, P)$, X takes the values $0, 1, 2, \ldots, 9$ with the distribution $P(X = j) = 1/10$. Hence, $E(X) = (1/10)(1 + 2 + \cdots + 9) = 4.5$. On $(B, \mathscr{A}_B, P_B)$, X takes the values $0, 1, 2, 3, 4$ with $P_B(X = j) = P(X = j \mid x < 1/2) = P(X = j, x < 1/2)/P(x < 1/2) = 2P(X = j, x < 1/2) = 2/10$. Thus, $E_B(X) = (1/5)(1 + 2 + 3 + 4) = 2$.

2. $P(X = j \mid X + Y = k) = P(X = j, Y = k - j)/P(X + Y = k)$. The numerator is either 0 or $1/36$ for given j and k, and $P(X + Y = k) = r/36$, where $r = $ the number of pairs (i, j) with $i + j = k$, $1 \leq i \leq 6$, $1 \leq j \leq 6$.

3. With the total probability rule, $P(\text{wins}) = \sum_{n \geq 0} P(\text{wins} \mid N = n)P(N = n) = \sum_{n=2}^{+\infty} \binom{n}{2}(1/6)^2(5/6)^{n-2}10^n e^{-10}/n! = (10/6)^2 e^{-10/6}/2$.

4. $P(A_1 \cap A_3 \mid A_2) = P(A_1 \cap A_2 \cap A_3)/P(A_2) = P(A_1 \mid A_2 \cap A_3)P(A_2 \cap A_3)/P(A_2) = P(A_1 \mid A_2 \cap A_3)P(A_3 \mid A_2) = 0.32$.

Section 1.7

1. One needs the infinite product space of the original probability space. Let B_k be the event that A occurs exactly n times in $2k$ repetitions of the experiment, but fewer than n times in $2k - 1$ repetitions. Then $P(B_k) = p\binom{2k-1}{n-1}p^{n-1}(1-p)^{2k-n}$ and the desired probability is the sum of $P(B_k)$ over $n/2 \le k < +\infty$.

2. Let X take the values $x_1, x_2, \ldots, x_s$, with distribution $p_1, p_2, \ldots, p_s$. Then $(X_1 + X_2 + \cdots + X_n)/n = x_1 f_{1,n}/n + x_2 f_{2,n}/n + \cdots + x_s f_{s,n}/n$, where $f_{j,n}$ is the frequency of x_j among $X_1, X_2, \ldots, X_n$. By Section 1.5, $f_{j,n}/n \to p_j$, except perhaps on a set D_j with $P(D_j) = 0$. Thus, on $\Omega \cap (\bigcup_{j=1}^s D_j)^c$, whose probability is one, $(X_1 + X_2 + \cdots + X_n)/n \to x_1 p_1 + x_2 p_2 + \cdots + x_s p_s = E(X)$ as $n \to +\infty$.

3. By the assumptions, if C is the event that $S_0, S_1, \ldots, S_n$ satisfy a particular condition A, say, $P(C) = N_n(A)/2^n$, where $N_n(A)$ is the number of $L_n(x)$ having property A. Now, if A is the property that every $S_j > 0$ and $S_n = k$, then $N_n(A)$ can be counted as all $L_n(x)$ with $L_n(1) = 1$ and $L_n(n) = k$ minus those which touch or cross the x-axis somewhere between 1 and n but $L_n(1) = 1$ and $L_n(n) = k$. For this latter kind of $L_n(x)$, let j be the first integer with $1 < j < n$ and $L_n(j) = 0$. Reflect the segment $L_n(x), 1 \le x \le j$, of such an $L_n(x)$ with respect to the x-axis, but continue with $L_n(x)$ for $j \le x \le n$. This way we can count the second kind of $L_n(x)$ as all those starting at $L_n(1) = -1$ and ending with $L_n(n) = k$. This yields the first equation. The second equation follows from $P(A \cap B) = P(A)P(B|A)$, valid for arbitrary events A and B, which we use with $\{S_1 = 1\} = A$ in the first term and $\{S_1 = -1\} = A$ in the second.

4. In order to have all $S_j > 0$ and $S_{2m} = 0$, we must have all $S_j > 0$, $S_{2m-1} = 1$ and $X_{2m} = -1$. By independence, the last requirement can be separated from the rest by multiplying the rest by $P(X_{2m} = -1) = 1/2$. Next, apply Exercise 3 with $n = 2m - 1$ and $k = 1$. Since $P(S_{2m-2} = 0) = \binom{2m-2}{m-1} 2^{-(2m-2)}$ and $P(S_{2m-2} = 2) = \binom{2m-2}{m} 2^{-(2m-2)}$, the identity

$$\binom{2m-2}{m-1} - \binom{2m-2}{m} = \frac{1}{m}\binom{2m-2}{m-1}$$

entails the claimed formula.

Section 1.8

1. Define $F^{-1}(y) = \inf\{x : F(x) > y\}$. Then $(Y \le z)$ and $\{X \le F^{-1}(z)\}$ differ only in a set of probability zero. Hence, X and Y are random variables at the same time. Also, for continuous F, $F(F^{-1}(z)) = z$ for $0 \le z \le 1$.

Hence, $P(Y \le z) = P(X \le F^{-1}(z)) = F(F^{-1}(z)) = z$. Finally, $P(Z \le u) = P(Y \ge e^{-u}) = 1 - e^{-u}, u > 0$.

2. By induction, $G(x_1 + x_2 + \cdots + x_n) = G(x_1)G(x_2)\cdots G(x_n)$. Substitute, first, $x_j = 1$ and, then, $x_j = 1/m$ for all j, and obtain $G(m/n) = G^{m/n}(1)$ and $0 < G(1) < 1$ whenever F is nondegenerate. If x is irrational, approximate by rationals and conclude from the monotonicity of G that $G(x) = G^x(1) = e^{-ax}$, $a = -\log G(1)$, for all x. The last part, by the definition of conditional probabilities, reduces to the first one.

3. Let x or y tend to infinity to get the marginals. The actual verification of (1.37) for the listed $G(x, y)$'s is not simple. A better way of concluding that they are distribution functions is to observe that they all have appeared in the literature.

4. $F(x, +\infty) = 1 - e^{-2x}, x > 0$, and $F(+\infty, y) = 0$ or $1/2$ or 1, depending on whether $y < 0, 0 < y < 1$, or $y > 1$; so they are distribution functions. But $F(\log 6, 2) - F(\log 6, 1/2) - F(\log 4, 2) + F(\log 4, 1/2) = -1/144$, i.e., (1.37) fails.

5. In the equivalent form $P(x^* < X \le x, y^* < Y \le y) = P(x^* < X \le x) \times P(y^* < Y \le y)$, all $x^* < x$ and $y^* < y$, of the definition of independence, use appropriate x^*, x, y^*, and y, and use the fact that x_j and y_j are not dense sets. Now, with z and u nonnegative integers, $P(X + Y = z) = \sum_{u=0}^z P(X = u, Y = z - u) = \sum_{u=0}^z P(X = u)P(Y = z - u)$. In the case of Poisson distributions for both X and Y, this last sum becomes $(e^{-a-b}/z!) \times \sum_{u=0}^z \binom{z}{u} a^u b^{z-u} = (a + b)^z e^{-a-b}/z!, z \ge 0$ integer.

6. $L_n(z) = 1 - P(W_n > z) = 1 - P$ (all $X_j > z) = 1 - (1 - F(z))^n$; $H_n(z) = P$ (all $X_j \le z) = F^n(z)$. Hence, for $F(x) = 1 - e^{-ax}$, $L_n(z/n) = F(z)$ and $H_n((z + \log n)/a) = (1 - e^{-z}/n)^n \to \exp(-e^{-z})$.

CHAPTER 2

Section 2.1

1. (i) Since

$$\sum_{j=n}^{+\infty} a^j e^{-a}/j! < e^{-a}(a^n/n!) \sum_{k=0}^{+\infty} (a/n)^k = \frac{e^{-a}a^n}{n!(1 - a/n)}$$

for all $n > a$, and $\log n! = \sum_{k=1}^n \log k > \int_1^n \log x \, dx = n \log n - n$, we have $nP(X \ge n) \to 0$ as $n \to +\infty$. Furthermore, for all large n, $P(k2^{-n} \le X < (k + 1)2^{-n}) = 0$ unless $k = j2^n, j > 0$ integer. Hence, by (2.8), $E(X) = \lim_{n \to +\infty} \sum_{j=0}^n jP(X = j) = \sum_{j=0}^{+\infty} a^j e^{-a}/j! = a$.

(ii) Since $nP(X \ge n) = ne^{-an} \to 0$ as $n \to +\infty$ and $P(k2^{-n} \le X < (k + 1)2^{-n}) = e^{-ka/2^n} - e^{-(k+1)a/2^n} = e^{-ka/2^n}(a/2^n + O(2^{-2n}))$, then, in the sum appearing in (2.8), we recognize a Riemann sum and get $E(X) = \int_0^{+\infty} axe^{-ax}\, dx = 1/a$.

(iii) Argue as in part (ii) for both X^+ and X^- and conclude that (2.8) leads to the improper integral $E(X) = (1/\sqrt{2\pi}) \int_{-\infty}^{+\infty} x \times \exp(-x^2/2)\, dx = 0$.

2. $E(X^+) \geq \sum_{n=1}^{N} 2^n (1/2^{n+1}) = N/2$ for all $N \geq 1$, i.e., $E(X^+) = +\infty$. Similarly, $E(X^-) = +\infty$.

3. Start with (ii). Set $Y_k = \sup_{n \geq k} X_n$ and $U_k = \inf_{n \geq k} X_n$. Clearly, $\{Y_k \leq z\} = \bigcap_{n=k}^{+\infty} \{X_n \leq z\}$ and $\{U_k > z\} = \bigcap_{n=k}^{+\infty} \{X_n > z\}$. Hence, if all X_n are measurable, so are Y_k and U_k. Now, by $\lim\inf_{n=+\infty} X_n = \inf_{k \geq 1} Y_k$ and $\lim\sup_{n=+\infty} X_n = \sup_{k \geq 1} U_k$, (ii) follows.

(i) First, let X and Y be measurable, each taking a finite number of values x_i and y_j, respectively. Then, for every z, the sets $\{X + Y \leq z\}$ are finite intersections of sets of the type $\{X = x_i\}$ and $\{Y = y_j\}$, so $X + Y$ is measurable. Now, if X and Y are arbitrary measurable functions, approximate X^+, X^-, Y^+ and Y^- by simple sequences of the kind which led to (2.8) and apply part (ii). The argument for XY is similar, while $\max(X, Y)$ and $\min(X, Y)$ have already been covered by part (ii). Note that the operations in (i) lead to finite values when X and Y are finite. However, in part (ii), if $P(X_n = n) = P(X_n = -n) = 1/2$, then both the sup of X_n and the limsup of $X_n = +\infty$ with probability $1/2$, so these operations do not necessarily result in random variables.

Section 2.2

1. By Ex. 1 of Section 2.1, $E(X) = a$ and $E(Y) = 1/b$. Hence, $E(3X - 2Y) = 3E(X) - 2E(Y) = 3a - 2/b$.

2. Let X take the values $x_1, x_2, \ldots$ with distribution $p_1, p_2, \ldots$. Then, from (2.9), $E(X) = \sum x_i p_i$ and $E(X^2) = \sum x_i^2 p_i$, whenever these are absolutely convergent. Hence, $V(X) = \sum (x_i - E(X))^2 p_i = \sum x_i^2 p_i - (\sum x_i p_i)^2$.

   (i) $x_i = i$, $p_i = \binom{n}{i} p^i (1-p)^{n-i}$, $1 \leq i \leq n$, $0 \leq p \leq 1$. Use $i\binom{n}{i} = n\binom{n-1}{i-1}$, $i \geq 1$, and $i(i-1)\binom{n}{i} = n(n-1)\binom{n-2}{i-2}$, $i \geq 2$; in calculating $V(X)$, write $i^2 = i(i-1) + i$. Hence, $E(X) = np$ and $V(X) = np(1-p)$.

   (ii) $x_i = i$ and $p_i = a^i e^{-a}/i!$, $i \geq 0$, $E(X) = a$ (see Ex. 1 of Section 2.1); after writing again $i^2 = i(i-1) + i$, we get $V(X) = a$.

   (iii) $x_i = i$ and $p_i = \binom{M}{i}\binom{N-M}{n-i}/\binom{N}{n}$, $0 \leq i \leq \min(M, n)$. Use the identities of (i). $E(X) = nM/N$ and $V(X) = (nM/N)(1 - M/N)[1 - (n-1)/(N-1)]$.

3. Apply L'Hospital's rule to $\lim[1 - F(x)]/(1/\pi x)$ as $x \to +\infty$ to conclude that $nP(X \geq n) \to 1/\pi$. Thus, by Theorem 3, X is not integrable. Since

the distribution function $F(x)$ is symmetric, it means that both X^+ and X^- fail to have finite expectation, so $E(X)$ is not defined.

4. By (2.10), $E(X)$ is finite if and only if $a > 1$. However, (2.11) is valid for all $a > 0$.

5. $E(X^u)$ is finite if and only if $E(|X|^u)$ is. Now, for $0 < u \lesssim r$ and arbitrary $c > 1$,

$$E(|X|^u) = E(|X|^u I(|X| < c)) + E(|X|^u I(|X| \geq c))$$
$$\leq E(|X|^u I(|X| < c)) + c^{-(r-u)} E(|X|^r) < +\infty.$$

6. From the concavity inequality of $\log z$ quoted in the exercise, $uv \leq u^r/r + v^s/s$ for $u > 0, v > 0, r > 1$, and $1/r + 1/s = 1$. We get Hölder's inequality by $u = |X|/E^{1/r}(|X|^r)$ and $v = |Y|/E^{1/s}(|Y|^s)$ and by taking the expectation.

(i) Apply Hölder's inequality to the random variables $|X|^{(a-b)/2}$ and $|X|^{(a+b)/2}$, $b \leq a$, with $r = 2$, and take logarithms.

(ii) Choose $r = b/a$ and $s = c/a$ such that $1/a = 1/b + 1/c$ and apply Hölder's inequality to $|X|^a$ and $Y \equiv 1$ (recall that $r > 1$).

Section 2.3

1. $E(|X - Y|^p) \geq E[|X - Y|^p I(|X - Y| \geq a)] \geq a^p E(I(|X - Y| \geq a))$.

2. We know that $E(X) = V(X) = 2$. By Chebyshev's inequality, we get the upper estimate $2/a^2$, which equals $1/2$ if $a = 2$ and $2/25$ if $a = 5$. For $a = 2$, the Poisson distribution gives $P(X = 0) + \sum_{k \geq 4} P(X = k) < 0.29$; for $a = 5$, we get $\sum_{k \geq 7} P(X = k) < 0.005$.

3. (i) Assume $P(X > X^*) > 0$. Then there is an integer $m > 0$ such that $P(X - X^* \geq 1/m) = p_m > 0$. Let $0 < \varepsilon < 1/2m$. We would have $1 - p_m \geq P(|X - X^*| < 2\varepsilon) \geq P(|X - X_n| < \varepsilon, \quad |X^* - X_n| < \varepsilon) \geq 1 - P(|X - X_n| \geq \varepsilon) - P(|X^* - X_n| \geq \varepsilon) \to 1$ as $n \to +\infty$.

(ii) Let $\varepsilon > 0$ be arbitrary. Then, by uniform continuity, there is a $\delta > 0$ such that, by assumption, $P(|g(X_n) - g(X)| \geq \varepsilon) \leq P(|X_n - X| \geq \delta, |X| \leq k) + P(|X| > k) \to 0$ as $n \to +\infty$, and then $k \to \infty$.

(iii) Similar to part (ii).

Note: Exercise 3 is trivial if Ex. 4 is proved first.

4. Assume that X_n does not converge to X in probability. Then, for a subsequence X_{n_k} of X_n, $P(|X_{n_k} - X| \geq \varepsilon) \geq q > 0$ for all $k \geq k_0$. But then no subsequence of X_{n_k} can converge to X.

Section 2.4

1. Note that $|XI_n| \leq |X|$, assumed integrable, and $XI_n \to 0$ in probability. The dominated convergence theorem applies.

2. Apply the mean value theorem of calculus to $(e^{-TX} - e^{-tX})/(T - t)$ and the dominated convergence theorem.

3. First, compute $U(y)$ in the three explicit cases. Evidently, $0 < y < 1$. Straight computation yields (i) $U(y) = (-\log y)^{1/2}$, (ii) $U(y) = -\log y$, and (iii) $U(y) = \log[(1 - y)/y]$. The limit of the stated ratio, as $t \to 0$, is $(\log u)/\log v$ in each of the three cases. In order to set the requested lower bound, first note that the expression behind liminf is the integral with respect to s, from $1/2$ to 1, of the fraction whose limit we have just determined. Hence, Fatou's lemma entails the lower bound $\int_{1/2}^{1} [(\log s)/\log v] \, ds = -(1 - \log 2)/2 \log v$.

4. First, one has to show that $X_{\omega_1}(\omega_2) = X(\omega_1 \omega_2)$ is a random variable on $(\Omega_2, \mathscr{A}_2)$ for each $\omega_1 \in \Omega_1$. This can easily be obtained when $X(\omega_1, \omega_2)$ is the indicator of a cylinder set (recall the construction leading to (1.33)). Since the set of indicators for which $X_{\omega_1}(\omega_2)$ is $\mathscr{A}_2$-measurable is closed under complementation and taking unions, $X_{\omega_1}(\omega_2)$ is, in fact, $\mathscr{A}_2$-measurable for all indicators. Now, use Exercise 3 of Section 2.1 and conclude that $X_{\omega_1}(\omega_2)$ is $\mathscr{A}_2$-measurable for simple functions; then for $X(\omega_1, \omega_2) \geq 0$, and, finally, for arbitrary random variables. Next, by applying the preceding steps, show that $\int X(\omega_1, \omega_2) \, dP_2$ is $\mathscr{A}_1$-measurable. Finally, establish the stated equations step by step for indicators, simple functions, nonnegative functions, and for arbitrary random variables.

Section 2.5

1. For a fixed $N > 0$, let $X_{j,N} = X_j$ if $X_j < N$ and $X_{j,N} = N$ if $X_j \geq N$. Then $X_{j,N}$, $1 \leq j \leq n$, are i.i.d. with $E(X_{j,N}) < +\infty$ but $E(X_{j,N}) \to +\infty$ as $N \to +\infty$ by Theorem 14. Now, $(X_1 + X_2 + \cdots + X_n)/n \geq (X_{1,N} + X_{2,N} + \cdots + X_{n,N})/n \to E(X_{1,N})$ as $n \to +\infty$. Let $N \to +\infty$.

2. This is a difficult exercise, but its content is significant for the concept of expectation. Let us first show that $\sum_{n=1}^{+\infty} P(A_n) = +\infty$. Let $0 = x_0 < x_1 < \cdots$ be real numbers such that $F(x_{k+1}) - F(x_k) \leq \delta$, where $F(x)$ is the common distribution function of the X_j. Clearly, $\sum_{n=1}^{+\infty} P(A_n) = \sum_{k=0}^{+\infty} \sum_{n=1}^{+\infty} P(A_n \cap \{x_k < X_n \leq x_{k+1}\}) \geq \sum_{k=0}^{+\infty} P(x_k < X_1 \leq x_{k+1}) \sum_{n=0}^{+\infty} P(S_{n-1} \leq \varepsilon x_k)$. Let $U(x) = \inf\{n : S_n \geq x\}$. Then $\sum_n P(S_{n-1} \leq \varepsilon x_k) = \sum_n P(U(\varepsilon x_k) \geq n) = E(U(\varepsilon x_k))$, so $\sum_n P(A_n) \geq \sum_k E(U \varepsilon x_k))[F(x_{k+1}) - F(x_k)]$, which, with some effort, can be shown to be infinity. Next, observe that $A_n \cap A_m \subset A_n \cap \{\sum_{k=n+1}^{m-1} X_k \leq \varepsilon X_m\}$ for $n + 1 < m$, so $P(A_n \cap A_m) \leq P(A_n)P(A_{m-n-1})$, where we used the fact that the X_j are i.i.d. Arguing similarly, one gets $P(A_n \cap (\bigcup_{j=n+k}^{+\infty} A_j)) \leq P(A_n)P(\bigcup_{j=k}^{+\infty} A_j)$, from which it follows that $P(A_n \text{ i.o.}) = 1$ (consider the events $A_n \cap (\bigcap_{j=n+k}^{+\infty} A_j)^c$ for fixed k and all n). For the second part, write $\liminf S_n/a_n \leq [\liminf S_n/$

$(1 + X_{n+1})]\limsup(1 + X_{n+1})/a_n$ and observe that the last term is finite by (ii), while the penultimate term is zero by (i).

3. Truncate each X_j at $n \log n$, obtaining, say, $Y_{j,n}$. Then $E(Y_{j,n}) \sim \log n$ and $V(Y_{j,n}) \le E(Y_{j,n}^2) \sim n \log n$. Setting $S_n(X) = X_1 + X_2 + \cdots + X_n$ and $S_n(Y) = Y_{1,n} + Y_{2,n} + \cdots + Y_{n,n}$ and applying Chebyshev's inequality to $S_n(Y)$, the obvious inequality $P(|S_n(X) - n \log n| \ge a_n) \le P(|S_n(Y) - n \log n| \ge a_n) + P(S_n(X) \ne S_n(Y))$ implies the desired limit. By Exercise 2, this limit cannot be strengthened to be a.s.

4. Since $\sum_{n=1}^{+\infty} P(|X_n| \ge n^{1/d}) = \sum_{n=1}^{+\infty} P(|X_1| \ge n^{1/d}) \le E(|X_1|^d) < +\infty$, it suffices to prove that $\sum (Z_n - E(Z_n))$ and $\sum |E(Z_n)|$ converge a.s., where $Z_n = X_n/n^{1/d}$ if $|X_n| \le n^{1/d}$ and $Z_n = 0$ otherwise. Now, with $A_j = \{(j-1)^{1/d} < |X_1| \le j^{1/d}\}$, the i.i.d. property of the X_j implies $\sum_{n=1}^{+\infty} E(Z_n^2) = \sum_{n=1}^{+\infty} \sum_{j=1}^{+\infty} n^{-2/d} \int_{A_j} X_1^2 \, dP$, which after interchanging the order of summations, is seen to be smaller than a constant times $E(|X_1|^d)$. Similarly, starting with $\sum_{n=1}^{+\infty} E(|Z_n|) \le \sum_{n=1}^{+\infty} n^{-1/d} \int_{|X_n| > n^{1/d}} |X_n| \, dP$, which is due to $E(X_1) = 0$, if we split the integral into integrals over A_j, $j \ge n$, replace X_n by X_1 for all n, and interchange the summations, we get $\sum_{n=1}^{+\infty} E(|Z_n|) < +\infty$. Now, apply Theorem 13.

Section 2.6

1. Use the result of Example 2.

2. $P(Z_n \ge 2) = P(X_1 \ge 2)$.

3. Set $S_n = X_1 + X_2 + \cdots + X_n$ and $A_n = \{S_n = 0\}$. Then $A_n \cap (\bigcup_{j=n+k}^{+\infty} A_j) \subset A_n \cap (\bigcup_{j=n+k}^{+\infty} \{S_j - S_n = 0\})$. Since the X_j are i.i.d., $P(\bigcup_{j=n+k}^{+\infty} \{S_j - S_n = 0\}) = P(\bigcup_{j=k}^{+\infty} A_j)$. Hence, by arguing as in Exercise 2 of the preceding section, $\sum P(A_n) = +\infty$ implies $P(A_n \text{ i.o.}) = 1$. Now, in the case of the simple random walk, $P(A_{2n}) = \binom{2n}{n} 2^{-2n} \sim cn^{-1/2}$ by Stirling's formula (see the introductory volume), i.e., $\sum P(A_n) = +\infty$. In the general case, note that $\{S_n = 0\} = \{|S_n| < 1/2\}$, because the X_j are integer valued. Hence, first prove that, for arbitrary k and m, $\sum_{n=1}^{+\infty} P(|S_n| < 1/2) \ge \sum_{n=1}^{km} P(|S_n| < 1/2) \ge (1/2k) \sum_{n=1}^{km} P(|S_n| < k/2)$ by decomposing $P(|S_n| < k/2) = \sum_{t=-k}^{k-1} P(t/2 < S_n < (t+1)/2)$ and $P(t/2 < S_n < (t+1)/2) = \sum_{r=0}^{n} P(U_t = r, t/2 < S_n < (t+1)/2)$, where $U_t = \inf\{a : t/2 < S_a < (t+1)/2\}$. Now, estimate this last probability by $P(U_t = r) \times P(|S_n - S_r| < 1/2) = P(U_t = r)P(|S_{n-r}| < 1/2)$ and interchange summations. From the inequality just established, we get $\sum_{n=1}^{+\infty} P(|S_n| < 1/2) \ge (1/2k) \sum_{n=1}^{km} P(|S_n| < n/2m) = (m/2)(1/km) \sum_{n=1}^{km} P(|S_n/n| < 1/2m) \to m/2$, as $k \to +\infty$, because $S_n/n \to 0$ by the law of large numbers. Since m is arbitrary,

$$\sum P(S_n = 0) = +\infty.$$

Section 2.7

1. Routine to set up the integrals. Note that not all these integrals are expressible by elementary functions.
2. Write $P(X \leq x) = P(Y \leq (\log x - m)/\sigma)$, $x > 0$, and differentiate. For $E(X^r)$, use (2.36) with $g(x) = \exp(rm + r\sigma x)$ and $F(x)$ the standard normal distribution function.
3. Use Riemann–Stieltjes sums to show that $\int_{-\infty}^{+\infty} x \, dF(x) = \int_0^{1/2} x \, dx + (F(1/2) - F(1/2 - 0))(1/2)$. In general, if $x_j, j \in J$, are the points of discontinuities and $f_j(x) = F'(x)$ for $x_j < x < x_{j+1}$, then the same method yields $\int_{-\infty}^{+\infty} x \, dF(x) = \sum_{j \in J} x_j(F(x_j) - F(x_j - 0)) + \sum_{j \in J} \int_{A_j} x f_j(x) \, dx + \int_{-\infty}^a x F'(x) \, dx + \int_b^{+\infty} x F'(x) \, dx$, where $A_j = (x_j, x_{j+1})$, $a = \inf J$, and $b = \sup J$.
4. Separate the contributions of the discontinuities to both integrals and construct Riemann–Stieltjes sums between the discontinuities.
5. It suffices to show that T preserves the measures of the intervals $[0, a]$. Since $T^{-1}[0, a] = \bigcup_{k=1}^{+\infty} [1/(k + a), 1/k]$, we have only to verify the equation $\int_0^a (1/(1 + x)) \, dx = \sum_{k=1}^{+\infty} \int_{1/(k+a)}^{1/k} (1/(1 + x)) \, dx$, which clearly holds.
6. Let $a \leq y < z \leq b$. First show that $v_f(a, z) = v_f(a, y) + v_f(y, z)$. It then follows that both $F(z) - F(y) = (1/2)v_f(y, z) + (1/2)(f(z) - f(y)) \geq 0$ and $G(z) - G(y) = (1/2)v_f(y, z) - (1/2)(f(z) - f(y)) \geq 0$ because $|f(z) - f(y)| \leq v_f(y, z)$. Hence, $f(x)$ takes the form of $U(x)$ in the last paragraph of p. 75.

Section 2.8

1. (i) Since $F_n(x) = 0$ if $x < n$, as $n \to +\infty$, $F_n(x) \to 0$ for all x. However, $1 = F_n(+\infty)$ does not converge to 0.
   (ii) $0 = F_n(x) \to F(x) = 0$ for $x < 0$. If $x > 0$, $F_n(x) = 1/2$ plus a finite number of times $1/2n$, so $F_n(x) \to F(x) = 1/2$ for $x > 0$. But $1 = F_n(+\infty) \not\to F(+\infty) = 1/2$.
2. $F_n(x) \to F(x)$, where $F(x) = 0$ for $x < 0$, $1/2$ for $0 < x < 1$ and 1 for $x > 1$. Since the integrand is continuous, the Helly–Bray lemma applies, yielding $\int_{-3}^2 e^x/(1 + x^2) \, dF(x) = 1/2 + (e/2)(1/2)$ for the desired limit.
3. Argue as in the first part of the proof on p. 82.
4. Let $Y_n = (f_n - np)/[np(1 - p)]^{1/2}$. Then

$$E(Y_n^k) = \sum_{r=0}^n (r - np)^k [np(1 - p)]^{-k/2} \binom{n}{r} p^r(1 - p)^{n-r}$$

Apply Stirling's formula to the terms $r = np + x_r[np(1 - p)]^{1/2}$ and obtain that the approximate value of the rth summand is $(2\pi)^{-1/2} x_r^k \times$

$\exp(-x_r^2/2)\,\Delta x_r$, where $\Delta x_r = x_{r+1} - x_r = [np(1-p)]^{-1/2}$. Thus, $E(Y_n^k)$ is asymptotically a Riemann sum to the integral $E(U^k)$, U being standard normal. Now, apply the method of moments.

5. Instead of finding the limit of $E(f_n^k)$, find the limit of $E[f_n(f_n - 1)\cdots(f_n - k + 1)] = \sum_{r=0}^{n} r(r-1)\cdots(r-k+1)\binom{n}{r}p^r(1-p)^{n-r}$. Apply repeatedly $t\binom{T}{t} = T\binom{T-1}{t-1}$, $t \geq 1$, and show that the above sum is asymptotically $(np)^k \sim \lambda^k = E(X(X-1)\ldots(X-k+1))$, X Poisson with parameter λ. Finally, apply the method of moments.

Section 2.9

1. Use the factorial form for binomial coefficients and observe that the same expressions result on the two sides.

2. For the recursive formula use the identities $\binom{T}{t} = \binom{T}{T-t}$ and $(T-t) \times \binom{T}{T-t} = T\binom{T-1}{T-t-1}$ with $T = M+1$ and $t = k$ on the left hand side and $T = R - M$ and $t = n - k$ on the right hand side. In order to get the dual, replace M by n and n by M.

3. The first part is straightforward. For the second part, use the definition of conditional probability and observe that all powers of p and $1 - p$ cancel out. The remaining binomial coefficients form a hypergeometric distribution.

4. By the approximation of Exercise 3 of Section 1.4, the true hypergeometric distribution of the number X in favor of A among those n selected can be replaced by the binomial distribution with parameters n and $p = P(A) = M/R$, M being the unknown number of those in the population in favor of A. Next, we use the normal approximation to the binomial in the form $P(c_1 n^{-1/2} \leq (X/n) - p \leq c_2 n^{-1/2}) \sim N(b) - N(a)$, where $c_1 = a\sqrt{p(1-p)}$ and $c_2 = b\sqrt{p(1-p)}$. Now, our record is that $X = k = 576$ and $n = 1,600$, or $X/n = k/n = 0.36$, so the unknown p must be close to it. In fact, if we choose $a = -b$, and $N(b) - N(a) = 0.95$, say, then a table for $N(x)$ yields $0.95 = N(b) - N(-b) = 2N(b) - 1$, i.e., $b = 1.96$. Therefore, with the chosen probability 0.95, $|(X/n) - p| \leq n^{-1/2}$, or $|p - 0.36| \leq 1/40 = 0.025$, since $b\sqrt{p(1-p)} < 1$. The correct interpretation of the result, therefore, is that $p = 0.36$, plus or minus 0.025, with probability 0.95. Note that only b changes with choosing another probability of confidence in place of 0.95, and its change is very minor which is clear from a look at a large portion of the table for $N(x)$.

5. Given that $X = n$, the number Y of marked items is binomial with parameters n and p. Hence, by the total probability rule,

$$P(Y = k) = \sum_{n=k}^{+\infty} \binom{n}{k} p^k (1 - p)^{n-k} \frac{\lambda^n e^{-\lambda}}{n!}$$

$$= \frac{(\lambda p)^k e^{-\lambda}}{k!} \sum_{n=k}^{+\infty} \frac{[\lambda(1 - p)]^{n-k}}{(n - k)!} = \frac{(\lambda p)^k e^{-\lambda p}}{k!}$$

where $\lambda > 0$ is the parameter of the distribution of X.

6. Use the summation formula for positive integer valued random variables X and Y: if X and Y are independent, then $P(X + Y = n) = \sum_{k=1}^{n-1} \times P(X = k) P(Y = n - k)$. First, let both X and Y be geometric, and then, in the step of induction, X is geometric and Y is negative binomial.

7. Substitute $x = r \cos \vartheta$ and $y = r \sin \vartheta$. Then $0 \le \vartheta \le 2\pi$ and $0 \le r < +\infty$; furthermore, the Jacobian of the substitution equals r. Hence, the double integral we started with becomes

$$\int_0^{+\infty} \int_0^{2\pi} r e^{-r^2/2} \, d\vartheta \, dr = 2\pi$$

8. Use the definition of conditional probability and the mean value theorem of calculus in the resulting integrals in the bivariate normal case. We get that $f(x|y)$ is the limit, as Δx and Δy both go to zero, of $f(u_1, u_2)/f_y(z)$ where $f(u_1, u_2)$ is the bivariate density of (X, Y) with $x < u_1 < x + \Delta x$ and $y < u_2 < y + \Delta y$ and $f_Y(z)$ is the (normal) density of Y with $y < z < y + \Delta y$. Carry out the details of computation with arbitrary $E(X)$, $E(Y)$, $V(X)$ and $V(Y)$ (not just zero expectations and unit variances). The result, by continuity, is $f(x, y)/f_Y(y)$, which is another normal density.

CHAPTER 3
Section 3.1

1. (i) $g(0) = -1$. (ii) $g''(0) = 2$, but we must have $g''(0) < 0$ for a characteristic function. (iii) $g(t) = 1 - t^4 + O(t^8)$ as $t \to 0$. If it were the characteristic function of X, we would have $E(X) = E(X^2) = V(X) = 0$. But then $X = 0$ a.s., whose characteristic function is $\varphi(t) = 1$.

2. Let φ_i be the characteristic function of F_i. Then $a_1 F_1 + a_2 F_2 + \cdots + a_k F_k$ is a distribution function whose characteristic function is $\sum_{j=1}^{k} a_j \varphi_j$. Finally, note that $\varphi_X(-t) = \varphi_{(-X)}(t)$ and $\operatorname{Re} \varphi(t) = (1/2)\varphi(t) + (1/2)\varphi(-t)$.

3. If X takes the values $a + jb$ with distribution p_j, $j = 0$, $j = \pm k$, $k \ge 1$, where $p_j = 0$ is possible for some j, then $\varphi_X(t) = e^{ita} \sum_{j=-\infty}^{+\infty} e^{itjb} p_j$. If

$b = 0$, any t can be t_0; if $b \neq 0$, take $t_0 = 2\pi/b$. Conversely, if $|\varphi_X(t_0)| = 1$, $t_0 \neq 0$, then $\int_{-\infty}^{+\infty} \exp(it_0 x) \, dF(x) = \exp(it_0 c)$, c real. Divide this equation by $\exp(it_0 c)$ and take the real part of the integral: $\int_{-\infty}^{+\infty} [1 - \cos(x - c)t_0] \, dF(x) = 0$. Since $1 - \cos(x - c)t_0 \geq 0$ and continuous, the last equation implies that $F(x)$ is a discrete function whose jumps are at the points where $1 - \cos(x - c)t_0 = 0$, i.e., $x = c + (2\pi/t_0)j$, j integer.

4. If there is a t such that $|\varphi(t)| \neq 1$, then $|\varphi(t)| < 1$ and $|\varphi(t)|^a > 1$ for $a < 0$, which is not possible. If $|\varphi(t)| = 1$ for all t, then $\varphi(t) = \exp(iu(t))$, $u(t)$ real, and, by the preceding exercise, the corresponding distribution is discrete with discontinuities at a lattice $a + bj$, j integer. We also know that $b = 2\pi/t_0$, where t_0 can be any number t in our case. This is possible only if $P(X = a + bj) = 0$ for $j \neq 0$ and $u(t) = a$ is a constant, i.e., X is degenerate.

5. (i) For $F(x) = 1 - e^{-ax}$, $x > 0$, $\varphi(t) = 1/(1 - it/a)$, so $|t| < a$.

   (ii) Use Exercise 2 of Section 2.7 and the elementary inequality $k! < \exp(k \log k)$ (because $\log k! < \int_1^k \log x \, dx < k \log k$) in (3.5) and conclude that this Taylor series converges only for $t = 0$. Finally, for the Pareto distribution $P(X \leq x) = 1 - x^{-A}$, $A > 0$, $x \geq 1$, $E(X^r) = +\infty$ for $r > A$, so (3.5) does not apply.

6. $(1 + t^2)^{-2} = (1 - t^2 + t^4/2 - \cdots)^2 = 1 - 2t^2 + O(t^4)$ as $t \to 0$, so, by (3.4), $E(X) = 0$ and $V(X) = E(X^2) = 4$.

Section 3.2

1. By the assumptions on $g(x, y)$, all integrals are finite. Form Riemann–Stieltjes sums and observe that the order of passages to the limit can be interchanged. (Compare with Exercise 3 of Section 2.4.)

2. Each X_j has the characteristic function $\varphi(t) = 1/(1 - it/a)$, so the characteristic function of $X_1 + X_2 + \cdots + X_n$ is $\psi(t) = \varphi^n(t) = (1 - it/a)^{-n}$. A gamma variable with density function $f(x) = x^{n-1} a^n e^{-ax}/(n - 1)!$, $x > 0$, has this same characteristic function (integrate by parts and use induction). Apply the uniqueness theorem.

3. Let $F(x)$ be the common distribution function. Then $P(X_{r:n} \leq x) = P$ (at least r $X_j \leq x$) $= \sum_{k=r}^{n} \binom{n}{k} F^k(x)[1 - F(x)]^{n-k}$. Thus, $(n - r)E(X_{r:n}) + rE(X_{r+1:n}) = nE(X_{r:n-1})$ for all $1 \leq r \leq n - 1$, $n \geq 2$. Since $r(1) = 1$, this recursive formula and any $E(X_{r(n):n})$, $n \geq 1$, uniquely determine all $E(X_{k:n})$. Hence, Example 10 (p. 110) can be used.

4. By the convolution formula, $|\varphi(t)|^2 = \varphi(t)\varphi(-t)$ is a characteristic function; then, by convolution again, so is $|\varphi(t)|^{2n}$ for every $n \geq 1$. Note that X, taking 0 and 1 with probabilities $1/4$ and $3/4$, respectively, has the characteristic function $\varphi(t) = (1/4) + (3/4)e^{it}$. Then $|\varphi(t)|^2$ is the

characteristic function of $X_1 - X_2$, where X_1 and X_2 are independent copies of X. If $|\varphi(t)|$ were also a characteristic function—of Y, say,—then $|\varphi(t)|^2$ would also be the characteristic function of $Y_1 + Y_2$, Y_i being independent copies of Y. By the uniqueness theorem, $X_1 - X_2$ and $Y_1 + Y_2$ must have the same distribution; since the first takes the values -1, 0, and 1, Y can take the values $1/2$ and $-1/2$ only. If $P(Y = 1/2) = p$, then $3/16 = P(X_1 - X_2 = 1) = P(Y_1 + Y_2 = 1) = p^2$ and $3/16 = P(X_1 - X_2 = -1) = P(Y_1 + Y_2 = -1) = (1 - p)^2$ leads to a contradiction.

5. We know that $\varphi_X(t) = [1 + p(e^{it} - 1)]^n$; thus, the characteristic function of $(X - np)/[np(1 - p)]^{1/2}$ is

$$\psi_n(t) = (1 + p\{\exp[it/(np(1 - p))^{1/2}] - 1\})^n \exp[-it(np/(1 - p))^{1/2}]$$

First use the Taylor expansion $e^z - 1 = z + z^2/2 + O(z^3)$ and then $\log(1 + y) = y - y^2/2 + O(y^3)$ in $\log \psi_n(t)$. For fixed t, as $n \to +\infty$, the asymptotic formula $\log \psi_n(t) = -t^2/2 + O(n^{-1/2})$ obtains. The combination of the continuity and uniqueness theorems yields the desired limit.

6. The characteristic function of $(Y - a)/a^{1/2}$,

$$\varphi_a(t) = \exp\{a[\exp(ita^{-1/2}) - 1]\} \exp(-ita^{1/2})$$

which, by the Taylor expansion of $e^z - 1$, reduces to $\varphi_a(t) = \exp[-t^2/2 + O(a^{-1/2})]$. Now, use the continuity and uniqueness theorems.

7. Let $C(x)$ be the convolution of $F(x)$ and $U(x)$, where $U(x)$ is the uniform distribution on $(-h/2, h/2)$. Then $u(x) = U'(x) = 1/h$ on $(-h/2, h/2)$, $\varphi_C(t) = \varphi(t)\varphi_U(t) = \varphi(t)(2 \sin th/2)/th$, and $C(z) = \int_{-\infty}^{+\infty} F(z - x)\, dU(x) = (1/h) \int_{-h/2}^{h/2} F(z - x)\, dx$. Now compute $C(z + h/2) - C(z - h/2)$ from the last formula and from the formula of Theorem 4. In the latter case, note that, because it converges absolutely, the limit becomes an improper Riemann integral and that $2 \sin^2(th/2) = 1 - \cos th$. The special case is a straightforward calculation.

8. Since $e^{ita} + e^{-ita} = 2 \cos ta$ for real a,

$$\varphi(t) = 1/2 + \sum_{k=0}^{+\infty} 4(2k + 1)^{-2}\pi^{-2} \cos(2k + 1)\pi t$$

which, by elementary Fourier theory, is the Fourier series of $1 - |t|$, $|t| < 1$, and equals $1 - |t|$. (Simply evaluate the integrals $\int_{-1}^{1} (1 - |t|) \times \sin(2k + 1)\pi t\, dt = 0$ and $\int_{-1}^{1} (1 - |t|) \cos(2k + 1)\pi t\, dt = 4(2k + 1)^{-2}\pi^{-2}$; the absolute convergence of the series in $\varphi(t)$ ensures equality to $1 - |t|$ on $|t| < 1$.) Clearly, $\varphi(t + 2) = \varphi(t)$.

9. Let $F^*(x)$ be the distribution function determined by $\varphi(-t)$. If $F(x) - F(x - 0) > 0$, then $F^*(-x) - F^*(-x - 0) > 0$. Hence, for the convolution $G(x)$ of F and F^*, $G(0) - G(0 - 0) = \sum_{j \in J} p_j^2$. Now, since $|\varphi(t)|^2 =$

$\int_{-\infty}^{+\infty} e^{itx}\, dG(x)$, $(1/2U) \int_{-U}^{U} |\varphi(t)|^2\, dt = \int_{-\infty}^{+\infty} (\sin Ux)/Ux\, dG(x)$. Split the integral at $|x| < \varepsilon$, where $\varepsilon > 0$ is chosen so that, for a given $\delta > 0$, $|(\sin Ux)/Ux - 1| < \delta$. For $|x| \geq \varepsilon$, use the fact that $(\sin Ux)/Ux \to 0$ as $U \to +\infty$ and apply the dominated convergence theorem.

10. By the triangular inequality, $|\varphi(t) - \varphi(0)| \leq |\varphi(t) - \varphi_n(t)| + |\varphi_n(t) - \varphi_n(0)| + |\varphi_n(0) - \varphi(0)|$. By the uniformity of convergence, the first and the last terms on the right-hand side become small for all $|t| < T$ if n is large. The middle term is small for such an n if t is close to zero by the continuity of $\varphi_n(t)$.

Section 3.3

1. (i) Write $G_n(x) = P(a_n X_n + b_n \leq x) = P(X_n \leq x + \delta_n)$, where $\delta_n = x(1/a_n - 1) - b_n/a_n \to 0$ as $n \to +\infty$. Hence, for large n, $F_n(x - \varepsilon) \leq G_n(x) \leq F_n(x + \varepsilon)$, where $\varepsilon > 0$ is arbitrary. Choose $\varepsilon > 0$ so that both $x - \varepsilon$ and $x + \varepsilon$ are continuity points of $F(x)$. First let $n \to +\infty$ and then $\varepsilon \to 0$.

   (ii) Write

$$G_n(x) = P(X_n \leq x, X_n = Y_n) + P(Y_n \leq x, X_n \neq Y_n)$$
$$= P(X_n \leq x) - P(X_n \leq x, X_n \neq Y_n) + P(Y_n \leq x, X_n \neq Y_n)$$

   and note that each of the last two terms $\leq P(X_n \neq Y_n) \to 0$.

   (iii) Let $D_n = X_n - Y_n$. Then $G_n(x) = P(X_n \leq x + D_n, |D_n| \leq \varepsilon) + P(X_n \leq x + D_n, |D_n| > \varepsilon)$, yielding $P(X_n \leq x + D_n, |D_n| \leq \varepsilon) \leq G_n(x) \leq P(X_n \leq x + \varepsilon) + P(|D_n| > \varepsilon)$. The left-hand side is further decreased by $P(X_n \leq x - \varepsilon, |D_n| \leq \varepsilon) = P(X_n \leq x - \varepsilon) - P(X_n \leq x - \varepsilon, |D_n| > \varepsilon)$. The terms involving $|D_n| > \varepsilon$ tend to zero by assumption. Now, first let $n \to +\infty$ and then $\varepsilon \to 0$.

2. In determining X_n, one divides the interval $(0, 1)$ into 10^n equal parts and, on one out of every consecutive 10 such subintervals, $X_n = j$. Hence, $P(X_n = j) = 1/10$. Similarly, for $1 \leq n_1 < n_2 < \cdots < n_k$, $P(X_{n_t} = j_t, 1 \leq t \leq k) = 1/10^k$, since the event in the parentheses occurs on 10^{n-k} of the 10^n subintervals, where $n = n_k$. That is, the random variables X_n, $n \geq 1$, are independent and identically distributed and so are the $Y_n(j)$ for any given j. The rest of the exercise is straightforward.

3. Write $2X = (X + Y) + (X - Y)$. By independence, $\varphi(2t) = \varphi_{X+Y}(t)$ $\varphi_{X-Y}(t) = \varphi^2(t)\varphi(t)\varphi(-t) = \varphi^2(t)|\varphi(t)|^2$. Hence, $|\varphi(2t)|^2 = \varphi(2t)\varphi(-2t) = |\varphi(t)|^8$, i.e., $\psi(2t) = \psi^4(t)$. By iteration, just as in the proof of Theorem 14, $\psi(t) = \exp(-t^2)$. Thus, from the first equations, $\varphi(2t) = \varphi^2(t)\exp(-t^2) = \exp(-2t^2 + ig(t))$, where we wrote $\varphi^2(t) = |\varphi(t)|^2 \exp(ig(t))$, $g(t)$ real. From these last two formulas, we get the equation $\exp(ig(t)) = \exp(2ig(t/2))$. Since $\varphi(t)$ is not affected if we add multiples of 2π to $g(t)$, we have

$g(t) = 2g(t/2)$ and, by iteration, $g(t) = tg(t/2^n)/(t/2^n)$. Let $n \to +\infty$; $g(t) = tg'(0)$ obtains ($g'(0)$ exists because V is finite) and $g'(0) = 0$ follows from $E(X) = 0$.

4. Let $\varphi(t)$ be the characteristic function of X. Set $s_n^2 = n \log n$ and write $\varphi(t/s_n) = 1 + \int_{-s_n}^{s_n} [\exp(itx/s_n) - 1] f(x)\, dx + \int_{|x| > s_n} [\exp(itx/s_n) - 1] f(x)\, dx$. By Taylor's formula, $e^z - 1 = z + z^2/2 + O(z^3)$, which gives $(-t^2/2s_n^2) \times (2 \log s_n) + O(1/s_n^2) = -t^2/2n + o(1/n)$ for the first integral (note that $\int_{-a}^{a} xf(x)\, dx = 0$ for $a = s_n$), while the second integral is majorized by $2(F(-s_n) + 1 - F(s_n)) = O(1/s_n^2) = o(1/n)$ (we used $|e^{ia} - 1| \le 2$ for real a). Thus, the characteristic function of $S_n/(n \log n)^{1/2}$, $\varphi^n(t/s_n) = (1 - t^2/2n + o(1/n))^n \to \exp(-t^2/2)$. The continuity and uniqueness theorems complete the solution.

Section 3.4

1. By the inequality of Theorem 16(ii), $|\varphi(t)| = 1$ for all t. Hence, by Exercise 4 of Section 3.1, $\varphi(t) = e^{ita}$, a real.

2. By Theorem 16(ii), $1 - |\varphi(t)|^2 \ge 4^{-n}[1 - |\varphi(2^n t)|^2]$. Now, by assumption, $|\varphi(2^n t)|^2 \le a^2$ for $b/2^n \le |t| < b/2^{n-1}$, where the last inequality can also be written as $4^{-n} > t^2/4b^2$. All these combined yield $|\varphi(t)|^2 \le 1 - t^2(1 - a^2)/4b^2$ for $0 < |t| < b$. In particular, $|\varphi(t)| < 1$ for $t \ne 0$ whenever there is a $b > 0$ such that $|\varphi(t)| \le a < 1$ for $|t| > b$. Apply Exercise 3 of Section 3.1.

3. Since $\varphi(t) = \int_{-\infty}^{+\infty} e^{itx} f(x)\, dx$, the Riemann–Lebesgue lemma (p. 131) applies.

4. Let X_n be binomial with parameters n and p and with characteristic function $\varphi_n(t)$. Refine slightly the estimates in Exercise 5 of Section 3.2 and establish that $|\varphi_n(t) - \exp(-t^2/2)| \le (c_1/n^{1/2})|t|^3 \exp(-t^2/2)$ for $|t| \le U = c_2 n^{1/2}$. Hence, the second inequality of Theorem 18 yields the estimate $c_3/n^{1/2}$.

Section 3.5

1. Set $W = \min(X, Y)$, $Z = \max(X, Y)$, and $d = Z - W = |X - Y|$. Let the characteristic function of (W, d) be $\varphi(t, u)$. Then $\varphi(t, u) = \int_0^{+\infty} \int_0^{+\infty} \{\exp[it \min(x, y) + iu|x - y|]\} e^{-x-y}\, dx\, dy$. Split this (double) integral into two parts: for $x < y$ and for $x \ge y$. Write $y = x + s$ in the first one and $x = y + s$ in the second, where $s \ge 0$. We get $\varphi(t, u) = 2 \int_0^{+\infty} \int_0^{+\infty} e^{itx + ius} e^{-2x-s}\, dx\, ds$, which is the product of two integrals; one the integral of $2e^{itx - 2x}$; the other, that of $e^{ius - s}$. Hence, by Theorem 20, W and d are independent. Note that the integrals representing the characteristic func-

tions of W and d are both exponential. By the uniqueness theorem, both W and d are exponential with $E(W) = 1/2$ and $E(d) = 1$.

2. Since independence is not affected by adding a constant or multiplying by one, we may assume that $E(X_j) = 0$ and $V(X_j) = 1$. Let $\varphi(t, u)$ be the characteristic function of (M_n, S_n^2). Then $\varphi(t, u)$ is the n-fold integral of the product of $\exp\{it(x_1 + x_2 + \cdots + x_n)/n + iu(n-1)^{-1} \sum_{j=1}^n [x_j - (x_1 + x_2 + \cdots + x_n)/n]^2\}$ and $(2\pi)^{-n/2} \exp[-(x_1^2 + x_2^2 + \cdots + x_n^2)/2]$ over $-\infty < x_j < +\infty$, $1 \le j \le n$. Substituting

$$y_1 = (x_1 + x_2 + \cdots + x_n)/n^{1/2}$$

and $y_j = \sum_{k=1}^n a_{k,j} x_k$, $2 \le j \le n$, where the coefficients $a_{k,j}$ satisfy the sum in k of $a_{k,j} = 0$, of $a_{k,j}^2 = 1$, and of $a_{k,j_1} a_{k,j_2} = 0$ for $j_1 \ne j_2$. Then the Jacobian of this substitution is one, $\sum_{j=1}^n x_j = y_1 n^{1/2}$, $\sum_{j=1}^n x_j^2 = \sum_{j=1}^n y_j^2$, and $\sum_{j=1}^n [x_j - (x_1 + x_2 + \cdots + x_n)/n]^2 = \sum_{j=2}^n y_j^2$. After this substitution, $\varphi(t, u)$ becomes the product of two integrals, one of which is in y_1 and involves t only and the other of which is an $(n-1)$-fold integral in $y_2, y_3, \ldots, y_n$ and involves u only. Hence, by Theorem 20, M_n and S_n^2 are independent. By carrying out these computations, one gets, via the uniqueness theorem, that M_n is normal and S_n^2 is a gamma variable.

Section 3.6

1. By (3.39), $p_n = (1/n!) \int_0^{+\infty} e^{-ax}(ax)^n \, dF(x) = \int_0^{+\infty} P(U_{ax} = n) \, dF(x)$, where U_{ax} is a Poisson variable with $E(U_{ax}) = ax$. Hence, $p_n \ge 0$ and $\sum_{n=0}^{+\infty} p_n = \int_0^{+\infty} dF(x) = 1$ under the conditions of the present section. Let $V(a)$ be a random variable with distribution $\{p_n\}$. Then $\mathrm{Lap}_{V(a)}(z) = \sum_{n=0}^{+\infty} p_n e^{-nz} = \sum_{n=0}^{+\infty} (-1)^n \mathrm{Lap}^{(n)}(a)(ae^{-z})^n/n! = \mathrm{Lap}(a - ae^{-z})$, the last equation by Taylor's formula. Now, $\mathrm{Lap}_{V(a)/a}(z) = \mathrm{Lap}_{V(a)}(z/a) = \mathrm{Lap}(a - ae^{-z/a}) \to \mathrm{Lap}(z)$ as $a \to +\infty$ (use $e^{-z/a} = 1 - z/a + O(a^{-2})$). So, by the continuity theorem for Laplace transforms, $P(V(a)/a \le x) \to F(x)$. This is (3.44).

2. By the total probability rule and by the independence of N and the X_j, $P(Y_N \le x) = \sum_{k=1}^{+\infty} P(Y_k \le x)P(N = k)$, where $Y_n = X_1 + X_2 + \cdots + X_n$. Hence, by (3.43), $\mathrm{Lap}_{Y_N}(z) = \sum_{k=1}^{+\infty} (1 + z/a)^{-k} p(1-p)^{k-1} = 1/(1 + z/ap)$, where $1/a = E(X_j)$. Apply Theorem 24.

3. Set $Y_n = (X_1 + X_2 + \cdots + X_n)/n$ and assume $E(X_j) = 1$. By independence, $\mathrm{Lap}_{Y_n}(z) = \mathrm{Lap}_{X_1}^n(z/n)$. By (3.40) and finite Taylor expansion, $\mathrm{Lap}_{Y_n}(z) = (1 - z/n + o(1/n))^n \to e^{-z}$ as $n \to +\infty$. Apply Theorem 25.

Section 3.7

1. $G_p(z) = \sum_{k=1}^{+\infty} p(1-p)^{k-1} z^k = pz/[1 - (1-p)z]$. Hence, $G_{X_1 + \cdots + X_n}(z) = G_p^n(z) = (pz)^n/[1 - (1-p)z]^n$, from which, by Taylor expansion (easily

obtained from the $(n-1)$th derivative of $G_p(z)/pz$, $P(X_1 + X_2 + \cdots + X_n = k) = \binom{k-1}{n-1}p^n(1-p)^{k-n}$. If $k = n + t$ with t fixed and $n \to +\infty$, together with $1 - p \to 0$ and $n(1-p) \to a$, then we get the Poisson limit (from the Poisson approximation to the binomial distribution).

2. Note that, in order to get to 2, the random walk must pass through 1. Also, when it has reached 1 in k steps (k is odd), the problem of reaching 2 from this point is the same as that of a new random walk to reach 1 in $n - k$ steps. Hence, $t_n^{(2)} = t_1 t_{n-1} + t_3 t_{n-3} + \cdots + t_{n-1} t_1$ and $t_1 = p$. Now, if the first step is -1 and 1 is reached for the first time in $n - 1$ steps, then we get $t_n = (1-p)t_{n-1}^{(2)}$. Clearly, $t_n^{(2)} = 0$ if n is odd. Upon multiplying the stated recursive formulas by s^{2k} and summing, we get the following relations for the generating functions $T_2(s)$ and $T(s)$ of $t_n^{(2)}$ and t_n, respectively: $T_2(s) = T^2(s)$, $t_1 = p$, and $T(s) = (1-p)sT_2(s) + ps$. These have two solutions for $T(s)$, only one of which has nonnegative terms in its Taylor expansion. This solution is

$$T(s) = \{1 - [1 - 4p(1-p)s^2]^{1/2}\}/2(1-p)s$$

Taylor expansion yields $t_{2k+1} = \binom{1/2}{k+1}(-1)^k(2p)^{k+1}[2(1-p)]^k$. $T(1) = 1$ if and only if $p = 1/2$, i.e., it reaches 1 i.o. with probability 1 only in the symmetric case.

3. For the random walk to be at zero in $2n$ steps, it must be at 1 or -1 in $2n - 1$ steps. Hence, $f_{2n} = (1-p)t_{2n-1} + pt^*_{2n-1}$, which yields $F(s) = (1-p)sT(s) + psT^*(s)$. Using the form for $T(s)$ obtained in the preceding exercise (also applicable to $T^*(s)$ by changing p and $1 - p$), we get $F(s)$ as stated. From p. 143, $Q(s) = 1/[1 - F(s)] = 1/(1-s^2)^{1/2}$ if $p = 1/2$.

Section 3.8

1. By definition, there is a Y whose distribution is normal and $X = \exp(Y)$. Now, $(e^{uY} - 1)/u = Y + 0(u)$ as $u \to 0$ if the sample point, and thus Y, is kept fixed. Hence, a.s. $(X^u - 1)/u \to Y$ as $u \to 0$, entailing the claimed asymptotic normality.

2. The lognormal case follows from the definition of lognormality and a property of normal distributions: write $X = e^U$ and $Y = e^V$, where U and V are independent normal variables. Then $XY = e^{U+V}$ and $X/Y = e^{U-V}$, from which the Example of Section 3.5 entails that XY and X/Y are independent. In the case of the gamma distributions we utilize that $X + Y$ and X/Y are independent if, and only if, $X + Y$ and $X/(X+Y) = 1/(1 + Y/X)$ are. Now, by the mean value theorem of calculus, calculate the probability that, concurrently, $X + Y$ is between u and $u + \Delta u$ and

$X/(X + Y)$ is between v and $v + \Delta v$, where Δu and Δv tend to zero. This means that X is "close to" uv and thus Y is "close to" $u - uv$. By substituting these values into the explicit form of gamma densities, the independence of $X + Y$ and $X/(X + Y)$ follows.

3. Clearly, $P(X \leq x) = P(Y \leq \log(e^x + 1)) = 1 - 1/(e^x + 1) = 1/(1 + e^{-x})$.
   (i) $E(S_n(x)) = nP(X_1 \leq x)$ and $V(S_n(x)) = np(1 - p)$, where $p = P(X_1 \leq x)$. Now, if $dP(X_1 \leq x)/dx = cP(X_1 \leq x)[1 - P(X_1 \leq x)]$, where c is a constant both in n and x, then with $p = p(x)$ above and $g(x) = 1/p(x) - 1$, the differential equation $g'(x) = -cg(x)$ results, whose solution is the exponential function $g(x) = e^{-cx+d}$, where d is a constant. Consequently, $p(x)$ is logistic. The converse is straightforward. (ii) If X is logistic then $P(X \geq -x | X \leq x)$ is exponential by the definitions involved (by the assumed symmetry about the origin, the location parameter of the distribution of X is zero). Conversely, if $P(X \geq -x | X \leq x) = 1 - e^{-cx}$, $x > 0$, $c > 0$ constant, then, by symmetry, the distribution function $F(x) = P(X \leq x)$ satisfies $2F(x) - 1 = F(x)(1 - e^{-cx})$, whose only solution in $F(x)$ is the logistic distribution function.

4. $F_1(x) = \lim_{y = +\infty} F(x, y) = \exp(-e^{-x})$ (use the branch $y > x$ of $F(x, y)$). $F_2(y) = \lim_{x = +\infty} F(x, y) = \exp(-e^{-y})$ upon using the branch when $x > y$. Thus the dependence function $D(u_1, u_2)$ is the solution of $F(x, y) = D(\exp(-e^{-x}), \exp(-e^{-y}))$. We have $D(u_1, u_2) = u_1 u_2^{1/2}$ if $0 \leq u_1 \leq u_2 \leq 1$. For $0 < u_2 < u_1 < 1$, write $\exp(-e^{y-2x}/2) = [\exp(-e^{-x})]^{e^{-x}e^y/2}$ and $e^y = -1/\log u_2$, $e^{-x} = -\log u_1$, yielding $D(u_1, u_2) = u_2 u_1^{(\log u_1)/2 \log u_2}$. In either case we find $D^k(u_1^{1/k}, u_2^{1/k}) = D(u_1, u_2)$ for any $k > 0$. (For the significance of this equation, see Section 6.7 and Exercise 5 in that section.)

CHAPTER 4

Section 4.1

1. The density of $Y_{1:n} \leq \cdots \leq Y_{n:n}$, $f(y_1, y_2, \ldots, y_n) = n! \exp(-\sum_{j=1}^n y_j)$ if $0 \leq y_1 < y_2 < \cdots < y_n$ and equals 0 otherwise. Let $\varphi(t_1, t_2, \ldots, t_n)$ be the characteristic function of $Y_{1:n}$, $Y_{2:n} - Y_{1:n}, \ldots, Y_{n:n} - Y_{n-1:n}$. Then $\varphi(t_1, t_2, \ldots, t_n)$ is the n-fold integral of $[\exp(i \sum_{j=1}^n t_j u_j)]f(u_1, u_1 + u_2, \ldots, u_1 + \cdots + u_n)$ with respect to $u_1, u_2, \ldots, u_n$ over $0 \leq u_j < +\infty$, $1 \leq j \leq n$, which, by the substitution of the value of f, becomes the product of $(n - j + 1) \int_0^{+\infty} \exp[it_j u_j - (n - j + 1)u_j] du_j$, $1 \leq j \leq n$. Hence, the differences $Y_{j:n} - Y_{j-1:n}$, $Y_{0:n} = 0$, $1 \leq j \leq n$, are independent exponential variables with expectation $1/(n - j + 1)$. Use this result and the identity $Y_{n:n} = \sum_{j=1}^n (Y_{j:n} - Y_{j-1:n})$ to get the characteristic function of $Y_{n:n}$. Since

the last sum for $Y_{n:n}$ has the same distribution as $\sum_{j=1}^{n} Y_j/j$ and because $\sum_{j=1}^{+\infty} (Y_j - 1)/j$ converges a.s. (Theorem 2.13), the asymptotic formula $\sum_{j=1}^{n} 1/j = \log n + c + o(1)$, where c is Euler's constant, and the limit $P(Y_{n:n} - \log n - c \leq x) \to H(x + c)$ yields $\int_{-\infty}^{+\infty} e^{itx} dH(x) = e^{itc} \prod_{j=1}^{+\infty} [\exp(-it/j)]/(1 - it/j)$.

2. Let $\psi(t) = \prod_{n=1}^{+\infty} |\varphi_n(t)|^2$. It is zero if $\varphi_n(t) = 0$ or if $\sum_{n=1}^{+\infty} (1 - |\varphi_n(t)|^2) = +\infty$. On the other hand, when the last sum converges, which is entailed by $\sum |1 - \varphi_n(t)| < +\infty$, $\psi(t) \neq 0$. So, by Theorem 1, $\psi(t)$ is a characteristic function. Hence, by Lemma 5, there are constants c_n such that $\sum (X_n - c_n)$ converges a.s., implying that $\prod_{n=1}^{+\infty} \varphi_n(t) \exp(-itc_n)$ is a characteristic function. But by assumption, $\varrho(t) = \prod_{n=1}^{+\infty} \varphi_n(t)$ is convergent on a set of positive Lebesgue measure. Thus, deduce that $\sum c_n$ is convergent and that $\varrho(t)$ itself is a characteristic function, so $\sum X_j$ converges a.s. (Theorem 3). Since, under the conditions of the exercise, $\sum X_j$ and $\sum (X_j - E_j)$ converge (a.s.) for all rearrangements of the summands, $\sum E_j$ converges after all rearrangements as well. That is, it is absolutely convergent.

3. Follow the proof of Lemma 4 with the indicated changes.

4. Apply Theorem 2 and conclude that (i) $\sum X_j$ converges a.s. if and only if $\sum a_j^2 < +\infty$ and (ii) $\sum (X_j + a_j)$ converges a.s. if and only if $\sum a_j$ and $\sum a_j^2$ converge. Now, if $a_j = 1/2^j$, the characteristic function of X_j is $\cos t/2^j$, so the characteristic function of $\sum X_j$ is $\prod_{j=1}^{+\infty} \cos t/2^j = (\sin t)/t$. To prove the last equation, work backwards: $(\sin t)/t = 2[\cos(t/2) \times \{\sin(t/2)]/t = [\cos(t/2)][\sin(t/2)]/(t/2)$ and use $\sin T = 2 \sin(T/2) \times \cos(T/2)$ repeatedly.

Section 4.2

1. Clearly, $(\varphi(t))^3(\psi(t))^4 = \exp(-3t^2)$, which can also be written as $\varphi(t)\varrho(t) = \exp(-3t^2)$. Apply Theorem 5. Now write $\varrho(t) = \psi(t)\eta(t)$ and apply Theorem 5 again.

2. It follows immediately from the convolution formula that both X and Y must be discrete and then that X and Y take only nonnegative integers. Let $p_k = P(X = k)$ and $r_k = P(Y = k)$, $k \geq 0$, and set $G(z) = \sum p_k z^k$ and $H(z) = \sum r_k z^k$. By $X + Y = Z$ and $p_0 r_0 \neq 0$, $G(z)H(z) = \exp[a(z - 1)]$ and $a^n e^{-a}/n! = \sum_{k=0}^{n} p_k r_{n-k}$, where $a > 0$. We thus have $r_n \leq a^n e^{-a}/n!p_0$. So, for real $z \geq 0$, $H(z) \leq (1/p_0)\exp[a(z - 1)]$. Similarly, $G(z) \leq (1/r_0) \exp[a(z - 1)]$. Hence, $G(z)$ and $H(z)$ are finite for all (complex) z and regular (as given by power series). Furthermore, neither G nor H have any zeros (their product is exponential). Therefore, we can turn to logarithms. Since $\log p_0 + \log r_0 = -a$, then, by the inequality established above, $\log |H(z)| \leq \log H(|z|) \leq -\log p_0 + a(|z| - 1) = a|z| + \log r_0$. On

the other hand, from its power series, $\log|H(z)| > 0$ for $z > 1$. Hence, $[\log|H(z)| - \log r_0]/|z|$ (and similarly $[\log|G(z)| - \log p_0]/|z|$) is bounded, and therefore constant by Liouville's theorem. Thus, for $z \geq 0$ real, $H(z)/r_0 = e^{cz}$ and $G(z)/p_0 = e^{bz}$, from which $H(1) = G(1) = 1$ yields that $c > 0$ and $b > 0$, and $H(z) = \exp[c(z-1)]$ and $G(z) = \exp[b(z-1)]$. One can now easily drop the restriction $z \geq 0$ real, so both X and Y are Poisson.

Section 4.3

1. By the compactness of distribution functions, there is a distribution function $R(x)$ and a subsequence $R_{n_k}(x)$ such that $R_{n_k}(x) \to R(x)$ weakly. Hence, by Theorem 6, $L(R_{n_k}, R) \to 0$ as $k \to +\infty$. Use the triangular inequality and the Cauchy property of the sequence R_n and conclude that $L(R_n, R) \to 0$ as $n \to +\infty$. Apply Theorem 6 again.
2. In the first part, we have to determine the infimum of h such that $x - 2h \leq x^2 \leq x + 2h$ for all $0 \leq x \leq 1$. The upper inequality is true for all $h \geq 0$. For the lower inequality, one gets $h \geq 1/8$, so $L(F, G) = 1/8$. For the second part of the exercise, we have to show that, for every n and all x with $e^{-n} \leq x \leq 1$, there is an $h_n > 0$ such that $h_n \to 0$ as $n \to +\infty$ and $x - 2h_n \leq (1 + n^{-1}\log x)^n \leq x + 2h_n$. Apply the inequality $e^z \geq 1 + z$ to see that the upper inequality holds with any $h_n \geq 0$. Note that $(1 + n^{-1}\log x)^n \geq 0$, while $x - \exp(-n^a) < 0$ if $e^{-n} \leq x < \exp(-n^a)$, where $a > 0$ is arbitrarily small. Now, for $\exp(-n^a) \leq x \leq 1$, write $(1 + n^{-1}\log x)^n = \exp[n\log(1 + n^{-1}\log x)]$ and use Taylor expansions. We get $L(F_n, F) \leq n^{2a-1}$ with arbitrarily (small) $a > 0$.

Section 4.4

1. By Remark 1 of Section 4.1, both Y and Z are a.s. convergent. By Lemma 6 of Section 4.3, $E(Y) = E(Z) = 0$. Remark 5 of Section 4.1 (p. 160) now yields that Y and Z are identically distributed.
2. By the assumption of finite variances, $x^2[1 - F_{j,n}(x) + F_{j,n}(-x)] \to 0$ and $x^2[1 - N_{j,n}(x) + N_{j,n}(-x)] \to 0$ as $x \to +\infty$. Hence, by integrating by parts in (4.29) (after writing the right-hand side as in the formula following (4.29)), we get $|\psi_{n,1}(t) - \tau_{n,1}(t)| \leq \sum_{j \in J_n} |t| |\int_{-\infty}^{+\infty} (e^{itx} - 1 - itx)(F_{j,n}(x) - N_{j,n}(x))\, dx|$. Integrate separately for $|x| \leq \varepsilon$ and for $|x| > \varepsilon$. When $|x| \leq \varepsilon$, use the estimate $|e^{itx} - 1 - itx| \leq t^2 x^2/2 \leq (t^2\varepsilon/2)|x|$, while, for $|x| > \varepsilon$, use $|e^{itx} - 1| \leq |tx|$ (both come from the Taylor expansion). Now, by integrating by parts in the integral form of $\sigma_{j,n}^2$, we get $\int_{|x| \leq \varepsilon} |x||F_{j,n}(x) - N_{j,n}(x)|\, dx \leq 2\sigma_{j,n}^2$. All these combined yield $|\psi_{n,1}(t) - \tau_{n,1}(t)| \leq \varepsilon|t|^3 \times \sum_{j \in J_n} \sigma_{j,n}^2 + 2t^2 \sum_{j \in J_n} \int_{|x| > \varepsilon} |x||F_{j,n}(x) - N_{j,n}(x)|\, dx$. The proof is completed

by using (4.27), the estimate preceding (4.29), and the assumption of the exercise.

3. Use Theorem 9. Alternatively, use Theorem 8 and Lemma 4 of the previous section. The result shows that arithmetical means do not necessarily converge to a constant, as in the weak law of large number.

Section 4.5

1. Starting with third moments rather than with second ones, use estimates similar to those in the proof of Theorem 11 and the specific and simple distribution of the X_j.

2. Put $h(a) = \int_{|x| \geq a} x^2 \, dN(x)$. By integrating by parts, find $\lim h(a) a^{-1} \times \exp(a^2/2)$ as $a \to +\infty$. Also, by differentiating $g(a) = h(a) a^{-1} \exp(a^2/2)$, find $\sup g(a)$ over $1 \leq a < +\infty$. The larger of these values is c_6. Then $c_5 = 8.4 + 2c_6$.

Section 4.6

1. Since $P(X_j/n \geq \varepsilon) = 1/n\varepsilon$ does not depend on j, so X_j/n, $1 \leq j \leq n$, are UAN. However, $P(\max_{1 \leq j \leq n} X_j \geq \varepsilon n) = 1 - P$ (all $X_j < \varepsilon n) = 1 - (1 - 1/\varepsilon n)^n \to 1 - \exp(-1/\varepsilon) \geq 1 - e^{-3} = 0.95$ for $\varepsilon \leq 1/3$.

2. Use the fact that the product of two characteristic functions is a characteristic function.

3. Note that every summand of a Riemann–Stieltjes sum to $\log \psi^*(t)$ is the logarithm of a Poisson-type characteristic function. So, when the integral in $\psi^*(t)$ is replaced by a Riemann–Stieltjes sum, we get a product of (Poisson-type) characteristic functions and the limit $\psi^*(t)$ of such a product is a characteristic function because it is continuous (for all t). Now, with $\psi(t)$ in Theorem 13(iii), note that $\psi(t)/\psi(at)$ is of the form of $\psi^*(t)$ with $M(x) = M_a(x)$ for $x > 0$ and $M(x) = N_a(x)$ for $x < 0$. The contribution of $x = 0$ is of the form $\exp(-Vt^2)$, i.e., normal. Now apply Theorem 12.

CHAPTER 5
Section 5.1

1. Use (5.2a) with $S_{(n)} = X$ and $G(x) = F(x)$. Now, if $|X| < A$, then $|X_{jn}| < A/n$, so the variance of X_{jn} does not exceed A^2/n^2. But then $V(X) \leq A^2/n$ for all $n \geq 2$, i.e., $V(X) = 0$. So, F must be degenerate.

2. If $\psi(t) = \exp[\lambda(e^{it} - 1)]$, $\lambda > 0$, then

$$\psi(t)/\psi(at) = \exp[\lambda(e^{it} - e^{iat})]$$

$$= \exp[\lambda(\cos t - \cos at) + \lambda i(\sin t - \sin at)], \quad 0 < a < 1$$

But $\cos t - \cos at > 0$ for $2\pi/3 \le t < \pi$ and $a = 3/4$, say, so $|\psi(t)/\psi(at)| > 1$, that is, it is not a characteristic function. Apply Theorem 4.12.

3. Let $\varphi(t)$ be the characteristic function of $G(x)$. Then $\varphi(t) = 0$ for $|t| > 1$, so it cannot be i.d.

Section 5.2

1. The new representation is equivalent to (5.9) if and only if (i) $m(x)$ is nondecreasing for $x < 0$, $m(-\infty) = 0$, and $\int_{-a}^{0} u^2 \, dm(u)$ is finite for all finite $a > 0$; (ii) $M(x)$ is nondecreasing for $x > 0$, $M(+\infty) = 0$, and $\int_{0}^{a} u^2 \, dM(u) < +\infty$ for finite $a > 0$; and (iii) $\sigma^2 \ge 0$.

2. Since $V(X) < +\infty$, $\psi(t)$, and thus $\log \psi(t)$ as well, is twice differentiable. Hence, (5.9) yields that $\int_{-\infty}^{+\infty} (1 + x^2) \, dK^*(x) < +\infty$. Set $K(x) = \int_{-\infty}^{x} (1 + y^2) \, dK^*(y)$ and $L_1 = L + \int_{-\infty}^{+\infty} x \, dK^*(x)$, by which (5.9) becomes Kolmogorov's formula. This also establishes the relation between $(L_1, K^*(x))$ and $(L, K(x))$. Exercise 3 of Section 4.6 gives that $\psi(t)$, which is given by Kolmogorov's formula, is i.d. and differentiation yields that $V(X) < +\infty$.

3. Note that $\varphi_p(t) = (1 - 1/p^\sigma)(1 - 1/p^{\sigma+it})^{-1} = (1 - 1/p^\sigma) \sum_{k=0}^{+\infty} p^{-k\sigma - kit}$ is the characteristic function of a geometric distribution (on the values $-k \log p$, $k \ge 0$), which is i.d. (see p. 208). Since $\psi(t)$ is the product of $\varphi_p(t)$ over all primes p, Theorem 3 yields that $\psi(t)$ is i.d.

4. First establish that the integral in $\log \psi(t)$ is absolutely convergent under the given conditions (use the facts that $|\log(1 - it/y)|$ is of the magnitude $|\log y|$ if y is close to zero and, by the Taylor expansion, $|\log(1 - it/y)|$ has the same magnitude as $1/y$ for large y). Now, the general term in a Riemann–Stieltjes sum for $\log \psi(t)$ is $-(U(y_{j+1}) - U(y_j)) \log(1 - it/y_j)$, which is the logarithm of a gamma characteristic function. Since the gamma characteristic functions are i.d. (p. 200), Theorem 3 implies that $\psi(t)$ is i.d.

5. Start with the fact that $\varphi_1(t) \equiv 1$ is a characteristic function. Then use Exercise 2 of Section 3.1, the convolution rule for characteristic functions, and the continuity theorem ($\varphi(t)$ is continuous at $t = 0$) to conclude that $\psi(t)$ is a characteristic function. But then $\psi^{1/n}(t)$ is a characteristic function because it is of the same form as $\psi(t)$ for every $n \ge 1$. So $\psi(t)$ is i.d.

6. Because $-u(t)$ has the Kolmogorov representation of Exercise 2, $\psi(t) = \exp[-u(t)]$ is an i.d. characteristic function with finite variance. In particular, with $F'(x) = 1/\pi(1 + x^2)$, i.e., $\varphi(z) = \exp(-|z|)$, $\psi(t) = \exp(-\int_0^t \int_0^v e^{-z} \, dx \, dv) = \exp(1 - |t| - e^{-|t|})$ is an i.d. characteristic function. By Exercise 5, so is $\exp(e^{-|t|} - 1)$. The product of these two characteristic functions is the Cauchy characteristic function $e^{-|t|}$, so

$X = Y + Z$, Y and Z are independent, and $V(Y) < +\infty$. Clearly, $V(Z)$ cannot be finite.

7. Recall that, in (5.15), $F_n(x)$ is the distribution function determined by $\psi^{1/n}(t)$, so it is a gamma distribution. Hence, (5.3), (5.5), (5.15), and the fact that $L = \lim L_n$ and $K^*(x) = \lim K_n^*(x)$ as $n \to +\infty$ (p. 207) lead to the desired expressions upon using the facts that $u\Gamma(u) = \Gamma(u + 1)$ (integrate by parts in (5.4)) and that $\lim n/\Gamma(\gamma/n) = \lim \gamma/(\gamma/n)\Gamma(\gamma/n) = \lim \gamma/\Gamma(\gamma/n + 1) = \gamma$ $(n \to +\infty)$. Now, use the relations of Exercise 2, which yield $L_1 = \gamma/a$ and $K(x) = (\gamma/a^2)(1 - (1 + ax)e^{-ax})$ for $x > 0$ and $K(x) = 0$ for $x > 0$.

Section 5.3

1. For every median m_{jn}, $P(X_{jn} \geq m_{jn}) \geq 1/2$. Now, under UAN, $\min_j P(|X_{jn}| < \varepsilon) = 1 - \max_j P(|X_{jn}| \geq \varepsilon) \geq 1/2$ for $n \geq n_0$. Hence, $\max_j |m_{jn}| < \varepsilon$.

2. From the definition of $F(x)$ and UAN condition it easily follows that $\max\{|c_{jn}| : 1 \leq j \leq n\} \to 0$ implies the UAN property. For simplicity, let $c_{jn} > 0$. We apply Theorem 7, for which L_n of (5.23) and $K_n^*(x)$ of (5.24) must be computed. Now, since $a_{jn} = 0$ by symmetry (see the top of p. 211), $F_{jn}^*(x) = F_{jn}(x) = F(x/c_{jn})$, which has density $c_{jn}/2x^2$ if $|x| \geq c_{jn}$ and zero otherwise. Thus, $L_n = 0$ by symmetry again and, for $x < 0$, $K_n^*(x) = (1/2)(\pi/2 + \tan^{-1} x)\sum_x c_{jn} + (\pi/4)\sum_x^c c_{jn}$, where $\sum_x$ signifies summation for those j, $1 \leq j \leq n$, for which $x < -c_{jn}$, and in $\sum_x^c$ we sum for the rest (complement) of j. Combined with the UAN condition, we have that the convergence of $\sum_{j=1}^n c_{jn}$ guarantees the convergence of $K_n^*(x)$ (both for $x < 0$ and $x > 0$).

3. Use (5.9). It turns out that $\psi(t)$ is the product of four characteristic functions, three of which are Poisson type and one of which is degenerate. So $G(x)$ is the distribution of $X_1 + 2X_2 + 3X_3 + Y$, where the summands are independent, the X_j are Poisson, and Y is constant. The actual computation yields that $-1 < Y < 0$, so Y does not play any role in $G(2)$ (because the rest is integer valued). Hence, $G(2) = P(X_1 = 1, X_2 = X_3 = 0) + P(X_2 = 1, X_1 = X_3 = 0) + P(X_1 = X_2 = X_3 = 0) + P(X_1 = 2, X_2 = X_3 = 0)$, which splits into products of the appropriate Poisson probabilities. To find $H(2)$, apply Theorem 8.

4. Write $\int_{-\infty}^{+\infty} (x/(1 + x^2))\, dF_{jn}^*(x) = \int_{|x|<r} x\, dF_{jn}^*(x) - \int_{|x|<r}(x^3/(1 + x^2))\, dF_{jn}^*(x) + \int_{|x|\geq r} (x/(1 + x^2))\, dF_{jn}^*(x)$ and choose r so that r and $-r$ are continuity points of $K^*(x)$. The sum over j of the second term on the right-hand side equals $\int_{|x|<r} x\, dK_n^*(x)$, which converges to $\int_{|x|<r} x\, dK^*(x)$ (Helly–Bray lemma). Similarly, the sum of the last term on the right-hand side is $\int_{|x|\geq r} (1/x)\, dK_n^*(x)$, which converges to $\int_{|x|\geq r} (1/x)\, dK^*(x)$.

Finally write $|\int_{|x|<r} x \, dF_{jn}^*(x)| = |\int_{|x-ajn|<r} (x-a_{jn}) \, dF_{jn}(x) - \int_{|x|<r} (x-a_{jn}) \times dF_{jn}(x) + a_{jn} \int_{|x|\geq r} dF_{jn}(x)| \leq (2r + A_n) \int_{r-A_n\leq |x|\leq r+A_n} dF_{jn}(x) + A_n \times \int_{|x|\geq r} dF_{jn}(x)$, the sum of which tends to zero as a consequence of the UAN condition, where $A_n = \max |a_{jn}|$ (see the top of p. 211 and p. 217).

Section 5.4

1. The geometric distribution is i.d. and has finite variance, so its characteristic function has the Kolmogorov representation. Indeed, by the Taylor formula, $\log\{pe^{it}/[1-(1-p)\,e^{it}]\} = it + \sum_{k=1}^{+\infty} (1-p)^k(e^{itk}-1)/k = \int_0^{+\infty} (e^{itx} - 1 - itx)(1/x^2) \, dK(x) + it/p$, where $K(x) = \sum_{k=1}^{x} k(1-p)^k$, $x > 0$ (and zero for $x < 0$). Since $K(x)$ is not continuous, the geometric distribution is not in class L (Theorem 9). A similar argument also applies to the Poisson distribution.

2. In (5.9), $(1 + x^2) \, dK^*(x) = dx$. Let $t > 0$ and let us first integrate for $x \geq 0$. The real part of (5.9) is $\int_0^{+\infty} (\cos tx - 1)/x^2 \, dx = -\pi t/2$ (we had this integral several times in the text) and the imaginary part of (5.9) is the limit, as $\varepsilon \to 0$, of $\int_\varepsilon^{+\infty} (\sin tx - tx/(1+x^2))(1/x^2) \, dx = \int_\varepsilon^{+\infty} (\sin tx)/x^2 \times dx - t \int_\varepsilon^{+\infty} 1/x(1 + x^2) \, dx$. Substitute $tx = u$ in the first integral, obtaining $t \int_\varepsilon^{+\infty} [v^{-2} \sin v - 1/v(1 + v^2)] \, dv - t \int_\varepsilon^{\varepsilon t} u^{-2} \sin u \, du$. The first integral is finite and the second is asymptotically the same as $\int_\varepsilon^{\varepsilon t} (1/u) \, du = \log t$. So the integral for $x \geq 0$ in (5.9) equals $-\pi t/2 + Ait - it \log t, t > 0$. With the same steps as above but for $x < 0$, we get $-\pi t/2 - Ait + it \log t$. Hence, (5.9) gives $-\pi t$ for $t > 0$. By changing t to $-t$, the previous calculations yield πt for (5.9) when $t < 0$. The two cases can be combined to $-\pi|t|$, so $\psi(t)$, or $G(x)$, is Cauchy. Since $e^{-|t|}/e^{-a|t|}, 0 < a < 1$, is again a Cauchy characteristic function, the Cauchy distribution belongs to class L. (Theorem 9 is also easy to apply in the light of the first part of the exercise.) For the limiting distributions of the extremes, apply Theorem 8. By $(1 + u^2) \, dK^*(u) = du$, $M(x) = -1/x$, $x > 0$, and $m(x) = -1/x$, $x < 0$. So, $H(x) = \exp(-1/x)$, $x > 0$, and $L(x) = 1 - \exp(1/x)$, $x < 0$.

3. Use the result of Exercise 7 of Section 5.2 and restate Theorem 7 for this particular case. Note that, by Theorem 9, the gamma distributions are in class L. Restate Theorem 11.

Section 5.5

1. Let $Y_j = 1/X_j$, where $X_j > 0$. $G_j(x) = P(Y_j \leq x) = P(X_j \geq 1/x) = 1 - F_j(1/x)$, whenever $x > 0$ and $1/x$ is a continuity point of $F_j(x)$. Clearly, $P(W_n \geq 1/x) = P(Z_n \leq x)$, where W_n is as defined in the text for the X_j, and $Z_n = \max(Y_1, Y_2, \ldots, Y_n)$. Hence, in conditions, $F_j(x)$ is to be replaced by $1 - G_j(y)$, $x \to 0$ is to become $y \to +\infty$, the normalizing constants b_n and B_n in $\lim P(W_n \leq b_n z) = L(z)$ and $\lim P(Z_n \leq B_n z) = H(z)$ have

similar properties through the relation $b_n = 1/B_n$, and the limiting distributions L and H are related by $H(z) = 1 - L(1/z)$. Now, work out the details for each statement in the section.

2. Use (5.43) and the elementary relation $\sum_{j=1}^{n} 1/j \sim \log n$. One gets that, with $b_n = 1/\log n$, both (5.40) and (5.41) hold, the latter with $a(z) = z$. Hence, $L(z) = 1 - e^{-z}$, $z > 0$.

CHAPTER 6
Section 6.1

1. By induction, $G(x_1 + x_2 + \cdots + x_n) = G(x_1)G(x_2)\cdots G(x_n)$, $x_j \geq 0$ integer. Choose $x_j = 1$ for each j: $G(n) = G^n(1) = \exp(n \log G(1))$, since $G(1) > 0$ by assumption. This equation also holds for $n = 0$, since $G(x + 0) = G(x)G(0)$ and $G(0) > 0$.

2. Let $F(x) = P(X_j \leq x)$ and $G(x) = 1 - F(x)$. Then $P(W_n > x) = G^n(x) = G(nx)$, the former equation by independence and the latter by assumption. This is the same equation as (6.6a) and (6.5) combined, so we have $G(r) = G^r(1)$ for $r > 0$ rational. Since $G(x)$ is monotonic and $G(+\infty) = 0$, $G(1) \neq 0$ (otherwise $F(x)$ would be degenerate) and $G(1) \neq 1$. So, $c = \log G(1) < 0$ and finite, and $G(r) = e^{cr}$ for rational $r > 0$. Now, if x is arbitrary, choose y_{1r} and y_{2r} rationals such that $y_{1r} \leq x \leq y_{2r}$ and $y_{1r} \to x$, $y_{2r} \to x$ as $r \to +\infty$. Then, by monotonicity, $\exp(cy_{2r}) \leq G(x) \leq \exp(cy_{1r})$. Let $r \to +\infty$.

3. By elementary arithmetic, every integer $n \geq 1$ can be written as $n = \prod p^k$, where the product is over those primes p and integers k for which $e_{p,k} = 1$. Hence, upon applying (6.1) a finite number of times, we get $g(x) = \sum e_{p,k}g(p^k) = \sum kg(p)e_{p,k}$, as desired. If $g(p) = c \log p$, $g(p^k) = ck \log p = kg(p)$, and the sum above reduces to $g(x) = c \log x$. If $g(p) = 1$ for all p, then $ke_{p,k}$ "counts" p k times if p^k is the exact power of p that divides x; thus, the sum representing $g(x)$ counts the number of prime divisors of x with multiplicity.

Section 6.2

1. Let $0 < m_1 < m_2 < \cdots < m_k$ be even integers. Then $P(S_{m_1} = S_{m_2} = \cdots = S_{m_k} = 0$ and $S_j \neq 0$ for $0 < j < m_k$, $j \neq m_t) = P(Y_j = m_j - m_{j-1}, 1 \leq j \leq k)$, where $m_0 = 0$. However, the left-hand side also equals $\prod_{t=1}^{k} \times P(S_{m_t - m_{t-1}} = 0$ and $S_j \neq 0$, $0 < j < m_t - m_{t-1})$ by the assumption on the X_j. Hence, the Y_j are i.i.d. Now, $P(Y_1 + Y_2 + \cdots + Y_R \geq 2n) = P(\{\text{number of } m \leq 2n \text{ with } S_m = 0\} \leq R) = \sum_{r=0}^{R} \binom{2n-r}{n} 2^{-(2n-r)}$. By

Stirling's formula $K! = (2\pi K)^{1/2}(K/e)^K(1 + o(1))$, $K \to +\infty$, and, by the Taylor expansion of the logarithmic function, we get $\binom{2n-r}{n}2^{-(2n-r)} = (\pi n)^{-1/2} \exp(-r^2/4n) + O(r^3/n^2)$. Hence, the sum above is an approximate Riemann sum to the integral of $(2/\pi) \exp(-y^2/2)$, $0 \le y \le z$, if we choose $R = z(2n)^{1/2}$. In order to remove z from the index of Y_R, we reparametrize by choosing $R = N$; thus, $2n = R^2/z^2 = R^2x$, resulting in the limiting distribution $2(1 - N(x^{-1/2}))$ as desired.

2. Use the formula of Theorem 6 for $\psi''(t)$. Hence, in order for $S_n/b_n - a_n$ to have a limiting distribution, we can choose $b_n = n^{-1/\alpha}$ and $a_n = ns$ if $\alpha \ne 1$ or $b_n = 1/n$ and $a_n = sn - (2c\gamma/\pi)\log n$ if $\alpha = 1$. Now, in Exercise 1, $b_n = n^{-2}$, so $\alpha = 1/2$ for $G(x) = 2(1 - N(x^{-1/2}))$.

3. If $G(x)$ is degenerate at x_0, then $E(|X|^u) < +\infty$ for all u if $x_0 \ne 0$ or for $u \ge 0$ if $x_0 = 0$. Let $G(x)$ be nondegenerate. If its characteristic exponent is α, then $G(x)$ is normal for $\alpha = 2$, in which case $E(|X|^u) < +\infty$ for all $u \ge 0$. For $0 < \alpha < 2$, $E(|X|^u) = +\infty$ for $u > \alpha$, which we first show for $0 < \alpha < 1$. Let $G^{(s)}(x)$ be the distribution obtained from $G(x)$ by symmetrization, i.e., the one determined by $|\psi(t)|^2$, where $\psi(t)$ is the characteristic function of $G(x)$. Then, $\|\psi(t)|^2 - 1| = |\int_{-\infty}^{+\infty} (e^{itx} - 1)\,dG^{(s)}(x)| \le \int_{-\infty}^{+\infty} |e^{itx} - 1|\,dG^{(s)}(x)$. But $|e^{itx} - 1| = 2|\sin(tx/2)| \le 2|tx/2|^u$ with $0 < u \le 1$. So, if $E(|X|^u) < +\infty$, we would have $\|\psi(t)|^2 - 1| \le 2^{1-u}|t|^uE(|X|^u)$, contradicting Remark 2 on p. 247. The argument is similar for $1 < \alpha < 2$, using the additional fact that $\int_{-\infty}^{+\infty} x\,dG^{(s)}(x) = 0$; thus, we can replace $e^{itx} - 1$ by $e^{itx} - 1 - itx = ix\int_0^t (e^{iyx} - 1)\,dy$ in the integral above. This time, use the estimate $|e^{iyx} - 1| \le 2^u|yx|^{u-1}$, which leads to the same contradiction as above. Finally, we show that $E(|X|^u) < +\infty$ for $0 \le u < \alpha$, $0 < \alpha < 2$. By Remark 2 of p. 247, $1 - |\psi(t)|^2 = O(|t|^\alpha)$ as $t \to 0$. Hence, with some constant $c > 0$, $c|t|^\alpha \ge 1 - |\psi(t)|^2 = \int_{-\infty}^{+\infty} (1 - \cos tx)\,dG^{(s)}(x) \ge \int_{|x|>z} (1 - \cos tx)\,dG^{(s)}(x)$ as $t \to 0$, implying $\int_0^{1/z} \int_{|x|>z} (1 - \cos tx) \times dG^{(s)}(x)\,dt \le [c/(\alpha + 1)]z^{-\alpha - 1}$. Now, interchange the orders of integration and conclude that $(1/z)(1 - G^{(s)}(z) + G^{(s)}(-z)) \le c_1 z^{-\alpha - 1}$. But $G^{(s)}(z)$ is symmetric, so integration by parts yields $\int_0^{+\infty} x^u\,dG^{(s)}(x) = u\int_0^{+\infty} x^{u-1}(1 - G^{(s)}(x))\,dx < +\infty$, i.e., $E(|X_1 - X_2|^u) < +\infty$, where X_1 and X_2 are independent copies of X. This obviously implies that $E(|X|^u) < +\infty$.

Section 6.3

1. Every monotonic function is measurable. Now, if $s(x)$ is increasing and if $t_0 > 1$, then $1 \le s(tx)/s(x) \le s(t_0x)/s(x)$ for every $1 \le t \le t_0$ and for all sufficiently large x. Let $x \to +\infty$: $s(tx)/s(x) \to 1$. If $t_0 < 1$, then $s(tx)/s(x) \to 1$ follows as above for $t_0 \le t \le 1$. If $s(x)$ is decreasing, then

reciprocals lead to the preceding conclusions. Next, assume that $s(tx)/s(x) \to 1$ has been established for $a \le t \le b$, $a < b$. Then write $s(tx)/s(x) = [s(Ttx/T)/s(tx/T)]s(tx/T)/s(x)$ and conclude that $s(Ty)/s(y) \to 1$ whenever t and t/T belong to $[a, b]$. This extends $s(tx)/s(x) \to 1$ from $a \le t \le b$ to $a/b \le t \le b/a$. Repeating this argument n times, the extension will cover $(a/b)^n \le t \le (b/a)^n$, so, ultimately, we can choose any $t > 0$.

2. Write $[s_1(tx) + s_2(tx)]/[s_1(x)s_2(x)] = [s_1(tx)/s_1(x)][s_1(x)/(s_1(x) + s_2(x)) + [s_2(tx)/s_2(x)]s_2(x)/(s_1(x) + s_2(x))$ and let $x \to +\infty$.

3. Choose D such that, in Karamata's representation theorem, $|\varepsilon(y)| < c/2$, where $c > 0$ is a given constant. Then $x^c s(x) = \exp[u(x) + \int_B^x \varepsilon(y)/y \, dy + c \log x] \ge K \exp[-\int_D^x (c/2y) \, dy + c \log x] = K x^{c/2} \to +\infty$ with x. Use a similar estimate for $x^{-c} s(x)$, in which the last term is $-c \log x$, so $x^{-c} s(x) \le K x^{-c/2}$ is obtained, which tends to 0 as $x \to +\infty$.

4. Once again, Karamata's representation can be used. We obtain $|\log s(x)| \le K^* + (c/2) \log x$ (see the previous solution). Let $x \to +\infty$ and note that $c > 0$ is arbitrary.

5. Set $T(t, x) = s_2(tx)/s_2(x)$ and write $s_1(s_2(tx))/s_1(s_2(x)) = s_1(s_2(x)T(t, x))/s_1(s_2(x))$. Let $x \to +\infty$ and apply the uniform convergence theorem (p. 256).

Section 6.4

1. If $V(X) = \lim s(x)$ is finite, then (6.25) holds and $s(x)$ is slowly varying. So, let $V(X) = +\infty$. Starting with the displayed formula just before (6.25), we get the same lower inequality as appears there and $s(kx) \le s(x) + k^2 x^2 [F(kx) - F(x) + F(-x) - F(-kx)]$. These two inequalities yield $s(kx)/s(x) \to 1$ for $k > 1$ as $x \to +\infty$ if (6.25) holds. For $k < 1$, use the same limit relation with $z = kx$, i.e., $x = z(1/k)$, where $1/k > 1$.

2. Integrate by parts in both $\int_0^x y^2 \, dF(y)$ and $\int_{-x}^0 y^2 \, dF(y)$. We obtain $R(x) = -x^2 T(x) + 2 \int_0^x y T(y) \, dy = -x^{2-\alpha} s(x) + 2 \int_0^x y^{1-\alpha} s(y) \, dy$, where $s(y)$ is slowly varying. Apply Theorems 7 and 9.

3. From the previous solution, $R(x) = -x^2 T(x) + 2 \int_0^x y T(y) \, dy$. Set $u(x) = x^2 T(x)/\int_0^x y T(y) \, dy$. If $x^2 T(x)/R(x) \to (2 - \alpha)/\alpha$, then from the equation above, we obtain that $u(x) \to 2 - \alpha$. Thus, $\log\{[\int_0^{tx} y T(y) \, dy]/[\int_0^x y T(y) \, dy]\} = \int_1^x u(tz)/z \, dz = u(t) \int_1^x [u(tz)/u(t)](1/z) \, dz \to (2 - \alpha) \log x$ as $t \to +\infty$, which says that $\int_0^x y T(y) \, dy$ is regularly varying with exponent $2 - \alpha$. Hence, if we rewrite $x^2 T(x) = R(x)[x^2 T(x)/R(x)]$ in the first line of this solution, we obtain $R(x) = [-(2 - \alpha)/\alpha + o(1)]R(x) + 2x^{2-\alpha} s(x)$, where $s(x)$ is slowly varying. That is, $R(x)$ is regularly varying with exponent $2 - \alpha$.

4. If $\alpha = 2$, i.e., if $G(x)$ is normal, then $s(x) = \int_{-x}^x y^2 \, dF(y)$ is slowly varying. Thus, for $r < 2$ and $k > 1$, we have $\int_x^{kx} |y|^r \, dF(y) \le x^{r-2} \int_x^{kx} y^2 \, dF(y) \le$

$x^{r-2}s(kx) \leq x^{-c}$, $c > 0$, because $s(kx)/s(x) \to 1$ and $s(x)/x^a \to 0$ for any $a > 0$ as $x \to +\infty$ (see Exercise 3 of Section 6.3). Consequently, for x sufficiently large, $\int_x^{+\infty} y^r \, dF(y) \leq x^{-c} \sum_{j=0}^{+\infty} 1/k^{cj} \to 0$ as $x \to +\infty$. The estimate is similar for the negative tail. Next, let $\alpha < 2$. Integrate by parts in $\int_0^{b+0} x^r \, dF(x)$ and $\int_{-b}^0 x^r \, dF(x)$, $r > 0$, and conclude that $E(|X|^r) < +\infty$ if and only if $x^{r-1}T(x)$ (the notation is as in Theorem 12) is integrable on $(a, +\infty)$, $a > 0$. Apply Theorem 12; we have $x^{r-1}T(x) = x^{r-\alpha-1}s(x)$, where $x^{-c}s(x) \to 0$ for any $c > 0$ as $x \to +\infty$ (Exercise 3 of the previous section). Hence, $x^{r-1}T(x)$ is integrable for all $0 \leq r < \alpha$ on $(1, +\infty)$.

5. By substituting $y = ut$ in $1 - F(xt) = \int_{xt}^{+\infty} f(y) \, dy$, we immediately obtain $[1 - F(xt)]/[1 - F(x)] \to t^{-1.5}$ as $x \to +\infty$. Hence, by symmetry about the origin, Theorem 12 applies, yielding that $F \in D(G)$, where the characteristic exponent of G is $\alpha = 1.5$. By the "instruction" towards the end of the proof of Theorem 12 (p. 262), we choose b_n by the rule $(n/b_n^2) \int_{-b_n}^{b_n} y^2 f(y) \, dy \sim 1$. This gives $b_n = d(n \log n)^{2/3}$, where $d > 0$ is a constant. By the symmetry of $f(x)$, we can choose $a_n = 0$ and, for this same reason, $G(x)$ is symmetric. Hence, by Theorem 6 (see Remark 3 on p. 247), the characteristic function of $G(x)$ is $\psi(t) = \exp(-c|t|^{1.5})$, $c > 0$.

Section 6.5

1. Apply Theorem 8.
2. First, $g^{-1}(g(z)) \leq z$ because $g(x)$ is nondecreasing and $z \in \{x : g(x) \geq g(z)\}$. Next, $g^{-1}(g(z) + \varepsilon) \geq z$ because $g(x)$ is right-continuous, so $z \notin \{x : g(x) \geq g(z) + \varepsilon\}$. By letting $\varepsilon \to 0$, the assumed continuity of g^{-1} yields $g^{-1}(g(z)) = z$.
3. By integrating by parts twice as indicated, we get $\int_x^{+\infty} \exp(-u^2/2) \, du = (1/x) \exp(-x^2/2) + O[(1/x^3) \exp(-x^2/2)]$ as $x \to +\infty$. From this, the two stated asymptotic formulas follow. When integrating by parts above, note that the first main term is an upper bound, while the first two terms form a lower bound for $(2\pi)^{1/2}[1 - N(x)]$. Use these inequalities and the equation $1 - N(a_n) = 1/n$ to obtain a_n. Then use $b_n = M(a_n)$.
4. Apply Theorem 17 with a positive integer x. Then $P(X = x) = a^x e^{-a}/x!$, while $P(X > x) = 1 - F(x)$ equals

$$\sum_{k=x+1}^{+\infty} \frac{a^k e^{-a}}{k!} \leq \frac{a^{x+1}e^{-a}}{(x+1)!} \sum_{k=x+1}^{+\infty} \left(\frac{a}{x+2}\right)^{k-x-1}$$

where the last sum, for $x > a$, equals $1/(1 - a/(x + 2))$ which converges to one as $x \to +\infty$. Thus, $(F(x) - F(x - 0))/(1 - F(x))$ fails to converge to zero as $x \to +\infty$.
5. Apply the rule $\max(X_1, X_2, \ldots, X_n) = -\min(-X_1, -X_2, \ldots, -X_n)$.

6. Setting $g(x) = xf(x)/[1 - F(x)] = -x\{\log[1 - F(x)]\}'$, we have $1 - F(x) = C \exp\{-\int_a^x [g(y)/y]\,dy\}$, where C and a are suitable constants. Hence, $[1 - F(tx)]/[1 - F(x)] = \exp\{-\int_t^{tx} [g(y)/y]\,dy\} = \exp\{-\int_1^x [g(ty)/y]\,dy\}$. Since $g(ty) \to \gamma$ as $t \to +\infty$, the previous ratio converges to $\exp(-\gamma \log x) = x^{-\gamma}$ by the dominated convergence theorem. Thus, Theorem 13 applies.

7. The assumptions that $\omega(F) = +\infty$ and that Theorem 15 applies entail that we can apply property (b) on p. 274. Evidently, $a_n \to +\infty$ with n. Consequently, since $b_n = M(a_n)$ is a possible choice and all other choices are proportional to this choice, we have $b_n/a_n = O(M(a_n)/a_n) = o(1)$ as $n \to +\infty$.

Section 6.6

1. Note that $s_n^2 = np(1 - p)$ and $K = 1$. Choose $\varepsilon = as_n/n$.

2. Choose $\varepsilon = x[p(1 - p)/n]^{1/2}$ in Exercise 1. The right-hand side becomes $2\exp\{-x^2/2[1 + x/2np(1 - p)]^2\}$, while the central limit theorem gives the asymptotic value $2(1 - N(x))$. For small values of n, the present inequality is superior.

3. With $f_n^* = n(f_n - p)/[2np(1 - p)\log \log n]^{1/2}$, we have $\lim \sup f_n^* = 1$ and $\lim \inf f_n^* = -1$ a.s.

Section 6.7

1. Start with independent standard (i) normal, (ii) Cauchy densities $f(x)g(y)$ for (X, Y). Set $aX + bY = U$ and $cX + dY = V$. For calculating $P(U \le u, V \le v)$, invert the linear equations above to get $X = AU + BV$ and $Y = CU + DV$, and integrate $f(x)g(y)$ accordingly. Differentiate in u and v.

2. Here we mean proper distribution functions. The Exercise involves simple calculus only.

3. One can proceed with dependence functions, and test the requirements, but a simpler way is to simply calculate the limiting distribution of bivariate normalized maxima for (i) (X, Y) and (ii) (X, X), where X and Y are independent unit exponential variables.

4. Let $F(x, y)$ be a bivariate distribution function with marginals $F_1(x)$ and $F_2(y)$. The inequality $P(A \cap B) \le \min[P(A), P(B)]$, valid for arbitrary events, implies that $F(x, y) \le \min[F_1(x), F_2(y)]$. Hence, $F^n(a_n + b_n u, c_n + d_n v) \le \min[F_1^n(a_n + b_n u), F_2^n(c_n + d_n v)]$, from which the upper inequality of the Exercise follows. The lower inequality is quite special to extreme value theory. We start with $P(A^c \cap B^c) = 1 - P(A \cup B) \ge 1 - P(A) - P(B)$, once again valid for arbitrary events A and B. This yields $F(x, y) \ge 1 -$

$(1 - F_1(x)) - (1 - F_2(y))$. Now, if $F^n(x_n, y_n) \to H(u, v)$, where $x_n = a_n + b_n u$, $y_n = c_n + d_n v$, then both $F_1(x_n)$ and $F_2(y_n)$ are close to one, so the previous inequality entails

$$F^n(x_n, y_n) \geq [1 - (1 - F_1(x_n)) - (1 - F_2(y_n))]^n$$

Write the right-hand side as $\exp(n \log(1 - z)) = \exp(-nz + n\, O(z^2))$ and utilize that, under the assumption of the Exercise, $n(1 - F_1(x_n)) \to e^{-u}$ and $n(1 - F_2(y_n)) \to e^{-v}$. This yields the lower bound.

5. By passing to infinity with x or y, we get the marginals, which equal $\exp(-e^{-x})$ and $\exp(-e^{-y})$, respectively. We have computed the dependence function $D(u, v)$ of $H(x, y)$ in Exercise 4 of Section 3.8. We observed that $D^k(u^{1/k}, v^{1/k}) = D(u, v)$ for all $k > 0$, which is sufficient for $H(x, y)$ to be a limiting distribution function of normalized bivariate maxima (since its marginals are univariate extreme value distributions).

CHAPTER 7
Section 7.1

1. Let X be simple. Then (7.5) and (iii) lead to (7.6). If $X \geq 0$ with $E(X) < +\infty$ and $X_n \leq X_{n+1}$ are simple such that $X_n \to X$, then (ii), (iii), and (7.6) for simple variables yield that $E(X \mid Z) \geq$ the right-hand side of (7.6). With the new sequence $X_n + \varepsilon$, $\varepsilon > 0$ arbitrary, a similar argument yields $E(X \mid Z) - \varepsilon \leq$ the right-hand side of (7.6). Let $\varepsilon \to 0$. Finally, if X is a random variable with $E(|X|) < +\infty$, we can apply (7.6) to $E(X^+ \mid Z)$ and $E(X^- \mid Z)$.

2. $E(S_{n+1} \mid S_n) = E(S_n + X_{n+1} \mid S_n) = S_n + E(X_{n+1}) = S_n$, all steps a.s.

3. Use the classical definition of conditional probability and show that, for all possible values, $P(Y = x_k \mid Y + Z = u_t) = P(Z = x_k \mid Y + Z = u_t)$. Hence, $E(Y \mid Y + Z) = E(Z \mid Y + Z)$ a.s. Now, $2E(Y \mid Y + Z) = E(Y \mid Y + Z) + E(Z \mid Y + Z) = E(Y + Z \mid Y + Z) = Y + Z$ a.s.

4. Let Z_1 and Z_2 take the values $\{x_{1t}\}$ and $\{x_{2k}\}$ on $\{C_{1t}\}$ and $\{C_{2k}\}$, respectively. Since $\sigma(Z_1) \subset \sigma(Z_2)$, for every t there is a set T_t such that $C_{1t} = \bigcup_{k \in T_t} C_{2k}$ and every k is in some T_t. Now, $Y_1 = E(X \mid C_{1t})$ on C_{1t} and $Y_2 = E(X \mid C_{2k})$ on C_{2k}. Hence, $U_{12} = E(Y_1 \mid C_{2k}) = E(X \mid C_{1t})$ on C_{2k}, where $k \in T_t$. That is, $U_{12} = Y_1$ a.s. Let us show that $U_{21} = Y_1$ a.s., as well. U_{21} is discrete and constant on C_{1t} and so is Y_1. The value of Y_1 on C_{1t} is $E(X \mid C_{1t})$. The value of U_{21} on C_{1t} is

$$E(Y_2 I(C_{1t}))/P(C_{1t}) = \sum_{k \in T_t} E(Y_2 I(C_{2k}))/P(C_{1t})$$

But $Y_2 = E(X \mid C_{2k})$ on C_{2k}. By summing over k, we get $U_{21} = E(X \mid Z_1)$ on each C_{1t}, i.e., $U_{21} = Y_1$ (everything a.s.).

Section 7.2

1. If $P(A) = 0$, A consists of a finite number of points, so $u(A) = 0$. Since $u(A) = \int_A (du/dP)\, dP$, $du/dP = c_k$ on $B_k = (k/n,\ (k+1)/n)$ and $u(B_k) = c_k P(B_k) = c_k/n = (2k+1)/2n^2$. So, one variant of $du/dP = \sum_{k=0}^{n-1} [(2k+1)/2n^2] I(B_k)$.

2. By Theorem 1, $u_2(A) = \int_A X\, du_1$ and $u_3(A) = \int_A Y\, du_2$. Now, by assumption, if $u_1(A) = 0$ then $u_2(A) = 0$, which then implies that $u_3(A) = 0$. So u_3 is absolutely continuous with respect to u_1. Apply Theorem 2 (with its extension to σ-finite measures) and the uniqueness part of Theorem 1.

Section 7.3

1. Because $X = [Z]$ is $\sigma(Z)$-measurable, $E([Z] \mid Z) = [Z]$ a.s.
2. By the basic properties, $E(S_{n+1} \mid S_1, S_2, \ldots, S_n) = E(S_n \mid S_1, S_2, \ldots, S_n) + E(X_{n+1} \mid S_1, S_2, \ldots, S_n) = S_n + E(X_{n+1}) = S_n$ a.s. The argument is the same for $E(S_{n+1} \mid S_n) = S_n$ a.s.
3. First show that a density of X_1 is standard normal $f_1(x)$. Set $f(y \mid x) = f(x, y)/f_1(x)$. Show by simple integration that $f(y \mid x)$ is a conditional density of X_2, given $X_1 = x$. Compute $E(X_2 \mid X_1)$ by Theorems 2 and 4, which yields $\varrho(\sigma_2/\sigma_1) X_1$ a.s.
4. Show that the joint distribution of $(S_1, S_2, \ldots, S_n)$, given $S_{n+1} = y$, is the integral of $f_y(y_1, y_2, \ldots, y_n) = n! y^{-n}$ if $0 \le y_1 < y_2 < \cdots < y_n$ and zero otherwise. Hence, $f_y(y_1, y_2, \ldots, y_n)$ is a variant of the desired density.

Section 7.4

1. Let $m < n$. Use repeatedly $E(E(S_{n+1} \mid \mathscr{A}_n) \mid \mathscr{A}_{n-1}) = E(S_{n+1} \mid \mathscr{A}_{n-1})$ on the one hand, while $E(S_{n+1} \mid \mathscr{A}_n) = S_n$ a.s. on the other. In $n - m$ steps we get $E(S_{n+1} \mid \mathscr{A}_m) = S_m$ a.s. Similarly for the submartingale case.
2. By assumption, $E(S_n \mid \mathscr{A}_m) \ge S_m$ a.s. if $m < n$, $m, n \in I$. Now, since $\mathscr{B}_n \subset \mathscr{A}_n$, $E(E(S_n \mid \mathscr{A}_m) \mid \mathscr{B}_m) = E(S_n \mid \mathscr{B}_m)$, but the left-hand side $\ge E(S_m \mid \mathscr{B}_m) = S_m$, the last equation being due to the $\mathscr{B}_m$-measurability of S_m.
3. Note that $\{T \le n - 1\} = \{T(n) = T(n-1)\}$ and $\{T \ge n\} = \{T(n) = n\}$. Furthermore, $\{T \le n - 1\}$, $\{T \ge n\} \in \mathscr{A}_{n-1}$. Hence,

$$
\begin{aligned}
E(S_{T(n)} \mid \mathscr{A}_{n-1}) &= E(S_{T(n-1)} I(T \le n - 1) \mid \mathscr{A}_{n-1}) + E(S_n I(T \ge n) \mid \mathscr{A}_{n-1}) \\
&= S_{T(n-1)} I(T \le n - 1) + I(T \ge n) E(S_n \mid \mathscr{A}_{n-1}) \\
&= S_{T(n-1)} I(T \le n - 1) + I(T \ge n) S_{n-1} \\
&= S_{T(n-1)} \text{ a.s.,}
\end{aligned}
$$

because $S_{n-1} = S_{T(n-1)}$ if $T \ge n$. Now, apply Exercise 1.

4. $Y_n - Y_{n-1} = R_n^2 - M_n - R_{n-1}^2 + M_{n-1}$. Since Y_n and Y_{n-1} are $\mathscr{A}_{n-1}$-measurable, $Y_n - Y_{n-1} = E(Y_n - Y_{n-1}|\mathscr{A}_{n-1}) = E(R_n^2|\mathscr{A}_{n-1}) - E(M_n|\mathscr{A}_{n-1}) - R_{n-1}^2 + M_{n-1} = E(R_n^2|\mathscr{A}_{n-1}) - R_{n-1}^2$ by the martingale property of M_n. On the other hand, by the martingale property of R_n,

$$E((R_n - R_{n-1})^2|\mathscr{A}_{n-1}) = E(R_n^2|\mathscr{A}_{n-1}) - 2R_{n-1}E(R_n|\mathscr{A}_{n-1}) + R_{n-1}^2$$

$$= E(R_n^2|\mathscr{A}_{n-1}) - R_{n-1}^2,$$

i.e., $Y_n - Y_{n-1} = E((R_n - R_{n-1})^2|\mathscr{A}_{n-1})$ a.s.
5. Apply Jensen's inequality and the martingale property.
6. Apply Jensen's inequality and the submartingale property.
7. Repeat the proof of Theorem 10 for the martingale $\{S_{n-i+1}, \mathscr{A}_{n-i+1}, 1 \le i \le n\}$. Since $\{|S_{n-i+1}|, \mathscr{A}_{n-i+1}, 1 \le i \le n\}$ is a submartingale, $E(|S_n|) \le E(|S_{n-1}|) \le \cdots \le E(|S_1|) < +\infty$. The crucial inequality of Theorem 9 now takes the form

$$(b - a)E(T_n) \le E((S_1 - a)^+) - E((S_n - a)^+)$$

Section 7.5

1. Since X_n/n is bounded, it is in L_r, $r > 0$. Now,

$$E(|X_n/n - p|^r) = \sum_{k=0}^n |k/n - p|^r \binom{n}{k} p^k (1 - p)^{n-k}$$

$$= n^{-r} \sum_{k=0}^n |k - np|^r \binom{n}{k} p^k (1 - p)^{n-k}$$

Choose $r = 2s$, s integer. Then, by Exercise 4 of Section 2.8, or by derivatives of the characteristic function of $(X_n - np)/[np(1-p)]^{1/2}$, $\sum_{k=0}^n (k - np)^{2s} \binom{n}{k} p^k (1-p)^{n-k}/[np(1-p)]^s$ has a finite positive limit (the corresponding moment of the standard normal distribution). So, if we divide by n^{2s} rather than n^s in this sum, the limit is zero. The case $r = 1$ follows from the central limit theorem. Now, if $r > 1$ is arbitrary, argue with either $1 < r < 2$ or $2s < r < 2s + 2$, as the case may be.
2. Since $X_n \in L_1$ and the X_n are identically distributed, the X_n are uniformly integrable. Now, since $\int_{A_t} (S_n/n) dP = \sum_{j=1}^n \int_{A_t} (X_j/n) dP \le \max_{1 \le j \le n} \int_{A_t} X_j dP$, we get from (7.26) that S_n/n is uniformly integrable. So, Theorem 13 is applicable.

Section 7.6

1. Use the inequalities $|E(X_n I(A)) - E(XI(A))| \le E[I(A)|X_n - X|] \le E(|X_n - X|)$.
2. Note that $g(t)/t$ is the slope of the secant of the graph of $g(x)$, connecting $(0, g(0)$ and $(t, g(t))$. It is increasing for convex $g(t)$, so β in Theorem 15 on p. 337 is finite. Since we assume that Young functions have finite powers (p. 336), Theorem 15 applies.
3. Make your own choices of $g(x)$. For $g(x) = \log(x + 1)$, q is not finite.

Section 7.7

1. By exchangeability,

$$0 \le V(X_1 + X_2 + \cdots + X_n)$$

$$= \sum_{i=1}^{n} E(X_j^2) + 2 \sum_{i<j} E(X_i X_j) - \left(\sum_{j=1}^{n} E(X_j) \right)^2$$

$$= nE(X_1^2) + n(n-1)E(X_1 X_2) - n^2 E^2(X_1).$$

Divide by n^2 and let $n \to +\infty$.
2. One possible example: $m(n) = n/2$, $p_{k,n} = P(N_m^{(n)} = k) = c/n^3$ if $k \ne 2$, and $p_{2,n} = 1 - c(n + 1)/n^3$, $c > 0$ constant. Then $E(N_n) \sim 2$, $E(N_n(N_n - 1)) \sim 2 + c/3$. So, (7.48) (use also the Corollary on the same page) is satisfied if $c = 6$, but direct computation from the corollary yields $P_n(N_m = 0) \to 1/4$, $P_n(N_m = 1) \to 1/2$, and $P_n(N_m = 2) \to 1/4$. Note that $P_n(A_1^{(n)}) = 2/n + o(1/n)$, $P_n(A_1^{(n)} \cap A_2^{(n)}) = 4/n^2 + o(n^{-2})$, and $P_n(A_1^{(n)} \cap A_2^{(n)} \cap A_3^{(n)}) = 3/2n^2 + o(n^{-2})$.
3. Use the asymptotic normality of the hypergeometric distribution and obtain mixtures of the normal and some distribution $U(x)$ as limiting distribution, where $U(x)$ is the limiting distribution of N_n when properly normalized.

CHAPTER 8

Section 8.1

1. Let $0 < u < s$. $\{X_{t+s} - X_t = 0\} = \{X_{t+s} - X_{t+u} = 0, X_{t+u} - X_t = 0\}$, so $G(s) = G(s - u)G(u)$, or, with $s - u = v$, $G(u + v) = G(u)G(v)$ for all u, $v \ge 0$. This is Cauchy's functional equation and, since $G(s)$ is non-increasing, its unique solution is $G(s) = e^{-as}$, $s \ge 0$, $a > 0$. By the assumption $X_t \not\equiv 0$, $G(s) \not\equiv 1$. Note that $G(s) \not\equiv 0$ is not possible, because we would then have, by a decomposition similar to the one above,

$P(X_{t+s} - X_t = 1) = 0$, and, by induction, $P(X_{t+s} - X_t = k) = 0$ for all $k \geq 1$.

2. Using the density formula for multivariate normal variables, $E(X_{t+s}X_t) = 0$ implies that the coefficient of the mixed term $x_1 x_2$ in the density is zero, i.e., the joint density of X_{t+s} and X_t has the form $C \exp(-d_1 x_1^2 - d_2 x_2^2)$, where C, d_1, and d_2 are positive constants. This is the independence property. If $X_1, X_2, \ldots$ are from an infinite sequence of exchangeable variables, then, by Exercise 1 of Section 7.7, the correlation of any two of the sequence is nonnegative. In several instances, the additional assumption of zero correlation leads to independence.

3. By the basic properties of conditional expectations, if the condition implies that $X_u = z$ and Z is independent of the condition, then $E(I(Z \leq x + X_u) | \mathscr{A}) = E(I(Z \leq x + z) | \mathscr{A}) = F_Z(x + X_u)$. So $P(Y_1 + Y_2 + \cdots + Y_n \leq x | X_1, X_2, \ldots, X_u) = E(I(Y_1 + Y_2 + \cdots + Y_n \leq x) | \sigma(Y_1, Y_2, \ldots, Y_u)) = E(I(X_n - X_u \leq x - X_u) | \sigma(X_1, X_2, \ldots, X_u)) = F_{nu}(x - X_u)$ a.s.

4. Let $A = \{j\}$. Then, with $u = t - 1$ and $s = t - 2$, (8.6) becomes $P(X_t = j | X_{t-2} = k) = \sum_{i=0}^M P(X_t = j | X_{t-1} = i)P(X_{t-1} = i | X_{t-2} = k)$. Also, by the total probability rule, $P(X_n = j) = \sum_{i=0}^M P(X_n = j | X_{n-1} = i)P(X_{n-1} = i)$. Now use induction over n.

5. Let u_j be a dense sequence in T. Then $\lim_{n=+\infty} \inf\{X_{u_j} : |u_j - t| < 1/n\} \leq X_t \leq \lim_{n=+\infty} \sup\{X_{u_j} : |u_j - t| < 1/n\}$ for all ω except $\omega \in B_t$ with $P(B_t) = 0$, $t \in T$. However, for t_j satisfying (8.7) and (8.8), if $\omega \notin B \cup (\bigcup B_{t_j})$, $\inf\{X_{u_j} : u_j \in I \cap T\} \leq \inf\{X_{t_j} : t_j \in I \cap T\} = \inf\{X_t : t \in I \cap T\} \leq \inf\{X_{u_j} : u_j \in I \cap T\}$; thus, the equations hold. Similarly for suprema.

Section 8.2

1. Let $0 < t < t_0$. Then $X_{t_0} = X_t + (X_{t_0} - X_t)$, where X_t and $X_{t_0} - X_t$ are independent and not identically zero. Thus, by Exercise 2 of Section 4.2, X_t and $X_{t_0} - X_t$ are Poisson. Now, if $t > t_0$, then $X_t = X_{t_0} + \sum_{k=1}^m (X_{t_0+s_k} - X_{t_0+s_{k-1}})$, where $0 = s_0 < s_1 < \cdots < s_m = t - t_0$ are such that each $s_k - s_{k-1} \leq t_0$. Thus, by stationarity and the first part, $X_{t_0+s_k} - X_{t_0+s_{k-1}}$ is Poisson and the summands are independent. But the convolution of Poisson variables is Poisson.

2. By Theorem 2, we have to prove (8.20). We know that the density of S_k is $f_k(x) = a^k x^{k-1} e^{-ax}/(k-1)!$, $k \geq 1$, $x \geq 0$. Hence, $P(X_t = k) = P(S_k \leq t, S_{k+1} > t) = \int_0^{+\infty} P(S_k \leq t, S_{k+1} > t | S_k = x) f_k(x) \, dx = \int_0^t P(S_{k+1} > t | S_k = x) f_k(x) \, dx = \int_0^t P(D_{k+1} > t - x) f_k(x) \, dx = (at)^k e^{-at}/k!$.

3. Let $0 < x_1 < x_2 < \cdots < x_n \leq t$ be fixed points. Let $y_j > 0$ be such that even $x_j - y_j \geq x_{j-1}$, $1 \leq j \leq n$, where $x_0 = 0$. Now, if $B_j = \{x_j - y_j < S_j \leq x_j\}$, then $P(B_1 \cap B_2 \cap \cdots \cap B_n | X_t = n) = P(B_1 \cap B_2 \cap \cdots \cap B_n \cap \{X_t = n\})/P(X_t = n)$, where the denominator is Poisson. On the other hand, the

event in the numerator means that there are no points in the intervals $(x_{j-1}, x_j - y_j)$, $1 \le j \le n + 1$, where $x_{n+1} - y_{n+1} = t$, and there is exactly one point in each of the intervals $(x_j - y_j, x_j)$. Since these intervals are nonoverlapping, the basic properties (i)–(iv) and Theorem 1 yield (note that $\sum_{j=1}^{n} (x_j - y_j - x_{j-1}) + \sum_{j=1}^{n+1} y_j = t$) that the numerator above equals $e^{-\lambda t} \prod_{j=1}^{n} [\lambda y_j e^{-y_j}]$. Divide by $y_1 y_2 \cdots y_n$ and let each $y_j \to 0$. The desired density $(n!t^{-n})$ follows, because when $x_j \le x_{j-1}$ for j, the joint density of the S_k, $1 \le k \le n$, given $X_t = n$, is zero.

4. Choose T as the positive real line or a finite subinterval of it. Then (i) and (ii) trivially hold. Now, since $\lambda(B)$ is an additive set function, (b) implies that the events $X(B) = 0$ are independent for disjoint sets B. Hence, by arguing as in Theorem 2, (c) implies (iii). Condition (iv) is obvious from the properties of integrals. Hence, Theorem 1 and (8.11) hold, $P(S_1 > t) = P(X(0, t) = 0) = \exp[-\int_0^t g(x)\, dx]$, and $P(S_1 \le x, S_2 > y) = P(X(0, x) = 1, X(x, y) = 0) = P(X(0, x) = 1)P(X(x, y) = 0) = [\int_0^x g(u)\, du] \exp[-\int_0^y g(u)\, du]$, $x < y$, $x, y \in T$.

5. Set $r(t) = S_N - t$, where $N = N(t)$ as defined for t_0 in the exercise. Then $P(r(t) > x) = \sum_{k=0}^{+\infty} P(r(t) > x, X_t = k) = \sum_{k=0}^{+\infty} P(r(t) > x, S_k < t \le S_{k+1}) = \sum_{k=0}^{+\infty} P(S_k < t, S_{k+1} \ge t + x)$. Now apply $P(S_k < t, S_{k+1} \ge t + x) = \int_0^t P(S_{k+1} \ge t + x | S_k = y) f_k(y)\, dy = \int_0^t P(D_{k+1} \ge t + x - y) f_k(y)\, dy$, where D_{k+1} is exponential and $f_k(y)$ is the familiar gamma density, so we have, for the kth term in the above sum, $e^{-\lambda(t+x)}(\lambda t)^k / k!$, $k \ge 1$, which is obviously the proper form for $k = 0$ as well. Summation over k results in the stated exponential form.

Section 8.3

1. Let $T(x)$ be another solution of (8.25) that is bounded over finite intervals and $T(x) = 0$ for $x < 0$. Then $R(x) - T(x) = D(x)$, where $R(x)$ is the renewal function (8.22), satisfies $D(x) = D(x) * F(x)$, $x > 0$. Hence, by repeated convolutions, $D(x) = D(x) * F^{(n*)}(x)$. But, as shown on p. 367, $F^{(n*)}(x) \to 0$ for all $x > 0$ as $n \to +\infty$, so $D(x) = 0$ for all $x > 0$.

2. By $P(X_t \ge n) = P(S_n \le t) = F^{(n*)}(t)$,

$$E(X_t^k) = \sum_{n=1}^{+\infty} n^k [P(S_n \le t) - P(S_{n+1} \le t)],$$

where we used $P(X_t = n) = P(X_t \ge n) - P(X_t > n + 1)$. Use the estimate on the n-fold convolution, developed on p. 367, and conclude that the sum above is convergent. A rearrangement of the terms of the above sum gives the desired formula.

3. Since $P(t + x > S_n > t, S_{n+1} > t + x) = \int_t^{t+x} P(S_{n+1} > t + x | S_n = y)\, dF^{(n*)}(y)$

and $P(S_{n+1} > t + x \mid S_n = y) = P(D_{n+1} > t + x - y) = 1 - F(t + x - y)$, summation over $n \geq 1$ (note that $0 < t \leq y$) yields the desired formula.

4. Use the formula of Exercise 3.

5. We know that $P(S_n - n\mu > x(nV)^{1/2}) \rightarrow 1 - N(x)$, where $N(x)$ is the standard normal distribution function. Hence, with $t = n\mu + x(nV)^{1/2}$, $P(X_t < n) \rightarrow 1 - N(x)$. Solving $t = n\mu + x(nV)^{1/2}$ for n as a function of t and x, we get $n = t/\mu - x(Vt)^{1/2} \mu^{-3/2} + O(1)$, because we had to observe that $t > n\mu$ for $x > 0$ and $t < n\mu$ for $x < 0$ in the two solutions of the above equation. Now, $1 - N(x) = N(-x)$ gives the result.

6. Let $E(w(t)) = \int_0^{+\infty} P(w(t) \geq x) \, dx = E$. Note that $P(w(t) \geq x) = \int_0^{+\infty} P(w(t) \geq x \mid S_1 = y) \, dF(y) = \int_0^t P(w(t - y) \geq x) \, dF(y) + \int_{t+x}^{+\infty} dF(y)$, where, for $y \leq t$, the origin is shifted to y, so the point t becomes $t - y$, while, for $y > t$, we observed that $P(w(t) \geq x \mid S_1 = y) = 0$ if $t < y \leq t + x$ and $= 1$ if $y > t + x$. Substituting this recursive formula into the integral giving E, we obtain $E[1 - F(t)] = \int_t^{+\infty} [1 - F(y)] \, dy$. Since $F(\cdot)$ is continuous, the right-hand side is differentiable; differentiation and the solution of the resulting differential equation gives the exponential distribution, since $F(0) = 0$ is a basic assumption in this section.

7. Let C_1 be the first time that a customer makes a purchase and $0 < S_1 < S_2 < \cdots$ be the arrival times of the customers. Then, by the total probability rule, $P(C_1 \leq x) = \sum_{k=1}^{+\infty} P(C_1 \leq x \mid C_1 = S_k) P(C_1 = S_k)$. By the assumptions, $P(C_1 = S_k) = (1 - p)^{k-1} p$ and $P(C_1 \leq x \mid C_1 = S_k) = F^{(k*)}(x)$. Thus, $P(C_1 \leq x) = p \sum_{k=1}^{+\infty} (1 - p)^{k-1} F^{(k*)}(x) = F_1(x)$, say. Similarly, if C_m is the mth time that a purchase is made, $P(C_1 \leq x, C_2 - C_1 \leq y) = \sum_{k,m} P(C_1 \leq x, C_2 - C_1 \leq y \mid G_{k,m}) P(G_{k,m})$, where $G_{k,m} = \{C_1 = S_k, C_2 = S_m\}$, $1 \leq k < m$. We easily get $P(C_1 \leq x, C_2 - C_1 \leq y) = F_1(x) F_1(y)$ upon observing that $P(G_{k,m}) = p^2(1 - p)^{m-2}$ and $P(C_1 \leq x, C_2 - C_1 \leq y \mid G_{k,m}) = F^{(k*)}(x) F^{(m-k,*)}(y)$. By induction we get that the points C_j form a renewal process with interval distribution $F_1(x)$. If the time scale is changed by a factor $1/p$, the interval distribution becomes $F_1(x/p)$. The actual formula for $F_1(x)$ reduces to $F(x)$ if $F(x)$ is exponential.

Section 8.4

1. Use the inequality on p. 377.

2. (i) If the parameter is λ, then $\lambda G(z) = G'(z)$; (ii) if the parameters are n and p, then $npG(z) = (1 - p + pz)G'(z)$; and (iii) if the parameters are n and p again, $npG(z) = [1 - (1 - p)z]G'(z)$. Now, upon differentiating $G_n(z) = G(G_{n-1}(z))$, we obtain $G_n'(z) = G'(G_{n-1}(z))G_{n-1}'(z)$. Apply the assumed differential equation for $G(z)$, yielding $G_n'(z) = [(a + bG_n(z))/(c + dG_{n-1}(z))]G_{n-1}'(z)$. Multiply by $c + dG_{n-1}(z)$ and differentiate k

times. The formula $P(X_{n+1} = k + 1) = [1/(k + 1)!]G_n^{(k+1)}(0)$ gives the desired recursive formula.

3. Clearly, $E(X_n | X_{n-1}) = \alpha X_{n-1}$ a.s. Since, by (8.36), $E(X_n) = \alpha^n$, $X_n / E(X_n)$ is a martingale. By $E((X_{n+1} - \alpha X_n)^2 | X_n) = V(X_n)$ a.s., each $X_n \in L_2$. Apply Theorem 7.14 to obtain the a.s. convergence of $X_n / E(X_n)$.

Section 8.5

1. Note the special role of the state 0 in Example 2: $\{0\}$ is a closed set, and 0 can be reached from all states. All other pairs of states intercommunicate. The features of Example 3 are that the chain there is finite and all pairs intercommunicate. Pay attention to periodicity.

2. Use the identity $P(A \cap B | C) = P(A | B \cap C)P(B | C)$ and the Markov property on both sides of the equation.

3. Since $p_{ij} > 0$ for all i and j, the chain is a single closed set. By $p_{11} > 0$, state 1 is aperiodic, so all states are aperiodic. The states must be persistent since the chain is finite. Apply Theorem 8 and the double stochastic property to conclude that $\mathbf{P}^n \rightarrow (r_{ij})$, where $r_{ij} = r$, independent of both i and j. Hence, $r = 1/N$ if $\mathbf{P}$ is an N by N matrix. Finally, apply Theorem 9.

4. Use (8.45) as we did on p. 388.

5. Since intercommunicating persistent states form a closed set, we would have a submatrix $\mathbf{P}_1$ of $\mathbf{P}$ that is stochastic and all states corresponding to $\mathbf{P}_1$ would be persistent with $E_j = +\infty$. If their common period is d (recall p. 387), then we can apply Theorem 8 to $\mathbf{P}_1^d$, by which all entries of the limit of $\mathbf{P}_1^{dn}$ would be zero. This is not possible for a finite stochastic matrix.

6. Use (8.53) and (8.42) with $t = 1$. Conclude by induction that the distribution of X_n, $n \geq 1$, is the initial distribution.

7. Since a finite chain must have a persistent state, state j is persistent. Let C be the set of states intercommunicating with j. Then either $C = S$ or $S = C \cup T$, where C is closed and the members of T are transient. Since $p_{jj} > 0$, j is aperiodic (and so are all members of C). If $\mathbf{P}_1 = (p_{it})$, $i, t \in C$, then $\mathbf{P}_1^n$ has a limit (r_{it}) by Theorem 8. Furthermore, $p_{it}^n \rightarrow 0$ for $t \in T$ and $p_{ti}^n \rightarrow r_{ti}$ if $t \in T$ and $i \in C$ (use the same estimate as on p. 388 and establish that $f_{ti}^* = 1$ since the chain is finite).

8. Apply Theorem 8 separately to $\mathbf{P}_1$ and $\mathbf{P}_2$, corresponding to the states C_1 and C_2, respectively. Since 4 is transient, $p_{j4}^n \rightarrow 0$. However, $\lim p_{4j}^n$ ($j \neq 4$, $n \rightarrow +\infty$) has to be determined with care. First, use the argument of p. 388 and then determine f_{4j}^* for $j \neq 4$. Note that $f_{40}^* = f_{41}^*, f_{42}^* = f_{43}^*$, and $f_{40}^* + f_{42}^* = 1$, because if the chain starts at 4, it will ultimately reach C_1 or C_2 and stay within C_i once it is reached. In addition, within C_i, either member will be reached from the other with probability one.

Section 8.6

1. Let $s_1 = u/v$ and $s_2 = m/h$ be two rational numbers, i.e., u, v, m, and h are positive integers. Let $n' = Kvh$, where K runs through the positive integers. Then $n's_1$ and $n's_2$ have a common subsequence, N, say, and P_N converges to r_{s_1} as well as to r_{s_2},' i.e., $r_{s_1} = r_{s_2}$ for arbitrary rationals s_1 and s_2. But then, whatever s, by uniform continuity, $P_{ns} - P_{ns_1}$ can be made arbitrarily small with an appropriate rational s_1. For large n, $P_{ns_1} - r$ (the limit for rational s) is small even with varying s_1, and thus P_{ns} converges to r, a value not dependent on s. Finally, let $t \to +\infty$. Fix an s, and choose n such that $ns \le t < (n + 1)s$. By uniform continuity again, $P_t - P_{ns}$ can be made arbitrarily small by choosing s small. Hence, $P_t \to r$.

2. Let s and t be positive numbers. Let $0 \le s_1 < s_2 < \cdots < s_n < s$. Then, for $i_k \in S$, $1 \le k \le n + 2$, compute $P(Y_{t+s} = i_{n+2} \mid Y_s = i_{n+1}, Y_{s_k} = i_k, n \ge k \ge 1)$ by the definition of conditional probability and by the meaning of Y_u in terms of X_v. If one writes $P(X_{T-s_k} = i_k, 1 \le k \le n, X_{T-s} = i_{n+1},$ $X_{T-s-t} = i_{n+2}) = P(X_{T-s_k} = i_k, 1 \le k \le n \mid X_{T-s} = i_{n+1}, X_{T-s-t} = i_{n+2}) \times$ $P(X_{T-s} = i_{n+1}, X_{T-s-t} = i_{n+2})$ and, in the denominator, $P(X_{T-s_k} = i_k,$ $1 \le k \le n, X_{T-s} = i_{n+1}) = P(X_{T-s_k} = i_k, 1 \le k \le n \mid X_{T-s} = i_{n+1})P(X_{T-s} = i_{n+1})$, by the Markov property of X_v, the ratio of the two expressions becomes $P(X_{T-s-t} = i_{n+2} \mid X_{T-s} = i_{n+1}) = P(Y_{t+s} = i_{n+2} \mid Y_s = i_{n+1})$.

3. In $G_{s+t}(i, j) - G_t(i, j) = P_{s+t}(i, j) - P_t(i, j) + q_i \int_t^{s+t} P_x(i, j)\, dx$, substitute $x = y + t$, and in both $P_{s+t}(i, j)$ and $P_{y+t}(i, j)$, use the Chapman–Kolmogorov equation, and note that the new expression equals $\sum_{k \in S} G_s(i, k)P_t(k, j)$.

4. Use the Chapman–Kolmogorov equation in $[P_{t+s}(i, j) - P_t(i, j)]/s$ and the fact that $P_s(i, k)/s$ for $i \ne k$ and $[1 - P_s(i, i)]/s$ converge as $s \to 0$ (to q_{ik} and $-q_{ii}$, respectively). This yields $P'_t(i, j) \ge \sum_{k \in S} q_{ik} P_t(k, j)$. Upon summing over $j \in S$, one can observe that strict inequality above cannot occur for any j under the condition of the Exercise.

5. By Theorem 10, $P_t(j, j) > 0$ and $P_t(i, j)$ is uniformly continuous in t for all i, $j \in S$. Hence, any s-skeleton of the chain is aperiodic which entails (see the preceding section) that $P_{ns}(i, j)$ converges as $n \to +\infty$. By Exercise 1, this limit does not depend on s, and by results for the discrete time situation (i.e., for the skeletons) the limit does not depend on i either. Now, by the discrete time case, the mentioned limits p_j are positive if the states are positive persistent (and if one state is such then so are all states) and clearly $\sum p_j \le 1$. Let $s \to +\infty$ in $P_{t+s}(k, j) = \sum_{i \in S} P_s(k, i) P_t(i, j)$. We get $p_j \ge \sum_{i \in S} p_i P_t(i, j)$. Summing these over j, we get that equation must hold for all $j \in S$ and all $t \ge 0$. Hence, by letting $t \to +\infty$ in $p_j = \sum_{i \in S} p_i P_t(i, j)$, the dominated convergence theorem yields that the equation remains to hold, and thus $\sum_{i \in S} p_i = 1$. Finally, if $r = (r_1, r_2, \ldots)$ with

Index

Milton Keynes UK
Ingram Content Group UK Ltd.
UKHW020009071024
449327UK00031B/2711